W9-CTT-387

Instrument Flying Handbook

Thomas P. Turner

McGraw-Hill

New York   San Francisco   Washington, D.C.   Auckland   Bogotá
Caracas   Lisbon   London   Madrid   Mexico City   Milan
Montreal   New Delhi   San Juan   Singapore
Sydney   Tokyo   Toronto

Library of Congress Cataloging-in-Publication Data

Turner, Thomas P.
    Instrument flying handbook / Thomas P. Turner.
      p. cm.
    Includes index.
    ISBN 0-07-136198-7
    1. Instrument flying—Handbooks, manuals, etc. I. Title.

TL711.B6T87   2000
629.132'5214—dc21   00-045581

McGraw-Hill

*A Division of The **McGraw·Hill** Companies*

  2 3 4 5 6 7 8 9 0   DOC/DOC   0 9 8 7 6 5 4 3 2 1

ISBN 0-07-136198-7

The sponsoring editor for this book was Shelley Carr, the editing supervisor was Sally Glover, and the production supervisor was Pamela Pelton. It was set in Garamond by Joanne Morbit and Michele Pridmore of McGraw-Hill's desktop publishing department, Hightstown, N.J.

Printed and bound by R. R. Donnelley & Sons Company.

McGraw-Hill books are available at special quantity discounts to use as premiums and sales promotions, or for use in corporate training programs. For more information, please write to the Director of Special Sales, Professional Publishing, McGraw-Hill, Two Penn Plaza, New York, NY 10121-2298. Or contact your local bookstore.

 This book is printed on recycled, acid-free paper containing a minimum of 50% recycled, de-inked fiber.

Contents

Introduction *xvii*

Section 1: Instrument Flight Basics   *1*

1   Basic Instrument Flight   *3*
Control, Primary, and Support Instruments   *4*
Trim, Pressure, and Control Displacement   *7*
Instruments That Lie—When and Why   *9*
Magnetic Compass Turns   *12*
Instruments   *15*
Visual Scan and Instrument Correlation   *24*
Turn Coordinator and Clock   *25*

2   Aircraft Control   *27*
Attitude Control   *27*
Altitude Control   *28*
Heading Control   *30*
Airspeed Control   *31*
Airspeed Transitions   *32*
Power   *33*
Trim in a Turn   *33*
Standard-Rate Turns   *34*
Straight-and-Level Flight   *36*
Climbs and Descents   *38*
"Two, Two, and Twenty"   *40*
Overcontrolling   *41*

3 Scan and Control *43*

Scan *43*

Distractions *44*

Instrument Scan and Interpretation *44*

Proper Instrument Scanning *45*

Instrument Errors *47*

Airspeed Control *48*

4 Phase of Flight *51*

Instrument Takeoffs *51*

Climbs by Reference to Instruments *52*

Straight and Level Flight and Turns *53*

Descent by Reference to Instruments *53*

Constant Airspeed Climbs *54*

Departure Climbout *55*

Constant Rate Climbs *56*

Chasing the Needle *57*

Climb Level Off *57*

Descents *58*

Constant Speed Descents *58*

Constant Rate Descents *59*

Descent Level-Off *60*

Approach Descents *60*

High-Speed Final *61*

Obstacle Clearance Concerns *62*

Aircraft Performance Restrictions *66*

5 Departure Procedures *69*

Getting an IFR Clearance *69*

Use of a Climb Profile *71*

Instrument Departure Procedures *72*

6 Practice Maneuvers and Emergencies *73*

"Pattern A" *73*

Minimum Controllable Airspeed *74*

"Vertical S" *77*

"Pattern B" *77*

"Pattern C"   *79*
Practicing Stalls   *80*
Stall Recovery   *80*
Steep Turns   *81*
Unusual Attitudes   *83*
Recovery Procedures   *84*
Partial Panel Procedures   *85*
Partial Panel Control Instruments   *87*
Loss of Radio Communication   *88*
VOR Tracking and Time/Distance Problems   *90*

7 Navigation   *95*

Cross-Country Flights   *95*
Preflight Planning   *96*
VOR   *98*
VOR Procedures   *102*
Heading Indicator Errors   *102*
VOR Proficiency   *104*
16-Point Orientation   *105*
VOR Time/Distance Check   *108*
Intercepting a Bearing or Radial   *109*
Wind Corrections   *111*
Common Interception Mistakes   *112*
"Cleared Direct"   *113*
Reference Heading   *113*
Bracketing   *115*
Chasing the Needle   *115*
Station Passage   *116*
Practice Patterns   *116*
NDB Procedures   *117*
ADF Orientation   *118*
ADF Time/Distance Checks   *119*
Intercepting a Bearing   *120*
Homing is Unacceptable   *122*
Tracking and Bracketing   *123*
Outbound Bearings from the NDB   *126*
Practice Patterns   *127*

ADF Holding Patterns   *128*
GPS   *129*
GPS Approval   *129*
GPS Demystified   *130*
Great Circles   *130*
Relationship of Arc to Distance   *134*
Primer of Celestial Navigation   *135*
Timing is Crucial   *138*
GPS Errors   *139*
Additional Radioo Navigation Aids   *140*

Section 2: IFR Flight Planning   *143*

8   Preflight Planning   *145*

The Flight Kit   *146*
IFR Aircraft Requirements   *148*
Icing and Thunderstorm Hazards   *149*
Airways and Altitudes   *151*
Fuel and Alternate Airport Requirements   *153*

9   Preparing for an Instrument Flight   *157*

GPS En Route Navigation   *158*
The Flight Log   *159*
Planning the Route   *159*
Standard Instrument Departures (SIDs)   *162*
Preferred Roues and TECs   *162*
Approach Planning   *163*
En Route Fixes   *169*
Communications Frequencies   *171*
Field Elevation   *172*
Airport Services   *174*

10   Minimum Equipment Lists and Approved Flight Manuals   *177*

Operations with a Minimum Equipment List   *178*
Operations without an MEL   *182*
Applying the Four-Step Test   *191*
Ramp Checks   *192*

11 Weather 195

Weather and Decision Making   *195*
Thunderstorms   *196*
Turbulence   *200*
Icing   *201*
Fog   *203*
Destination Minimums   *205*
The "One, Two, Three" Rule   *208*
Selecting an Alternate   *208*
Personal Minimums   *209*
Weather Factors Reviewed   *211*
Weather Briefings and Flight Plans   *212*
Computer Weather Services: The "Big Picture"   *212*
DUATS   *213*
TAFs and METARs   *214*
Forecast Reliability   *217*
Faster Service   *220*
Transcribing the Weather   *221*
Go or No-Go?   *226*
Estimated Climbout Time   *228*
Wind and Ground Speed   *228*
Flight Plan to Alternate   *229*
In-Flight Notations   *232*
Filing the Flight Plan   *234*
Abbreviated Briefings   *234*
Outlook Briefings   *236*
The "One-Call" Technique   *236*
Total Time En Route   *237*

12 Airplane, Instrument and Equipment Checks 239

Fuel Quantity   *239*
ATIS   *240*
Altimeter and Airspeed Errors   *241*
Electrical Equipment   *241*
VOR Checks   *242*
Cockpit Organization   *243*
VOR Checks with VOT   *245*
Ground and Airborne VOR Checks   *246*

ILS Check   *246*
ADF Check   *247*
Transponder Check   *247*
Tips to Reduce Cockpit Confusion   *247*
Gyro Instruments   *248*

13 Clearances and Communications   *249*

Practice Clearances   *249*
Clearance Shorthand   *250*
Handling Amended Clearances   *250*
Obtaining Clearances   *252*
Unacceptable Clearances   *253*
Remote Communications Outlets   *254*
Void Time Clearances   *254*
Taxi Checks   *256*
Rolling Engine Run-Up   *256*
Runway Checks   *257*
IFR Communications   *257*
Standard Phraseology   *258*
"Who, Who, Where, What"   *258*
Calling Ground Control   *259*
When You Hear Nothing Further   *260*
Managing Frequencies   *260*
Required Reports   *261*
Canceling IFR   *262*
Radio Contact Lost   *263*

Section 3: Instrument Charts and Procedures   *265*

14 En route Charts   *267*

Road Maps in the Sky   *271*
Minimum Altitudes, Fixes, and Reporting Points   *278*
Other Symbols   *283*
Fitting All the Pieces Together   *284*

15 Departure and Arrival Procedures   *291*

The Class B Trap   *295*
SIDs, STARs, and Profile Descents   *297*

Looking at a Profile Descent   *302*
Charted Visual Approaches   *304*

16 Approach Chart Plan View   *309*
Creating the Charts   *312*
Howie Keefe's Update Concept   *312*
Filing the Charts   *318*
The Bird's-Eye View   *321*
Transitioning for the Approach   *327*
The Rest of the Way In   *329*
Odd and Ends   *331*

17 Approach Chart: Profile View   *333*
Nonprecision Approaches   *333*
Inbound From the FAF   *335*
Visualizing the VDP   *336*
Precision Approaches Cut a Fine Line   *337*
DH, MAP, and Other Acronyms   *340*
CAT II and CAT IIIA   *341*
Missed Approach   *344*
Landing Minimums   *345*
Circling Minimums   *347*

18 Airport Charts   *351*
New Symbols   *352*
The Airport as Seen From the Air   *355*
Additional Runway Information   *355*
What You See Ain't Always What You Get   *357*
Takeoff Minimums   *360*
Alternate Minimums   *362*
Expanded Airport Charts   *366*

19 NOS Approach Charts   *369*
ILS by NOS   *369*
A Lot of Page Turning   *372*
Basic Takeoff and Alternate Minima   *375*
Picking a Chart Apart   *383*
Airport Lighting   *384*

NOS Procedure Turns *388*
Climb and Descent Gradients *393*

Section 4: Arrival Procedures   397

20 Radio Technique and Holding Patterns   *399*
Proper Position Reports *400*
Required Reports to ATC *402*
Communications Confusion *403*
More Communication Failure Procedure *404*
The Holding Pattern *407*
Holding Pattern Entries *408*
STARs *410*

21 More on Holding Patterns   *413*
"Five Ts": Time, Turn, Twist, Throttle, Talk *413*
Wind Corrections *414*
En Route Holding *415*
Holding Pattern Entry *416*
Choosing the Correct Entry *418*
Importance of Altitude Control *419*
Holding Pattern Variations *420*
Intersection Holds *421*
DME Holding Patterns *422*

22 Approach Angles   *425*
Getting It Down Safely *426*
In the Beginning *427*
Precision and Nonprecision Approaches *428*
Approach Minimums *429*
Aircraft Categories and Descent Minimums *430*

23 Mastering Minimums   *435*
Visibility *435*
Visual Approach *436*
Contact Approach *438*
Straight-In and Circling Minimums *440*

Circling Approach   *441*
Runway Not in Sight   *443*

24 Approach Basics and NDB Approaches   *447*

Nonprecision Approaches   *447*
Precision Approaches   *448*
Altitude Minimums   *449*
Adjustments to MDA   *450*
Operation Below MDA   *457*
Visibility Minimums Required for Landing   *458*
Missed Approach Planning   *459*
NDB Approaches   *461*
Radar Vectors   *461*
The Full Approach Procedure   *463*
Procedure Turns   *463*
Approach Speeds   *466*
Approach Communications   *467*
Flying the NDB Approach   *468*
When to Descend   *468*
Timing the Approach   *469*
Final Approach Course   *470*
Missed Approaches   *471*
Circling Approaches   *472*
NDB on Airport   *477*

25 VOR, DME and GPS Approaches   *481*

Flying the VOR Approach   *481*
DME and DME Arc Approaches   *484*
Flying the DME Arc   *486*
GPS Approaches   *490*
GPS Approach Basics   *492*
GPS Approach Planning   *492*
Tips on Flying Approaches   *497*

26 ILS, Localizer, and Radar Approaches   *499*

Needle Sensitivity   *499*
Flying the ILS   *501*

Analyzing an ILS Approach   *502*
Decision Height   *504*
Marker Beacons   *506*
ILS Tips   *507*
Back Course Approaches   *508*
Localizer, LDA, and SDF Approaches   *511*
Radar Assists   *514*
ASR Approaches   *516*
No-Gyro Approaches   *518*
PAR Approaches   *518*
Visual and Contact Approaches   *519*
Instrument Takeoffs   *520*
Flying "Away" From the Needle   *523*
Vectors to Final—A Warning   *526*

27 Multiengine Aircraft Approaches   *527*
One-Engine Inoperative Instrument Approach   *528*
Real-World Training   *531*

Section 5: The Human Factor   *535*

28 What Causes IFR Accidents?   *537*
The IFR Accident Record   *537*
Thunderstorms   *538*
Turbulence   *538*
Airframe Ice   *539*
Reduced Visibility   *539*

29 The Pilot   541
Preflighting the Pilot   *541*

30 Physiological Factors   *557*
Legal Drugs   *557*
Alcohol, Cigarettes, and Hypoxia   *558*
Illegal Drugs   *561*
Vertigo Awareness   *562*
The Inner Ear   *563*
Believing the Instruments   *565*

31 Physiology at Night   569
Background Information   572
Technique   594
Skills to Practice   600
Further Reading   601

32 High-Altitude Physiology   603
Respiration   603
Hypoxia   604
Other High-Altitude Sickness   614
Causes and Effects of Gas Expansion and
  Bubble Formations   614
Vision   616

33 Pilot Judgment   617
Three Mental Processes of Safe Flight   618
Five Hazardous Attitudes   619
Accident Prevention   620

34 Stress   623
What is Stress?   623
Flying Stress   624
Physical Factors   627
Effects of Stress   627
Nonflying Stress   628
Flight Test Stress   630

35 Cockpit Resource Management   633
Working Together As a Flight Crew   634
In-Flight Cockpit Management   635
Flight Watch   639

Section 6: Instrument Training Syllabus   643

36 Instrument Training Syllabus   645
Flight Lesson 1: Introduction to IFR   645
Background Briefing 1-2: Introduction to Basic
  Instruments   646

Flight Lesson 2: Maneuvering Solely by Reference to
    Instruments—Part I   *648*

Flight Lesson 3: Maneuvering Solely by Reference to
    Instruments—Part II   *649*

Flight Lesson 4: VOR Tracking and Bracketing   *650*

Flight Lesson 5: VOR Holding Patterns   *651*

Flight Lesson 6: Unusual Attitudes, Partial Panel   *652*

Flight Lesson 7: ADF Orientation, Tracking,
    and Bracketing   *653*

Flight Lesson 8: ADF Holding   *654*

Background Briefing 8-9: Instrument Approach
    Procedures   *655*

Flight Lesson 9: NDB Approaches—I   *657*

Flight Lesson 10: NDB Approaches—II   *658*

Flight Lesson 11: VOR Approaches—I   *659*

Flight Lesson 12: VOR Approaches—II   *659*

Flight Lesson 13: VOR Approaches—III   *660*

Flight Lesson 14: ILS Approaches—I   *661*

Flight Lesson 15: ILS Approaches—II   *661*

Flight Lesson 16: ILS Back Course, Llocalizer, LDF, SDF,
    and Radar Approaches   *662*

Background Briefing 16-17: IFR Cross-Country
    Procedures   *663*

Flight Lesson 17: Long IFR Cross-Country Flight   *665*

Flight Lesson 18: Progress Check   *666*

Flight Lesson 19: Flight Test Preparation   *667*

Background Briefing 19-20: Preparation for the
    Instrument Flight Test Oral Exam   *667*

Flight Lesson 20: Flight Test Recommendation Flight   *672*

Flight Instructor Endorsements   *673*

37 Long IFR Cross-Country   *675*

The Value of Actual IFR   *676*

Uncontrolled Airports   *676*

Void Time Clearances   *677*

Partial Panel   *677*

Fuel Management   *678*

Logging the Flight   *678*

Obtaining Weather Information in Flight   *680*

Lost Rradio Contact   *682*
Two-Way Radio Communications Failure   *684*
Importance of Logging Times   *685*
Emergency Altitudes   *685*
Complete Electrical Failure   *686*
IFR Cross-Country Tips   *687*

38 The Instrument Pilot's Library   *689*

Afterword   *691*

Index   *693*

Introduction

Improved safety. Increased flexibility and convenience of travel. A step toward a professional flying career. Reduced aircraft insurance premiums.   These are all good reasons to get your instrument rating, maintain Instrument Flight Rules (IFR) currency, and exercise the privileges of your instrument rating.

Whether you're just learning the art of flight by reference to instruments, a rated instrument pilot wanting to refresh and expand your skills, or an instrument instructor wanting to better educate your students, this volume provides a wealth of information and expertise you can use to meet your goal. Some of the finest general-aviation authors have written extremely helpful and informative books on different aspects of instrument instruction and flight. This handbook unites the best of their work into a single text you can use to develop and refine your instrument flying skills.

We'll review their combined expertise in six distinct areas. In Section 1 we'll explain basic attitude flight as the focus of instrument flight.

We'll start with a review of the instruments themselves, then cover flying solely by reference to instruments in normal and partial panel regimes. We'll apply those skills to the various phases of flight (takeoff, climb, cruise, descent, and approach), and discuss exercises for practice and dealing with emergencies. Then we'll add instrument navigation, getting from here to there without outside visual references.

In Section 2 we'll look at the instrument flight planning process. We'll discuss general flight planning technique, obtaining and evaluating weather information, conducting an instrument preflight check, and obtaining an instrument clearance.

Section 3 is our review of instrument charts and approach plates.

In Section 4, we'll concentrate on instrument let-down and arrival procedures, including holding patterns, precision and nonprecision approaches, and missed approach procedures.

Section 5 goes over the most important element in flying safely, the human factor. We'll look at the instrument and weather-related accident record, physiological factors, pilot attitude and mental processes that affect judgment and decision-making, and techniques for managing cockpit workload and resources.

In Section 6 we'll lay out an instrument flight training syllabus and recommendations for continued study as part of a regular, ongoing instrument pilot education.

Of course, none of this is possible without the superb source material provided by the following authors:

Lewis Bjork, *Piloting at Night*. 1998. Lewis Bjork is a pilot for Skywest Airlines. A specialist in aerobatics with thousands of instruction hours, he has flown more than 130 types of airplanes. As an experimental aircraft builder and test pilot, Bjork refined the handling characteristics of microlight twin and a pressurized Lancair IV. He is the author of several McGraw-Hill Aviation titles, including *Piloting for Maximum Performance* (McGraw-Hill).

Paul A. Craig, *Multiengine Flying*. Second Edition, 1997. Paul A. Craig is the chief flight instructor at the Middle Tennessee State University Aerospace Department in Murfreesboro, Tennessee. He holds the Gold Seal Multiengine and Instrument Flight Instructor Certificate, as well as an airline Transport Pilot Certificate. He is a recipient of the North Carolina Flight Instructor of the Year award and is an FAA safety program counselor. Craig holds a doctorate in education, specializing in pilot decision-making and flight training. He also has a Masters of Aerospace Education degree and a Specialist in Education degree. He is the author of numerous McGraw-Hill Aviation titles.

David Frazier, *The ABCs of Safe Flying*. Fourth Edition, 1999. David Frazier (Jackson, MI) is currently Director of Aviation at Jackson College; prior to this, he taught aviation at Vincennes University (Indiana) for 20 years. He is a multi-rated pilot with over 10,000 hours of instructing experience, in both aircraft and simulators, Frazier has been an FAA Designated Pilot Examiner since 1982. He is the author of *AG Pilot Flight Training Guide*, and *How to Master Precision Flight*.

Michael C. Love, *Flight Maneuvers*. 1999. Michael C. Love (New Glarus, WI) is a flight instructor, aerobatic pilot, and holds a commercial pilot's rating. He is the author of *Better Takeoffs and Landings and Spin Management and Recovery* in McGraw-Hill's Practical Flying Series.

Henry Sollman and Sherwood Harris, *Mastering Instrument Flying*. Third Edition, 1999. Henry Sollman is an award-winning flight instructor who has been flying for more than 50 years. He developed the highly successful 10-day instrument certification program on which this book is based. Sherwood Harris is a flight instructor with 30 years' experience who is the author of several books and numerous articles on aviation.

J.R. Williams, *The Art of Instrument Flying*. 1996. J.R. Williams is a retired commercial airline pilot whose flying career also included service in the Persian Gulf War. A contributing editor to *Private Pilot* magazine for seven years, Williams has written several aviation books and numerous articles.

My thanks to them for their significant contribution to aviation safety!

Section 1

Instrument Flight Basics

1

Basic instrument flight

A common misconception among instrument pilots is that the primary emphasis is on navigation. The rapid advance of computerized navigation equipment like Global Positioning System (GPS) receivers and moving-map displays makes the pilot's job more and more one of system management. Autopilots, to which pilots can delegate some flying tasks, remove us even farther from the basics of instrument flight.

The natural urge in instrument flight training is also to teach, and to master, the navigational tasks associated with instrument flight. Many Certified Flight Instructors—Instruments (CFIIs) inadvertently reinforce this navigational emphasis by teaching instrument approach procedures before the student is really ready to handle them.

Certainly, as a motivational tool, CFIIs can and perhaps should demonstrate an arrival procedure early in the instrument syllabus, perhaps even on the first instrument flying lesson. But the student should not begin serious practice of departure, en route, and approach navigation until he or she is capable of simply flying the airplane solely by reference to the instruments.

For that's the misconception—that instrument flying is solely a matter of navigation along precise approach procedures or along a departure procedure. Instrument flying, instead, consists primarily of the ability to control the airplane's heading and attitude without outside visual references. It's a matter of using what at first may not be familiar or intuitively user-friendly gauges to replace the sensations of pitch, roll, and yaw that we've grown up with. Instrument flying is the process of interpreting the gauges and guiding the airplane, almost subconsciously, as *if* we can see the outside world.

We fly on instruments by flying basic attitude flight. The attitude indicator, heading indicator, altimeter, airspeed, quality-of-turn instrument, vertical speed indicator, and aircraft power, flaps, and landing gear controls tell us how the airplane is oriented and configured. They accept our inputs to make the airplane go where we want it to go. Only after mastering the basics of flying solely by reference to instruments can we add the additional tasks of navigation and systems management.

Control, primary, and support instruments

The attitude indicator provides pitch information during the step maneuver; it is the central control instrument or simply the *control instrument.*

Primary information on the quality of control comes from the altimeter. It shows a direct, almost instantaneous reading on whether or not the quality is good enough to stay within ±20 feet.

Here's a good way to determine the primary instrument: It indicates the most pertinent information about how well you are doing and *does not move* when flying precisely. The altimeter is the *primary instrument* in maintaining level flight. Does it meet the test? You bet it does. It indicates the most pertinent information regarding altitude. And if you're doing a good job, the altimeter will not move.

The vertical speed indicator (VSI) has neither a control nor a primary role maintaining level flight. But it is an additional source of information on the rate of climb or descent and may be used for cross-checking. The VSI is thus a *support instrument* during straight and level flight.

Figure. 1-1 shows the primary, support, and control instruments for all phases of instrument flight. Note how the instruments just discussed are used to control pitch in level flight. I will classify instruments as control, primary, and support when discussing other instrument flight maneuvers. Refer to this chart from time to time to better understand the roles these instruments play during each phase of flight. The primary instruments will have added importance later during partial panel practice because they become control instruments during simulated failure of the vacuum-driven attitude and heading indicators.

If you can maintain altitude within ±20 feet for 20 seconds, we'll try it for 40 seconds, then carry it on to the next step, one minute. When

CONTROL		PRIMARY	SUPPORT
STRAIGHT and LEVEL			
Pitch	Attitude Indicator	Altimeter	VSI (Rate of Climb)
Bank	Attitude Indicator	Heading Indicator	Turn Coordinator
Power		Airspeed	RPM/MP
SPEED CHANGES			
Pitch	Attitude Indicator	Altimeter	VSI
Bank	Attitude Indicator	Heading Indicator	Turn Coordinator
Power		Airspeed	RPM/MP
STANDARD RATE TURN			
Pitch	Attitude Indicator	Altimeter	VSI
Bank	Attitude Indicator	Turn Coordinator	Sweep Second Hand
Power		Airspeed	RPM/MP
MINIMUM CONTROLLABLE AIRSPEED			
Pitch	Attitude Indicator	Airspeed	RPM/MP
Bank	Attitude Indicator	Heading Indicator	Turn Coordinator
Power		Altimeter	VSI
CLIMB ENTRY			
Pitch	Attitude Indicator	Attitude Indicator	VSI
Bank	Attitude Indicator	Heading Indicator	Turn Coordinator
Power		RPM/MP	
CONSTANT AIRSPEED CLIMB			
Pitch	Attitude Indicator	Airspeed	VSI
Bank	Attitude Indicator	Heading Indicator	Turn Coordinator
Power		RPM/MP	
CONSTANT RATE CLIMB			
Pitch	Attitude Indicator	VSI	Altimeter, Sweep Second Hand
Bank	Attitude Indicator	Heading Indicator	Turn Coordinator
Power		Airspeed	RPM/MP

CONTROL	= Main reference instrument
PRIMARY	= Key quality instrument*
SUPPORT	= Back-up or secondary instrument

Variable Power (cruise, etc.) — "Power to the speed, pitch to the altitude."
Constant Power (min. controllable airspeed, etc.) — "Pitch to the speed, power to the altitude."
Power + Attitude = Performance
*The Primary Instrument is always the instrument that gives the most pertinent information and is not moving when flying precisely.

Fig. 1-1. *Control, primary, and support instruments for all the basic regimes of flight.*

LEVEL OFF (to cruise from climb or descent)

Pitch	Attitude Indicator	Altimeter	VSI
Bank	Attitude Indicator	Heading Indicator	Turn Coordinator
Power		Airspeed	RPM/MP

CONSTANT AIRSPEED DESCENT

Pitch	Attitude Indicator	Airspeed	VSI
Bank	Attitude Indicator	Heading Indicator	Turn Coordinator
Power		RPM/M	

CONSTANT RATE DESCENT

Pitch	Attitude Indicator	VSI	Altimeter Sweep Second Hand
Bank	Attitude Indicator	Heading Indicator	Turn Coordinator
Power		Airspeed	RPM/MP

Fig. 1-1. (*Continued.*)

you get a good grasp of altitude control with this method I will cover up the altimeter, wait 30 seconds or so, then uncover it to see how far up or down the airplane has drifted. With a little practice you will be amazed at how well you can fly without looking at the altimeter. This also serves as a painless little introduction to partial panel work, which we will take up on later flights.

Instructor note. It is a good idea to raise the hood every now and then so the student can compare what is seen on the instruments with the real horizon and other visual references. For example, when the nose is pitched up on the attitude indicator, raise the hood and show that the nose is also pitched up with reference to the actual horizon. This sounds obvious but it's not obvious to the beginning instrument student unless actually confirmed. The beginning instrument student needs all available help at this point.

Before moving on to turns, climbs, and descents, let's work on straight and level until it becomes automatic. Concentrating on one element at a time, you will succeed in learning very quickly to work within tight tolerances. The control pressures necessary to maintain altitude in straight and level flight are minimal, the tendency to overcontrol is much reduced, and even the heading will remain relatively constant because wings are level. And best of all, you will soon do this without hardly thinking about it!

Most readers recall getting a bicycle for a birthday or Christmas. Within the first week or two you probably wore the skin off your knees learning to ride by trial and error. Then after two weeks, suddenly the secret of balancing came and you were in full control. You could command that bicycle to turn left or right merely by leaning or shifting weight on the seat of the bicycle. And it wasn't long before you were riding along gleefully in front of friends with arms folded, just leaning left and right to make turns. Turning had become an automatic reflex—not reacting, but acting. That's the secret of straight and level flight.

Students develop an automatic reflex so that if the long needle of the altimeter is to the left of the big zero, low, apply back pressure to move the miniature airplane one-half bar width above the horizon line. And if the long needle is to the right, apply forward pressure to lower the miniature airplane one bar width below the horizon line.

I practice this with students until they are making corrections automatically for sudden updrafts and downdrafts without pausing to analyze what needs to be done. Then it will be time to move on to heading control. This isn't the end of altitude control. I'll have much more to say when considering climbs and descents to accomplish altitude changes.

Trim, pressure, and control displacement

When learning to fly, every student has a difficult time with two techniques: separating control pressure from control displacement and the proper use of trim tabs.

Trim is the easier of the two to explain. An explanation of trim: The proper use of trim, in each trim-equipped control axis, is to remove the existing pressure from the associated cockpit flight control in such a manner that the aircraft will continue in the desired flight path or attitude with little or no assistance from the pilot.

It is also easy to explain how to trim to accomplish this. Manipulate the cockpit controls to elicit the desired flight path or attitude from the aircraft. Then operate the trim control for each trim-equipped axis, one at a time, until the pressure being exerted on the control is eliminated. To double check, remove your hand or foot from the control. If the aircraft continues as desired, the trim is proper. Note that you do not fly the aircraft with the trim; you merely use it to remove the pressure from the controls; to try and fly the aircraft with the trim will usually result in the trim and the aircraft ending up 180 degrees out of phase. The proper use of trim is very important to good instrument flight.

Differentiation of control pressure and control displacement is difficult to explain and to teach, and I have found that many people never actually learn it. Properly learned and executed, this differentiation is what separates the aircraft drivers from the aircraft pilots. Many times, control pressure will not result in any noticeable control displacement; it is merely a light touch in the direction you want the control to move.

You cannot feel these pressures if you're gripping the yoke tightly. If you do grip tightly, you deaden some of the nerve endings that are necessary for this sense of feel: RELAX. That's one of the most important words for any pilot. Turn loose of the yoke and flex your hand a few times to restore the blood circulation, then place your hand—lightly—back on the wheel. Don't clamp it tightly; allow some space between your hand and the yoke all the way around. That way, if the aircraft is out of trim, you will feel the pressure immediately.

Although many times the aircraft "driver" will deliver a very precise flight, the aircraft "pilot" will give the smoothest flight, with precision as well: RELAX—that's the word. Remember, everyone reading this has carried passengers or will someday. When you do, although your primary consideration has to be safety, your secondary consideration should be passenger comfort. Your passengers should not know when your aircraft leaves the blocks and should have minimal sense of taxiing to the runway, braking, takeoff, cruise, descent, landing, taxiing onto the ramp, and stopping on the blocks. This goes for J-3 Cubs as well as 747s. It's the mark of a good pilot.

This is not to say that you don't displace the controls, for you certainly do. The slower you fly, the more displacement is necessary to achieve the same pressure, and when recovering from unusual attitudes, you use a lot of displacement and throw pressure out the window.

Although some of your instruments have lags related to them, a good instrument pilot should imagine that he has a string connected between the controls and the instrument he wants to move and that proper control pressure will move the instrument in concert with the pressure, even eliminating the inherent lag.

Instruments that lie—when and why

Murphy had a great gift for simplifying technical problems in an unforgettable way. The first law: "If a part can be installed wrong, it will be." Here is Murphy's law as it applies to instrument flying: "If an instrument can fail, it will."

Here, in summary form; is a list of instruments and problems to prepare for.

Attitude indicator. This instrument is driven by gyros powered by the vacuum system. The indicator fails when there is a failure in the vacuum system, the result, usually, of an engine-driven vacuum pump failure. Failure of the attitude indicator shows up gradually. The instrument doesn't just roll over and die but begins to drift off slowly at first as its gyros wind down. For a good idea of what this looks like, watch the way an attitude indicator behaves after engine shutdown at the end of a flight.

If you have trouble maintaining straight and level, stabilized climbs and descents, or smooth standard rate turns, suspect failure of the attitude indicator. Cover it up to prevent scan distraction and switch to partial panel operation.

The attitude indicator might also show erroneous information if it has been set incorrectly. There is only one situation in which the attitude indicator can be set correctly. And that is straight and level unaccelerated flight. That is the only time the miniature airplane can be accurately matched to the horizon line.

If you reset the attitude indicator on the flight line, it will be incorrect due to the nose-up attitude of the parked airplane. Other errors will occur if a reset is attempted in turns, climbs, and descents.

Heading indicator. Like the attitude indicator, the heading indicator is driven by the vacuum system. It will fail if the vacuum system fails, and it can also fail when the vacuum system is operating normally. Like the attitude indicator, a failure of the heading indicator is rarely dramatic. If you suspect a failure, cover the instrument and switch to partial panel operation.

Make sure that the heading indicator shows the same heading as the magnetic compass. Students make three common mistakes when setting the heading indicator.

First, they fail to reset the heading indicator when lined up on the runway centerline for takeoff. This is the best time to reset a heading indicator because you know the runway heading. If it's an instrument runway, the heading is often shown to the last degree on an approach chart instead of rounded to the nearest 10° interval.

Second, because of precession the heading indicator slowly drifts off. It should be reset every 15 minutes, or after maneuvers that involve a lot of turns in a short time, such as holding patterns.

Third, the heading indicator cannot be reset accurately unless the airplane is in straight and level, unaccelerated flight. The magnetic compass is the culprit here. It is accurate only when stabilized in straight and level, unaccelerated flight. Meanwhile, resist the temptation to reset the heading indicator in a turn. I have seen students make errors of as much as 30° while trying to match heading indicator and magnetic compass in a turn.

Magnetic compass. Let's review a few points learned in VFR training:

- In calculating headings, account for *variation* due to the earth's magnetic field, and *deviation* due to magnetic influences on a specific compass because of its location in the airplane.
- Turning toward the north, the compass lags behind the turn due to dip error, turning toward the south, the compass leads the turn.
- On easterly and westerly headings, acceleration produces an indication to the north, deceleration produces an indication to the south. Remember ANDS (Acceleration North Deceleration South).

Altimeter. The altimeter will read erroneously if not set to the correct barometric pressure at all times. It will also read erroneously if the static port is clogged. Insects, ice, and dirt can clog a static port. The problem becomes apparent when airborne and the altimeter needles don't move.

An alternate static source aboard an airplane can restore the altimeter to normal operation if the static port has become clogged: However, the altimeter will read *higher* than normal.

If you do not have an alternate static source aboard, create one by breaking the glass face of the VSI. This vents the pitot-static system to the cabin, the same as an alternate static source. Again, the altimeter will read higher than normal. (Breaking the glass face of the VSI usually damages the needle and renders the instrument inoperative.)

Airspeed indicator. A blockage of the pitot tube will render the airspeed indicator useless. As is the case with the altimeter, you won't know this until airborne. Pitot heat will prevent ice from clogging the pitot tube. That's why I recommend turning on the pitot heat before takeoff on every instrument flight. When ice has clogged the tube, pitot heat might melt it too slowly. (This is developing good habit patterns.)

Insects love to nest in the pitot tube; use the pitot tube cover. Insects can clog those tiny air passages in just a few minutes.

Blockage of the static source also causes erroneous airspeed readings. An alternate static source will produce an indicated airspeed a knot or two faster than normal.

Vertical speed indicator (VSI). The most important thing to remember about the VSI is that it only gives an accurate reading when the needle has been stabilized for 7 seconds or longer. If the needle is moving, forget it.

Another quirk of the VSI is that when you first raise the nose of the airplane to begin a climb, the VSI needle initially shows a descent. The reverse is true in the beginning of a descent when the needle will momentarily show a climb.

The needle of the VSI should point to zero when the aircraft is sitting on the ground. If not, the needle can usually be zeroed by turning a small screw at the lower left corner of the instrument case. If this adjustment cannot be made, add or subtract the error for an accurate reading in flight.

Turn coordinator. The turn coordinator is powered by electricity; it will continue to operate even if there is a failure of the vacuum-powered attitude indicator and heading indicator. If the turn coordinator fails, the needle won't move. It will remain fixed in an upright position. That's why it's important to check the movement of the turn coordinator while taxiing out. It's pretty rare, but I have also seen the ball of the turn coordinator get stuck in the tube.

Fuel gauges. Here's another Murphy's law: "On land, air, and sea, the second half of the tank always empties faster than the first half."

rust any fuel gauges. There is no way to judge how accurate
e and most of them are fairly crude. Always note takeoff time
e time en route from each major position fix, then calculate fuel
consumption based upon airplane performance figures. Ask "What if
my fuel gauges failed completely? Am I keeping track of fuel calculations well enough independently of the gauges to know exactly how
much more flying time I have left?" (It is *time* in your tanks.)

Oil pressure and temperature gauges. Engine instruments should
be scanned every few minutes. The main concerns are low oil pressure and rising oil temperature. When these symptoms appear, a serious problem is developing in the engine oil system. Land the
airplane as soon as possible. Don't stop to think about whether or
not the gauges are functioning properly.

Low oil pressure with no rising oil temperature indicates either an
instrument error or an incorrectly set pressure relief valve. Keep an
eye on this situation. As long as the indications remain stable, the
flight can be continued. But if the oil pressure drops and the oil temperature rises, land.

High oil pressure with normal oil temperature usually means that the
pressure relief valve has been set incorrectly. As long as engine indications remain normal there is no reason to discontinue the flight.

Magnetic compass turns

Note how the magnetic compass starts to come into play as a primary instrument (Fig. 1-2). The magnetic compass is one of the most
familiar and perhaps least understood instruments in the cockpit. It
is extremely reliable, even though it bounces around a lot, and it is
the only source of heading information that operates completely independently of all electrical, vacuum, and pitot-static systems. Pilots
who understand the behavior of the magnetic compass and make a
point of practicing with it frequently can achieve amazing flying precision with this instrument.

However, the magnetic compass "lies." You must know when and
why. VFR training taught you how to use magnetic compass lag and
lead to roll out on headings accurately. Here is a review:

1. When on a northerly heading and you start a turn to the east,
   the magnetic compass will initially show a turn to the west
   and will gradually catch up as the turn progresses to give an

CONTROL		PRIMARY	SUPPORT
STRAIGHT and LEVEL			
Pitch	Airspeed Indicator	Altimeter	VSI (Rate of Climb)
Bank	Turn Coordinator	Magnetic Compass	ADF
Power			RPM/MP
STANDARD RATE TURN			
Pitch	Airspeed Indicator	Altimeter	VSI
Bank	Turn Coordinator	Sweep Second Hand	ADF
Power		Airspeed Indicator	RPM/MP
CONSTANT AIRSPEED CLIMB			
Pitch	Airspeed Indicator	VSI	Altimeter
Bank	Turn Coordinator	Magnetic Compass	ADF
Power			RPM/MP
CONSTANT AIRSPEED DESCENT			
Pitch	Airspeed Indicator	VSI	Altimeter
Bank	Turn Coordinator	Magnetic Compass	
Power			RPM/MP
CONSTANT RATE DESCENT (ILS)			
Pitch	Airspeed Indicator	Glide Slope Needle	VSI
Bank	Turn Coordinator	Localizer Needle	ADF
Power		Airspeed Indicator	RPM/MP

CONTROL	= Main reference instruments
*PRIMARY	= Key quality instrument
SUPPORT	= Back-up or secondary instrument

*The Primary Instrument is always the instrument that gives the most pertinent information and is not moving when flying precisely.

Fig. 1-2. *Control, primary, and support instruments for partial panel. Simulating loss of vacuum affecting attitude indicator and heading indicator.*

accurate indication when passing through 090°. Conversely, when on a northerly heading and you start to turn to the west, the compass will initially show a turn to the east, then gradually catch up.

2. When on a southerly heading and you start a turn toward the east, the magnetic compass will initially indicate a turn to the east, but will exaggerate the turn, gradually reducing the error and will be accurate when passing through 090°. When turning to the west, the compass behaves similarly by exaggerating the amount of turn initially, but in the correct direction.

This lagging and leading behavior of the magnetic compass in 1. and 2. above is *dip error* caused by the earth's magnetic lines of force and their effect on the magnetic compass when it is not precisely level. (A rough formula: dip error is equal to the closest latitude in degrees.)

3. When flying on an easterly or westerly heading, aircraft acceleration results in a northerly turn indication; deceleration results in a southerly turn indication.

Remember this by the acronym ANDS:

Accelerate

North

Decelerate

South

Magnetic compass procedures and exercises

1. When turning to a heading of south, maintain the turn until the compass passes south the number of degrees of dip error (degrees of latitude) minus the normal rollout lead (one-half the angle of bank).

2. When turning to a heading of north, lead the compass by the amount of dip error (latitude) minus the normal rollout lead.

3. When turning to a heading of east or west, anticipate the rollout by the normal method.

The main point to remember is that on southerly headings, the magnetic compass precedes or leads the actual turn. On northerly headings, the magnetic compass lags behind the turn. Figure 1-3 shows the procedures for making magnetic compass turns in graphic form.

Turns to a heading based upon indications from the magnetic compass are imprecise at best; however, you should practice magnetic compass turns on instruments until you can roll out on a specified heading ±10° using the magnetic compass alone. But don't waste time trying to go beyond this. Concentrate on learning and recognizing the errors affecting the magnetic compass and never believe it unless you are straight and level and in stabilized flight.

This is especially important during the departure and approach phases of an instrument flight when the heading indicator must be accurately set. Do not adjust the heading indicator unless the air-

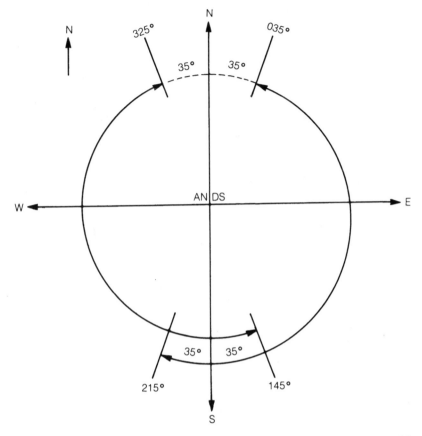

Fig. 1-3. *Magnetic compass leads (to the south) and lags (to the north) assuming a latitude of 30° for the dip error and one-half a bank angle of 10° for leading the roll in, with acceleration/deceleration errors on east and west headings.*

plane is in stable and straight and level flight. In a normal approach, the last opportunity for determining heading indicator accuracy might be the outbound (1 minute) leg of the procedure turn.

Instruments

In most general-aviation instruments there are several mechanisms that are normally placed in standard positions relative to each other. With that said, you will probably find that the airplanes you fly have some variation of the instrument panel configuration that we will be using here. Having the instruments placed in a different position in the panel is not a real problem, but know where the instruments are so that you don't waste time looking for them and are able to scan the instrument panel effectively. As you will see, there are normally ways to cross-check instruments using the proper scanning techniques that can help you quickly determine if you have a malfunctioning instrument. Knowing where to look on the panel can help you smoothly scan the panel and more easily maintain the proper attitude, heading, and altitude.

Figure 1-4 shows a basic instrument panel with a series of six instruments laid out in it. Across the upper row from left to right are the airspeed indicator, the artificial horizon, and the altimeter. On the second row left to right you will find the turn and bank indicator, the directional gyro, and the vertical speed indicator. Each of these instruments has a significant piece of information to offer in reference to the airplane's status. Let's work our way across the panel and look at each instrument's function and relation to the other instruments.

Airspeed indicator

The airspeed indicator is a pressure-sensitive instrument that uses pressure differences to determine the speed of the airplane. The pitot tube and static system measure the pressure of air and use the differences between them to calculate the airspeed. As the airplane flies through the air, the pitot tube measures the pressure of air entering the pitot tube, which is placed along the wing on most

Fig. 1-4 *Instrument layout.*

general-aviation aircraft. Figure 1-5 depicts a typical static system. As you can see, the pitot tube connects to the airspeed indicator, which then connects to the static vent and alternate static vent. The altimeter and vertical speed indicator are also connected to the static system. The pitot is positioned so that the airflow impacts it and flows into it, generating pressure. The airspeed indicator is marked with several colored arcs that show ranges for normal flight, flap extension, stall speeds, and the never-exceed speed.

But you can also use the airspeed indicator to cross-reference with the artificial horizon and vertical speed indicator to verify that they are also working correctly. If you are in a dive, you will notice that the airspeed will increase. In a climb it will decrease as the nose rises. If you keep power settings constant, a drop in airspeed could mean the nose is rising. Cross-referencing this information with the artificial horizon, altimeter, and vertical speed indicator will help you determine if the instrument is accurate. If the other gauges give an indication that the plane is in level flight but the airspeed indicator is decreasing or increasing, it could be a sign of a malfunctioning airspeed indicator.

While flying under instrument conditions, the airspeed indicator is used to assure that in climbs or descents the plane's airspeeds are maintained within the correct ranges. During a climb or descent, the attitude indicator can be used to control the pitch, but the airspeed indicator is used to refine the airspeeds to the correct values. In a climb you want to use the correct airspeed for the situation, such as

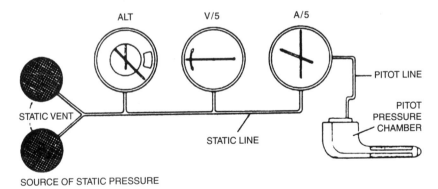

Fig. 1-5 *Static system.*

V_x or V_y, while in a descent you want to also be sure you are not exceeding recommended speeds. When pilots become disoriented under instrument conditions, it is not uncommon for the nose to drop and the airspeed to build rapidly. By monitoring the airspeed indicator, this problem can be more easily avoided.

Artificial horizon

The artificial horizon (AH) is one of the most useful, and most abused, instruments in the panel. Both bank attitude and pitch angles are available from the artificial horizon, which is extremely important when flying on instruments. But many pilots have a tendency to focus almost exclusively on the AH to the point of excluding the other instruments. Illustrated in Fig. 1-6, the artificial horizon is normally made up of a "ball" that depicts the sky and ground, the intersection of those two areas being the horizon. While the color of the sky and ground will be different for various AH instruments, they will normally use blue or white to represent the sky, and brown, black, or other darker tones for the ground. Graduated lines show pitch up and down angles on each of these areas. This central "horizon" portion of the instrument moves up and down to represent pitch angles and rolls left and right to show bank angles. Graduated marks along the sides of the artificial horizon are used to measure the bank angle of the airplane. In the center of the instrument there is usually some representation of the airplane that allows you to more easily visualize the pitch and bank attitudes for the plane.

Fig. 1-6 *Artificial horizon.*

The artificial horizon is a vacuum-driven gyroscopic instrument on most general-aviation aircraft. Essentially, the vacuum pump on the airplane's engine generates a flow of air through the instrument that spins up a gyroscope that maintains a constant attitude relation to the earth. Figure 1-7 shows a typical pump-driven vacuum system and the instruments normally attached to it. In most general-aviation planes, the directional gyro, artificial horizon, and the turn and bank indicator are driven by the vacuum system. As the plane banks or changes pitch angle, the gyroscope pivots on its mounts, trying to stay level. The central horizon is attached to the gyroscope and also maintains a relative position to the earth. As the plane changes pitch or bank angles, the gyroscope holds its position, which allows the instrument to show the bank and pitch angles.

The artificial horizon is useful, but it should also be cross-checked against the altimeter, the airspeed indicator, the vertical speed indicator, and the directional gyro. For instance, if the AH shows the plane is in a bank, but the directional gyro shows the heading is constant, there may be a problem with one of the instruments. A further cross-check with the compass will confirm if it is in sync with the directional gyro. If the two match, there is a strong likelihood that the artificial horizon is inaccurate. On the other hand, if the compass is showing a turn and the directional gyro (DG) is not moving, the DG could be malfunctioning. You can begin to see from just this little discussion that flying on instruments demands that you understand the relation of the instruments to each other and know how to rule out specific instruments when one or the other is malfunctioning. It also means that attempting to fly on instruments when you are not rated to could be more demanding than you might have anticipated.

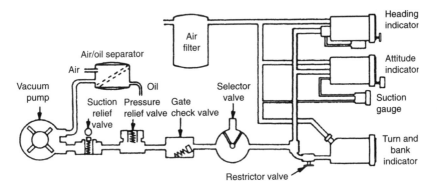

Fig. 1-7 *Pump-driven vacuum system.*

Altimeter

The altimeter is used to determine the altitude of the airplane through the measurement of barometric pressure. The altimeter is adjusted to the correct pressure by setting the value in the small window to the current barometric pressure. The altimeter shown in Fig. 1-8 shows an altimeter setting of 30.34. Like the airspeed indicator, it uses the static system as the source for its measurements. Altimeters can be equipped with one, two, or three needles used to indicate altitude. On an altimeter equipped with three needles, the long needle indicates hundreds of feet, the next shorter needle indicates thousands of feet, and the shortest needle shows tens of thousands of feet. If the barometric pressure in the window is incorrect, the altitude the instrument indicates will also be incorrect. Assuming your altimeter is working correctly, you can also determine if you are climbing or diving based on whether it shows your altitude is increasing or decreasing. This can also be useful for cross-checking with the artificial horizon and the vertical speed indicator to determine if all instruments show the same information regarding climbs or descents.

Turn and Bank Indicator

Figure 1-9 shows a turn and bank indicator, also known as a turn and slip indicator. You can see that it is made up of a vertical white bar, called a needle, and a black sphere in a glass tube, called the ball. We have already covered in great detail the fact that the ball is used to indicate whether the plane is in coordinated flight. Figure 1-10A shows the needle and ball when the plane is in a slip, while

Fig. 1-8 *Altimeter.*

Fig.1-9 *Turn and bank indicator.*

Fig. 1-10A *Turn and bank: slip.*

Fig. 1-10B *Turn and bank: skid.*

1-10B shows the plane in a skid. In both cases the ball is not centered in the tube, indicating that the plane is in uncoordinated flight. The phrase "step on the ball" is a general rule of thumb that says you should step on the rudder to the side that the ball is toward. In 1-10A, this would mean you need to push on the right rudder, while in 1-10B you need more left rudder.

The needle indicates the direction of the bank, which is to the right in both examples. Located above the needle are three figures. The middle white mark above the needle is the center marker, while on either side the dog-house shaped marks are used to indicate the plane is in a standard rate turn. If the ball is centered and the needle is located on the center mark, the plane is in a wings-level attitude. But if the needle is to the left or right of the center marker, the plane is banked. This can be a useful indicator of the plane's attitude and a good cross-check of the artificial horizon. If the needle is on the left or right "dog-house," and the ball is centered, the plane is making a turn in that direction at 3 degrees per second of heading change. When flying under instrument conditions, turns are normally made at the standard rate. The timing and layout for most instrument approaches are set up for standard rate turns.

Like the artificial horizon, the turn and bank indicator is driven by a gyroscope. This gyro is driven by the vacuum system on many aircraft, while on others it is an electrically powered gyroscope. The reason for having the turn and bank on a separate drive system is that if the vacuum pump fails, you will not lose all gyroscopic instruments. In the unfortunate event that the vacuum system does have problems, you could still maintain control of the plane by using the turn and bank indicator—in conjunction with the compass, altimeter, and vertical speed indicator—to maintain level flight at the correct heading. Granted, this requires a little more work on the pilot's part, but the redundancy of instruments gives you a fallback in the event of major system problems.

Directional gyro

A directional gyro (DG) is depicted in Fig.1-11. As you can see, it has the headings of a compass on it, with a small plane in the center of the instrument. As the name implies, this is a gyroscopically driven instrument that allows you to determine the heading of the plane. The DG is basically a dumb instrument; it does not inherently know magnetic directions. Before you fly you must set the initial heading on the DG to match the compass. After the heading is set, the gyroscope in the DG attempts to maintain its alignment with that initial setting. When the plane turns, the card with the compass headings rotates within the DG, showing the airplane's current directional heading.

It should quickly be apparent that if the initial heading that the DG is set to is inaccurate, the headings that it shows will also be in error. The gyroscope in the DG also tends to precess, or shift on its axis as

Fig. 1-11 *Directional gyro.*

time passes. In order to be sure the DG shows an accurate heading, you will need to periodically reset the heading it shows to match the heading on the compass. Depending on the condition of the DG, it may be necessary to reset the heading as often as every ten minutes. The advantage the DG has over the compass is that it will maintain a steady indication of the plane's heading and is not subject to the bobbing and weaving that the compass is prone to in rough air. Anyone who has attempted to hold an accurate heading while flying in turbulent air by using only the compass will attest to the problems this creates. The directional gyro is not affected by this type of movement and allows you to more accurately hold a heading.

When a DG fails, it can slowly precess, the card that shows the compass points slowly turns, or it can fail more completely. In a severe failure of the DG it will often spin around in circles at an amazing rate. Major failures are easy to pick up; the instrument is making so much nonsense that it is very easy to spot the behavior. But in a slow failure the heading can slowly drift off from the correct value. If you do not catch this you might find that you are turning unconsciously as you attempt to hold the correct heading on the DG. While you may think you are heading in the right direction, this situation can put you significantly off the correct one. For this reason you should frequently cross-check the DG with the compass.

Vertical speed indicator (VSI)

The vertical speed indicator is the last instrument in the basic cluster we will review. Figure 1-12 shows the instrument, which is another that is attached to the static system. The instrument shows the rate of altitude change in hundreds or thousands of feet per minute. The VSI is

Fig. 1-12 *Vertical speed indicator (VSI).*

useful for setting up descents at the proper rate for a number of instrument approaches. They are often calibrated for 500-feet-per minute descents, which provide a comfortable descent angle and manageable airspeeds during the approach. They are also useful for cross-checking the altimeter and artificial horizon to assure that they are accurate.

Visual and instrument correlation

A current, and ongoing, question among instructor pilots is how much instrument integration instruction a student pilot should receive, and how soon it should be integrated.

My personal feeling on this is that the student pilot should be taught to use the instruments and to correlate them with his or her external visual references from the first lesson. At no time, however, should the necessity of adequate head-up flying be minimized. I have always felt that no student should be soloed until he or she can take off, fly the pattern, and land with a simulated blockage of the airplane's pitot-static system. As much stress should be placed on the full use of a pilot's senses (sound of wind and engine, and visual height separation) as in the days of open-cockpit biplane trainers.

From the beginning, though, the student should be shown the similarities between the instrument and the visual cues, such cues as the attitude indicator showing wings level, the outside horizon being level, and the like.

He or she should also be shown how to interpret this information. For example, if your aircraft is flying wings-level with the nose on the horizon, and your attitude indicator shows the same, does this mean that you are actually flying straight and level? The answer is "Of course not." If you encounter an updraft or downdraft (*wind shear*), or if your power is too high or too low, your aircraft can climb or descend accordingly and still be in a level flight attitude. Visually, the ground will appear to come nearer or farther away. On instruments you will notice the vertical speed indicator deflect up or down, with a corresponding change in the altimeter. If the instrument changes are due to changes in power setting, you will also notice improper airspeeds and a change in engine sound, and the power instruments will show a difference in RPM and/or manifold pressure.

Here's another example. Does a nose-high attitude indicate a climb? Not unless you have added sufficient power. In slow flight the aircraft has to assume a nose-high attitude to maintain altitude.

All of these relationships should be pointed out to the student from the beginning. What we should be pointing out to the students, in other words, is that one visual cue, or the indication of one instrument, is not enough. You must derive information from as many sources as possible in order to accurately interpret the flight path and attitude of the aircraft.

Turn coordinator and clock

The turn coordinator is the primary, or quality, instrument. In a standard rate turn, the symbolic airplane tilts and its wings line up with left and right benchmarks (Fig. 1-13). The ball remains centered throughout the turn with proper rudder input. (In standard rate turns with an older turn and slip indicator, the needle is displaced one needle width in the direction of the turn.)

The sweep-second hand of the clock may be included in the scan during standard rate turns. It is an invaluable support instrument because it shows whether you're turning faster or slower than 3° per second. No matter where the second hand is when you commence a turn, it should make half a sweep for every 90° of turn, and a complete sweep for every 180° of turn.

If you have turned 90° and the second hand is more than halfway around the clock face, you know the rate of turn is too slow; speed

Fig. 1-13 *Standard rate turn to the left at 100 knots. The wings of the miniature airplane match the benchmark on the turn coordinator. But the angle of bank falls between marks on the attitude indicator because the airspeed is 100 knots.*

up the rate of turn by increasing the angle of bank slightly. Likewise, if you have turned 90° and the second hand has not reached halfway, the rate of turn is too fast. Decrease the rate of turn by decreasing the angle of bank slightly. Be careful, however, not to fixate on the clock. That steadily advancing sweep-second hand is a powerful attention-getter. Make a conscious effort to direct attention away from the clock and return to your normal scan, resuming, as usual, with the attitude indicator.

Practice level standard rate turns in both directions for a full 360°, then try turns of 180° and 90°. To hit the rollout heading exactly, anticipate the target heading by one-half the degrees of bank, just like VFR turns to a heading. If the angle of bank in the turn is 20°, for example, begin the roll out 10° before reaching the target heading. A 10° angle of bank requires a lead of 5°.

2

Aircraft control

Now that you understand the basics of instrument scan and operation, you can begin practice on techniques and procedures for flying solely by reference to the instruments.

Attitude control

To make the aircraft perform these maneuvers the pilot must maintain or change the attitude. The attitude indicator was specifically developed as the basic reference for maintaining or changing attitude in instrument flight. The other flight instruments are vital, of course, and lifesavers if the attitude indicator does fail. But for normal IFR flying, the attitude indicator is the star player, and the other flight instruments are the supporting cast.

Straight and level

I have found it best to work first on straight and level flight. Most of the time an instrument cross-country is in straight and level flight. And, more often than not, the other basic maneuvers begin and end with a return to straight and level.

This basic maneuver is divided into two elements: altitude control and heading control. It seems to work better to practice altitude control first, then work on heading control.

Choose an odd or even thousand-foot altitude—such as 3,000, 4,000, or 5,000 feet—for practicing straight and level. It's interesting to note in passing that on an IFR flight, most of the time you will be flying at an odd or even thousand foot altitude with the long hand of the altimeter pointing straight up to zero.

Set the power for cruise. Trim the airplane. Trim is important because the closer you can trim the airplane to fly "hands off," the easier it will be to make the small corrections that are so vital to precise instrument flight.

Altitude control

In level flight, the representative wings on the attitude indicator should superimpose the horizon line to form one line (Fig. 2-1). If it doesn't look like this in level flight, reset the attitude indicator.

Try maintaining altitude within ±20 feet for 20 seconds with reference to the attitude indicator primarily, using the altimeter as a cross-check. If altitude drops, use slight back pressure on the control yoke to pitch the nose up slightly. Raise the wings of the miniature airplane to one-half bar width above the horizon line, keeping the wings level (Fig. 2-2). (Bar width is the thickness of the wings of the miniature airplane represented on the attitude indicator.) When you get back to the desired altitude, relax the back pressure and align the wings with the horizon line.

If altitude increases, apply a slight forward *pressure* on the yoke to pitch the nose down slightly. Lower the miniature airplane to one-half bar width below the horizon line, keeping the wings level (Fig. 2-3). Cross-check with the altimeter. Level off at the desired altitude. Relax the forward pressure on the yoke and align the wings with the horizon line again. *Pressure* (not movement) is the key to smooth control.

Step climbs and descents

The "step" climb and descent is particularly good for developing an automatic reflex for making altitude corrections. The maneuver is very helpful to the student trying to reach the goals for climbs and descents and controlling altitude. Figure 2-4 shows the maneuver from a starting altitude of 3,000 feet.

Instructor note. The step climb and descent is very useful for another reason: It can be introduced early in the syllabus without overloading the student.

After a student has demonstrated an ability to hold some convenient altitude consistently for two or three minutes on a constant heading, practice the step maneuver next—climb 100 feet, level off, and hold

Fig. 2-1. *In straight and level flight, the miniature airplane on the attitude indicator should be set so that the top of the "wings" make a straight line from left to right with the horizon line.*

Fig. 2-2. *When below a prescribed altitude by 100 feet or less, correct by increasing pitch one-half bar width.*

Fig. 2-3. *When above a prescribed altitude by 100 feet or less, correct by decreasing pitch one-half bar width.*

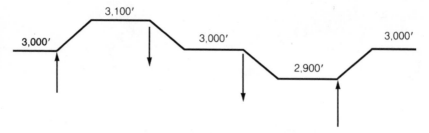

Fig. 2-4. *Step climbs and descents for practicing altitude adjustments.*

that altitude for one minute. The miniature airplane on the attitude indicator is held one-half bar width above the horizon line with back pressure alone. No power changes are needed for a change of less than 100 feet.

After the minute is up, descend to the original altitude, level off, and hold that for one minute. The miniature airplane should be pitched one-half bar width below the horizon line for the descent. Again, there should be no change in the power setting.

One-half bar width on the attitude indicator is a very small adjustment, but with a little practice, it can be easily achieved every time. One-half bar width is also an excellent rule of thumb: For altitude changes of 100 feet or less, use no more than a one-half bar width correction. Remember: *No power changes.*

The remainder of the exercise consists of one more descent and one more climb performed the same way—establishing pitch with the attitude indicator and making no power changes. The wings should remain level according to the attitude indicator.

The exercise should be continued until altitude can be controlled within ±20 feet of what is desired. Students who master step climbs and descents usually have little difficulty with altitude changes of more than 100 feet.

Heading control

As an expert at maintaining altitude in straight and level flight, it's time to exercise the same finesse to control and maintain heading. Set up straight and level on a convenient heading. With wings level on the attitude indicator, apply control pressure equally and very gently to both rudder pedals. It's not necessary to be heavy-footed; just keep both feet in contact with both pedals and balance the pressure.

Constant heading with wings level is maintained by small adjustments in the balance of pressure on the rudder pedals. Without constant pressure on the rudder pedals, the airplane will drift off heading if it yaws due to poorly adjusted rigging or constantly changing air currents.

Make minor adjustments in the pressure on the rudder pedals to compensate for heading changes caused by these factors. For heading adjustments of 5° or fewer use rudder pressure only, holding wings level; flight will be uncoordinated momentarily, but you will avoid the tendency to overcontrol.

Overcontrolling is an impulse that all students seem to have and they need to work consciously to avoid it. Think about it. If you want to make a 2° heading change and you roll in 20° of bank, what happens? Suddenly the heading has changed 20° or 30°, and you are powerfully tempted to roll into a steep bank in the opposite direction.

Airspeed control

Achievement of precise altitude and heading for longer and longer periods in straight and level flight leads to precise airspeed control. Recall from VFR flying that to increase airspeed in straight and level flight it is necessary to increase power—to decrease airspeed, reduce power. The same principles apply to IFR flying, of course, but the principles are applied more precisely than before.

Precision will demand very small changes in power settings—increments of 100 RPM or 1″ of manifold pressure in a high-performance airplane, unless a specific power setting is required. Light control pressure makes it easier to hold altitude and heading.

Likewise, small power changes will also make it easier to control the airplane within the goal of 2, 2, and 20. Large changes in power, or "throttle jockeying," is a form of overcontrolling and must be eliminated or else precise instrument flight cannot be attained.

A power change of 100 RPM or 1″ of manifold pressure yields a change in airspeed of $7\frac{1}{2}$ knots. Strangely enough, this rough rule of thumb can be applied to any propeller-driven airplane you are likely to fly, Cessna 152, Cherokee Arrow, Beech Baron, or whatever. For example, if cruising at 100 knots and you want to reduce the airspeed to 95 knots, the 5-knot change would equate to a reduction of 75 RPM or $3/_4$" manifold pressure. To increase or decrease airspeed by 10 knots, you would increase or decrease RPMs by 125 or change manifold pressure by $1^1/_4$". Remember that this rule of thumb is not set in concrete. Minor adjustments are almost always necessary. It is something to start with. (I will offer many rules of thumb throughout this book; they are valuable guides to precision IFR flight and you won't have to waste time reinventing them—don't waste time trying to rediscover old knowledge.)

Airspeed transitions

Let's take our discussion of airspeed control one step further and talk about airspeed transitions—how to change from one airspeed to another with facility and precision.

First, establish a specific speed for each flight condition for the type of airplane flown. For a Cessna 172 in level flight, for example, consider 110 knots as normal cruise, 90 knots as slow cruise, and 70 knots as slow flight.

Next, establish the power settings to maintain these selected speeds in a Cessna 172. Maintain 110 knots with 2450 RPM, 90 knots with 2100 RPM, and 70 knots with 1900 RPM.

To slow down from 110 to 90 knots, for example, reduce power to the setting established for that speed, 2100 RPM.

Maintain level flight by slowly raising the nose of the miniature airplane on the attitude indicator, with wings level. Increase back pressure on the yoke as the airspeed bleeds off, and trim out the pressure as the desired airspeed approaches.

Cross-check with the primary instruments: altimeter for the best information on pitch, heading indicator for bank, and airspeed for power. Include the support instruments in your scan: VSI for pitch, turn coordinator for bank, and tachometer for power.

To speed up and resume 110 knots, reverse the procedure. Increase power to 2450 RPM and maintain level flight by slowly lowering the nose with reference to the horizon line on the attitude indicator. Increase forward pressure on the yoke, then trim out the pressure near 110 knots. Cross-check the same primary and support instruments.

Power changes induce airspeed changes, while adjusting attitude changes altitude. In straight and level cruising flight, it's "power to the speed, pitch to the altitude." You'll hear instructors say this over and over as you make airspeed changes in level flight.

Power

There are only two basic power conditions for flight: variable (adjustable) and not variable (either by choice or by accident).

Variable power is used in level flight to control airspeed (power to the speed, pitch to the altitude). Variable power is also used in climbs and descents at a specified rate, such as 500 feet per minute, and on the glide path of an ILS approach. You will see how this works later when these maneuvers are analyzed.

The conditions under which power is not variable occur when using full throttle during takeoff, when the throttle is closed or the engine fails, or when the power is in transit during a transition from one maneuver to another, such as intercepting a glide slope.

In these situations remember "pitch to the speed, power to the altitude" from VFR training and flight at minimum controllable airspeed.

Trim in a turn

A level turn requires back pressure on the control wheel to maintain altitude. Students frequently ask me whether or not they should apply nose-up trim during a turn to reduce the back pressure needed to keep the nose up. This is an individual matter and there is no strict rule for trimming. I think you will find that a little extra back pressure on the yoke during the turn will work very well in counteracting the tendency of the nose to descend. But it's really up to you; find the method that works best for you and stick with it.

Cross-check attitude indicator, altimeter, and VSI to make sure you are using the correct amount of back pressure to maintain constant altitude. In a full 360° turn, which requires two minutes, it *is* helpful to add a quarter of a turn of nose-up trim. Remember that this trim will have to be removed after completing the turn or the nose will rise.

Standard-rate turns

After you perfect this technique, you should develop the ability to execute standard-rate turns. A standard-rate turn for normal general aviation aircraft is 3 degrees per second. This means that it will require 2 minutes (120 seconds) to execute a 360-degree circle. For slower aircraft this turn rate will require a bank angle of 10–25 degrees. High-performance aircraft require steeper banks. High speeds can even require a 4-minute (or half-standard-rate) turn in order to keep the bank angle from becoming excessive. The maximum bank angle you'll want to make on instruments is 30 degrees, and most flight command systems will limit the aircraft to 27 degrees.

One instrument that will help maintain a constant standard-rate turn is the turn coordinator or its equivalent. Most general aviation aircraft use a 2-minute turn indicator. Because all instruments have certain errors, it becomes necessary to determine the exact needle position on your turn coordinator that will result in a standard-rate turn.

Roll smoothly into a turn and place the low wing of the airplane figure on the turn index mark. Check the bank angle on the attitude indicator. Fly through 360 degrees, timing your turn as you pass through a few of the cardinal headings; continue through another 360 degrees or so, readjusting the angle of bank as necessary to complete a 360-degree turn in 2 minutes at your normal cruising airspeed.

Once you have established the exact needle position and attitude indicator position for your aircraft, you should be able to execute a 3-degree-per-second turn every time. Time the turn coordinator in both directions.

Timing the turn through a few cardinal headings can be explained. If you are turning at a rate of 3 degrees per second, you should turn through 30 degrees in 10 seconds, 45 degrees in 15 seconds, and 60 degrees in 20 seconds; so, as you are turning, you won't have to fly the entire 360 degrees to see if your angle of bank is resulting in the desired rate of turn. But once you have adjusted and readjusted your bank angle to get the desired results, it is still necessary to go through a complete 360-degree turn to ascertain that it is correct.

Once you have accomplished this experiment, you should be able to establish a standard-rate turn by using either the turn coordinator or the attitude indicator. Naturally, when both instruments are working, you will constantly use one as a cross-check against the other, but if one fails, you should still be able to complete the turn. Not only that, but by rolling into and out of turns smoothly and slowly, always at the same roll rate, you should be able to roll out on any heading— or be able to make a turn of any number of degrees—without reference to the heading indicators, just by timing the turns. This ability has saved many pilots when their respective magnetic compasses have sprung a leak.

To execute a timed turn properly, begin timing just as you begin rolling into the turn, and begin rolling out again at the end of the timing period.

The ball on the turn coordinator will help you keep the turn coordinated. It should always be centered. If the ball is on the high side of the turn, it indicates a *skid*. If it is on the low side, it indicates a *slip*. In either case, "step on the ball" to center it. First, take all pressure off the rudder pedals, which will usually eliminate 70 percent of the skid. Then apply just enough rudder pressure to center the ball, pressing the rudder pedal on the same side to which the ball is deflected.

As explained in Chapter 1, apply smooth pressure to the aileron and rudder to roll slowly into a coordinated turn. As the aircraft reaches the desired angle of bank as indicated both by the turn coordinator and the attitude indicator, the speed will decay somewhat due to some of the lift being used to turn the aircraft. As it occurs, the nose will begin to drop as evidenced by the attitude indicator. About this time, the vertical speed indicator will indicate a descent, and if you wait much longer, the altimeter will begin unwinding.

Apply back pressure to the yoke to hold the proper nose attitude on the attitude indicator and then trim off the back pressure, cross-checking the compass and heading indicator, turn coordinator, and attitude indicator for the proper rate of turn. It will be necessary to add a small amount of power to maintain the desired speed due to the increased drag produced by the increased angle-of-attack.

As you approach the desired heading, apply smooth, coordinated pressure to the opposite aileron and rudder to roll out of the turn on the proper heading. As you near a wings-level attitude, the upward lift vector will increase and the nose attitude will tend to rise. Stop this by applying forward pressure on the yoke and then by trimming off the pressure. It will also be necessary to ease off the extra power applied when entering the turn.

The most common errors committed while rolling into and out of turns are:

- Rolling in and out of turns at varying roll rates.
- Omitting the attitude indicator from the scan, allowing the nose to lower after entering the turns or rise after rolling out.
- Anticipating an altitude loss entering turns or a gain when rolling out and reacting too soon, causing the opposite results.

Practice should consist of perfecting turns using the magnetic compass, perfecting timed turns, and rolling in and out of turns smoothly and at a consistent roll rate.

Straight and level flight

Combine the pitch instruments with the heading instruments and try to fly straight and level.

The first thing to do is to get the attitude indicator (AI) to show a wings-level, nose-level attitude with the power set for cruise. You must program both a heading and an altitude in your mind. As you scan the instruments, if you find the airspeed is higher or lower than normal for the power you have set, check the altimeter and vertical speed indicator (VSI) for any deviation. If none is noted, check your power control to determine if it has slipped in or out.

If you do have a change of altitude, make a slight pitch change, referencing the AI, to try to stop the unwanted trend. It will be important to remember the new pitch position of the AI. After making the slight pitch change, you should scan the directional gyro (DG) and needle/ball (or turn coordinator, as the case might be). If the needle is upright (or the airplane figure is wings-level), the ball is centered, and the AI indicates wings-level, you should have no change in the heading.

If your heading has changed and is still changing, the turn needle will be deflected and the ball will probably be off center. To stop the turn, apply aileron pressure opposite the direction of the turn needle deflection, and use pressure on the rudder pedals to keep the ball centered.

If you apply slow, smooth pressures, you might imagine that the turn needle (or airplane figure) is directly connected to the yoke or stick, and you can almost "will" it to the center by the light pressure you are exerting on the yoke. When the needle centers, glance at the AI to note the wing position. That will be your new wings-level reference for straight flight, at least momentarily, until any precession has worked out of the gyro.

Now, return your scan to the pitch instruments. The altitude deviation should have stopped due to your corrective pitch change, and now you can make another pitch change to bring the aircraft back to your desired altitude. This pitch change should result in a vertical speed equal to twice the altitude change required, up to a maximum of 500 fpm. In other words, if the altitude is off by 100 feet, you should climb or descend by 200 fpm until you return to the desired altitude. This 100-foot climb or descent will take 30 seconds, during which time you will go back to the heading indicators and begin a slight correction back to the desired heading.

Straight and level flight is probably the hardest to master, perhaps because most pilots become complacent and feel that it's too easy. Once you're able to fly fairly straight and level at cruise power and speed, start varying your speed as you did before, only this time keep your heading by scanning the heading instruments.

From straight and level flight at cruise power, reduce to approach speed, maintaining both heading and altitude. Slowly and smoothly reduce the power to slightly less than required for approach speed. As the airspeed begins to decrease, you will notice a tendency for the pitch attitude to lower, accompanied by the VSI showing a descent. You should apply back pressure smoothly, raising the pitch attitude on the AI at a rate necessary to maintain a zero indication on the VSI and altimeter. This back pressure should be trimmed off as necessary.

At the same time, the ball will tend to move to the left and the nose will tend to yaw to the right. This can be corrected by either easing the pressure off the right rudder pedal or by applying sufficient pressure to the left rudder pedal, as the case might be, to keep the ball centered. Properly executed, this maneuver should result in no change of heading.

When the airspeed reads 3–5 knots above the approach speed, the power should be advanced smoothly to the approach speed power that you learned earlier. The pitch attitude and power setting will be a close approximation and might have to be adjusted slightly to maintain the proper airspeed and altitude.

Now, increase airspeed back up to cruise, while at the same time maintaining heading and altitude. To do this, increase the power smoothly to the climb power setting and reverse the previous procedures. Keep the ball centered by using slight right rudder pressure as the power is increased. When the airspeed reaches cruise, smoothly reset cruise power.

Climbs and descents

Once you are able to do these maneuvers fairly well, try a climb, straight ahead. To initiate a climb, apply slight back pressure to set a climb attitude indication on the AI. As the airspeed decreases, the increasing back pressure on the yoke should be trimmed off. When the airspeed reads 5 knots above climb airspeed, slowly and smoothly advance the power lever to the climb power setting. This procedure will result in a smooth transition from level flight to a climb attitude.

Throughout this maneuver, the right rudder pressure will have to be slowly increased to keep the ball centered and to hold the desired heading. If the aircraft has a rudder trim tab, this pressure should be trimmed out. Pitch attitude should be readjusted as necessary to establish and maintain a 500-fpm rate of climb on the VSI.

During the last 100 feet of climb, the pitch attitude should be gradually changed to level off at the desired altitude. As the airspeed increases, the pitch control pressure should be removed through the use of trim. When the airspeed reaches cruise speed, the power should be smoothly reduced to cruise power, and the rudder pressure should be removed to keep the ball centered.

So far, so good. Now for descents. There are at least three different airspeeds commonly used during descents. You should learn to use them all.

The first and easiest is merely to lower the pitch attitude and retrim as needed to establish a 500-fpm rate of descent on the VSI, allowing the airspeed to increase, and reducing power only if the airspeed approaches the redline V_{NE}. This is a good descent procedure to use in calm air. But instrument flight is seldom conducted in calm air, so a slower descent, your second choice, is made by merely reducing the power sufficiently to establish a 500-fpm descent at cruise airspeed. In this case, very little pitch trim will be necessary. Close to the desired altitude, slight nose-up pressure (trimmed out) and reestablishment of cruise power should hold the airspeed constant.

The last type of descent is made at approach speed. From level flight and approach speed/power, you can lower the gear and/or flaps and retrim to establish the required descent rate, reduce power as you did in the preceding method, or use a combination of these methods. The important thing is to first reduce to approach speed and level flight. The fewer attitude, speed, and configuration changes you have to make at one time, the easier it becomes to do it correctly.

If you have to use drag devices (gear and/or flaps) to establish the descent, when the aircraft is close to the desired altitude, maintain the altitude by retracting the drag devices or increasing power. Establishing the new altitude by the retraction of drag devices requires a longer lead time than by establishing the altitude with power changes, and usually tends to confuse or alarm passengers.

All of these procedures should be practiced first in straight flight and then combined with turns to specific headings until they can be accomplished smoothly with little or no noticeable changes of gravity forces and with minimum instrument deviation. The turns should be practiced using the directional gyro, the magnetic compass, and time, separately and in combination. Altitude changes should be practiced using altimeter and time.

In short, practice, practice, practice, until the procedures become second nature and your instrument scan becomes automatic, avoiding the most common errors:

- Omission of one or more instruments from the scan.
- Gazing too long at one instrument, called *instrument fixation.* This is sometimes caused by staring at the instrument while waiting for a change to take place.

- Anticipating the need for corrective pressures and applying them before they are needed. This is especially prevalent when rolling into or out of turns.

- Misinterpreting the instruments, resulting in such errors as unnecessary power changes. For example, you see the airspeed is too high, so you reduce power without noticing that the increased airspeed is due to a loss of altitude and that simply leveling off by applying back pressure will slow the airspeed.

- Muscle tension on the controls that hampers your ability to sense unwanted control pressures. From time to time, flex your hands and feet to maintain sensitivity and blood circulation.

- Erratic and excessive control displacement, resulting in excessive instrument deflections, instrument lag, and gravity forces, leading to vertigo.

- Improper trim techniques.

"Two, two, and twenty"

From the very first flight I teach students to maintain heading within ±2°, airspeed within ±2 knots, and altitude within ±20 feet. This is necessary to become a member in good standing of the "2, 2, and 20 Club." This might sound corny, but it makes students think about goals to aim for right from the beginning.

The FAA allows tolerances on the instrument flight test of ±10° on heading, ±10 knots on airspeed, and ±100 feet on altitude. Stop and think about it; these are actually very large deviations—from one extreme to another, up to 20° on heading, 20 knots on airspeed, and 200 feet in altitude.

If a pilot is trained from the beginning to hold 2, 2, and 20, there will be no problem on the flight test. Even more important, the pilot will have better control of the airplane. If your personal maximum altitude deviation is 20 feet, you will use much smaller control adjustments to maintain it, compared to a 100-foot deviation limit. Larger tolerances invite larger corrections, and the larger the correction, the greater the tendency to overcontrol.

Overcontrolling

Remember the problem from VFR training? Overcontrolling occurs when a pilot uses too large a change in attitude to make a correction: too much bank in a turn, nose too high or too low in a climb or descent, for example. Soon after the correction, the pilot corrects again in the other direction to avoid overshooting the heading or altitude. The flight of an overcontrolled airplane is a wobbling, bobbing affair in which it is impossible to hold headings and altitudes with any degree of accuracy.

Overcontrolling is, then, a problem of attitude control. And what is the best way to control attitude, avoid overcontrolling, and achieve the goal of 2, 2, and 20? Concentrate on the fundamentals of controlling the aircraft by—the answer—reference to instruments.

3

Scan and control

Scan

An efficient instrument scanning technique will develop while practicing altitude and heading control in straight and level flight by means of control, primary, and support instruments—it happens almost automatically. The scan develops even further to include the support instruments.

Students often ask what is the best way to scan an instrument panel: up and down, left to right, clockwise or counterclockwise? It doesn't really make much difference as long as you adhere to two very important principles:

- Don't fixate on one instrument
- Always return to the attitude indicator, the control instrument, after checking each of the other instruments

Teaching scan

I have found that an ideal method of teaching the scan is *attitude, heading, altitude.* Repeat these three words while flying to help guide your scan.

Look first at the attitude indicator to determine if any attitude corrections are required. Next, check the heading indicator to see if any drift in heading has occurred. Glance at the altimeter to determine if any correction in altitude control is required. Go back to the attitude indicator and make any adjustments dictated by the glance at heading and altitude.

Repeat this process continually throughout the flight. On every 10th scan, include the VOR or the ADF, depending on which is used for navigation. Every couple of minutes, include all the instruments in your scan, including engine gauges. Adopt a system and stick with it.

Distractions

The human eye is constructed to immediately respond to any movement it picks up. This is a well-known trait from ancient forebears who had to respond quickly to movement of any kind in their environment; movement either indicated the presence of something that they would eat or the presence of something that would eat them.

The implication for the modern instrument pilot is that we descendants still tend to respond with greatest interest to something that moves. In bumpy weather our attention tends to become fixed on the oscillations of the turn needle or the rapid up-and-down movement of the VSI needle instead of the attitude indicator. Our attention is riveted on the wanderings of the VOR needle near a station while neglecting heading and altitude.

Make an effort to tear your eyes away from an instrument that is showing rapidly changing indications and methodically scan from one instrument to another in whatever sequence is most comfortable.

The control instrument for all phases of instrument flight is the attitude indicator. Center the scan pattern on the attitude indicator to maintain control of the airplane on instruments. When the eyes move away from the flight instruments—tuning a new communication or navigation frequency—always return to the attitude indicator to resume the scan.

You will quickly see small changes in attitude develop and make the small automatic adjustments of yoke and rudder pedals that are so important in smoothly controlling the airplane.

Instrument scan and interpretation

If you were able to see the flight path of an aircraft, you would see that it is maintained by a series of corrections to the controls to make the aircraft follow the desired track. The smaller and more frequent these corrections are, the closer the aircraft will conform to the desired path.

In order to make small corrections, you must receive a large input of information in a short period of time, interpret it correctly, and then make the desired control response in a timely fashion.

How do you do this? You learn to scan the instruments and to interpret what you see. As you practice this procedure, you learn to give
bare notice to those instruments with a proper indication, while de-
voting most of your attention to those instruments that have deviated
because they will stand out like a soldier out of step.

The two major errors most commonly made in instrument scanning
are omitting necessary instruments from the scan and spending too
much time looking at one instrument. An instrument is omitted from
the scan because the pilot is spending too much time worrying
about another one, so these errors are usually interrelated.

At one time or other, we are all guilty of having unwanted heading
changes of 15–20 degrees or unwanted altitude deviations of
±100–150 feet. How much time would such inadvertent changes
take? Unless your aircraft encounters exceptional trim changes, wind
shears, updrafts, downdrafts, or power changes, the normal rate of
climb (or descent) in an unpressurized aircraft is rarely more than
500 fpm. At this rate, it would take 10–12 seconds for your altimeter
to change by 100–150 feet. This means that for 10–12 seconds you
have omitted the vertical speed indicator, altimeter, and attitude indicator from your scan. Ten to 12 seconds of daydreaming is a long
time, but we all do it. How can we avoid it? By practice and hard
work, of course. Keep your eyes constantly moving over the instruments: SCAN.

Proper instrument scanning

You must have a systematic approach to your scan. In most IFR situations the attitude indicator will be your prime instrument, just as
in most VFR situations the outside horizon will be your prime visual
aid. Some people will argue with this idea, but I feel it's easier to
learn to fly on instruments by keeping one of the instruments as a
primary one that you can always fall back on.

Your scan will almost always begin with the attitude indicator and
proceed to the supporting instruments for the aspect of flight that
you're focusing on. You will scan the other instruments as well, but
your major concern will be those that support the flight characteristics that you are most interested in at the time.

For example, in level flight your scan would begin with the attitude
indicator, and then, while attempting to hold a constant attitude,
you would glance at the altimeter, the vertical speed indicator, the

airspeed indicator, the power setting, and back to the attitude indicator. Hopefully, by the time you get back to the attitude indicator, it will still be indicating the same as the first time you glanced at it. You will then repeat the scan of the instruments mentioned above.

Provided that your attitude and power have remained constant, one of three trends will have been established: descent, level flight, or climb. One cue that you will look for is the airspeed. Is it higher, lower, or normal for your power setting? Notice that the position of the attitude indicator has not been mentioned; it is enough that you hold the attitude indicator constant. At this point a lower airspeed at a given power setting would indicate a climb (or flight through a strong downdraft). This will be confirmed by the second scan of the vertical speed indicator and the altimeter. The second and third scans determine the trend. If the trend indicates a climb or descent when you want level flight, you will have to correct it. This correction should be accomplished in two *smooth* steps (notice the stressed word). The first step is to stop the unwanted trend. The second step is to begin a trend back to the desired flight situation.

In the example above, you began to climb slightly. If your power setting is proper for level flight, you will stop your climb by a slight change of the pitch attitude shown on the attitude indicator. You do this by relaxing enough back pressure (or by adding enough forward pressure) on the yoke to cause the imaginary string leading to the vertical speed indicator and altimeter to move the needles to a level-flight indication.

Although an experienced pilot would actually be doing all of these things simultaneously, as a student you will have to try to control one instrument at a time while learning. Because the attitude indicator has the least lag, it is the easiest instrument to control.

After you have made your slight change, you again try to hold the attitude indicator constant at the desired attitude and scan the other instruments again. Your change will have accompanied one of three things:

- If you did not change the pitch attitude sufficiently, you will continue climbing, but at a reduced rate.
- If you made an overcorrection, you will descend and your airspeed will increase.

- If you have changed the pitch the proper amount, your altimeter and vertical speed indicator should indicate level flight, and your airspeed will begin to increase. After you have stopped the unwanted trend, you can apply another small correction to reverse the previous trend and get your aircraft back to its desired flight path or attitude.

As you learn to control the vertical speed better through smooth corrections, and to coordinate that with the changes made to the attitude indicator, you will find that more of your corrections will be of the latter type.

The smaller and smoother you can make your corrections, the less chance you will have of developing vertigo (spatial disorientation), the smoother ride you will afford your passengers, and the more accurately you will fly your aircraft.

Instrument errors

Earlier explanations intentionally made no mention about the initial attitude indication. This is because you must think of the attitude indicator as a *relative* instrument rather than as an absolute indicator of the aircraft performance. Your aircraft (and its attitude indicator) can be nose-low or nose-high, and depending on loading (center of gravity) and speed, still be in normal flight.

The attitude indication on instruments will also change due to precession of the gyro as well as from acceleration and deceleration errors. For example, if you have been climbing for a long while, the attitude indicator gets used to indicating a nose-high attitude, and it will continue to indicate slightly more nose-high than normal for a few minutes after you have leveled off and established cruise airspeed; therefore, all attitude corrections are made to the previous indication, not to the actual attitude.

Gyro precession and acceleration errors are inherent gyro instrument errors that you learn to live with, and with proper scanning procedures, they become relatively unimportant. Nongyro instruments are also subject to errors. You have all undoubtedly heard of altimeter and vertical speed instrument lag. By horsing the controls abruptly you can get a momentary instrument reaction opposite to the aircraft's reaction, but if you make slow, smooth corrections as mentioned earlier, this lag becomes barely noticeable.

Many good books thoroughly explain instruments and their construction and the reasons for their errors. My purpose is only to tell you how to live with them.

Airspeed control

The remainder of this chapter concentrates on changing airspeed while maintaining level flight. The purpose of this is to teach the relationships among power setting, pitch attitude, airspeed, and trim.

One nice thing that you will learn about trim is that it is closely related to airspeed, and that once trimmed for a given speed and center of gravity (CG), the aircraft will tend to remain at the same airspeed regardless of the power setting. When you reduce the power, the airspeed will want to decrease, but if you are trimmed for the higher speed, the nose will lower instead and you will begin to lose altitude. If your plane is fairly stable, it will porpoise a few times, and eventually the speed and the rate of descent should stabilize.

From this point on in these discussions you will be attempting to maintain exact altitudes, airspeeds, headings, and rates of climb or descent. Here the allowable margin of error will be ±0. I realize that the FAA Practical Test Standards are more lenient, but you will not achieve a high level of competence by training for minimum performance. You should be striving for perfection from the start. You'll never achieve it, but you should try.

Let's begin with this thought. In level flight, any given power setting should result in a specific airspeed and corresponding pitch attitude (instrument error disregarded). Ideally then, at cruise power and airspeed, your attitude indicator should show precisely level flight.

Because the four forces working on an aircraft (lift, thrust, drag, and gravity) must always be in balance, anytime one of them varies, the other three will change to bring all four back into equilibrium.

If you increase power and maintain pitch, the thrust will overcome drag and your speed will increase until the drag and thrust balance each other out. At the same time, the increased speed (with no deliberate change of attitude) will increase lift. This, in turn, will cause your aircraft to climb until the lift is balanced out by gravity again.

To counteract this climbing effect, as the airspeed increases, you must slowly lower your pitch attitude to maintain level flight. To do this you will apply forward pressure on the yoke: enough pressure to keep the vertical speed indicator at zero and the altimeter at the required altitude. The more the speed increases, the more forward pressure you will have to exert on the yoke. The opposite occurs if you reduce power while trying to maintain altitude.

John Doster, past director of the FAA GADO at Allentown-Bethlehem-Easton Airport in Pennsylvania, has always felt that the level turn was the best training maneuver to teach aircraft control. He would begin by asking each of his thousands of students and flight certificate candidates what caused an aircraft to turn. There was and is only one correct answer: lift.

As the ailerons deflect to roll into a turn, the lowered aileron on the up-wing causes increased drag, which turns the nose of the plane slightly in the direction opposite the turn. Rudder pressure has to be added to counteract this yawing tendency.

As the aircraft banks, part of the lift that up to now has balanced out gravity (to maintain level flight) is vectored to the inside of the bank, lifting the aircraft into the turn.

Inertia is a factor for a little while. The aircraft maintains level flight until the vertical component of lift decreases enough to allow gravity to take over, causing the aircraft to descend. As this occurs, back pressure has to be applied to the yoke to increase the angle of attack on the wing, thereby inducing enough added lift to maintain the desired altitude. Unfortunately, this added lift also increases drag, and the plane begins to slow down unless enough thrust is applied to overcome the increased drag and keep all four forces in balance.

The opposite situation occurs when rolling out of the turn. First, the rudder has to be applied again to counteract yaw. As the aircraft rolls to a wings-level attitude, inertia again works on it momentarily. Then the plane wants to begin climbing, due to the increased vertical component of lift.

This climbing tendency can be overcome by forward pressure on the yoke (decreasing the angle of attack) but when you do this, the power you added previously begins to overcome drag, and your airspeed increases unless you reduce power again.

John Doster feels that the best training maneuvers for any pilot are turns of various degrees of bank and duration, with zero permissible altitude and airspeed fluctuation.

As mentioned above, when your airspeed changes, so will the control pressures, and here's where it's very important to retrim the aircraft. Remember that the goal is to have the aircraft trimmed in such a manner that it will maintain its desired path and attitude if you momentarily take your hands off the controls.

By learning how to maintain level flight while varying pitch and power settings, you also learn how to slow down to approach speed before beginning your descent. This will be of importance when you get into the approach phase of instrument flight.

During all phases of instrument flight, try to keep changes as simple as possible. When beginning an approach, try not to go from cruise speed/attitude/power to approach speed/attitude/power all at once. You can control your aircraft easier if you first slow to approach speed/attitude while still maintaining level flight and then begin a descent by easing off a little power, lowering some flaps, or extending the landing gear.

For practice, try flying level, with and without the hood, varying the airspeed from 5 knots above stall speed to 5 knots above cruise speed. Trim the aircraft as the speed changes so that once your speed has stabilized, the aircraft will fly practically hands-off. When you do this without the hood, compare your aircraft's nose and wing tip positions (their relationship with the horizon) to what is presented on the attitude indicator. Naturally, you should have an instructor or safety pilot with you to scan outside at all times.

Develop your instrument scan by covering one pitch instrument at a time and flying with the remaining pitch instruments to learn their interrelationship. With the attitude indicator covered, try to develop a touch and smoothness on the yoke that will minimize the instrument lag inherent in the vertical speed indicator and the altimeter.

4

Phase of flight

Now that you have a firm idea of how to fly basic maneuvers by reference to instruments, let's take a beginning look at a typical instrument flight. We'll break it down into stages, or phases of flight, to better see how to apply basic instrument flight maneuvering to the "real world."

Instrument takeoffs

The instrument takeoff is basically a training maneuver. If the weather is really zero/zero, not even the birds will take off, let alone intelligent pilots. After all, where could you go if you had to make an emergency landing shortly after takeoff?

A pilot would be justified in making an actual ITO in only a few instances, perhaps if the airport is socked in by a strictly local condition, such as a ground fog where the tops are only a few hundred feet high and there's another field nearby that is open. Even then you'd be taking a chance, but at least it would be well calculated.

In making an ITO, taxi onto the runway, line up with the centerline, and let the plane roll a few feet to make sure that the nosewheel or the tailwheel is rolling straight, then stop. Set the DG to the painted runway heading, even though the magnetic compass might be reading differently. You can and will always reset the DG later, but on takeoff it's easier to steer to a specific mark on the face of the DG.

Next, set the attitude indicator to the proper aircraft attitude, which is approximately nose level with a tricycle gear but nose high (near the normal climb attitude) in a taildragger.

Be sure that the gyros are uncaged and up to speed. Remember that it takes up to 5 minutes for gyros to spin up to a reliable speed.

Finally, advance the throttle(s) smoothly to maximum allowable power (takeoff power) while keeping all pressure off the brakes. Do not attempt to steer with the brakes. As in visual takeoffs, you will be correcting the heading with rudder pressure. There will be a psychological tendency to overcontrol your heading at first because you will think that you are about to tear out all of the runway lights. Pay more attention to the action of the DG on your next visual takeoff to see the normal reactions and to help you understand them better during an ITO.

As your airspeed approaches climb speed, establish a climb attitude on the AI. This will be about the same attitude whether you are in a tailwheel or nosewheel aircraft. In a taildragger, you will have to lift the tailwheel off the ground first, but just bring it off slightly, nowhere near the amount you do on a normal visual takeoff. For this reason, you should practice this technique a few times under visual conditions to see what it feels like and looks like before trying it under the hood.

Remember, as the aircraft lifts off, the AI will precess and show a slightly higher nose attitude. Scanning the AI and VSI should be quick because you're holding the heading with the DG and the needle/ball. Improper interpretation of the AI at liftoff can result in touching back down. The first 150 feet of altitude will be fairly critical. Above that, you will be well out of ground effect and the instrument precession and lag should have pretty much recovered.

ITOs are a lot of fun, and while they might not be the most practical maneuver, they do a lot of good toward teaching good instrument scanning techniques. They also help pilots build confidence.

Climbs by reference to instruments

ATC may have you initially climb after you contact them to reduce the chances of encountering hazards on the ground. If that is the case, you will want to use the same climb power settings and pitch angles that you would during a VFR climb. The climb should be made at a constant airspeed, which means that you need to maintain a constant pitch. To begin the climb, add power to climb settings, then use the artificial horizon to establish a gradual nose-up pitch. For most airplanes, a 10-degree pitch-up angle from level on the artificial horizon would be a good place to start. Watch the airspeed as the nose pitches up, and if the airspeed drops below the correct value, slightly reduce the backpressure on the control yoke. Keep the inputs small to avoid overcontrolling the airplane. Once the airspeed is established at the correct climb value, hold a constant pitch angle to maintain the airspeed. You will also need to monitor the altimeter and directional gyro to make sure you do not overshoot the desired altitude or stray from the heading ATC has assigned you. Don't forget to keep the ball centered through proper rudder use during the climb. As you approach the assigned altitude, slowly lower the nose to a level attitude by using the artificial horizon. As the speed builds, reduce power to cruise settings and trim for level flight at that altitude. Keep scanning the gauges; don't focus on just the artificial horizon or the altimeter.

Straight and level flight and turns

In order to maintain straight and level flight, you should establish the correct power settings, then trim the plane to hold a constant altitude with neutral pressure on the elevator control. Proper use of trim can help reduce your workload and fatigue factor, which is very important in this situation. Use the artificial horizon as the primary indicator of your pitch and bank attitudes, but also scan the directional gyro for proper heading, and check the altimeter for the altitude. The turn and bank indicator is useful when you need to turn the airplane and should be used to establish coordinated, standard-rate turns. If you use standard-rate turns whenever you change headings, you will avoid becoming so steeply banked that you lose orientation. Use of standard-rate turns will also reduce the chances that your turning rate is so fast that you turn past the heading you are turning to. If you do make turns, remember to use a slight amount of backpressure on the control yoke to maintain a constant altitude during turn. It is quite common for new instrument students to overbank the plane during turns and allow the nose to drop as they enter the turn. This allows loss of altitude and can cause a significant increase in the airspeed. It cannot be stressed enough. Keep the control inputs small and smooth, and scan the gauges.

Descent by reference to instruments

If you need to descend on instruments, you will once again use the artificial horizon to initially set up the descent angle. If necessary, reduce the engine power setting to avoid gaining too much airspeed. Use the vertical speed indicator in conjunction with the airspeed indicator to establish a safe rate of descent. If you notice the airspeed is becoming too high, raise the nose of the plane and/or reduce power. If you use approximately a 500-foot-per-minute descent rate and power settings at the bottom end of the green arc, you should avoid excessive airspeeds. However, every plane is different, and it may be necessary to use different techniques for the plane you fly.

Like the climb, a descent should be at a constant airspeed and rate of descent. Using a constant pitch angle and power settings is necessary to establish the descent. You should decide on the altitude you want to descend to before you start the descent, then begin raising the nose to a level attitude and increasing power as you approach the target altitude. If you are setting up for landing, it can be a good idea to get the airplane into the proper landing configuration before you begin the descent to reduce the workload during the descent. Continue to scan the airspeed indicator, artificial horizon, altimeter, and directional gyro during the descent. Again, it is common for pilots to focus on just one instrument during descents. This can vary from the artificial horizon to the altimeter, but if you exclude the other instruments from your scan you will not be in complete control of the plane.

Constant airspeed climbs

In VFR flying you have already become accustomed to constant airspeed climbs. After takeoff, climb at a speed that will give the best rate of climb, usually 80 knots for a Cessna 172 or Cherokee 180. In IFR flying, just as in VFR, the most important thing is to pick the best rate of climb speed (V_y) and stick to it.

Full power is used for climbs, so climbs are situations in which power is "constant," not "variable" and you will "pitch to the speed." Climb speed is thus established by setting the pitch attitude. You will set and control the pitch attitude by reference to the attitude indicator. Establish the climb by pitching up to that first index line in the blue "sky" section of the attitude indicator (Fig. 4-1). If entering the climb from straight and level, add full power and simultaneously increase pitch.

Pitch adjustments

Make adjustments from that first reference line, as necessary. If airspeed is too high and climb performance is poor, correct by a slight

Fig. 4-1. *Establish a climb by pitching up to the first index line above horizon line on the attitude indicator.*

additional pitch up adjustment to reduce the airspeed; if airspeed is too low, make a slight pitch down adjustment. Hold the pitch constant on the attitude indicator and the airspeed will remain constant.

Pitch adjustments on the attitude indicator—like all adjustments in instrument flight—are small and deliberate, not more than a quarter or half a bar width at most. Bar width is the thickness of the miniature airplane "wings" on the attitude indicator. The smaller the adjustments, the smoother the flight.

Note that when entering a climb, the attitude indicator serves a triple purpose: the control instrument and the primary instrument for pitch and the control instrument for bank. The main concern in climb entry is establishing the correct pitch; the attitude indicator gives the best information on the "quality" of pitch, as well as serving as the reference by which you control the airplane in pitch and bank.

When the climb is stabilized at the best rate of climb speed, the airspeed indicator becomes the "quality" instrument for pitch. The primary and support instruments for both climb entry and stabilized climb are the same.

Departure climbout

Be prepared on the first instrument training flight to climb out on instruments after takeoff. Under ATC clearance, adhere to standard instrument climbs, turns, and descents. Make full use of this excellent opportunity for practicing instrument climbs and you won't have to use valuable time later in the flight to hone climb skills.

Let's see how this works in the real world of IFR. The clearance from ATC contains climb instructions; ATC frequently requires a step or two in the climbout before reaching cruise altitude for better traffic separation. A typical clearance might be "Maintain three thousand feet, expect further clearance to five thousand feet in ten minutes."

Two climbs are in this clearance. First is the initial climb to 3,000 feet after takeoff. Then there is a stretch of level flight until ATC calls back 10 minutes later with further clearance to climb to 5,000 feet. How, in practical terms, do you handle this type of climb clearance?

As a rule of thumb, climb at the highest practical rate up to the last 1,000 feet before the assigned altitude. Then reduce the rate of climb to a constant 500 feet per minute for the last 1,000 feet to avoid overshooting the assigned altitude when leveling off.

In the previous example, you would take off and climb to 2,000 feet with full power at the best rate of climb airspeed. Then you would adjust the attitude to produce a 500 foot-per-minute climb from 2,000 to 3,000 feet. Then level off until ATC clearance to resume the climb to 5,000 feet, approximately 10 minutes later.

Constant rate climbs

Upon reaching 2,000 feet you would switch from a stabilized *constant speed* climb to a stabilized *constant rate* climb at the rate of 500 feet per minute. (You will find this less complicated in flight than it sounds in print.) Most single-engine planes can't manage much more than a 500 foot-per-minute climb at full throttle anyway. To attain a constant rate climb, simply reduce pitch slightly, leaving the throttle at full power.

The VSI is the primary instrument for "quality" information on the rate of climb.

You will have to wait approximately 7 seconds for the VSI to stabilize at the new pitch. In constant rate climbs and descents the attitude indicator continues to be the instrument by which you control the airplane. The VSI becomes the primary instrument for attitude.

If the VSI stabilizes at more than 500 feet per minute, lower the nose to decrease the rate of climb; if less than 500 feet per minute, raise the nose to increase the rate of climb. Always remember that you cannot get an accurate indication from the VSI until it has stabilized for 7 seconds.

Chasing the needle

When I give instrument checkrides I can tell very quickly if candidate pilots do not understand the limits of the VSI because they "chase the needle." If the needle is rising fast, they push forward on the yoke to slow it down; if it is descending fast they pull back. This produces a form of porpoising, which is a sure tip-off that the candidate has developed neither a good scan nor an understanding of instrument control. Instead, the pilot's attention remains fixed on one instrument too long.

Instructor note. A student who develops the VSI needle chasing symptom has an instrument fixation problem. Return to level flight and teach the student to forcefully shift his or her focus from one instrument to another, starting and stopping each scan cycle with the attitude indicator.

Climb level off

Leveling off from a climb is a simple procedure, but it's the source of a common problem for beginning instrument students. That problem is overshooting or undershooting the target altitude. To avoid this, anticipate reaching the target altitude by 10 percent of the rate of climb. If the VSI shows a rate of climb of 600 feet per minute, for example, begin leveling off 60 feet before the target altitude.

Lower the miniature airplane to the horizon line on the attitude indicator when reaching the desired altitude and allow the airspeed to build up to cruise before reducing power. Exert forward pressure on the yoke to prevent ballooning above the assigned altitude as airspeed builds up, and trim out the excess control pressure as it becomes heavier.

It is very important to have a predetermined idea of what the cruise power setting should be. You can automatically reduce power to that specific cruise power setting in one motion upon reaching the cruise speed. Any further adjustments of power and trim will then be relatively minor.

Include the sweep-second hand of the clock in your scan. No matter when you start the constant rate climb, you know that the airplane should gain 500 feet by the time the sweep-second hand returns to its starting position. The altitude gain should be 250 feet when the sweep-second hand is 30 seconds from its starting position.

Descents

Overshooting the target altitude when descending is a more critical matter. Most precision instrument approaches go down to 200 feet above the runway, where there's no room for error. You must be able to control descents so there is never a question of coming too close to the ground or obstacles in the airport vicinity.

It's worth noting that airspeed control during descents becomes very important when making VOR and other nonprecision approaches. Understand that the *missed approach point* (MAP) on a nonprecision approach is frequently based upon how long it takes at a given airspeed to fly from the final approach fix to the airport.

If you can't control airspeed on a descending final approach course, you might as well throw the stopwatch away. Elapsed time on that final leg to the airport will be meaningless if the airspeed on which it is calculated is not constant. If the time is off, the airplane might end up off course. If there are hills or obstacles around the airport, poor timing due to poor airspeed control might be disastrous.

Constant speed descents

In a constant speed descent use "pitch to the airspeed" like the constant speed climb. While in straight and level flight, reduce power to slow cruise. Set 1900 RPM in a Cessna 172 or Cherokee 180, approximately 90 knots for a comfortable and efficient descent speed.

"Pitch to the airspeed" again. Hold the nose up in level flight until speed bleeds off to 90 knots, then gently allow the nose to pitch down to one line below the horizon line on the attitude indicator (Fig. 4-2). That should produce a 90-knot descent at about 500 feet per minute. (It might vary slightly with different types of airplanes.)

"Pull the plug" if a faster descent is necessary. Reduce power to 1500 RPM, slow to 90 knots, then lower the nose to maintain 90 knots.

Fig. 4-2. *Establish a descent by pitching down to first index line below horizon line on attitude indicator.*

This will produce a rate of descent of about 1,000 feet per minute at a somewhat steeper angle.

Some high-performance airplanes can descend 2,000 feet per minute by reducing power and lowering the landing gear and flaps. The *dirty* descent (gear and flaps down) can be made without reducing power excessively. This procedure avoids the shock of sudden over-cooling, which can damage the engine.

Constant rate descents

Let it be known here and forever carved in stone that a change of 100 RPM or 1" of manifold pressure will produce a change of 100 feet per minute (fpm) in the rate of descent. Suppose you are in a 400-feet-per-minute descent with power at 1900 RPM, and airspeed 90 knots. How do you change the rate of descent to 500 feet per minute? Simply reduce power by 100 RPM to 1,800. Power adjustments of 100 RPM or 1" in descents will produce changes of 100 feet per minute in almost every plane that you are likely to fly.

This rule of 100 RPM or 1" = 100 fpm will become vitally important later on as you learn to fly the ILS approach with its very sensitive electronic glide slope. But the rule also applies to normal descents, and it will help avoid overshooting that critical altitude.

ATC will expect a descent at 1,000 fpm until reaching 1,000 feet above the target altitude. Then you are expected to reduce the rate of descent to 500 fpm. In other words,

- A change to a lower altitude begins with a *constant speed* descent at 90 knots and approximately 1,000 fpm
- At 1,000 feet above the target altitude, switch to a *constant rate* descent at 500 fpm

This is similar, of course, to the situation in a climb. The first part of a climb is at a *constant speed*—the best rate of climb speed. At 1,000 feet below the target altitude, the descent becomes *constant rate* at 500 fpm. The first part of a descent is at a *constant speed*. At 1,000 feet above target altitude, change to a *constant rate* descent of 500 fpm.

Again, the sweep-second hand of the clock should be included in the scan to ascertain whether you are ahead or behind in the climb.

Descent level off

The target descent altitude should be anticipated by 10 percent of the rate of descent—the same as in a climb. If descending at a rate of 500 fpm, begin to level off 50 feet prior to reaching the target altitude.

Instrument students sometimes fixate on the target altitude and don't start the transition until they reach it. That is a sure way to overshoot the altitude and, on an instrument approach to go below the minimums for that approach. Don't do it! "Busting the minimums" is hazardous to your health when close to the ground. And on an instrument checkride, busting minimums is an automatic failure, no matter how brilliant the rest of the checkride might have been.

To level off, simultaneously increase power to the cruise setting and raise the nose and set it on the horizon bar of the attitude indicator. Adjust trim as the speed builds up to prevent ballooning above the level-off altitude. Scan the attitude indicator and the primary and support instruments.

Approach descents

Before leaving the subject of descents, I want to cover two slightly modified descents used on instrument approaches.

The first is a constant airspeed descent with the addition of 10° of flaps. Let's call them approach flaps because the setting might be more or less than 10° on different airplanes. Most airplanes handle better at slow speeds with a notch of flaps down. This is the configuration used on most instrument approaches on the final approach leg just before landing.

Start from cruise speed, reduce power, and extend approach flaps. Allow the airspeed to bleed off while in level flight, then "pitch to the airspeed." Set the nose of the miniature airplane on the first black line below the horizon on the attitude indicator. Adjust power for 500 fpm, remembering that 100 RPM or 1" = 100 fpm, or "power to the altitude."

If flying a high performance airplane with retractable landing gear, reduce power, set approach flaps, and allow the speed to decrease to the desired descent airspeed while remaining in level flight. When you reach the descent speed, extend the gear; the nose dips automatically to just about the right pitch for a descent when the gear is lowered.

High-speed final

More and more these days ATC might say: "Keep your speed up on final." This is frequently followed by something interesting such as like "727 overtaking." If you cannot comply with this request, do not hesitate to tell ATC right away. But it is best to cooperate with this request whenever possible, for obvious reasons. ATC won't let the separation between you and that big jet get too narrow. If a jet is behind you coming in 10 or 20 knots faster and chewing up the distance in between, guess who is most likely to be ordered to go around, you or the big jet? ATC might request a 90° left or right turn to let the jet pass before vectoring you back for the approach.

Practice constant rate descents at cruise speed, or faster, as well as at the normal descent speed. Some pilots automatically make high-speed final approaches whenever they fly into airports with a lot of jet traffic. This certainly makes it easier for ATC.

To set up a high-speed descent, lower the nose of the miniature airplane to the first black line below the horizon line on the attitude indicator. Reduce power 500 RPM to set up a 500-fpm descent. When stabilized, "pitch to the airspeed, power to the altitude" to maintain cruise airspeed at a 500-fpm descent. Leave flaps and landing gear up during a high-speed approach. If jets are on the approach behind you, you know the runway will be plenty long enough to slow down and extend flaps and gear when the runway is in sight and landing is assured.

Obstacle clearance concerns

It might be a good idea here to take a look at how the FAA devises
obstacle clearance specifications, explained by Jeppesen:

> *Obstacle clearance is based on the aircraft climbing at 200
> feet per nautical mile, crossing the end of the runway at 35
> feet AGL, and climbing to 400 feet above the airport elevation
> before turning unless otherwise specified in the procedure.*
> [This is the basic obstacle clearance specification.] *A slope of
> 152 feet per mile, starting no higher than 35 feet above the
> departure end of the runway, is assessed for obstacles. A min-
> imum of 48 feet of obstacle clearance is provided for each
> mile of flight. If no obstacles penetrate the 152 feet per mile
> slope, IFR departure procedures are not published.* [So far so
> good. If nothing is specified on the SID or in the takeoff sec-
> tion of the airport plan chart, then all you need do is meet
> the above climb criteria.] *If obstacles penetrate the slope, ob-
> stacle avoidance procedures are specified. These procedures
> may be: a ceiling and visibility to allow the obstacles to be
> seen and avoided; a climb gradient greater than 200 feet per
> mile; detailed flight maneuvers; or a combination of the
> above. In extreme cases, IFR takeoff may not be authorized
> for some runways.* [Unless you have some really bad prob-
> lems, any aircraft capable of IFR flight is able to meet the
> standard obstacle climb gradient.]

> *Climb gradients are specified when required for obstacle
> clearance. Crossing restrictions in the SIDs may be estab-
> lished for traffic separation or obstacle clearance. When no
> gradient is specified, the pilot is expected to climb at least 200
> feet per mile to MEA unless required to level off by a crossing
> restriction.* [Perhaps we should get our heads together here.
> We are used to thinking in terms of climbing and descend-
> ing at a certain feet per minute rate. These climb gradients
> are based on climbing so many feet per mile. To meet these
> restrictions, the rate of climb will be a function of both air-
> speed and groundspeed. The faster you fly across the
> ground, the higher your rate of climb will have to be to meet
> the climb gradient.] *Climb gradients may be specified to an
> altitude/fix, above which the normal gradient applies. Some
> procedures require a climb in visual conditions to cross the*

airport (or an on-airport NAVAID) at or above an altitude. The specified ceiling and visibility minimums will be enough to allow the pilot to see and avoid obstacles near the airport. Obstacle avoidance is not guaranteed if the pilot maneuvers farther from the airport than the visibility minimum. [This is a very important point. If you are given an IFR clearance with a two-mile visibility restriction, it is your responsibility to stay within two miles of the airport until above the ceiling specified in the same clearance. You must remain in visual conditions to see and avoid any obstacles.] *That segment of the procedure which requires the pilot to see and avoid obstacles ends when the aircraft crosses the specified point at the required altitude. Thereafter, standard obstacle protection is provided.*

Take a look at an actual SID. Figure 4-3 is the DUMBARTON FIVE DEPARTURE from San Francisco, California. You can see that it is labeled SID in the upper right hand corner so as not to be confused with an approach plate or a STAR (*standard terminal arrival route*). As a further identification aid, Jeppesen assigns SIDs the code number of 10-3 followed by an alphabetical suffix if there is more than one SID for the airport, as opposed to the code number 10-2 for STARs, and 10-1 for area charts.

As procedures are changed, ATC changes the numerical suffix of the SID. In this case, the Dumbarton Five Departure has updated the Dumbarton Four Departure, so you must make sure that you have the correct numerical suffix for the SID that is specified in the clearance.

(Reminder: By the time you read these words, many of the charts appearing in this book will be obsolete. *Do not attempt to use any chart in this book for actual navigation.* It is your responsibility to obtain the current charts for the routes you plan to fly.)

The Jeppesen SID shows the SID both textually and pictorially so you can follow it more easily. Remember that you must have at least the textual portion in your possession.

The text of the SID is self-explanatory, giving headings and altitudes as well as the direction of all turns for each takeoff runway that you might use. If there is a runway at the airport that is not mentioned in the SID, you are not allowed to use it for an instrument takeoff when using that specified SID.

SID

JEPPESEN 31 DEC 93 (10-3) Eff 6 Jan **SAN FRANCISCO, CALIF**
BAY Departure (R) 120.9 | SAN FRANCISCO INTL

DUMBARTON FIVE DEPARTURE (DUMB5.BARTN) (PILOT NAV)
(RWYS 10L/R and 19L/R)

This SID requires the following minimum climb gradients for obstacle clearance:
Rwy 19L: 480' per NM to 1400'.
Rwy 19R: CAT A & B, 480' per NM to 1400', CAT C & D, 530' per NM to 1800'.

Gnd speed-Kts	75	100	150	200	250	300
480' per NM	600	800	1200	1600	2000	2400
530' per NM	663	883	1325	1767	2208	2650

TAKE-OFF
Rwys 10L/R and 19L/R: Turn LEFT (Rwys 19L/R departures turn LEFT as soon as practicable due to steeply rising terrain to 2000' immediately south of airport) and climb via SFO R-090 to Bartn Int. Thence via (transition) or (assigned route). Expect further clearance to filed altitude 10 minutes after departure.

TRANSITIONS
— **Linden (DUMB5.LIN):** From Bartn Int to **LIN VOR:** Via OSI R-028 and LIN R-229. Cross OSI R-028 D24 at or above 11000'. **MAINTAIN** assigned altitude.
— **Red Bluff (DUMB5.RBL):** From Bartn Int to **RBL VOR:** Via OSI R-028 and RBL R-152. Cross OSI R-028 D24 at or above 11000'. **MAINTAIN** assigned altitude.
— **Sacramento (DUMB5.SAC):** From Bartn Int to **SAC VOR:** Via OSI R-028 and SAC R-177. Cross OSI R-028 D24 at or above 11000'. **MAINTAIN** assigned altitude.
— **Woodside (DUMB5.OSI):** From Bartn Int to **OSI VOR:** Via OSI R-028.

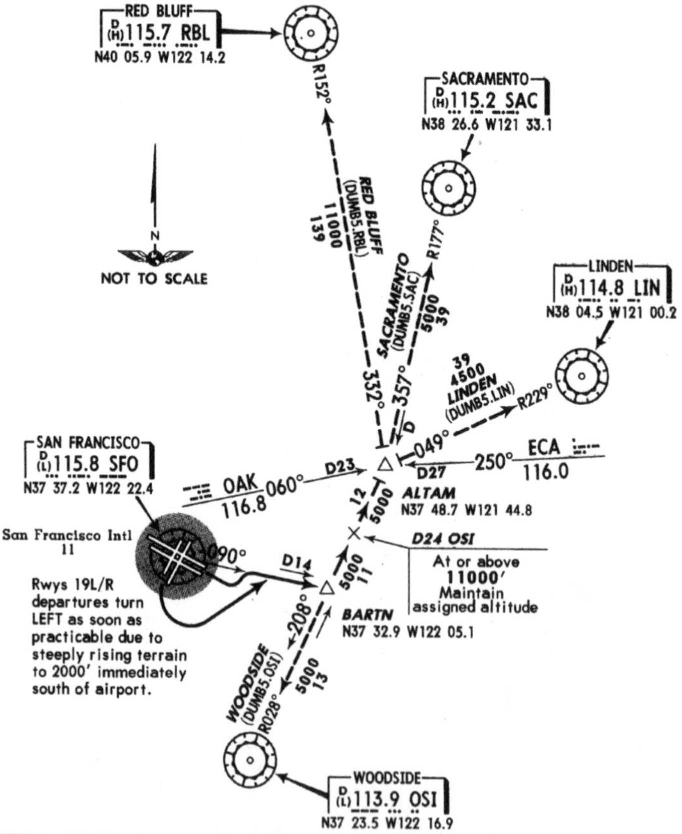

NOT TO SCALE

RED BLUFF
(D)(H) 115.7 RBL
N40 05.9 W122 14.2
R152°
RED BLUFF (DUMB5.RBL) 11000 139

SACRAMENTO
(D)(H) 115.2 SAC
N38 26.6 W121 33.1
R177°
SACRAMENTO (DUMB5.SAC) 5000 39
332° 357°

LINDEN
(D)(H) 114.8 LIN
N38 04.5 W121 00.2
39 4500 LINDEN (DUMB5.LIN) R229°
049° 250° ECA 116.0
D27

SAN FRANCISCO
(D)(L) 115.8 SFO
N37 37.2 W122 22.4
OAK 060° D23 116.8
12 5000 ALTAM N37 48.7 W121 44.8

San Francisco Intl 11
090°
D14 5000 11
X D24 OSI
At or above 11000' Maintain assigned altitude

Rwys 19L/R departures turn LEFT as soon as practicable due to steeply rising terrain to 2000' immediately south of airport.
208°
BARTN N37 32.9 W122 05.1
WOODSIDE (DUMB5.OSI) R028° 5000 13

WOODSIDE
(D)(L) 113.9 OSI
N37 23.5 W122 16.9

CHANGES: Alcoa, Bebop, & Clukk Ints deleted.

Fig. 4-3 *Reproduced with permission of Jeppesen Sanderson, Inc. Not for use in navigation.*

Following along with the Dumbarton Five SID, you can see that, after following the takeoff instructions for the runway you are using, you fly to the Bartn Intersection, and then fly whatever transition has been assigned. That will get you on your way in good shape. You can see that on two of the runways, 19L and 19R, you have to climb as soon as practicable because of the steeply rising terrain immediately south of the airport. Departures from these runways will require specific climb gradients as mentioned earlier. The climb gradient chart at the upper left of the SID illustrates how ground speed affects the rate of climb in order to achieve these specific gradients.

An important point to learn that should be reiterated: There will be much less radio congestion and much less chance of error in copying, reading back, and complying when you receive a simple clearance using a SID. In this case you might hear something such as:

> *Speedy 50X is cleared to the Lake Tahoe Airport, Dumbarton Five Departure, Sacramento Transition, then as filed, maintain 11,000 feet.*

Isn't this a lot better than the clearance you'd receive if you didn't use the SID? Without the SID, the same clearance would be complicated:

> *Wordy 20F is cleared to the Lake Tahoe Airport as filed, maintain 11,000 feet. After takeoff, climb via the San Francisco 090-degree radial to Bartn Intersection. Then via the Woodside 028-degree radial and the Sacramento 177-degree radial. Cross the 24-mile DME on the Woodside 028-degree radial at 11,000 feet.*

Filing for a specific SID is straightforward. The computerized world of the ATC system has codes for all of its intersections, fixes, SIDs, STARs, approaches, and the like. If you look again at the DUMBARTON FIVE DEPARTURE, you will notice that the computerized code is (DUMB5.BARTN).

When you file your flight plan, you will use this code if you use the SID and don't care about which transition you fly. On the other hand, if you look down at the transitions listed, you will see that each one has its own specific code. The Sacramento Transition mentioned earlier has the specific code (DUMB5.SAC), and if you wanted to file that specific SID and transition, it would be filed as that code. Computers work better when they're given specifics, and it eases the workload on an already overworked ATC system.

Aircraft performance restrictions

You won't always be able to use a specific SID due to operating limitations of your aircraft. An obvious example can be seen in the SID we have been examining. Although this is not calculated on the chart, the SID requires a rate of climb that will average out to about 400 feet per mile or greater in order to cross the Woodside R-028/24 DME fix at or above 11,000 feet because that fix is only 25 miles from the airport.

If you are heavy or have a low-performance aircraft, you might not be able to make it. For example, if your best-rate-of-climb airspeed will average out at 90 knots ground speed, you will need an average rate of climb of about 650 feet per minute all the way to 11,000 feet. Will your low-powered aircraft be able to sustain that rate of climb up to 11,000 feet? If your best-rate-of-climb airspeed averaged out at a ground speed of 120 knots, you would need an average rate of climb of 915 feet per minute. It is something to contemplate because the faster your airspeed in climb, the less time you will have to make the altitude, so the higher your rate of climb will have to be. If you are unable to comply with the requirements of a SID, you naturally cannot accept it, and most of the time you are the only one who knows the performance capabilities of the aircraft. ATC might give you a clearance utilizing a SID you can't conform with and it's up to you to advise them of the appropriate capabilities.

I might as well explain what the Woodside R-028/24 DME fix is. Although it looks a little complicated at first, it is simply identifying the fix by giving the VOR involved (Woodside), the radial off of Woodside (028-degree radial), and the distance away from the VOR on that radial (24 DME). It really shouldn't be necessary to mention this, but as Murphy's Law is always lurking just around the corner, remember that the radial mentioned is just that, a radial *from* the VOR.

In some cases it's quite obvious that the SID has been designed for high-performance aircraft. A good example of this is the 10-3G SID for San Francisco, SHORELINE NINE DEPARTURE. It isn't necessary to show the actual SID to get a point across. The Melts Transition for this departure calls for crossing the Linden R-240/18 DME fix at or above 16,000 feet, crossing the Linden VORTAC at or above Flight Level 200 (approximately 20,000 feet; above 18,000 feet altimeters are set to 29.92 inches and the altitude is referred to as a flight level, which is the altitude reading sans the last two zeros), and then crossing Melts Intersection at FL 230. Those last two requirements, high altitudes, put all but transport and very-high-performance general aviation aircraft out of the picture.

Because of mountains to the west of SFO, there is also a note on the SHORELINE NINE DEPARTURE that says, in part, when departing Runway 28 L/R, the weather conditions must be 2,000-foot ceiling and 3 miles prevailing visibility with 5 miles to the west and northwest, so here you see another form of obstruction-clearance instructions.

Refer back to the climb gradient chart in the illustration. It is laid out to comply with the climb restrictions off Runway 19 L/R. The note says climb is required at a minimum of 480 feet *per nautical mile* to 1,400 feet, if you are taking off from Runway 19L. If you are taking off from Runway 19R and are operating a CAT A or B aircraft, climb is required at a rate of 480 feet per nautical mile to 1,400 feet; a CAT B or D aircraft rate of climb must be 530 feet per nautical mile to 1,800 feet. Notice the high rate of climb required for high-performance aircraft.

The climb gradient chart has ground speeds across the top and the required rates of climb listed below. It would appear that many general aviation aircraft might be hard pressed to comply with the 480 feet-per-nautical-mile gradient required off Runway 19 L/R, especially with strong south winds dumping down off the mountains.

This points out again the need to study all of the information available before you file a flight plan. As a rule, the people at ATC have no concept of the operational capabilities of your aircraft. ATC is trying to help you out, and by trying to do so, it might issue a clearance that you have asked for; however, you could very well find yourself in the air, IFR, and unable to comply with the limitations of the SID or some other part of the clearance. You have to study it all beforehand because you won't have much time once you're in the air.

If you are unable to comply with the requirements of one SID, you might still be able to find one that will be compatible with the operating capabilities of your aircraft. If you can't find one that you can use, or if you don't feel comfortable with any, simply file, "NO SID" in the remarks section of the flight plan. It's really much easier for all concerned to use the SIDs. Remember the old saying, "A picture is worth a thousand words."

There's another phrase in your clearance that you need to look at: cleared as filed. This is an abbreviation that is used by ATC when the route is essentially the same as what you filed. If only one small portion has been changed, you might get a complete readout, or you might receive, "Cleared as filed, *except* . . ." with a readout of only the portion that has been changed.

One very important point that must be emphasized here is that the cleared-as-filed clearance does not include the altitude filed for. Examine earlier examples in this chapter to see what I mean. The altitude must be given to you in addition to the cleared-as-filed portion.

If the original clearance has been substantially changed by you, your company, or ATC, you will definitely receive a full readout of the new clearance. If you don't receive it automatically, then you should request it.

For an example of an abbreviated clearance with a slight change, suppose that you have filed to Lake Tahoe from San Francisco, and have asked for a "Dumbarton Five Departure, Direct . . . FL 180." You might get a clearance that reads:

> *Speedy 50X is cleared to the Lake Tahoe Airport, Bartn Transition, then as filed; maintain flight level 180, except, climb on runway heading to 2,000 feet before turning intersecting the San Francisco 090-degree radial.*

In order to ensure the success of the program, as well as the accurate relaying of information, you should:

- Include specific SID/transitions and preferred routes in your flight plans whenever possible.
- Avoid making changes to a filed flight plan just prior to departure.
- Most especially, *request route/altitude verification or clarification from ATC if any portion of the clearance is not clearly understood.* You should never be put in a position to have to say, "Well I thought you meant. . . ."

5

Departure procedures

Getting an IFR clearance

File a flight plan at least 30 minutes prior to departure to give ATC ample time to process the clearance and to fit it into the system. Once a flight plan has been filed, how do you go about getting your clearance?

This will depend upon which type of airport you are operating from. You will normally find yourself in one of three situations:

- At an airport with a tower and a *discrete clearance delivery* frequency, you will simply call clearance delivery directly prior to calling *ground control* for taxi clearance; at airports where pretaxi clearance procedures are in effect (as indicated in the *Airport/Facility Directory*), you will get the clearance not more than 10 minutes prior to taxi; at airports where this procedure is in effect, you will then know if you are to expect a flight delay prior to starting the engine(s).

- At an airport with or without a tower or flight service station, but at least within radio range of one of the above, you will get the clearance over the radio from whichever facility you are able to contact, the tower being the first choice. One exception to this rule is when you are unable to contact any of the above, but are able to establish contact with an *air route traffic control center* (ARTCC) frequency, in which case you can call them directly.

- At an airport with no ATC facilities, and one from which you are unable to contact a control facility by radio while on the ground, you may take off in VFR conditions and pick up the IFR clearance once airborne (and still VFR) and within radio range of a facility. If you are unable to take off in VFR conditions, you will have to telephone the facility with which you filed your flight plan. He or she will normally ask what time you expect to depart, and you will usually be given a clearance containing a "clearance void" time, which means that if you are not airborne by that time, your clearance will be voided and you will have to call for a new one. If something happens that makes it look as if you won't be able to make it off by the clearance void time, your best bet is to get on the horn again and ask for an extension of the void time. In any case, you must let ATC know within 30 minutes that you will not be able to comply. This will help prevent further delays as well as costly reroutes to other traffic that have been issued clearances based on your departure.

You should have a good working knowledge of all of the facilities and basic airways in the vicinity of the departure airport. This is because it isn't unusual to receive a change to the clearance shortly after becoming airborne. If you know the primary airways and the VOR frequencies, it will enable you to comply with the changes at once, double-checking the charts as time permits, rather than trying to fumble with the charts while still trying to get the gear up or maintaining control in turbulence.

These course changes are especially prevalent in radar environments. You will be advised of any expected radar vectors prior to takeoff, normally advised of the initial heading and the reasons for the vectors. By listening to the instructions, you merely fly basic instrument maneuvers (climbs, turns, etc.) until you are established on course and told to "resume normal navigation."

Sometimes, *departure control* will hand you off to a center frequency where you will receive further vectors. You must not allow yourself to become too complacent when following these simple directions. You must still monitor the navigation receivers and be continuously aware of the aircraft position with respect to your requested route of flight. You never know when the radios might decide to take a holiday and leave you with no air/ground communications.

Before going any further, an emphatic statement: *You should read back every clearance you receive.* If the controller read it wrong or if you copied it wrong, this will be your last chance to correct the error. This holds true for all clearances, including taxi, takeoff, climb, descent, approach, landing, or whatever. Also, don't try to be a speed reader because the faster you read back the clearance, the more chance there is for an error to go undetected. Be sure that you include your complete aircraft number or flight number in the read back. *EVERY clearance.*

Use of a climb profile

Many aircraft types are uncomfortable at best-angle-of-climb speeds for very long. The typically low airspeeds and high power settings may combine to produce high engine temperatures. The high deck angle during the climb presents a problem in forward visibility, and for many pilots the speed itself is uncomfortably close to stall. With this in mind, a more practical departure might employ a variety of configurations, power settings, and speeds to enhance climb performance and obtain a good safety margin on departure. A horizontal, or profile, view of this climb path will often describe several steps where configuration changes and accelerations take place as part of the climb procedure. The entire procedure is known as a *climb profile.*

A typical climb profile in the Canadair Regional Jet as used by one local airline involves a configuration set to 20-degree flaps and rotation off the runway, retracting the landing gear, and an attitude such that the airplane climbs at single-engine best angle of climb plus 10 to 20 knots until 400 ft, a flap retraction to 8 degrees maintained until 1000 ft, flaps retracted and acceleration to best rate of climb until 1500 ft, whereupon the thrust is reduced to a more nominal climb setting. If that sounds involved, it becomes more so out of airports that are surrounded by high terrain, such as Butte, Montana. The southbound Butte departure involves a modified climb profile, where 20-degree flaps are retained until the airplane is much higher, and a course reversal in the climb, using 15 degrees of bank that keeps the airplane over the airport area until it reaches an altitude where a potential engine failure would be less of a problem.

Climb profiles such as this are developed with much careful study of the airplane's climb performance characteristics on one engine. The profiles plan for an engine failure, then write the procedure such that the airplane can maintain terrain separation while climbing IFR, until reaching altitudes where normal instrument procedures apply.

If you fly a multiengine airplane, you would do well to consider its single-engine climb performance in selecting your initial flight path on a night departure, planning for the potential failure and your subsequent need to continue a climb without hitting any unseen obstacles nearby. If you are flying single-engine, select a procedure and flight path that would take you over terrain that you might possibly use for an emergency runway. Your best choice could easily be the airport runway you just departed, which would be possible if your flight path in the climb positioned your airplane over or near the field.

Instrument departure procedures

A good place to look for guidance in planning your night departure is the IFR departure procedures for your airport. Although many airports do not have IFR departure procedures, most airports with much night traffic do. A few dollars will get you some National Oceanic Service (NOS) charts for your airport, and they will outline a suggested route for IFR traffic. The route will often specify a minimum climb gradient and flight path to follow, assuming that outside references are invisible. It will even suggest minimum altitudes before continuing on course, based upon your direction of flight. As a night VFR pilot, you are certainly not bound to follow these IFR-specific procedures, but they may offer excellent help in avoiding obstacles in the area while you climb.

You may wish to create your own departure procedure. If so, the IFR departure might be a good place to start, then use a VFR chart, identify well-lit areas that should offer good outside reference and plan something that would allow you to climb safely and conveniently to altitude.

6

Practice maneuvers and emergencies

Now that you've mastered the basics of instrument flight, it's time for some practice instrument maneuvering in normal conditions and in an emergency where one or more of your instruments has failed. This will prepare you for the ultimate goal of adding departure, en route, and arrival navigation to your instrument flying skills.

Pattern A

It's time to assemble all the elements covered so far: straight and level, speed changes, and standard rate turns. A good exercise for this is Pattern A, shown in Fig. 6-1. It also contains all the maneuvers required for a full-scale instrument approach, except for the descents. You will fly this pattern, or portions of it, on every instrument approach. Make a copy of Pattern A and attach it to your clipboard as a ready reference.

Set up the exercise in straight and level flight at normal cruise speed. Start the pattern on 360° the first few times you try it. As soon as you are comfortable with the pattern, vary the initial headings and start on 090°, 180°, or 270°. This will be good practice to learn which roll-out headings will be with different initial headings.

Start timing at the beginning of the exercise with the sweep-second hand at 12 o'clock. Time each leg consecutively; each new leg starts when the time for the old leg has expired and control pressure is applied to adjust for the new leg.

Pattern A is an excellent exercise to practice under the hood without an instructor, but with a safety pilot in the right seat (Appendix B: FAR 91.109 (b)(1)). Don't continue practicing these patterns if

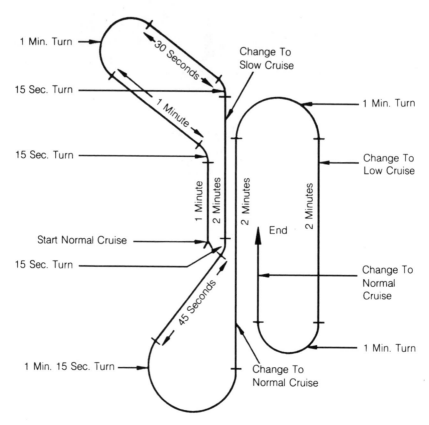

Fig. 6-1. *Pattern A practice maneuver.*

problems develop. It might be frustrating and you might unknowingly develop bad habits in scanning or procedures. As any instructor will tell you, it takes a lot of extra time and effort to break bad habits. Try again another day with an instructor.

Minimum controllable airspeed

I like to introduce minimum controllable airspeed under the hood early in the training syllabus. This surprises a lot of students. They are just beginning to master straight and level flight under the hood and I ask them to try flight at minimum controllable airspeed!

Introducing minimum controllable airspeed early serves several important purposes. First, it will improve your handling ability on instruments because you will learn to control the aircraft throughout a wide range of pitch and power changes. And you will learn to anticipate the major changes in control pressures that accompany these transitions. A student who learns to handle the plane well in minimum controllable airspeed under the hood, always finds it easier to reach the goal of 2, 2, 20 when flying straight and level and other fundamental maneuvers.

Second, the student learns to extend and retract flaps (and landing gear if so equipped) while under the hood. With very little practice extending and retracting flaps and gear on instruments becomes an automatic reflex. This is very important on instrument approaches. The last thing you want to happen while descending on final approach is to break your scan while fumbling with flaps and landing gear.

Simulation of missed approach

The third reason for practicing minimum controllable airspeed under the hood also has to do with instrument approaches. Imagine this situation: You have flown a perfect approach down to the ceiling minimum and the runway is not visible; it's time to execute a missed approach. The airplane is low and slow, the nose is pitched up, and flaps are extended. To stop the descent, you must add full power, level off, and start to retract the flaps.

The airplane might be as low as 200 feet above the runway. You must be in absolute, positive control of the airplane to increase airspeed rapidly and establish a climb to avoid obstructions and put some distance between yourself and the ground. The transitions during recovery from minimum controllable airspeed are identical to those during a missed approach: full power, increase speed, retract flaps, maintain altitude.

Missed approaches will be introduced and practiced in later lessons. Minimum controllable airspeed provides several fundamental building blocks that must be mastered early in the course to become a precise instrument pilot who can handle all maneuvers with absolute safety. The best way to master minimum controllable airspeed on instruments is to think that it is an extension of the second fundamental item, speed changes, except that the changes are carried to greater limits.

Entering the maneuver

Set up the maneuver from straight and level flight on a convenient cardinal heading. Start reducing power; for the Cessna 172 or Piper

Cherokee 180, you will find that reducing power to 1500 RPM will be adequate. As the nose starts to feel heavy, and the airspeed slows to the flap extension range, apply the first notch of flaps, or 10°. As airspeed decreases further, lower the flaps another notch. Adjust the trim as necessary throughout the maneuver. As the nose gets heavy again, extend full flaps. Do not attempt to drop full flaps all at once because the changes in attitude and control pressures will be so abrupt and heavy that you will have difficulty controlling the airplane by reference to instruments.

As minimum controllable airspeed approaches, the instructor will request power applications (perhaps full power) to maintain altitude at the lowest possible speed without stalling. Aim for the airspeed at which altitude control becomes difficult. That will be the minimum controllable airspeed for this exercise and it will not place the aircraft in an imminent stall situation.

Another important lesson to be learned from minimum controllable airspeed is when flying with full power, the power is "constant" and not "variable." Recall and implement the memory aid "pitch to the speed, power to the altitude." In other words, maintain airspeed by adjusting nose attitude and maintain altitude by adjusting power.

Once again, the attitude indicator is the instrument used to control the airplane. The airspeed indicator is the primary, or "quality" instrument for pitch. The heading indicator is primary for bank and the altimeter is primary for power. Include VSI, turn coordinator, and tachometer or manifold pressure in your scan as supporting instruments for pitch, bank, and power respectively.

Instructor note. The student must do clearing turns prior to minimum controllable airspeed. Make sure there are no airplanes in the area because the high nose-up attitude in this maneuver will limit your vision as safety pilot. My recommendation is that in the beginning the student should do standard rate turns left and right 90° prior to reducing power. As the student gains more experience, the clearing turns may be made while reducing power and lowering flaps. This is good practice because in the real world of IFR there will be many times during approaches when the plane will have to be slowed and flaps extended while in a turn.

Vertical S

When students file and depart IFR on every flight, they will usually have more than enough opportunity to practice climbs and descents in the real world of IFR. So there isn't much point in practicing additional vertical maneuvers. An instructor has to avoid a natural tendency to teach mechanics of the maneuver rather than the goal of the maneuver.

The Vertical S (Fig. 6-2) and its variations, the S-1 and S-2, are excellent exercises for an instrument student to practice with a safety pilot. The Vertical S consists of climbs to 500, 400, 300, and 200 feet with reversals at the top of each climb and descents back to the original altitude before climbing to the next altitude in the series. The Vertical S can also be a series of descents.

The Vertical S-1 is a combination of the Vertical S and a standard rate turn. Make a standard rate turn each time you return to the original altitude. Alternate turns to the left and to the right.

The Vertical S-2 differs from the S-1 in that the direction of turn is reversed with each reversal of vertical direction.

Pattern B

The Vertical S, S-1, and S-2 are recommended maneuvers in the FAA's *Instrument Flying Handbook*. However, I have found that

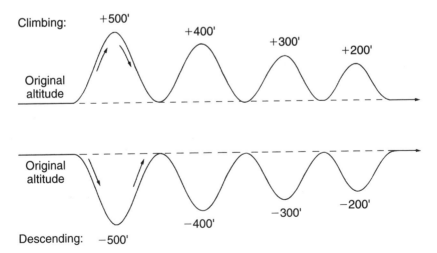

Fig. 6-2. *Vertical S practice maneuver.*

Pattern B is much more effective in teaching students how to combine the fundamentals of instrument flight: straight and level, speed changes, standard rate turns, climbs, and descents. It's an excellent maneuver for "putting it all together."

The turns and straight stretches in Pattern B (Fig. 6-3) are the same as those in Pattern A. But B adds speed changes and includes a descent and an emergency pull-up to simulate an approach and missed approach.

Roll out on headings regardless of time passage. The turn to the final leg is a descending standard rate turn. Note that a prelanding checklist is included, then a little later you extend ¼ flaps as if commencing the final "approach." If you are flying an airplane with retractable gear, also lower the landing gear at this point.

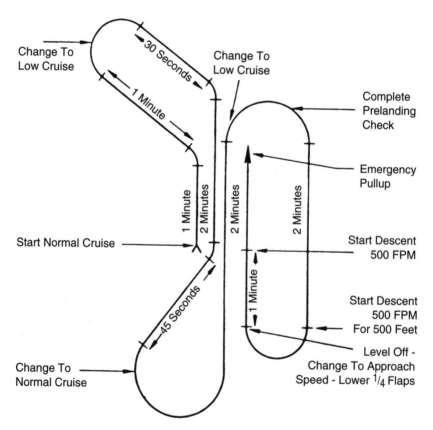

Fig. 6-3. *Pattern B practice maneuver.*

At the emergency pull-up, don't forget to retract approach flaps and landing gear, if so equipped. Does something seem familiar here? Right! It's the recovery from minimum controllable airspeed. The pieces indeed come together at this point. Maybe not perfectly, but the goal of 2, 2, and 20 is in sight.

Pattern C

Don't worry, Pattern C isn't required! But you will feel a great sense of achievement if you can do it. It has been called a basic airwork "graduation exercise." (Fig. 6-4.)

If you can fly C with its nonstandard climbs and descents and maintain 2, 2, and 20, you will have certainly mastered the fundamentals of attitude instrument flight. Patterns B and C are good exercises to practice with a safety pilot. Break off practice if the pattern work is not going well, otherwise you might unconsciously develop bad habits. Work with an instructor on whatever is causing the problem before any bad habits have a chance to take hold.

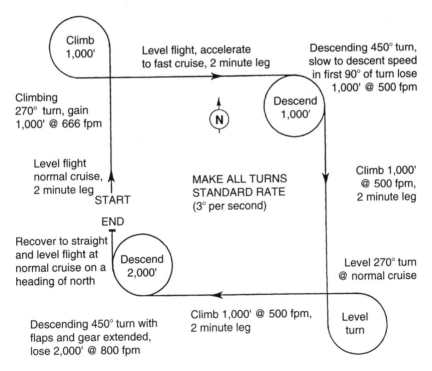

Fig. 6-4. *Pattern C practice maneuver.*

Master Pattern C on full panel, then try it on a partial panel. It is a sure cure for overconfidence; it is also instant insanity. Some dedicated instrument students have done this. I think they were former military pilots who had partial-panel Pattern Cs inflicted on them by sadistic military instructors. Civilian instructors, of course, would never pull a stunt like that. Flying is supposed to be fun!

Practicing stalls

You have been practicing flight at minimum controllable airspeed since Flight Lesson 2. Stall work will simply be an extension of minimum controllable airspeed. In practicing the power-off stall, fly at minimum controllable airspeed with power set at 1500 RPM or 15 inches of manifold pressure and flaps and gear down if applicable.

Instructor note. Pick a convenient safe VFR altitude for practicing stalls and *always* do clearing turns. The student should do the clearing turns under the hood while setting up for minimum controllable airspeed. This will provide valuable additional practice in handling the airplane on instruments through a wide range of changing control forces. The instructor has sole responsibility for collision avoidance.

When straight and level with flaps and gear down and power at 1500 RPM, reduce power to idle. Cross-check with the turn coordinator and add rudder pressure to keep the ball centered. Allow the airspeed to decrease while holding altitude constant until a full stall occurs.

Stall recovery

As the airplane stalls, effect the stall recovery by applying full power. Reduce back pressure to reduce the angle of attack. Don't push forward on the yoke—you will lose too much altitude if you do. Reduce back pressure to pitch down slightly (as seen on the attitude indicator) then return to straight and level flight promptly without inducing a secondary stall. Use the attitude indicator to maintain straight and level. Use rudder pressure, not ailerons, to hold heading while the airspeed is low. (Smoothness is very important, as always!)

In a stall, one wing will frequently drop. The reaction of many pilots in critical situations and at critical airspeeds is to use the ailerons to raise that low wing. This is incorrect in most airplanes and might make things worse. The proper procedure is to use opposite rudder to add a little speed to the slower descending wing and give it lift.

Instructor note. Because of deficient earlier VFR training, students might require extra practice and instruction in the use of rudder rather than ailerons to raise a low wing during a stall.

As the power becomes effective, start raising the flaps in increments. When climb airspeed (V_y) is attained in a straight and level attitude, gradually pitch up to the first line above the horizon on the attitude indicator. (In a retractable gear airplane, delay raising the landing gear until a positive rate of climb has been established.)

Retracting the gear

Flaps are raised before the gear is retracted for two reasons. First, the flaps create much more drag than the landing gear at slow speeds. A climb can be established sooner when the flaps are raised first.

Second, when you execute the missed approach, you want to establish a positive rate of climb before retracting the landing gear. If you have erroneously allowed the airplane to drift down, you will touch the runway with the wheels, not the first 6 inches of the propeller blades!

This stall maneuver combines the power-off stall that might occur on the final of an instrument approach with the full-power, flaps-and-gear-down situation of a missed approach. The IFR stall maneuver should be practiced on instruments until you can quickly recognize the stalled condition when you begin to lose altitude, then effect a prompt recovery with minimal additional loss of altitude. After you become skillful at this, you should lose no altitude. Losing more than 25–50 feet in the recovery is unsatisfactory. Recognize the stall quickly and execute the recovery procedures promptly, correctly, and automatically.

Steep turns

The next maneuver was required to obtain a private pilot certificate: steep turns at 45° bank. The difference now is steep turns with the hood on, solely by instruments.

Instructor note. The traditional method is to establish a 45° bank and then complete a 360° turn in one direction, followed immediately by a 360° turn in the opposite direction, rolling out on the original heading.

Common errors are altitude control, especially during and immediately following the roll-out, and changing to the opposite direction. Students will usually lose altitude at 60° into the turn when executed to the left and gain altitude in a right turn because of "P" factor. During the change of direction, there is also a tendency to gain altitude due to the excess back pressure required. An excellent training maneuver to overcome this problem if it persists is to have the student do a series of entries to steep (45° bank) turns in opposite directions until the problem is solved.

Steep turns require a faster scan to make sure you absorb all the information that the instruments are showing you when you need it. You can't afford to fixate on any single instrument because everything will be happening quickly. You will use the same control and support instruments for steep 45°-bank turns as standard rate turns, but move your eyes around the panel faster using the same scan pattern—attitude, heading, altitude—but do it faster.

One problem with this maneuver is that the inner ear senses a 90° bank with each change in direction. Beware of nausea; when airsickness occurs, all learning ceases.

Altitude control

Cross-check with the altimeter to maintain altitude. On this maneuver ±100 feet is allowed. However, it is much easier to limit the variation in altitude to ±20 feet than to allow the altitude to vary by 100 feet. Continue the turn for a full 360°. When established in this steep turn, of course, the bank tends to increase. If you are not paying attention or your scan is too slow, expect a rapid loss of altitude and an ever-steepening bank. Correct this by reducing bank to 20° or 30° to recover the lift lost in the steep turn. It's almost impossible to regain the lost altitude unless you decrease the angle of bank.

Trim

Students almost always ask what to do about trim in steep turns. I recommend that the maneuver be performed a few times without adjusting the trim to feel what the control pressures are like and what it takes to cope with the control pressures solely by instruments.

Add a little power—100 RPM or so—in the turn to help keep the nose up. Add a touch or two of nose-up trim. With practice the combination of added power and nose-up trim will result in smoother, easier steep turns.

After completing a 360° turn in one direction, make a smooth transition to a steep turn in the opposite direction. Do not pause during the reversal and do not fly straight and level between one turn and the other.

Keep the attitude indicator dot right on the horizon as you roll and do not allow altitude to vary more than ±20 feet for an easier time managing the control pressures in the reversal. You have been applying so much back pressure (or adding power and nose-up trim) to maintain altitude during the 45° bank that when you roll into this reversal, it feels like you have to push the nose forward to keep from climbing. Also, lift increases when the wings roll through the level position. Fix that dot on the horizon line and visualize rolling around it from one direction to another.

Unusual attitudes

A steep turn with a bank that increases beyond 45°, coupled with a rapid loss of altitude, opens the realm of unusual attitudes. From the beginning, my students work on unusual attitudes with the attitude indicator covered up. It is advisable not to practice initially with a full panel. (This is the instructor's decision.)

Let's consider what happens when an attitude indicator begins to fail. The first point is that it takes time for the gyros to wind down. Even if there is a sudden failure of the vacuum system powering the gyros of the attitude indicator, the gyroscopes in the instrument will lose momentum slowly.

You might not be aware that a failure has occurred. The attitude indicator doesn't suddenly roll over and die at a dramatic angle; the indicator gradually drifts off. You might continue to use the attitude indicator as the control instrument while it is gradually leading you astray.

The first assumption dealing with an unusual attitude is that the attitude indicator has failed. Don't stop to analyze the failure. Assume that it has occurred, deal with the unusual attitude immediately, and when everything is under control again, try to figure out what went wrong.

Recovery procedures

The first instrument to check in an unusual attitude is the airspeed indicator. Its indication will determine what actions to take. If the airspeed is *increasing* the airplane is in a dive and might run out of altitude during an approach or departure, or exceed the redline airspeed (V_{ne}, never exceed) if flying at cruise altitude.

The recovery procedure for a *diving* unusual attitude is:

1. Reduce power to idle.
2. Level the wings with rudder and aileron. Center the needle of the turn coordinator to control the bank. Keep the ball centered.
3. Raise the nose to stop the descent. Refer to the airspeed indicator to increase pitch and stabilize the airspeed at cruise.
4. When cruise airspeed is attained, apply cruise power and establish straight and level flight on partial panel. (*See* the subsection regarding partial panel procedures in this chapter.)

Let's cope with an unusual attitude in which the airspeed is *decreasing*. The recovery from a *climbing* unusual attitude is:

1. Add full power to increase airspeed and reduce the risk of a stall.
2. Lower the nose.
   - ~ Don't run out of airspeed and get into a power-on stall.
   - ~ Decrease the angle of attack.
   - ~ Use the airspeed indicator to decrease pitch and return to cruise airspeed.
3. Level the wings with rudder and aileron.
   - ~ Stabilize the turn coordinator to control the bank.
   - ~ Keep the ball centered.
4. When the airspeed reaches cruise, reduce to cruise power and establish straight and level flight by partial panel.

Unusual attitude recovery procedures summary

Resumption of control is initiated by reference to airspeed, altimeter, VSI, and turn coordinator.

Nose low, airspeed increasing:
1. Reduce power
2. Level the wings
3. Raise the nose

Nose high, airspeed decreasing:
1. Increase power
2. Lower the nose
3. Level the wings

Common errors:
1. Fixation. Staring at the least important instrument. (Keep eyes moving.)
2. Improper trimming or continuously holding pressure against the trim. (Learn proper trimming.)
3. Cockpit disorganization. (Plan to take every step in proper sequence.)
4. Attempting recovery by the seat of the pants, which produces misleading sensory inputs. (Believe the instruments.)

Finally, as the famous violinist told the budding musician who inquired about how to get to Carnegie Hall: Practice! Practice! Practice! Practice!

Partial panel procedures

Recognition of instrument failure becomes the first step in partial panel work, regardless of whether the failure becomes apparent in an unusual attitude or in erroneous readings that gradually become more serious. Following recognition of the failure, make a positive decision to disregard erroneous instruments and turn to other instruments to supply the missing information (Figs. 6-5 and 6-6).

Some pilots find it helpful to cover a failed attitude indicator and heading indicator and they carry little rubber suction cup soap dishes on every flight just for that purpose. This is a good idea. In a tight situation

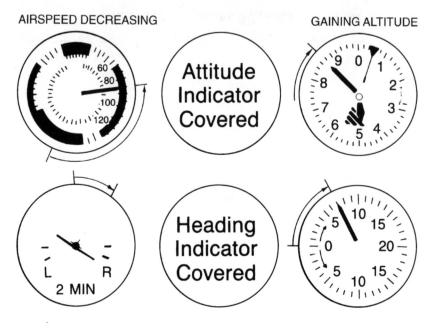

Fig. 6-5. *Nose-high unusual attitude on partial panel: steep right turn.*

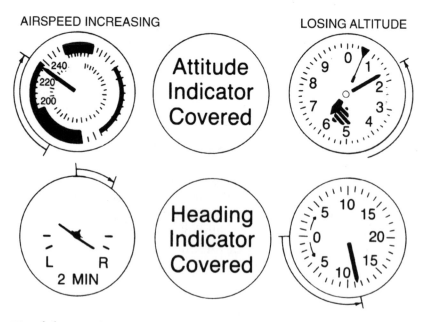

Fig. 6-6. *Nose-low unusual attitude on partial panel: steep right turn.*

you don't want to inadvertently include failed instruments in the scan. If you don't have any soap dish suction cups handy, just tear off pieces of paper and use them to cover the failed instruments.

The next step in the real world of IFR would be to land immediately after a major instrument failure. Don't hesitate to declare an emergency and get vectors from ATC to the nearest VFR airport. If no VFR field is available, head for the nearest IFR airport that has your personal minimums. There is absolutely no point in continuing an IFR flight with major instrument malfunctions. Things will only get worse as fatigue sets in.

Partial panel control instruments

Because you can't always count on going VFR immediately after an instrument malfunction, be prepared to fly on partial panel. Other instruments replace the functions of the attitude indicator and heading indicator when no longer available. As with full panel, a primary instrument provides the most pertinent information about how well you are doing and it does not move when flying precisely.

Use the same scan pattern for partial panel as full panel: attitude, heading, altitude. The only difference with partial panel is that you scan two instruments for attitude information rather than one.

The control function, formerly performed by the attitude indicator, is now divided between the airspeed indicator and the turn coordinator.

Pitch is controlled with the airspeed indicator. If the airplane is 100 feet below the desired altitude, use back pressure on the yoke to reduce airspeed 5–7 knots. This will pitch the nose up slightly and regain the lost altitude. Conversely, if the airplane is 100 feet above the desired altitude, increase airspeed by a very slight forward pressure on the yoke and pick up approximately 5 knots, indicating a slight nose-down pitch. This will gradually take the airplane back to the desired altitude.

When you reach the desired altitude and the airspeed stabilizes at the desired cruise speed, a minor trim adjustment might be required to help hold that altitude. Keep wings level with the turn coordinator. As with full panel, keep both feet on the rudder pedals and apply pressure if the coordinator leans steadily in one direction or the other. The turn indicator is very sensitive; don't attempt to make a correction every time it moves, or else you will begin chasing the needle and very quickly lose control of the heading. As always, keep the ball centered with rudder pressure.

The poor man's heading indicator

Here is an item I bet you won't find in any other book. Note that the ADF is listed as a support instrument for bank in straight and level and several other flight conditions. Why is this? Because the ADF is an excellent source of bank information. When an ADF is tuned to a station in front of the airplane, the ADF will very quickly indicate drifting left or right, the same way the heading indicator would if it were functioning. I call the ADF the poor man's heading indicator.

The ADF can also help you make standard rate turns. For example, if the ADF is tuned to a station ahead and you want to make a 30° turn to the right, simply turn and begin the rollout approximately five degrees before the ADF needle has moved 30° to the left. There's more to the ADF than meets the eye!

With practice you will automatically look to the correct instruments for control references to replace the attitude indicator. As you make corrections and enter climbs, descents, and turns, be careful to make minimal control movements. The greatest problem I find with instrument students flying partial panel is overcontrolling. Know exactly where to look for control references and minimize control movements and you will rapidly build up skills in the fine art of partial panel flying.

For small changes of heading, use a half-standard rate turn ($1\frac{1}{2}$° per second). Roll into the half-standard rate turn and count "one one thousand, two one thousand, three one thousand" up to the number of degrees of heading change that you wish to accomplish. This will also bring you close enough to make minor adjustments after the magnetic compass settles down.

Practice steep turns, unusual attitudes, partial panel, and magnetic compass turns from Flight Lesson 6 through the end of the program. The test standards require demonstration of these techniques during the instrument flight test.

Loss of radio communication

If you do lose radio contact, what will you do? We'll look into this subject much more deeply during the *en route* discussion, but if it happens during the departure phase of your flight, you will find yourself in one of three situations:

- Flying the clearance as received, in which case you keep right on complying with the clearance
- Flying an amended clearance
- Flying a radar vector

When flying either an amended clearance or a radar vector, you will have been issued instructions as to where the change of heading is taking you. The amended clearance might be explained:

50X, amendment to your clearance, now climb on the 030-degree radial until 8,000 feet and then turn left on course.

If you lose communications capability while climbing through 3,000 feet on the 030-degree radial, just comply with the clearance as it was given. Continue climbing to 8,000 feet, and then turn left to get on the flight planned route.

If you were on a westbound radar vector, you might have been told something like:

50X, radar contact, turn left now, heading 240 degrees to intercept the Spartan 276-degree radial.

If you were to lose communications at that point, you should turn farther left to intercept the radial at a 90-degree angle in order to comply with Part 91.185(c)(1)(ii):

If being radar vectored, [fly] by the direct route from the point of radio failure to the fix, route, or airway specified in the vector clearance.

During taxi to the active runway, give the instruments a good check. Prior to taxiing, you should have set the directional gyro to the magnetic compass heading. Then during turns while taxiing, you are able to check for proper movement of the DG.

The same thing holds true for the needle/ball, turn/slip, slip/skid indicator, or turn coordinator, whichever term is the vogue for this instrument at the time you read this. (I prefer to call it the needle/ball.) In any event, watch it while turning. The needle should deflect to the same direction the aircraft is turning.

Remember that the attitude indicator, needle/ball and DG are all gyro-driven, and should have at least five minutes to come up to speed to prevent excessive precession and bearing wear.

VOR tracking and time/distance problems

You have one last area to work on before departing on your first cross-country instrument flight. This is a procedure that will cover VOR tracking, radial intercepts, and time-and-distance problems. You can usually accomplish this problem successfully in one practice session using a single VOR facility (Fig. 6-7).

To begin with, cross the VOR and proceed outbound on the 135-degree radial for a period of 5 minutes, then turn right to a heading of 270 degrees. The problem now will be to intercept and track inbound on the 180-degree radial.

In addition, you will need to figure the time to the station from the point of intercept. This problem is primarily one of mental compu-

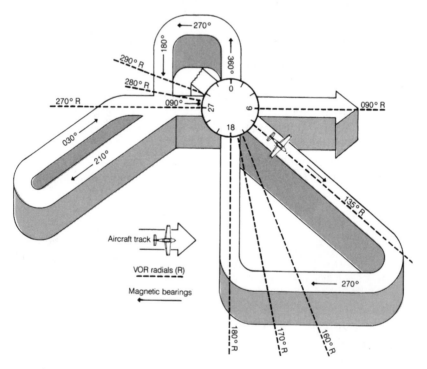

Fig. 6-7.

tation, devised to get your thought processes working independently of the motor processes that are actually flying the aircraft.

Once again, the pattern has been set up to keep you within a reasonable amount of airspace. You can set up any headings and intercepts you want, though the ones depicted here are easy to follow and will suffice to teach what you will need to know in actual instrument conditions.

A simple solution to a time/distance problem on an intercept is to position the aircraft 90 degrees to the inbound radial. When 20 degrees from the radial, begin timing through the next 10 degrees of bearing change, converting the time to seconds. Ten percent of the resulting time in seconds (just place the decimal point one digit in from the right) will provide the approximate time to the station in minutes.

For example, if you have flown through 10 degrees in 2 minutes (or 120 seconds), time to the station (in minutes) would be 10 percent of that, or 120 seconds × 10 percent = 12.0 minutes. Distance from the station would then be the ground speed (in nautical miles per minute) times the time to the station. So if you have a ground speed of 120 knots (or 2 nm per minute), you would be 24 nm from the station. Easy? You bet, but it seems much more difficult in a moving aircraft when you are trying to control it on instruments alone or with a partial panel.

You can then add one other step to the problem. In order to allow adequate time to pin down the inbound course before entering the cone of confusion around the VOR, you should make it a practice to turn inbound only if you are more than 3 minutes from the station. This means that it must take more than 30 seconds to cross that 10 degrees of bearing change that you timed.

If you find that you are 3 minutes or fewer from the station, you should turn 30 degrees away from the reciprocal of the inbound course, fly outbound for 1.5 minutes, and make a standard-rate 180-degree turn (in the direction away from the station) back toward the course. Once you intercept the course, fly inbound to the station. Take another look at the diagram in Fig. 6-7. You can see how this resembles a procedure turn. In addition to everything else, this pattern should also give you practice in procedure turns.

Returning to the explanation of the problem depicted in Fig. 6-7: You are heading 270 degrees, which is 90 degrees to your inbound course (the 180-degree radial of the VOR). To orientate the *omni bearing indicator* (OBI), or *needle*, properly, you should have it set to your inbound course, which will be 360 degrees TO. In this example then, the OBI should be deflected to the left, showing that you have not reached the course.

(Before everyone jumps on me for using the term *omni bearing indicator* instead of *course deviation indicator* (CDI), I prefer to use the term CDI in relation to flight directors, and OBI in relation to the simple VOR receiver. If you prefer the term CDI, feel free to substitute it whenever you see OBI.)

In order to time yourself through 10 degrees of bearing change and still have time to make decisions as to whether to fly inbound or to fly a procedure turn first, you should back up the omni bearing selector (OBS) 20 degrees from the inbound course. That means that in this example, you should set it to 340 degrees TO. Once again, the needle should be deflected to the left side of the indicator.

Before long the needle will move off its peg. When it centers, you should check the clock, noting both the minute hand and sweep-second hand positions or the indication of minutes and seconds on a digital clock. Reset the OBS to 350 degrees TO. When the needle centers again, take the time in seconds and compute the time and distance. For the first 30 seconds of time from 340 degrees TO through 350 degrees TO, it's useful to keep saying "procedure turn." After 30 seconds have passed, if I have not yet made it through the 10-degree bearing change, I automatically start saying "direct inbound."

If you flew the initial outbound leg for 5 or 6 minutes, you should be more than 3 minutes from the station, and you will make a right turn inbound when you get to the course. Naturally, after crossing the 350-degree course that was set on the OBS, you should immediately reset it to the desired inbound course of 360 degrees. All this time, you should be maintaining the proper heading, airspeed, and altitude.

A strong wind will play havoc with this procedure because it might screw up your estimates, but for the most part, the estimates should work out pretty close.

Now turn inbound, track to the VOR, and continue on the same course outbound for 1 minute. As you cross the VOR, you should make a simulated position report to ATC. Your new problem is to track inbound on the 270-degree radial.

In order to track that radial inbound, you must first be positioned on the proper side of the station; therefore, after flying the 1-minute outbound leg from the station, you should turn left to a heading of 270 degrees and fly outbound for 1 minute. Then turn south, which will be 90 degrees to your inbound course again.

Once again, set the inbound course on the OBS. This time it will be 090 degrees TO. The OBI should be deflected to the right. If it is, re-set the OBS to 110 degrees TO, which is 20 degrees from the desired inbound track. When the needle centers, check the time, and advance the OBS 10 degrees at 100 degrees TO. In this case, assuming that your time is less than 30 seconds (it should be), you should fly a procedure turn as depicted, and while turning outbound, you should once again set the inbound course on the OBS.

As you can see, you can set up many problems using this simple pattern, although with the exception of the procedure turn, it might not be of much use on an actual IFR flight. Still, it is an excellent pattern for building up mental prowess and your command of reciprocal bearings, while at the same time teaching you to fly the aircraft on instruments almost automatically.

Once you feel that you have mastered the instruments and are quite comfortable doing these problems, you will be ready for the first actual cross-country IFR flight.

7

Navigation

You can now control your airplane solely and naturally by instruments alone. You understand the relationship between attitude, power, and aircraft configuration, and how changing any one will create changes in the others as well as affect your altitude, rate of climb or descent, your heading, and your airspeed. By practicing a series of flight maneuvers, you've begun to develop an intuitive "feel" for your airplane and now you can predict precisely how it will respond to your changes in control input. At last, it's time to add the reason for instrument flying in the first place—point-to-point navigation.

Cross-country flights

Almost all that you have previously learned, plus a few new tips, will be put to use on a cross-country flight. To realize the full benefit from your flight training, you must be able to plan and execute flights to other areas. To safely accomplish a cross-country flight, you have to be familiar with your aircraft, sectional charts, trip planning, weather, and navigation in order to be able to navigate from one point to another.

A safe cross-country flight consists of two distinct phases: preflight planning and the flight itself. If your preplanning is thorough, you will be able to fly your trip with greater confidence and probably enjoy it a lot more. The more planning you do on the ground, the less busy you will be in the air. On your first few cross-country flights, you will have plenty of things to do to occupy your time and mind without trying to find that frequency, heading, or some other important detail that you should have already written down and put in a safe place.

Preflight planning

A really safe cross-country flight doesn't just happen; it is planned. I remember my first dual cross-country flight. My instructor called me and said he had to go to Macon, Missouri, and would I like to go along. Would I? We got into the aircraft, flew to Macon, and returned. It was over before I knew it. He said I had done just fine, and he endorsed my certificate for solo cross-country privileges. There was just one problem: I did not know where we were going, where I had been, or how we had gotten there and back. There was virtually no preflight briefing, no calls to the FSS, no line on a chart (which he held the entire time), and only a general direction in which I was told to fly. At that time, I thought wind was something that kept you cool in the summer and a course was something that you took in school.

The odd part of this little self-incrimination was that the man was a very good basic flight instructor. He just didn't place much stock in the little things such as headings, charts, and frequencies. But he had taught me to fly the basics so well that I had time to look around, do some figuring, and with some common sense, I made it. The kicker is that I believe this made me a better pilot in many respects because I was forced to learn or suffer the consequences. I'm not recommending this as a way to learn cross-country flight technique, only suggesting that it can be done. If you receive good instruction, cross-country can be one of the more pleasant portions of your flight training.

The first things you need to know before you go on your cross-country are where you are going, the time of day you will be flying, and the charts that will be necessary to safely complete the trip (Fig. 7-1). You also have to plan whether the trip will be made by VOR, pilotage, or a combination of the two. Most VFR flights are made by using a combination of VOR and pilotage. However, pilotage is the method you must use to plan and execute your cross-country flight for your private pilot checkride.

Pilotage

Pilotage means to navigate an aircraft by using a map and prominent landmarks on the surface of the earth. It is obviously the oldest form of navigation, dating back to the early 1900s. It is also one of the most certain methods for finding and maintaining your position relative to earthly landmarks since they appear around and below you in plain sight.

Fig. 7-1. *Check the course and distance as part of your initial preparation.*

Today's pilot can utilize U.S. government-supplied sectional charts that depict the landmarks in very minute detail. In fact, there is so much information on these sectional charts that in a heavily populated area there is sometimes almost too much information. For instance, these charts display all towers, tell you how high they are above both sea level and the ground, as well as almost any other identifiable object you might imagine. A short list of items displayed on these charts would contain drive-in movie theatres, rivers, lakes, towns, towers, roads, airports, hills, mountains, cliffs, valleys, power plants, race tracks, railroad tracks, and just about any other solid object you can imagine. If you could see it, hit it, or use it, it will be displayed on the government sectional charts.

I have a small but significant piece of advice for you new or would-be pilots concerning the proper method of using the sectional chart. As is the case with all maps I am aware of, sectional charts are oriented to north. That is the top of the map, when held so it is easily readable, is north. Okay, so that's not earthshaking of itself. But human nature is such that we all like to be able to read things easily and so it is with sectional charts. We like to hold the map so that the map is always right side up. It's just easier to read that way.

Trouble is, our flights aren't necessarily always going from south to north. Sometimes we are going to fly southwest or whatever. When this happens, if you continue to hold the map so that it's right side up, the objects appear out the window of the aircraft out of place. They aren't in their proper perspective. Say a lake shows itself to your right, or west, of your course on the sectional. If you are flying south while holding the map upright, it's easy to believe the lake should appear out the left window of your aircraft. Of course this is not the case, but when you hold the map in what amounts to be an upside down position, funny things can happen to your navigation. Ask Wrong-Way Corrigan.

The cure for this malady is to always hold the map with the course you are following pointed out the front windshield of your aircraft. Turn the map so the departure point is nearest your body and the destination is at the point farthest away from you. It's only when landmarks appear out the window in the same relative positions in which they appear on the charts that the potential problem of reverse orientation is most likely avoided.

VOR

Without getting too complicated, let's review the functions and normal use of the very high frequency omnidirectional radio range (VOR). The VOR is called the *omni*. *Omni* is Latin for "in all directions." It can be a very good friend or a frightening enemy, depending on your understanding of its use and purpose. Well understood, it is of invaluable assistance in navigation. If you try to fly the VOR without fully understanding its use, you might actually wind up going in the wrong direction.

From its ground base, the VOR transmits signals in all directions. Each VOR has its own frequency, and you tune it in on the navigation side of your radio. The following describe the features of your VOR receiver.

Frequency selector

The *frequency selector* is manually rotated to select any of the frequencies in the VOR range of 108.0 to 117.95 MHz.

Course selector

By turning the OBS (omni bearing selector), the desired course is selected. This usually appears in a window or under an index on the VOR receiver head on your instrument panel.

Course deviation indicator

(CDI) The *course deviation indicator* is composed of a dial and a needle. The needle centers when you are on the selected course or its reciprocal, regardless of your heading (Fig. 7-2). Full needle deflection from the center position to either side of the dial indicates the aircraft is 10 degrees or more off course (assuming normal needle sensitivity).

TO/FROM indicator

The *TO/FROM indicator* is also called *sense indicator* or *ambiguity indicator*. The TO/FROM indicator shows whether the selected *course* will take the aircraft to or from the station. It does not indicate whether the aircraft is heading to or from the station (Fig. 7-3).

Flags

Flags can be labeled as signal strength indicators. The device to indicate whether a signal is usable or an unreliable signal is called an OFF/VOR flag. This flag retracts from view or says OFF when signal strength is sufficient for reliable instrument indications. When the VOR signal is strong enough to give reliable navigation guidance, the flag switches to VOR. Insufficient signal strength might also be indicated by a blank TO/FROM window.

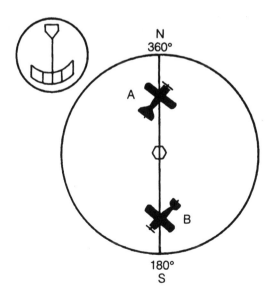

Fig. 7-2. *Course deviation indicator.*

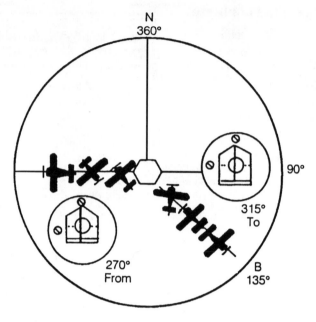

Fig. 7-3. *TO/FROM indicator.*

There are a couple of very important points to remember when using the VOR. The VOR transmits 360 possible magnetic courses to and from the station. These courses are called *radials.* They are oriented *from* the station (Fig. 7-4). For example, the aircraft at A, heading 180 degrees, is flying to the station on the 360 radial. After crossing the station, the aircraft is flying from the station on the 180 radial at 2A. Aircraft B is shown crossing the 225 radial. Similarly, at any point around the station, an aircraft can be located somewhere on a VOR radial. The important point is if you want to know what radial you are on, turn the OBS until the CDI centers and the TO/FROM indicator reads FROM. That is the radial you are on.

To properly utilize your VOR, you must first know where you are (what radial you are on), where you are going (what radial you have to intercept), and how to track the radial once you get there. This process utilizes three steps: orientation, interception of the radial, and tracking.

The following demonstration can be used to intercept a predetermined inbound or outbound track. The first three steps may be omitted if you turn directly to intercept a course without initially turning to parallel the desired course.

Turn to a heading to parallel the desired course. Turn in the same direction as the course to be flown.

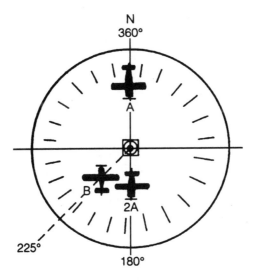

Fig. 7-4. *VOR courses called radials are oriented FROM the station.*

- Determine the difference between the radial to be intercepted and the radial on which you are located.
- Double the difference to determine the intercept angle not less than 20 degrees or more than 60 degrees.
- Rotate the OBS to the desired radial or inbound course.
- Turn to the interception beading (magnetic).
- Hold this magnetic heading until the CDI centers, indicating the aircraft is crossing the desired course.
- Turn to the magnetic heading corresponding to the selected course and track inbound or outbound on the radial.

VOR tracking also involves drift correction sufficient to maintain a direct course to or from a station. The course selected for tracking inbound is the course shown on the course index with the TO/FROM indicator showing TO. If you are off course to the left, the CDI is deflected right; if you are off course to the right, the CDI is deflected to the left. Turning toward the needle returns the aircraft to the course centerline and centers the needle.

To track inbound with the wind unknown, proceed using the following steps (Fig. 7-5). Outbound tracking is the same.

- With the CDI centered, maintain the heading corresponding to the selected course.

- As you hold the heading, note the CDI for deflection to the left or right. The direction of CDI deflection from centerline shows you the direction of the crosswind. Figure 7-5 shows a left deflection, therefore a left crosswind.
- Turn 20 degrees toward the needle and hold the heading until the needle centers.

Reduce drift correction to 10 degrees left of the course setting. Note whether this drift-correction angle keeps the CDI centered. Subsequent left or right needle deflection indicates an excessive or insufficient drift-correction angle—either add or remove some correction (heading). With the proper drift correction established, the CDI will remain centered until the aircraft is close to the station. Approach to the station is shown by a flickering of the TO/FROM indicator and CDI as the aircraft flies into the no-signal area (almost directly over the station). Station passage is indicated by a complete reversal of the TO/FROM indicator.

Following station passage and TO/FROM reversal, course correction is still toward the needle to maintain course centerline. The only difference is that now you are tracking *away from* the station instead of *to* the station. In the previously listed steps, you were tracking inbound on the 180 radial. After station passage, although your heading hasn't changed, you are tracking outbound on the 360 radial.

VOR procedures

Despite the great promise of GPS, VHF omnidirectional range stations (VORs) will remain the heart of the airway system for many years. The radials from a VOR are highways in the sky. VOR radials form most intersections, and VORs are used for more instrument approaches than any other type of facility.

VOR use requires positive identification of the facility. Obviously it can be fatal if you fail to make a positive identification and use the wrong frequency on an instrument approach. So get in the habit of automatically turning up the volume and checking the identification for every VOR.

Heading indicator errors

Another point to emphasize is the necessity of periodically checking the heading indicator and readjusting it to match the magnetic com-

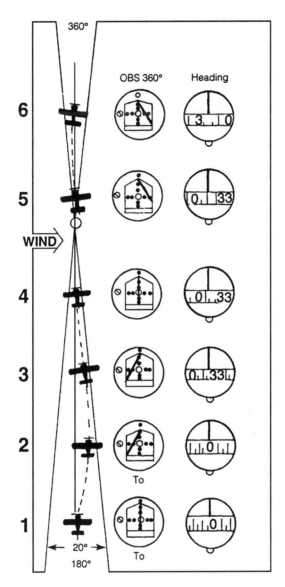

Fig. 7-5. *Tracking inbound with the wind unknown.*

pass. Maintaining an accurate course is difficult if not impossible if the heading indicator has drifted off.

The gyro of the heading indicator might precess a small amount due to bearing friction. The turns and reversals during instrument departures, approaches, and training maneuvers will produce additional precession errors. A heading indicator that precesses up to 3° every 15 minutes is within acceptable tolerances. I recommend checking the heading indicator at least every 15 minutes and prior to intercepting the final approach course on all approaches. Here are several other times the heading indicator must be checked or reset:

- When you line up on the centerline of the active runway for takeoff. This is the best opportunity to set the heading indicator with greatest precision: The gyro is up to speed, the plane is stable, and you know the runway magnetic course.

- When you begin every approach, even after a missed approach. This item must be on the approach checklist.

- On leaving a holding pattern and after practicing Patterns A, B, and C, holding, or any similar maneuvers that require numerous turns.

- After practicing unusual attitudes. These maneuvers might cause the heading indicator to wander off considerably or "tumble" because of the extra bearing friction produced by the maneuvers.

VOR proficiency

In the beginning of an instrument student's training, VOR skills must be determined. Does the student understand the basic principles of VOR orientation, and intercepting, tracking, and bracketing bearings and radials? Have any bad habits crept in since obtaining the private certificate? For example, does the student tend to get fixated on the course deviation indicator (CDI) needle and neglect other instrument indications? Remember, the goal remains 2, 2, and 20 in VOR work as well as other phases of instrument flight.

Don't spend expensive flight time to determine VOR proficiency. Make a quick pencil-and-paper check of the basic principles of VOR orientation by completing the VOR diagnostic exercise in Fig. 7-6.

Begin at the top line with the omni bearing selector (OBS) set at 030°. Then pick the correct VOR presentation for each lettered position. For example, with the OBS set at 030°, the most appropriate display for the A position is number 5. Proceed across through the G position, then drop down and complete the 090° line the same way, and so on.

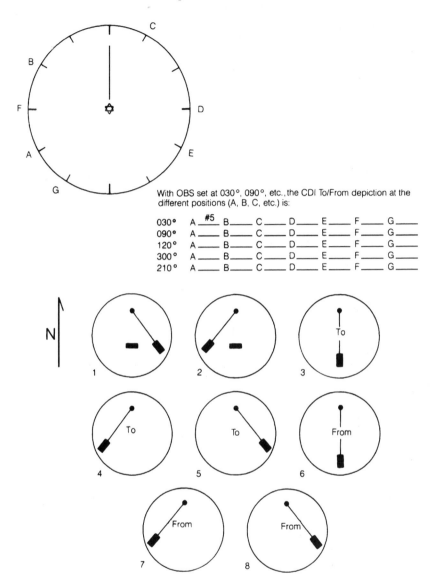

With OBS set at 030°, 090°, etc., the CDI To/From depiction at the different positions (A, B, C, etc.) is:

030°	A #5	B ___	C ___	D ___	E ___	F ___	G ___
090°	A ___	B ___	C ___	D ___	E ___	F ___	G ___
120°	A ___	B ___	C ___	D ___	E ___	F ___	G ___
300°	A ___	B ___	C ___	D ___	E ___	F ___	G ___
210°	A ___	B ___	C ___	D ___	E ___	F ___	G ___

Fig. 7-6. *VOR diagnostic test. Answers are on the next page.*

Answers are in Fig. 7-7. If any answer is wrong, review the exercise with your instructor. Proceed to the exercise in Fig. 7-8 for more VOR fundamentals.

16-point orientation

I developed this exercise many years ago. It works very well when teaching private pilot students exactly what happens around a VOR.

With OBS Set at:	CDI/To-From Depiction:						
030°	A #5	B #1	C #6	D #7	E #2	F #5	G #3
090°	A #4	B #5	C #8	D #6	E #7	F #3	G #4
120°	A #4	B #3	C #1	D #8	E #6	F #4	G #2
300°	A #8	B #6	C #2	D #4	E #3	F #8	G #1
210°	A #7	B #2	C #3	D #5	E #1	F #8	G #7

Fig. 7-7. *Answers to VOR diagnostic test on previous page.*

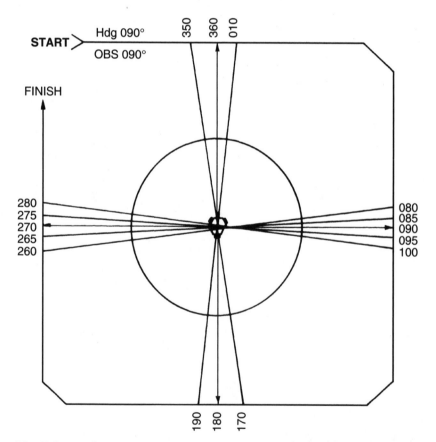

Fig. 7-8. *A 16-point VOR orientation exercise. Begin heading 0908 with OBS set on 090. You get 16 lines of position without changing OBS setting.*

The 16-point Orientation Exercise is a teaching exercise and a good diagnostic exercise because it will quickly reveal whether or not a student understands the basic principles of VOR. If not, the exercise can be repeated to bring the student up to par in short order.

Flying from west to east, with a VOR station to the south, one setting of the OBS will reveal 16 lines of position with precision while flying around the VOR station.

Most students are puzzled when I describe this exercise the first time. Sixteen lines of position from a single OBS setting? How can this be? It's really very simple when you understand what's going on, and when you understand, you will have mastered the basic principles of VOR work.

Enter the pattern on a heading of 090° (Fig. 7-8) with a convenient VOR station to the southeast. Set the OBS to 090°. As you begin the exercise, the TO-FROM indicator will show TO.

The 360° radial is the boundary between TO and FROM. The red flag will start to appear approximately at the 350° radial. When the red flag indicates OFF, you will be passing the 360° radial. That's the first precise line of position. When the red flag disappears, you will be approximately on the 010° radial. So the first leg of the exercise gives at least one precise line of position at 360° plus the lines of position 350° and 010° with lesser accuracy.

Continue on the 090° heading another 2 minutes or so, then turn right to a heading of 180°. The CDI needle will soon come alive and start moving in from right to left. When it reaches the outermost dot, you will be on the 080° radial. When the needle reaches the edge of the bull's-eye in the center, you will be at the 085° radial, and when the needle centers, you will be at the 090° radial. That's three lines of position.

As you continue on the 180° heading, the needle will pass the other edge of the bull's-eye at 095°, and the last dot on the left at 100°, adding two more lines of position.

You have produced three lines of position on the first leg and five on the second leg for a total of eight. If you continue the pattern as shown in Fig. 7-8, you will add three more lines of position on the heading of 270°, and five more on the 360° heading for a grand total of 16 for the full exercise—all without resetting the OBS.

If further practice is needed, enter the maneuver at other cardinal headings.

VOR time/distance check

Another excellent exercise in teaching the fundamentals of VOR orientation is the VOR time/distance check, or poor man's DME. Use this procedure to estimate out how long it will take to reach a VOR station in no-wind conditions. To tell the truth, I have never heard of a situation where someone has had to use the time/distance check to determine the time to a VOR on an actual flight; however, practicing the time/distance check will sharpen your VOR skills and, as we shall see later, introduce you to a similar procedure used in making a DME arc approach.

Turn toward a convenient VOR station and adjust the OBS so the CDI needle centers in the configuration. Note the heading to the station. Next, turn the airplane 80° right or left of the inbound course. Rotate the OBS in the *opposite* direction of the turn to the nearest increment of 10. In other words, if you turn right 80°, turn the OBS *left* (counterclockwise) to the nearest 10° increment (Fig. 7-9). You are flying a short tangent to an imaginary circle around the station.

Maintain the new heading. When the CDI needle centers, note the time. Continue on the same heading and change the OBS another 10° in the same direction as above. Note the number of seconds it takes for the CDI needle to center again. Divide the number of seconds by 10 to determine the time to the station in minutes. The formula is:

$$\text{Minutes to station} = \frac{\text{Time in seconds}}{\text{Degrees of bearing change}}$$

You can also calculate the distance to the station by using this formula:

$$\text{Distance to station} = \frac{\text{TAS} \times \text{minutes flown}}{\text{Degrees of bearing change}}$$

For the second leg, turn 20° *toward* the station and stop the turn. Turn the OBS 10° in the *same* direction as the turn. When the needle centers, note the time and turn another 10° in the same direction. When the needle centers again, note the elapsed time and calculate time to station.

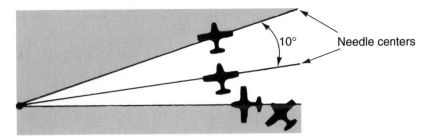

Fig. 7-9. *VOR time-distance check.*

Turn the airplane another 20° toward the station to start a new leg, and repeat the process. Continue these short tangents around the station as many times as you wish. Complete the exercise by selecting an inbound bearing to the station that lies ahead of the last tangent leg you plan to fly. Start turning inbound 10° before reaching the inbound course.

When are you 10° from the inbound course? Simple. Start turning inbound when the needle reaches the outermost dot on the CDI display. This will indicate that you are 10° away from the inbound course.

If you need more practice in either the 16-point orientation exercise or the time/distance check, use a simulator—you can get in more practice in an hour in a simulator because you don't have to copy clearances, take off, and fly to a practice area. Whenever you begin to have difficulty, you can stop the exercise and analyze what's causing the problem. Furthermore, an hour in a simulator costs much less than an hour in an airplane.

Intercepting a bearing or radial

The first practical application of VOR work will most likely be clearance to a VOR station after takeoff—cleared "direct" to the first VOR on the clearance or to intercept a specific radial. Clearance via a specific radial involves intercepting that radial and tracking inbound to the station. Two assumptions will simplify this procedure.

First, departure instructions and any radar vectors from ATC will always point in the right direction to make a quick, efficient intercept, unless you are being diverted away from traffic, higher terrain, or other obstacles. This is also true with en route clearance changes that direct you to intercept a specific radial.

Second, clearances always state *radials,* not magnetic courses to the VOR. Because a radial radiates *from* the station, you will fly *toward* the station on the reciprocal of the radial. A quick and easy way to determine the reciprocal is to refer to the heading indicator. Follow the radial from its number on the edge of the dial, through the center, and out in a straight line to the number on the opposite outer edge (Fig. 7-10). Set the assigned radial on the OBS and continue on the assigned heading. When the needle begins to move, the airplane is on the radial that lies 10° before the target radial.

Inbound turn

When you reach this 10° lead radial, turn to intercept the inbound bearing at an angle of 60°. When the needle reaches half-scale deflection, turn an additional 30° and maintain this 30° intercept heading until just before the needle reaches the bull's-eye, the small

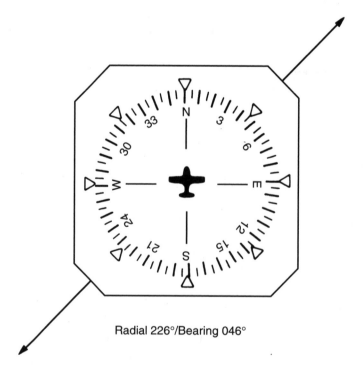

Radial 226°/Bearing 046°

Fig. 7-10. *Use heading indicator to visualize reciprocals.*

center circle on the CDI presentation. When the needle touches the bull's-eye, set the OBS to the inbound magnetic course.

Here's an illustration of how this works. Assume you have been assigned a heading of 320° and you have been cleared to the VOR via the 226° radial. As you reach the lead radial—216°—the needle comes alive. Turn right to a heading of 346° to set up a 60° intercept angle.

If you're not sure what heading will produce a 60° angle, refer again to the heading indicator, and count off 60° from the inbound bearing (Fig. 7-10). You will see this is 346°. With a little practice, you will be able to read reciprocals and intercept angles off the heading indicator at a glance.

When the needle reaches half-scale deflection, turn right again to 016°. Hold 016° until the needle reaches the bull's-eye, then steady up on the inbound heading of 046°, plus or minus whatever wind correction is necessary to hold the needle in the center.

Wind corrections

I'm always surprised when instrument students are unable to offer even an educated guess when asked, "Which way is the wind coming from?" Knowledge of the wind should almost be second nature by the time a person receives a private pilot certificate. If not, work on it during instrument training. An instructor should keep asking "which way is the wind?" until you begin anticipating and adjusting for the wind automatically.

Flight planning revealed wind forecasts at various geographical points and altitudes and you know exactly what the wind was at takeoff. It is a simple matter of deciding whether the wind is going to push to the left or to the right departing the airport toward the first VOR fix.

A tailwind will speed interception and a headwind will delay interception. And when turning onto the inbound heading, add a wind correction factor automatically, maybe 2°, 5°, or 10° according to your best estimate. Refine this correction by making adjustments en route toward the station.

There are two other interception techniques. If you believe you are close to the station when intercepting a radial, make the first turn

45° toward the station, rather than 60°. Hold that 45° interception course until the needle is about three-quarters of the way from full-scale deflection.

The needle reaches this position about $2\frac{1}{2}°$ from the assigned radial; turn to the inbound heading at this point. Add a correction for the wind when established on the inbound course.

How can you tell if you're close to the station? The more sensitive the needle, the closer you are to the station. When in close, the 45° intercept will put you on the inbound bearing quickly and at a greater distance from the station. The 45° intercept will provide time to adjust the inbound heading for the wind and it also gives you a better chance of being exactly on course over the station.

The second interception procedure is a reinterception technique utilized when off course and the needle is pegged at full deflection. This happens when a strong wind changes abruptly or when you are seriously distracted and drift left or right without correcting the problem.

In either case, make an en route correction to return to the desired radial or bearing. If the wind is from behind, use an intercept angle of 10° or 20° to return to course and avoid overshooting; if a headwind, use an intercept angle of 20° or 30°. The larger angles will get you back on the correct course sooner.

Common interception mistakes

The needle never centers. This indicates that (1) you turned to the inbound heading too abruptly or too soon or (2) a headwind was much stronger than anticipated. In either case, use the reinterception technique described above in the wind corrections subsection. Turn 10°, 20°, or 30° toward the needle and wait until the needle centers to resume the inbound heading.

The needle passes through the center and moves toward the opposite side. As in the first case, the inbound heading turn might have been faulty. You might have made the turn too slowly or waited too long to start the turn or the tailwind was stronger than anticipated. Use the reinterception technique to center the needle.

Another possibility is that you were so close to the VOR that the width of that 10° arc from full deflection to the center might have been only a few feet. A 45° intercept angle 1 mile from the VOR is almost impossible. The needle will peg with a FROM indication almost as soon as you turn to the inbound course. Ideally, you should have at least 5 miles before you get to the VOR to do a skillful job of intercepting a radial. The only solution is to steady up on the outbound heading and reintercept after the needle has settled down in the FROM position.

A good interception with the needle perfectly centered after the first turn to the inbound heading is not a matter of luck. With practice, and using the correct techniques, you will learn to judge the wind and turn so that the needle will center every time.

"Cleared direct"

In some cases you will be cleared "direct" to the first VOR on the route; ATC means directly to the VOR in a straight line. ATC expects you to establish a course to the station and to stay on that course—with the needle centered, which is the next VOR challenge.

To fly direct to a VOR, turn the OBS knob until the needle centers in the TO position and read the course in the window. That is the course toward the station and that is the course ATC expects you to fly.

You know whether the wind is from the left or right and approximately how strong it is. After establishing the course to the station, fly a trial correction, left or right, of 2°, 5°, or 10° to keep the needle centered. (See Fig. 7-11.)

Reference heading

Note the heading. This is a "reference heading" or "holding heading" because it is the heading that holds the airplane on the correct magnetic course with the needle centered. Make small corrections left and right of the reference heading and the needle should hover around the center.

The overall procedure for establishing a reference heading and adjusting it to keep the needle in the center is known as *tracking*. In addition to using tracking to stay on course directly to a VOR, tracking will also maintain a course along a prescribed airway.

Keep in mind that the wind will rarely remain constant for any length of time. So when you are established on your radial or bearing and

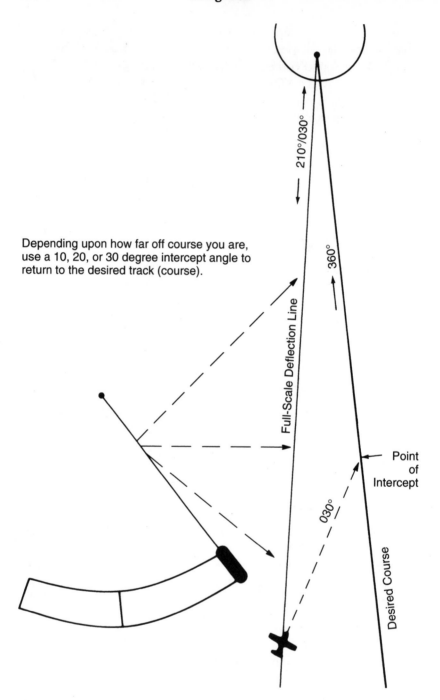

Depending upon how far off course you are, use a 10, 20, or 30 degree intercept angle to return to the desired track (course).

210°/030°

360°

Full-Scale Deflection Line

Point
of
Intercept

030°

Desired Course

Fig. 7-11. *En route correction to return to a VOR course.*

have worked out the holding heading, you will still have to adjust it a few degrees for slight changes in wind direction and strength. This, too, is part of tracking.

Bracketing

In the real world of IFR, you might not always be able to determine wind strength and direction, especially if you fly through a front or other rapidly changing weather conditions. Or you might become momentarily disoriented and unsure of what correction to make to stay on course and keep the needle centered. In either case, *bracketing* will get you back on course quickly and, at the same time, show what the holding heading should be to maintain that course.

Bracketing is a series of smaller and smaller turns from one direction to another across the desired course. Bracketing can be done on inbound and outbound legs. Start by turning to a heading that is the same as the desired magnetic course. Then make a turn 30° toward the needle. Hold the correction and make the needle move back to the center.

When the needle has returned to the center, cut that first 30° correction in half, and turn 15° toward the needle. Make it move back to the center again. When the needle has centered cut the correction in half and turn 7½° toward the needle. Again, make it move toward the center. You will quickly find a reference heading that will position the needle near the center and stop it from moving. Make minor adjustments left and right of that holding heading to keep the needle centered.

Chasing the needle

In describing bracketing, you will note that I was careful with each turn to say "make the needle move." What I meant by this is maintain the correction until achieving the desired result—in this case, returning the needle to the center.

Because the human eye is always quick to pick up motion, there is a great temptation to fixate on a VOR needle as soon as it starts to move and to turn toward it—to "chase" it. If you chase the needle making larger and larger corrections it will be impossible to predict when the needle will stop or reverse direction; pretty soon you will be way off course. If you chase the needle when the airplane is a half a mile from the station, you could make a 45° correction when only 50 feet off course and blow the station passage.

Set in a correction, hold it until the needle moves to the center, then adjust the correction. Don't start taking the correction out as soon as the needle starts to move.

Station passage

If you bracket and track the VOR properly, the adjustments to the holding heading will become smaller and smaller near the station. This is very important because the needle gets extremely sensitive closer to the station. If you are still making large heading changes close to the station, the airplane will pass way off to one side or the other. If the VOR that was just missed is the final fix on a VOR approach, execute a missed approach and try again.

Make your greatest efforts several miles out to establish the reference heading for perfect station passage. When you approach within a quarter of a mile of a station (and this is slant range), you enter a zone of confusion where none of the VOR instrument indications will hold steady.

Maintain the reference heading through this zone of confusion and note the time that the TO-FROM indicator flips to FROM. Maintain the reference heading—or turn to a new outbound heading and hold that—until the instrument indications settle down.

Don't chase the needle! As you maintain the reference heading, or turn to a new one, analyze whether that track is to the right or the left of the outbound course and then set up a reinterception or a bracketing procedure to get back on course. Then make minor adjustments to keep the needle centered.

Practice patterns

Patterns A and B are excellent for practicing VOR interception and tracking. Start the patterns over a VOR station, using a VFR altitude (odd or even thousand feet plus 500 feet) to avoid IFR traffic that might also be using the same VOR station. Start the pattern at the station and plan for each straight leg to return over the station.

While developing basic attitude instrument flying skills you will also hone VOR interception and tracking skills. And as noted before these patterns contain all the elements in an instrument approach and when practiced at a VOR, the effort becomes an introduction to VOR approaches.

After your instructor has introduced these patterns, practice them with a safety pilot. As always, discontinue practice if problems begin to crop up.

NDB procedures

The nondirectional beacon (NDB) is for everyone. I love it, you should too. Not too long ago many people thought the NDB was on its way out, a relic of the past as far as modern IFR flying was concerned. But this has not proven to be the case. NDBs have traditionally made it possible for small airports to have inexpensive and reliable instrument approaches when ILS and VOR installations are not feasible or too costly. (For example, the chief flight instructor for a school at Lincoln Park, New Jersey, obtained FAA approval for Lincoln Park's first instrument approach, an NDB approach based upon the compass locator for the Morristown ILS 23 approach.)

In addition to serving small airports, NDBs serve many other crucial functions at larger airports: fixes for holding patterns, procedure turns, missed approaches, and as compass locators for ILS approaches. For example, New York state has 65 airports with instrument approaches. Of these, 26 have NDB approaches. If the plane you fly on your instrument flight test has an ADF, chances are you will be asked to demonstrate an NDB approach.

For some reason NDBs and the airborne ADF receivers have been a mystery for many years. The first aviation direction finding equipment was very complicated, no doubt about it. You had to turn a wheel in the ceiling of the cockpit that rotated a loop antenna attached to the fuselage. In photographs of older planes—as well on the older airplanes in museums—this loop is quite prominent.

That was not an ADF loop, it was a *DF loop*. There was nothing automatic about it. It had a direction finding antenna, but the pilot had to turn it manually, listen to the signal build and fade, then interpret the signal to determine the direction to or from the station.

The introduction of the automatic direction finding (ADF) system was a great advance. Now the indicator of the direction finding instrument automatically points to the station, no matter where the airplane is and the ADF indicator is on the instrument panel not overhead or in back.

ADF orientation

This leads to the first concept to be stressed in ADF work: The needle always points to the station. This is an obvious point, but many people do not clearly understand its implications. First of all it means that when tuned to an ADF station for homing, tracking, intersections, holding, or an approach, you never have to touch the system. Unlike the VOR, there's no OBS to think about and no need to "twist" anything at station passage or at any other time. This makes ADF much simpler to use.

Second, with ADF you always know where you are in relation to the station. There is no TO-FROM to interpret, no confusion about radials and bearings, no way to set the wrong OBS numbers. ADF orientation is much simpler than VOR orientation. The head of the ADF needle always points to the station. With the azimuth set on 0 (zero)—straight ahead—the ADF needle will always indicate *relative bearing* to the station. (Relative bearing is the number of degrees that the station is from the nose of the airplane.)

To determine magnetic bearing to the station, simply add the *relative bearing* to the *magnetic heading* shown on the *heading indicator*, which equals the *magnetic bearing* (course) to the station. You undoubtedly learned this in your primary training, but let's do a quick review now for some hints to simplify the process.

If you are on a magnetic heading of 030° and the ADF needle is 90° to the right, the magnetic bearing to the station is 120°.

$$030 + 090 = 120$$

Turn to 120° and the ADF needle will point straight ahead.

You don't even need to make a turn to confirm this. Take a medium-length pencil and place it on the needle of the ADF, much in the same manner as a parallel ruler. Move it onto the heading indicator and the pencil will point to the magnetic bearing to the station, eliminating the arithmetic. This is one of the shortcuts used in flight to simplify a visualization of "where we are now."

An inexpensive feature on many ADF indicators is a third, even simpler method of determining magnetic bearing to the station: the rotating azimuth ring. Simply rotate this ring manually to line up the magnetic heading with the mark at the top of the ADF indicator and the ADF needle will automatically point to the magnetic bearing to the station.

Let's say you are on a heading of 300° and the needle is pointing to the right wing of the airplane at 090°. Apply the formula and add 300° plus 090° and come up with 390°. If the answer is more than 360°, all you have to do is subtract 360° from the total for the magnetic bearing to the station.

$$300 + 090 = 390 - 360 = 030$$

Be careful, as with a VOR, to tune the station correctly and verify its Morse code identifier. Turn up the audio sufficiently to hear the ADF signal faintly in the background—but not so loud as to interfere with communications—and keep the volume at that level.

When there is a disruption in VOR signals, warning flags appear on the face of the nav instrument; this does not happen on the ADF indicator. Continuously monitor the identifier to detect any signal disruptions. Keep the volume low enough to hear the signal in the background. This is the only way to be sure that the ground station has not gone off the air for some reason or that the ADF unit in the airplane has not malfunctioned.

Check the heading indicator against the magnetic compass at least every 15 minutes and reset it as necessary throughout all phases of IFR and VFR flight. This is critical in ADF work because it is impossible to determine the magnetic bearing if the heading indicator has drifted off. Prior to a magnetic bearing determination, verify accuracy of the heading indicator.

ADF time/distance checks

An exercise to develop NDB orientation awareness is the ADF time/distance check. Start in a simulator, which often reinforces the ability to orient around an NDB so well that a little additional practice in flight is needed.

Pick stations with the highest output for the best results. The needle might wobble with a commercial station, but this usually clears up closer to the station and it is not a problem with powerful stations.

Tune the station and determine the relative bearing. Turn the number of degrees necessary to place the ADF needle at either 090° or 270° relative—the right wing or left wing.

Note the time and fly a constant heading until the bearing changes 10°. Note the number of seconds it takes for the bearing to change 10° then divide by 10 for the time to the station in minutes.

$$\frac{\text{Time in seconds}}{\text{Degrees of bearing change}} = \text{Minutes to station}$$

You can determine distance to the station with this formula:

$$\frac{\text{TAS} \times \text{minutes flown}}{\text{Degrees of bearing change}} = \text{Distance to station}$$

Intercepting a bearing

Let's take an example from the real world of IFR. Suppose you have executed a missed approach and want to return to an NDB via a specified bearing to be in a good position to make another approach.

First, turn to the desired bearing.

Second, note the number of degrees of needle deflection to the left or right of the 0° position on the face of the azimuth card (ADF indicator) and double this amount to determine the *intercept angle*.

Third, turn toward the head of the needle the number of degrees determined for the *intercept angle*. As you turn toward the needle this predetermined number of degrees, the needle will pass through the 0° position and on *to the other side* of the 0° position on the face of the ADF indicator.

Wait a minute, the needle always points to the station. Why does it appear to move? The answer is that the needle doesn't move, the airplane moves and the ADF indicator face is attached to the airplane. In a turn, the needle continues to point to the station, but the airplane is moving under the needle.

A good, simple way to visualize this is to place a book or other object on the floor to simulate an NDB station. Stand a few feet away from the object, and point toward it so your arm simulates the ADF needle. Your nose becomes the zero point on the indicator face, matching the nose of the airplane.

Now turn your body to a new "heading" while continuing to point toward the "station." Your arm will behave the same way as the ADF needle, apparently moving away from your nose. But you will quickly see that it is really your body that is turning while your arm continues to point steadily at the "station."

Back to intercepting the bearing.

The fourth step is to maintain the new *intercept heading* until the needle is deflected on the opposite side of the nose the same number of degrees as the *intercept angle.*

Then turn to the desired heading, which is the magnetic course inbound. Hold this heading until you notice a drift of the needle, which indicates *wind drift.* The procedure outlined above may be repeated as often as necessary. As you become more proficient through practice, you will be able to determine a wind correction angle and make corrections as you proceed to keep on the magnetic course. (This is not as difficult as it seems at first. Have faith and the mystery gradually unfolds. This is also where thorough flight instructors are worth their weight in gold.)

Let's say you want to head inbound to the station on a 360° bearing (Fig. 7-12). Turn first to a heading of 360° and note the 15° deflection of the needle to the left (A).

Next, double the 15° to the left for an *intercept angle* of 30°. Now turn 30° toward the left (B). Note that the needle has now passed to the other side of the nose and the airplane is now heading 330°. Maintain this intercept heading until the ADF needle has deflected 30° to the right as at position (C).

Finally, turn inbound on a heading of 360°, the same as the inbound bearing to the station (D). (Lead the turn by 5° to avoid overshooting.)

With no wind, all you would have to do is continue the inbound heading to the station, but there is always wind. So let's proceed to examine the techniques used to correct for the effects of wind.

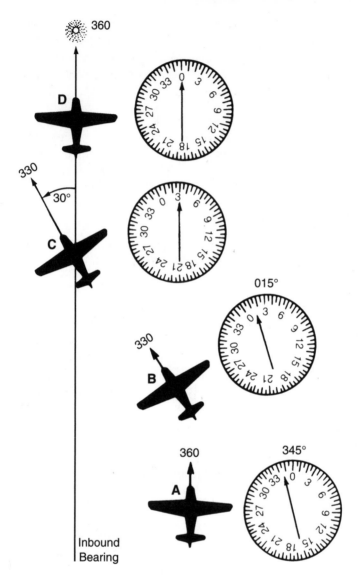

Fig. 7-12. *Intercepting an ADF bearing.*

Homing is unacceptable

Heading inbound you can "home" in to the NDB by placing the ADF needle on the nose and keeping it there with heading adjustments. With the wind constantly pushing from one side, you will have to constantly change the heading as you proceed toward the station to keep the needle on the nose.

Figure 7-13 is an illustration of the homing method of reaching an NDB. The airplane starts heading inbound on the 360° bearing at A. As the wind blows from the left, the heading has to be adjusted to maintain the needle on the nose (B and C). At D, the airplane has been blown so far off the inbound bearing that it is flying a heading of 315° instead of the desired 360° to keep the needle on the nose.

Homing is unacceptable for IFR navigation because the airplane strays too far from the intended course. The wide, looping course shown in Fig. 7-13 might lead into the side of a hill, a radio tower, or other obstruction at the minimum altitudes of an NDB approach. Figure 7-13 is not an exaggeration. Many poorly prepared instrument students do this on the instrument flight test. (This is a certain failure on a flight test!)

Tracking and bracketing

When you have intercepted the desired bearing, hold that heading and see what affect the wind has. Let's continue with the previous example—interception and tracking of the 360° bearing to the station—and see what happens (Fig. 7-14). As the airplane proceeds toward the station, the wind from the left blows it off the bearing. At B the airplane has drifted off the bearing by 15° relative. To get back on course, double the drift noted at point B and turn toward the needle that amount, in this case 30° (C). You are reintercepting the bearing.

The ADF needle will then swing over to a relative bearing of 015°. Hold the intercept heading (330°) until the relative bearing reads 030°. That puts the airplane back on the desired bearing to the station (D). Now reduce the corrections by half—15° in this case—to compensate for the wind (E). (You should also lead the turn back by approximately 5°.)

This method of correcting for wind drift is called *bracketing* and you might have to "bracket" several times to establish a reference heading that will remain on the desired bearing, especially if a long distance from the station. The initial wind correction might be too large or too small to stay on the bearing. If so, adjust the correction. In the example above, for instance, if a 15° correction proves to be too much, reintercept the bearing and try a 10° correction.

Chasing the needle is a common mistake in ADF intercepts, tracking, and bracketing, just as in VOR navigation. It is so tempting to follow that moving needle! Resist the temptation. Hold the heading steady until the needle reaches the relative bearing you want, then make the turn.

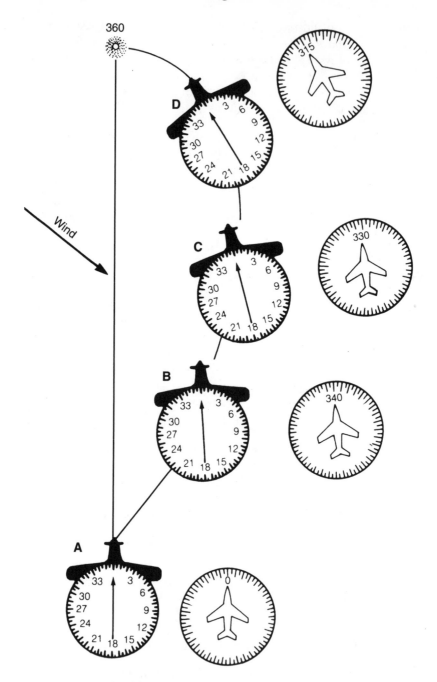

Fig. 7-13. *The problem with ADF homing is that it takes you off course to a degree that is unacceptable for IFR flying—especially on NBD approaches.*

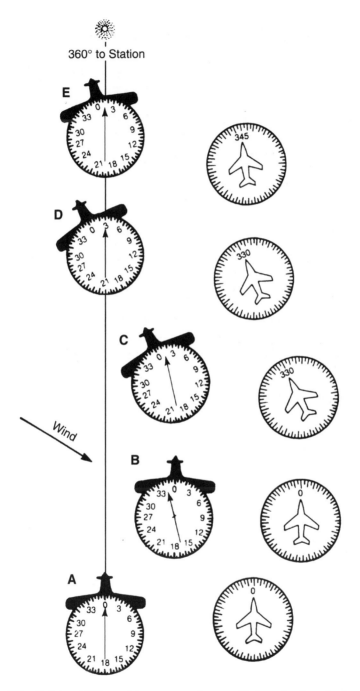

Fig. 7-14. *ADF tracking corrections utilizing the bracketing procedure.*

Near the station the ADF needle will become "nervous" and start oscillating, becoming more sensitive. Don't chase the needle, just fly the reference heading. Passing over the station the needle will commence a definite swing to the right or left. Note the time and start timing the outbound leg.

Wait until the needle has definitely swung around to verify station passage—at least 5–10 seconds past the ADF station. At this point you should turn to the outbound magnetic course to determine which side of the bearing you are now situated and how much you will have to correct. It can be fun!

Outbound bearings from the NDB

The procedures for intercepting, tracking, and bracketing outbound from the station are almost identical to procedures for the inbound magnetic bearing.

Turn to the outbound bearing and determine how many degrees you are off the bearing. Then double the error and turn toward the desired bearing by this amount. If off the bearing by 10°, turn toward the bearing 20°.

When the angle of the ADF needle off the tail and the intercept angle are the same, the airplane is on the desired bearing. Turn toward the outbound bearing and bracket outbound to determine the wind correction necessary to hold that bearing.

Remember that the needle always points to the station. Never put the needle on the tail by changing the heading. This will cause you to miss the bearing completely and as you will see during an NDB approach if you lose that outbound bearing you will miss the airport.

Try to visualize where you are at all times. If in doubt, turn to the outbound bearing and check whether you are to the left or to the right of course, and by how much. Then double this amount and reintercept the outbound bearing. Practice this at altitude until you become thoroughly proficient at tracking outbound before commencing NDB approaches.

Better yet, if you are having a problem tracking NDB bearings outbound, try the visualization exercise again. Place an object on the floor of a large room or parking ramp to represent the NDB and walk through the entire procedure of tracking a bearing inbound, then station passage, then tracking the same bearing outbound, using your right hand to always point to the NDB and your nose as the nose of the airplane. Believe me, this works!

I am very serious about practicing this visualization exercise until you understand clearly what the ADF needle is indicating, especially outbound from the NDB. This is the critical final approach leg on an NDB approach. One of the most frequent causes for failures on flight tests is confusion on the final leg of the approach to the airport. Time and again a candidate will track inbound to the NDB accurately, then turn the wrong way after passing the station.

In so many elements of IFR flight, visualization is the key to making the correct moves. The visualization exercise doesn't cost a cent, no matter how often you practice it.

Practice patterns

A good exercise for sharpening your skill at tracking and bracketing inbound and outbound is the simple pattern depicted in Fig. 7-15. It is a good idea to use a commercial broadcast station for practicing this and other patterns to avoid straying into busy airspace.

Pick a *cardinal heading* such as 270°, and intercept the bearing inbound to the station as shown. (A cardinal heading is one of the four directional points of a compass: north, east, south, west.) After station passage, track outbound correcting for the wind for 3–4 minutes.

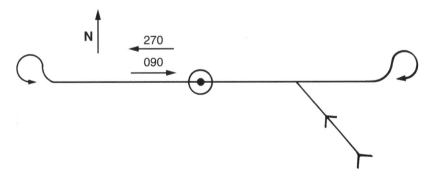

Fig. 7-15. *Pattern for inbound-outbound ADF tracking practice, with 90-270 course reversals.*

Then reverse course with a 90-270. Make a standard rate turn in one direction for 90°, then reverse smoothly into a standard rate 270° turn in the other direction. The 90-270 is a quick and efficient 180° change in heading. And if you have maintained a steady bearing outbound, the 90-270 will place you close enough to the inbound bearing so that only small corrections will be required.

Track inbound to the station, then outbound on the other side for the same amount of time. Do another 90-270, and repeat the process until you can maintain steady bearings with corrections. A standard procedure turn may be substituted for the 90-270 course reversal.

Other good ADF exercises are Pattern A and Pattern B set up on an NDB or commercial broadcast station. Start each pattern over the station and orient the pattern on cardinal headings, at least in the beginning. Plan for each straight leg to return over the station.

These two patterns will provide plenty of practice in intercepting bearings and tracking inbound and outbound. And because the patterns contain all elements of an instrument approach they are good introductions to NDB approaches. Patterns A and B are also good exercises to practice in a simulator.

ADF holding patterns

Nothing could be simpler or easier than an NDB holding pattern. (Procedures for entering a VOR holding pattern apply to entering an ADF holding pattern.)

Track inbound on the desired bearing, making wind corrections as needed to maintain the bearing. At station passage, begin a standard rate turn to the right (left in a nonstandard holding pattern). Roll out on the reciprocal of the inbound bearing and double the wind correction to account for the effect of the wind on the outbound leg and in the two turns.

Mark station passage when the ADF needle reaches the 90° position. Adjust the timing of the outbound leg (nonprecision side) to produce a 1-minute inbound leg. Now here's a good cross-check that many instructors overlook. After flying outbound 1 minute, the needle should point 30° off the tail: 30° off to the right in a standard holding pattern, 30° on to the left in a nonstandard pattern.

Completing the outbound leg, turn again, rolling out on the inbound heading. Note the bearing error and correct for it.

Once again, the quickest way to visualize ADF holding patterns is to run through the floor exercise described earlier in this chapter. Review VOR holding for methods of entering holding. The same procedures apply to entering ADF holding patterns.

GPS

The space age is really upon us. *The global positioning system* (GPS) that was merely a dream a short time ago is now up and fully operational. In 1991, although the system was incomplete and would be inoperable for up to 4 hours a day, I was able to use a hand-held unit to navigate a yacht under the Golden Gate Bridge in fog so thick that the first my passenger and I saw of the bridge was the span itself, *when we were directly underneath it.*

The most senior FAA navigational inspector told me in the early 1990s that it would be a long time before the FAA would accept GPS as being safe for aviation use. Recall that an air carrier demonstrated the accuracy of GPS in 1994, 5 years ahead of schedule, by making more than 170 fully coupled (autoland) landings using GPS for both vertical and horizontal guidance, and each landing touched down within 10 feet of each other.

It is generally accepted that the rapid completion of the GPS constellation was caused by the necessity of accurate navigation in the desert during the Persian Gulf War. In any event, it's here, and it looks like it's here to stay. Not only is GPS here to stay, but it is generally believed that within 10 to 20 years, GPS will make all other types of navigation obsolete, and the other types will go the way of the lighted and Adcock ranges.

GPS approval

Since the middle of 1993, the FAA has allowed GPS to be used for en route navigation as well as for "overlay" approaches to be flown using TSO C129 GPS avionics certified for nonprecision approaches, provided the procedure is retrievable from airborne navigation databases (such as the Jeppesen NavData Services). By the middle of 1994, the FAA had issued the first three "stand-alone" GPS approaches (Fig. 7-16). These approaches are completely new, and do not overlie existing approaches.

Figure 7-17 illustrates how Jeppesen identifies both a stand-alone and an overlay GPS approach. Another important illustration to notice is the way Jeppesen identifies its database identifiers (see the lower illustration in Fig. 7-17). Prior to October 1994, the identifiers were in italic type within parentheses. These have been changed to italic type with square brackets.

Figure 7-18 illustrates how GPS waypoints are incorporated into various portions of approaches. One thing to note is the Sensor FAF in the last two illustrations. An important fact here is that Jeppesen shows the distance from the Sensor FAF to the missed approach point.

GPS demystified

With the foregoing in mind, just what is GPS, and how does it work? Basically, GPS is made up of a constellation of 24 satellites. Twenty-one of the satellites are required to provide three-dimensional navigational capability, 24 hours a day, anywhere in the world. The other three birds are used as spares. These satellites are in fixed orbits at about 10,900 miles high. Each satellite circles the Earth twice daily; this places each one over a monitoring station twice a day. The monitoring station will in turn update the positions of the satellites.

Because the satellites will know where they are all the time, they will then be able to tell us where we are after the onboard computer has four of them in range. I tend to compare getting a satellite fix to getting a three-star fix in celestial navigation. Perhaps a short introduction to celestial navigation will help you understand the concept. Too many pilots fly transoceanic with no real conception of latitude and longitude nor their relationship to each other. This short introduction into celestial navigation might help you to better understand these concepts.

Great circles

Look at Fig. 7-19. This is a circle representing a great circle of the Earth. It has no relationship to compass directions because *a "great circle" is any circle that cuts through the center of the Earth*, so for the time being, do not confuse yourself with thinking in terms of north, south, east, or west.

Great circles   131

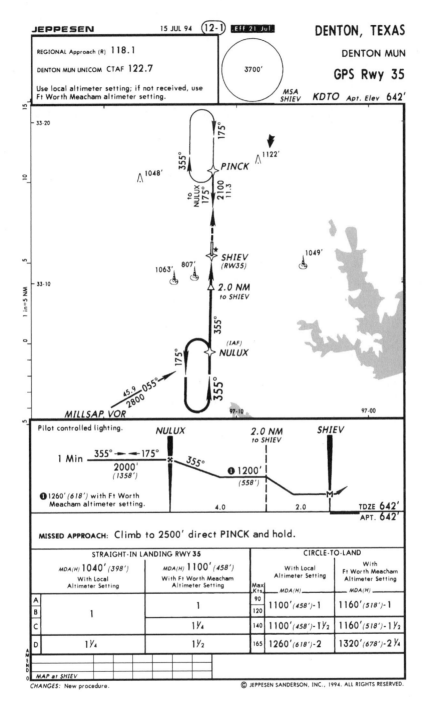

Fig. 7-16 *Reproduced with permission of Jeppesen Sanderson, Inc. Not for use in navigation.*

132   Navigation

APPROACH CHART LEGEND
GPS APPROACH CHARTS

This GPS Approach Chart Legend supplements the standard approach chart legend beginning on Introduction Page 101. Equipment requirements, database requirements, and requirement or non-requirement for monitoring conventional navaids are not addressed in this legend-Refer to Jeppesen Air Traffic Control (ATC) pages for this information. [For the United States, refer to the Jeppesen Navigation Aids pages of the Airman's Information Manual.]

STAND ALONE GPS APPROACH. Procedure is included in Jeppesen's NavData Service.

GPS APPROACH, overlies an established conventional navigation non-precision approach. Procedure is included in Jeppesen's NavData Service.

FREDERICK, MD

FREDERICK MUN

GPS Rwy 5

KFDK Apt. Elev 304'

OAKLEY, KAN

OAKLEY MUN

NDB or GPS Rwy 34

NDB 380 OEL

KOEL Apt. Elev 3044'

Airport identifier to assist in selecting the appropriate airport information from the database.

GPS OVERLAY, overlies an established conventional navigation non-precision approach. Note that GPS is not part of instrument approach procedure title. *(GPS)* indicates GPS approach information has been applied to the approach chart. Procedure is included in Jeppesen's NavData Service.

GENEVA, SWITZERLAND

COINTRIN

GVA VORDME Rwy 23

(GPS) VOR 114.6 GVA

Airport identifier to assist in selecting the appropriate airport information from the database.

LSGG Apt. Elev 1411'

Jeppesen database identifiers are always shown in italic type. They are enclosed within square brackets, as [D255G], or prior to October 1994 within parentheses, as (D255G).

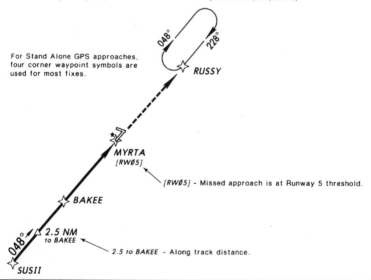

For Stand Alone GPS approaches, four corner waypoint symbols are used for most fixes.

048°   228°

RUSSY

MYRTA
[RW05]

[RW05] - Missed approach is at Runway 5 threshold.

BAKEE

048°   2.5 NM
to BAKEE

2.5 to BAKEE - Along track distance.

SUSII

Fig. 7-17 *Reproduced with permission of Jeppesen Sanderson, Inc. Not for use in navigation.*

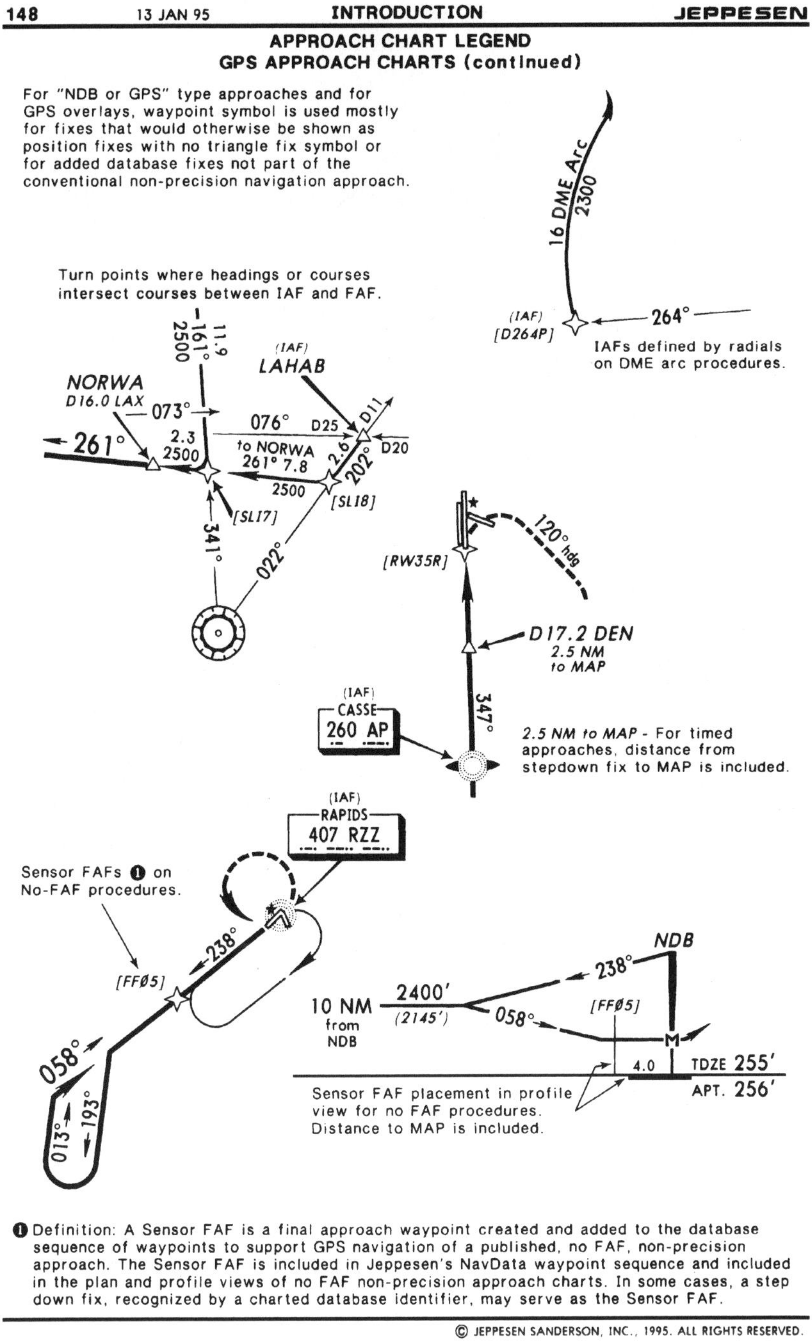

148 13 JAN 95 INTRODUCTION JEPPESEN

APPROACH CHART LEGEND
GPS APPROACH CHARTS (continued)

For "NDB or GPS" type approaches and for GPS overlays, waypoint symbol is used mostly for fixes that would otherwise be shown as position fixes with no triangle fix symbol or for added database fixes not part of the conventional non-precision navigation approach.

16 DME Arc
2300
(IAF)
[D264P]
264°
IAFs defined by radials on DME arc procedures.
Turn points where headings or courses intersect courses between IAF and FAF.
11.9
161°
2500
(IAF)
LAHAB
NORWA
D16.0 LAX
073°
261°
2.3
2500
076°
D25
D11
to NORWA
261° 7.8
2.6
202°
D20
2500
[SL18]
[SL17]
341°
022°
[RW35R]
120° hdg
D17.2 DEN
2.5 NM
to MAP
(IAF)
CASSE
260 AP
347°
2.5 NM to MAP - For timed approaches, distance from stepdown fix to MAP is included.
(IAF)
RAPIDS
407 RZZ
Sensor FAFs ❶ on No-FAF procedures.
238°
[FF05]
058°
013°
193°
NDB
238°
10 NM
from
NDB
2400'
(2145')
058°
[FF05]
M
4.0
TDZE 255'
APT. 256'
Sensor FAF placement in profile view for no FAF procedures. Distance to MAP is included.

❶ Definition: A Sensor FAF is a final approach waypoint created and added to the database sequence of waypoints to support GPS navigation of a published, no FAF, non-precision approach. The Sensor FAF is included in Jeppesen's NavData waypoint sequence and included in the plan and profile views of no FAF non-precision approach charts. In some cases, a step down fix, recognized by a charted database identifier, may serve as the Sensor FAF.

Fig. 7-18 *Reproduced with permission of Jeppesen Sanderson, Inc. Not for use in navigation.*

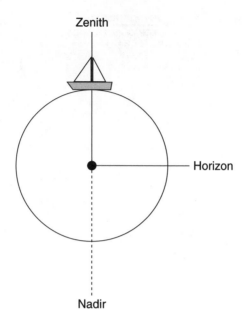

Fig. 7-19

All lines of longitude are great circles. The only line of latitude that is a great circle is the equator. All other lines of latitude are parallel to the equator; thus, they are called "parallels." An unlimited number of great circles cut the lines of latitude and longitude at various angles. Remember, any line that cuts through the center of the Earth, *no matter what the orientation to the compass*, is a great circle.

A great circle is the shortest route between two points on the Earth, and most long-range navigation systems will provide a pilot with great-circle routes. The nearer a route is to east and west, the more distance will be saved by flying a great-circle route. Until now, all of your flying has been between VORs, which has been flown at a constant compass heading (wind correction excepted). In a great circle, the compass heading is constantly changing, although for practical purposes, great-circle flight plans usually update headings at regular intervals such as each degree, 5 degrees, or 10 degrees, depending on aircraft speed.

Relationship of arc to distance

A circle, as in Fig. 7-19, is 360 degrees in circumference. Each degree of arc is made up of 60 minutes, which means that a full circle can be converted to 21,600 minutes. Give or take a few miles, the circumference of the Earth at the equator is 21,600 nautical miles. Recall that a statute mile is 5,280 feet in length. In instrument flying, you will be using the nautical mile, which is 1.15 times longer than a statute mile at 6,076 feet. For simplicity, we round that down to 6,000 feet.

One degree of arc along the equator (or along any other great circle on the Earth) is equal to 60 nautical miles; therefore, 1 minute of a degree of arc on a great circle equals 1 nautical mile. One minute of a degree on a line of latitude will vary in length from 1 nautical mile at the equator to zero at the North or South Pole where all of the lines of longitude converge. If you are flying along a great-circle route, you can calculate the distance in your head. If you are not on a great-circle route, you will have to use the conversion scale on the chart.

Primer of celestial navigation

Refer to Fig. 7-19 again. In order to determine a position on the Earth using celestial navigation, it is first necessary to determine the height of a celestial body above the horizon: Sun, Moon, Venus, Mars, Saturn, Jupiter, or any of 76 navigational stars. From any point on Earth (in this case, the boat in Fig. 7-19), the point directly overhead is called the *zenith* (Z). The point on the opposite side of the Earth is called the *nadir.* The horizon is where the Earth and sky join, which is 90 degrees from the zenith; this is applicable to all in all directions. The arc of this 90-degree circle is 5,400 nautical miles in length (90 × 60).

Examine Fig. 7-20. Let's use a sextant to determine that the height of a star is 30 degrees above the horizon. The illustration indicates that the angle is being measured from the center of the Earth because the distance to any celestial body is so much greater than the radius of the Earth that for all intents and purposes the center of the Earth can be assumed for our position. Corrections are made for this and other anomalies when the sights are reduced.

The point on the Earth directly underneath the celestial body is termed the *geographical position* (GP).

Subtracting the height of the body (30 degrees) from the 90-degree arc between the zenith and the horizon will result in the *zenith distance* (ZD). In this case, the ZD is 60 degrees of arc, which equals 3,600 miles (if converted into distance), which will be the distance from the GP to the boat.

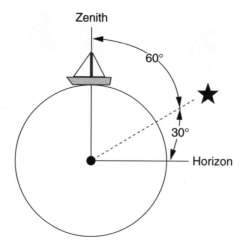

Fig. 7-20

The GP of the body at the exact second the shot was taken can be determined from the *Nautical Almanac*. If you had a long enough string, by knowing the exact latitude/longitude of the GP, you could pin one end of the string to the GP and scribe a circle. This circle would be called a *circle of equal altitudes* (COEA) because no matter where you were on that circle, the altitude of the body would be the same (30 degrees). The circle is also called a *circle of equal radius* (COER). The term used would depend on whether it is referring to the altitude of the body or the distance from the GP on the Earth. The boat, therefore, would have to be somewhere on that circle. The circle could also be referred to as a *circle of position* (COP), and if we were to transcribe a small portion of a circle of that magnitude onto a chart, the result would be a straight line, which would be called a *line of position* (LOP).

If this procedure were repeated using two celestial objects, two COEAs would result, as can be seen in Fig. 7-21. Note that these circles will cross at two points: A and B. Obviously, if the boat had to be somewhere on the first COEA, it would also have to be somewhere on the second, which means that it will have to be at either point A or point B, and because these points will usually be many miles apart, the navigator would know which point the boat was on.

A number of factors can affect the accuracy of that position. So to attain the greatest accuracy, the navigator will take a third shot on another object, which is called a *three-star fix* (Fig. 7-22).

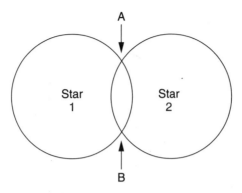

Fig. 7-21

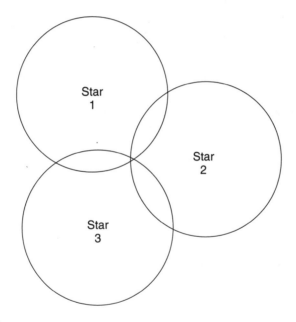

Fig. 7-22

Theoretically, these three COERs would all cross at one position, which, if the bodies shot were properly chosen, would cancel out all errors, resulting in a perfect fix.

This is rarely the case, however, and the navigator usually ends up with what is called a "cocked hat," as in Fig. 7-23. The navigator then applies rules of logic and her own experience to determine a more precise position.

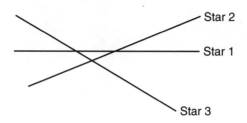

Fig. 7-23

Timing is crucial

Just as the navigator knows the GP of the various bodies from the tables in the *Nautical Almanac,* the GPS receiver knows where the satellites are because the receiver's computer "builds an almanac" from information received from the satellites themselves, which you recall are updated from the master ground stations twice daily.

By measuring the length of time it takes the satellite signal to reach the receiver, the receiver's computer knows how far away it is from each satellite. By knowing the distance from the satellite and knowing where the satellite is from the information in the "almanac," the computer "draws" a COER around the GP of the satellite. This is exactly the same thing the navigator does when she draws a LOP of a celestial shot.

It gets a little more complicated than that though. In order for the receiver to know how long the signal took to get down from the satellite, it has to know exactly when the signal was sent. After all, the signals travel at the speed of light, which is 186,000 miles per second, so we're speaking in terms of short time periods here. (Understanding time periods in nanoseconds is about like trying to really comprehend the size of the United States's deficit. Still and all, it's duck soup for the computers.)

The satellites have very expensive atomic clocks onboard to assure the proper timing of the signals. If the same expensive clocks were put in the receivers, only the government could afford them. For this reason, quartz clocks are used in the receivers, which requires adding one extra satellite to the equation to compensate for the difference in time measurements that the receiver's clock might end up with.

The way it works: If the receiver gets a signal from two satellites, it will figure out two LOPs, just as celestial navigators do from two stars.

In the case of the navigator, if her watch or clock has a 10-second error in it when she "shoots the stars," she will be more than 3 miles away from where the LOPs cross. This would occur because the Earth rotates at a rate of about 1 mile every 4 seconds. This error can be reduced by taking a three-star fix as previously mentioned.

The same thing is true of the satellites. To correct for the time error, the GPS receiver will use three satellites in the two-dimensional mode and four satellites in the three-dimensional mode. To simplify it, in the 2-D mode (latitude and longitude only), if there is an error in the clocks, the three LOPs won't cross in one spot. They'll end up with some sort of cocked hat. The receiver's computer then begins adjusting the times (by use of algebraic equations) until the LOPs do cross in one spot, which is the exact position. In the 3-D mode (which pilots are more interested in), the addition of a fourth satellite provides altitude information.

GPS errors

Other factors will affect the accuracy of GPS receivers. One of these is that the speed of light varies as it goes through the atmosphere. This might result in a receiver error of up to 12 feet. The clock error alone might add up to 2 feet. Built-in receiver errors might come to another 4 feet, the worst-case satellite-selection errors might add another 25 feet, and the ephemeris error will result in 2 more feet. All of this adds up to somewhere between 10 and 30 feet of error. One other error has until recently prevented GPS from being used for precision IFR approaches: *selective availability* (SA).

Two frequencies are used by the satellites. The *precision code* (P-code) is used by the military, and the *course acquisition code* (C/A-code) is used by the public. The military can arbitrarily induce an error into the C/A-code, which can impart an error of up to 300 feet without the user being aware of it. Naturally, an error of this magnitude would prevent the pilot from using GPS for a precision approach.

To overcome this, ground stations are being provided that "know" their position precisely. This ground station will then derive its own GPS position, recognize any error, and then transmit that difference to the user's receiver so that the receiver's computer will then be able to give a precise location with repeatable accuracy of within 3–10 meters (roughly 10–30 feet). This is called *differential GPS* (DGPS). It is this precision that is enabling aircraft to make fully-coupled autopilot landings, and which, someday, will make all other types of navigation obsolete.

Additional radio navigation aids

There are several other radio navigation aids available to a pilot. We will briefly cover those in this section, but the intent is only to make you aware of their existence and basic function. The aids we will cover include Radar Services and the Global Positioning System (GPS). Radar services are commonly used in both VFR and IFR flight and have been in service for several decades at many tower-controlled airports. Additionally, ATC uses radar to track the position of aircraft along routes of flight, providing heading and altitude guidance to pilots. GPS navigation has recently been made available to the general public and is radio navigation provided through the acquisition of data from satellites in orbit around the earth. Originally developed for the military, a small receiver can detect the transmissions from these satellites and establish the aircraft's position with a great deal of accuracy.

If you encounter IFR weather, Radar Services can be of immediate aid in providing heading and altitude directions to you. Center frequencies can be obtained from a variety of sources, but the easiest way to get help quickly is to use the emergency VHF frequency of 121.5 MHz. This frequency is monitored by Flight Service Stations and radar facilities, and the personnel will be able to quickly help you in the event of an emergency. When radar services are requested, the radar operator will initially ask the pilot to set a unique code into the plane's transponder. The transponder is a device that transmits and receives information with the radar facility and enhances the image of the plane on the radar operator's screen. This unique code identifies your airplane on the radarscope, helping the operator to quickly determine where you are. In the event your airplane is not equipped with a transponder, the operator may have you fly a series of turns until they can determine which plane is yours. The operator will also ask you a series of questions that include whether you are in VFR or IFR conditions, the amount of fuel you have, the number of people on board the aircraft, and if you are instrument rated. This provides the radar operator with information that can help them formulate the most efficient plan for getting you safely to an airport.

You should make it a point to occasionally use radar services even in VFR flight conditions. This will help you become familiar with the procedures and abilities of the service and teach you to communicate with them in an efficient manner. They can provide you with traffic information and help you avoid other aircraft. In the event of a true emergency, having this experience will allow you to quickly take advantage of the help they can provide.

The Global Positioning System receiver is a device that has enjoyed widespread use in recent years. At one time hailed as the end to the VOR system, the GPS provides an incredible navigation aid to pilots, boaters, the military, and just about anyone that wants to be able to find out where they are and the direction to head to get somewhere else. There are a number of GPS vendors that provide products with a wide variety of functionality. Basic aviation models will allow you to plot a course from point to point, then act somewhat like a VOR, letting you know if you are left or right of the desired course.

More advanced versions of the GPS include what is known as a moving map. This can be anything from a rough line map on a small screen to models that incorporate laptop computers with color maps that resemble sectional charts. In accidental incursions into IFR or low-visibility weather, many GPS systems have a NEAREST function. This takes your current position and determines the nearest airport to where you are. The direction, distance, and flying time are normally provided by the GPS unit in these situations. Many pilots use hand-held GPS units, while some aircraft have them mounted into the panel and tied into the autopilot. While it is not likely to be the demise of the VOR system, GPS receivers provide a tremendous amount of information to pilots and can be a worthwhile investment for any pilot that flies cross-country on a regular basis.

Instrument navigation

If you become immersed in instrument conditions, descending until you can see the ground and navigating via VFR procedures may not be an option. In severe weather cases, the clouds, rain, snow, or fog may completely obscure the surface and not allow you to descend. The same holds true for executing a 180-degree turn to get out of the weather, or climbing above it. In those instances ATC may direct you to navigate via instruments to an area of better visibility or to an airport that you can land at. This section will cover the basics of navigating by instruments. Like the previous section, this is not intended to be a self-taught instrument course, but information that may be useful in an emergency. You should find a qualified flight instructor to give you proper training in instrument flight, whether you intend to go on to get your instrument rating or are just interested in practicing flying by reference to instruments for proficiency.

Mastering the basics of instrument flight, then, starts with the understanding that instrument flying is a matter of airplane attitude, configuration and power, and the resulting effects of changing any or all of these basic elements. After you've gained confidence and proficiency in maneuvering the airplane without outside visual references, you can then add the task of navigation, still using basic attitude flying as the means of following a course or navigation signal.

Section 2

IFR flight planning

8

Preflight planning

You've probably already noticed that instrument flight works best when you think ahead. When you first started flying on instruments, it was easy to get engaged in the activity of the moment, to get so caught up in doing one task or procedure that others would go undone. How many times early in your instrument training did you suddenly realize that your heading was where you'd been working to keep it, but that your altitude was way off? Or that heading and altitude were fine, but that you'd strayed from the centerline of the airway?

With time and practice, though, you probably felt yourself opening up, scanning more rapidly, and noticing more and more of what was going on around you. Your stress level dropped dramatically, and deviations from altitude, heading, and navigational track became fewer and fewer. You could begin to anticipate what would happen next and make minor corrections not only to fix an immediate deviation from attitude, but also what would make the transition to the next attitude or heading smoother and easier.

Instrument flight planning is like that. It's the process of visualizing how the flight will progress, before you ever take off. It's becoming familiar with the headings and routes you'll fly, the minimum safe altitudes for each segment of your trip, and the names and approximate locations of the prominent landmarks and navigational fixes along the way. It's taking an informed look at the weather to make a proper go/no-go decision. It includes airplane, instrument, and equipment checks to verify that your airplane is safe, and it entails filing a flight plan and communicating with Air Traffic Control. "Plan the flight and fly the plan," goes the cliché—you'll find that it's far easier to keep on top of basic attitude flying and navigational tasks, as well as handling those common cases where your flight plan is altered en route by weather, Air Traffic Control direction, or for other reasons, if you've planned your flight beforehand. Let's look at some instrument flight planning skills.

Preflight planning to many people means nothing more than scanning the charts, making a quick call to the weather service or flight service to see if it's still VFR, and a cursory glance at the aircraft with perhaps a perfunctory kick at a tire.

Instrument pilots cannot afford this degree of nonchalance. Murphy's well-known law states that, "Anything that can go wrong, will go wrong." When we combine that with the fact that emergencies always seem to snowball, we tend to become a little more careful, or at least we should.

At least 50 percent of effective instrument flying is psychological; thus, a proper and careful preflight procedure will put us in a mental condition to properly handle the flight. In addition, this proper preflight procedure is required by the regulations. FAR 91.103 says:

> *Each pilot in command shall, before beginning a flight, familiarize himself with all available information concerning that flight. This information must include:*
>
> *(a) For a flight under IFR . . . weather reports and forecasts, fuel requirements, alternatives available if the planned flight cannot be completed, and any known traffic delays of which the pilot in command has been advised by ATC. . . .*

The flight kit

A fairly good place to begin this discussion is with the pilot's personal flight kit. Every instrument-rated pilot eventually puts together a flight kit that seems to work best for him or her. The items listed below make up a fairly comprehensive kit. As you gain more experience, you may add to it or subtract from it to suit individual desires. You'll notice that I have recommended more than one of some of the items; these are inexpensive items that don't take up space or add much weight to your flight kit. Believe me, you'll be willing to pay a king's ransom should one fail or wear out in flight and you end up without a spare.

- Airway charts, SIDs, STARs, and approach plates for the section of the country you'll be flying in. For most light aircraft, the en route low altitude charts will suffice for the airway charts.

- Sectional charts to cover the same area, and/or WAC charts, depending on the speed of your aircraft. These charts will better enable you to visualize the terrain you'll be flying over. There are many times when you'll get enough of a break in the clouds to catch a glimpse of the ground, and in many cases, the sectional will enable you to confirm your position. (One other use of the sectional shall be discussed later.)

- Two or more flashlights, with fresh spare batteries and extra bulbs, in case of an electrical power failure at night. If you, like some other pilots, prefer to put a red lens on one of the flashlights, you'll have to remember that red lines and markings on your charts will be invisible under the red light.

- Two or more pens; the felt tips work best. They write heavily and dark enough to show up well in both dim lighting conditions and in turbulence, and they are excellent for copying clearances. Black is the best color for maximum visibility.

- Six or so sharpened pencils (#2 lead seems to be the best) for working up flight plans and other miscellaneous figuring. Dull points are difficult to write with and read later, so you should carry a small hand-held pencil sharpener, like you'd find in a child's pencil box. In fact, a child's pencil box is a great container for all of these small items. And don't forget a good eraser. The erasers on the pencils wear out a lot faster than the pencils do, so it's a good idea to have spares.

- One pocket calculator, at least. There are many good computers on the market today; I happen to prefer the Jeppesen CR-2, which fits in a shirt pocket.

- Two or more plotters, one for sectional and WAC charts, and one for the IFR en route charts. Remember, unlike the WAC or sectional charts, the en route charts can be almost any scale. Always check the chart's scale so you will use the proper plotter scale.

- A pad or two of the latest FAA flight plan forms.

- Scratch pads or note paper. Here I find that the back sides of the small telephone message forms work just fine for clearances and miscellaneous figuring. They'll fit right in your shirt pocket along with your computer so they're always close at hand.

- You'll be working with at least two different times: local time for everyday life and UTC (Zulu) for filing flight plans and flying. Two inexpensive dual-time watches are ideal; wear one and leave the spare in the flight bag. The two watches permit easily shifting from local to UTC and from 12- to 24-hour timekeeping. These watches are available in all drugstores as well as catalogues.

IFR aircraft requirements

Just as you need special charts, your aircraft will need special instrumentation for IFR flight. Part 91.205 lists what you need on board to operate in all environments. The instruments and equipment required for IFR flight are, in addition to those required for VFR flight:

- A two-way radio communications system and navigational equipment appropriate to the ground facilities to be used
- Gyroscopic rate-of-turn indicator
- Slip/skid indicator
- Sensitive altimeter
- Clock showing hours, minutes, and seconds with a sweep-second pointer or digital presentation
- A generator or alternator of adequate capacity
- Artificial horizon (attitude indicator)
- Directional gyro or equivalent
- For flight above 24,000 feet MSL, distance measuring equipment (DME).

There are also special requirements for VOR receivers. Part 91.171 says this:

> *(a) No person may operate a civil aircraft under IFR using the VOR system of radio navigation unless the VOR equipment of that aircraft—*
>
> *(1) Is maintained, checked, and inspected under an approved procedure; or*
>
> *(2) Has been operationally checked within the preceding 30 days, and was found to be within the limits of the permissible indicated bearing error. . . .*

These permissible errors are ±4 degrees when using an FAA-operated or approved VOR *test signal* (VOT), or designated (usually by paint marks) surface VOR receiver checkpoint, or if dual VORs are checked against each other. The permissible error is ±6 degrees if monitored on an approved in-flight check.

An aircraft logbook entry must be made for each VOR check; however, it is not necessary to have a mechanic do the check. You can do your own check as the pilot, but don't forget to make the logbook entry according to FARs.

Other required items for IFR flight deal with external items such as lighting and won't be covered in this discussion.

Icing and thunderstorm hazards

After ascertaining that the aircraft and personal flight kit are in order, you should move to the weather briefing. After all, if the weather's too bad, you won't go. Remember, an instrument ticket is not a license to fly in any and all weather. Rather, it's proof that you've been taught to recognize which weather not to fly into, as well as which weather conditions you can handle.

For example, the regulations prevent you from flying into most icing conditions. Part 91.527, which concerns large and turbine-powered multiengine aircraft, states:

(a) No pilot may take off in an airplane that has—

(1) Frost, snow, or ice adhering to any propeller, windshield, or powerplant installation, or to an airspeed, altimeter, rate of climb, or flight attitude instrument system;

(2) Snow or ice adhering to the wings or stabilizing or control surfaces; or

(3) Any frost adhering to the wings, or stabilizing or control surfaces, unless that frost has been polished to make it smooth.

(b) Except for an airplane that has ice protection provisions that meet the requirements in [a regulation setting forth anti-ice and deice requirements not applicable to most light aircraft] no pilot may fly—

(1) Under IFR into known or forecast moderate icing conditions.

Although these restrictions are specifically written to apply to large and turbine-powered multiengine aircraft and there are no restrictions specifically written concerning light aircraft, it would be foolish to fly lightplanes where the heavies have been restricted. It could be considered a violation of FAR 91.13 (careless or reckless operation) to do so.

Ice is one of the greatest hazards a pilot can face, short of a wing falling off. It can add weight to the aircraft very quickly, which will increase the stalling speed. Additionally, it might build up to the point that the engine is no longer capable of carrying the load. It can build up in such a manner as to put the aircraft outside of CG limits, and especially with *rime ice*, it builds in such a nonuniform shape as to completely destroy the lift of the wings. The worst icing conditions that you can encounter are freezing rain and freezing drizzle, both of which are capable of pulling you out of the sky in just a few minutes.

Though rime ice can form at temperatures as low as −40°F (−40°C), the very hazardous *glaze* or *clear ice* is usually confined to temperatures of 14–32°F (−10–0°C).

Check the weather carefully for reports of icing, and avoid flying into IFR conditions when the outside air temperature (OAT) is in the range of 10–32°F. Find a different altitude that will result in a higher or lower OAT.

Even following these hints, you will eventually find yourself in icing conditions. You must remember that ice does increase the stalling speed, so be prepared to land at a higher speed than normal should you have ice on the aircraft.

One of the most insidious forms of icing is *carburetor ice*. If the humidity is high enough, this can form at any OAT from 14–70°F. A functioning carburetor heat control will usually melt the ice, provided it is used in the early stages of formation. Pulling the carburetor heat out for a few minutes every 10 or 15 minutes will usually prevent major carburetor icing.

The reason that carburetor ice will form at high temperatures is that as the intake air passes through the venturi of the carburetor, the air cools rapidly, and any moisture in the air can frost up the throat of the carburetor as well as the fuel nozzles. (I have had carburetor icing even over Hawaii.) This is why carburetor heat is pulled out on many aircraft when you close the throttle during power-off landings.

Another hazard to avoid like the plague is the thunderstorm. In addition to the icing hazard always associated with these awesome phenomena, you can be subjected to hail that can damage windshields and the leading edges of airfoils. They are also associated with severe *wind shears* that literally break an aircraft into pieces, and with *microbursts* that have even forced high-powered jet aircraft into the ground.

The thunderstorms usually associated with fast-moving cold fronts can normally be seen and avoided, but if they're hidden (*embedded*) inside the solid blanket of clouds associated with warm fronts, stationary fronts, or the warm side of occlusions, stay on the ground and fly another day.

Airways and altitudes

After checking the weather, you will have to check any NOTAMs that might apply to your flight, and, if you are flying into or out of large terminal areas, you will be looking for the possibility of preferred routes that have been established for the most orderly flow of traffic in and around those busy areas. These routes are listed in the back of the *Airport/Facility Directory (A/FD)*; the routes are either one-way or two-way.

Now, with all of this in mind, check your charts for a routing, using preferred routes if possible, that will avoid the thunderstorms and icing conditions. You might still have other restrictions to your choice of routes, such as altitude restrictions and obstructions that are beyond the operational limits of the aircraft.

It might be a good idea to review some of the terminology and definitions that you will run across. The lowest altitude you can ever fly on airways under instrument conditions will be determined from one of the following four restrictions as spelled out in the glossary section of the *Air Traffic Control Manual*:

- **Minimum En route IFR Altitude.** The lowest published altitude between radio fixes that assures acceptable navigational signal coverage and meets obstacle clearance requirements between those fixes. The MEA prescribed for a Federal Airway or segment thereof, area navigation low or high route, or other direct route, applies to the entire width of the airway, segment, or route.

- **Minimum Crossing Altitude.** The lowest altitude at certain fixes at which an aircraft must cross when proceeding in the direction of a higher minimum en route IFR altitude (MEA).

- **Minimum Obstruction Clearance Altitude.** The lowest published altitude in effect between radio fixes on VOR airways, off-airway routes, or route segments that meets obstacle clearance requirements for the entire route segment and that assures acceptable navigational signal coverage only within 25 statute (22 nautical) miles of a VOR.

- **Minimum Reception Altitude.** The lowest altitude at which an intersection can be determined.

If a specific route lists both MEA and MOCA, the pilot may operate below the MEA down to but not below the MOCA only when within 25 statute miles of the VOR concerned. This can be helpful when you have to descend to avoid or get out of icing conditions.

It is legal, and sometimes necessary, to operate off the airways. When the pilot decides to do so, the *ATC Handbook* states:

> *Pilots are reminded that they are responsible for adhering to obstruction clearance requirements on those segments of direct routes that are outside of controlled airspace. The MEAs and other altitudes shown on low altitude IFR en route charts pertain to those route segments within controlled airspace, and those altitudes may not meet obstruction clearance criteria when operating off those routes. When planning a direct flight, check your sectional or other VFR charts, too.*

Aha, there is the other use of the sectional charts previously mentioned in this chapter. Sectionals are a necessary part of your IFR flight kit.

Another reason to be concerned about MEAs and MOCAs is that in some sections of the country you have to cross high terrain and might run into the need for supplemental oxygen that might not be on board. The regulations stipulate that all required crewmembers be on oxygen for all flight time exceeding 30 minutes that the aircraft is above 12,500 feet MSL cabin altitude, up to and including 14,000 feet MSL. For flight at *cabin altitudes* above 14,000 feet MSL, crewmembers must be on oxygen for the entire period of time. In unpressurized aircraft, *aircraft altitude* and *cabin pressure altitude* are one and the same. Perhaps you will have to consider the alternative of roundabout routings if you do not carry oxygen.

When making up a tentative route, study the charts carefully to be familiar with all other routes that will be nearly parallel because it is not unusual to be cleared for a routing other than what you file for, or to have the routing suddenly changed after becoming airborne.

Fuel and alternate airport requirements

With all of this in mind, you can now prepare the flight plan. This should be done as accurately as possible, using all available wind and weather information. Even though you know that the conditions will seldom be exactly as forecast, especially in the lower altitudes, it's a lot easier in flight to make small corrections rather than major corrections. Besides, you base the required fuel load on the information on the flight plan. In order to see exactly what's figured in, in the way of IFR fuel, take another look at the regulations. You can find the fuel requirements in Part 91.167:

(a) . . . No person may operate a civil aircraft in IFR conditions unless it carries enough fuel (considering weather reports and forecasts, and weather conditions) to—

(1) Complete the flight to the first airport of intended landing;

(2) Fly from that airport to the alternate airport; and

(3) Fly after that for 45 minutes at normal cruising speed. . . .

(b) Paragraph (a)(2) of this section does not apply if—

(1) Part 97 of this subchapter prescribes a standard instrument approach procedure for the first airport of intended landing; and

(2) For at least 1 hour before and 1 hour after the estimated time of arrival at the airport, the weather reports or forecasts or any combination of them indicate—

(i) The ceiling will be at least 2,000 feet above the airport elevation; and

(ii) Visibility will be at least three miles.

Although part (b) might sound a little confusing at first, all it really means is that if the intended airport has an instrument approach procedure that has been approved, and if the weather is forecast to be VFR (actually, with the 2,000-foot ceiling, a little higher than VFR), you won't need to file an alternate airport, and your VFR fuel will suffice.

There is a built-in safety margin here if you look at it again. The fact that the field has an instrument approach should allow you to get in. The chances are quite remote that the weather will go from VFR all the way down to below IFR minimums without being forecast. On the other hand, if the weather is below VFR minimums or forecast to go below VFR minimums within the previously mentioned two-hour time period, you must file an alternate airport just in case the weather really does go down fast and closes the destination airport.

Under IFR alternate airport weather minimums, FAR 91.169:

> *(c) . . . Unless otherwise authorized by the Administrator, no person may include an alternate airport in an IFR flight plan unless current weather forecasts indicate that, at the estimated time of arrival at the alternate airport, the ceiling and visibility at that airport will be at or above the following alternate airport weather minimums:*

> *(1) If an instrument approach procedure has been published in Part 97 of this chapter for that airport, the alternate airport minimums specified in that procedure, or, if none are so specified, the following minimums:*

> *(i) Precision approach procedure: ceiling 600 feet and visibility 2 statute miles.*

> *(ii) Nonprecision approach procedures: ceiling 800 feet and visibility 2 statute miles.*

> *(2) If no instrument approach procedure has been published in Part 97 of this chapter for that airport, the ceiling and visibility minimums are those allowing descent from the MEA, approach, and landing under VFR.*

The alternate minimums are normally well above the primary landing minimums for the airport. This additional safety margin is thrown in because if the weather at the intended airport goes below IFR minimums and you have to divert to the alternate, chances are that the weather will still be above landing minimums at the alternate airport. But at the moment you divert, the alternate airport becomes your new airport of intended landing, and the published landing minimums are applicable for what is now your new destination, utilizing the facilities that are appropriate for the procedure.

Be aware that in determining your alternate airport and fuel loading requirements, even if the forecast for the intended destination airport indicates that conditions will only be "occasionally" (or "chance of") below the minimums specified in 91.167(b) (2) and 91.169(b), an alternate is still required.

Similarly, you cannot select an alternate that is forecast to be "occasionally" (etc.) below the minimums in 91.169(c).

ATC advises filing a flight plan at least 30 minutes prior to estimated time of departure; otherwise, expect a delay up to 30 minutes long. If you file before the aircraft preflight check, you should have the clearance in hand and be ready for departure by the time you're finished with the preflight and the runup.

9

Preparing for an instrument flight

Planning an IFR flight seems to be a large order for the beginning instrument student. Unfamiliar aeronautical charts must be mastered, the weather analysis is more complex compared to VFR analysis, and procedures for the departure, en route, and approach phases of the flight might be unclear.

It's only natural to be a little puzzled at first; however, many things can be done to master the process more quickly and make it more interesting. Believe it or not, flight planning can even be fun!

The single most important factor in taking the mystery out of IFR flight planning is to plan and file an IFR flight plan on every training flight, including the very first one.

Even if you are only going out to practice in the local area, you should still plan and file IFR to a destination 50–75 nautical miles from the home airport. You should also work out an IFR flight plan for the return trip, whether you expect to use it or not.

In the beginning, the instructor will suggest two or three nearby destinations. Get your instructor's ideas well ahead of the flight so you can do your planning at home when you have more time and are under less stress.

Plan for all destinations the instructor has suggested, not just one. This will give you more practice, expose you to a variety of situations, and will not cost you a cent. Students who show up at the airport without their homework finished waste their time and their instructor's time.

You will be surprised how much you can accomplish at home. You can map out the flight route, review departure procedures, and go over the destination approach charts and airport information in great detail. You can get preliminary weather briefings, check the NO-TAMs, and even file your flight plan from home. In fact, you can make several calls if the weather is changing rapidly, or there is something you didn't catch clearly on the first call. This is particularly helpful in the beginning of your IFR training.

ATC will often make amendments to your clearance as the IFR flight proceeds. When you carefully work out your flight at home, familiarize yourself with other VORs and airways between your departure and destination airports and pick out all those obscure intersections that ATC might use for clearance limits, rerouting, or holding fixes. That way there will be no unpleasant surprises!

The cockpit is not the place for basic research. A newly rated instrument pilot departed one of the New York airports in actual IFR weather flying a light twin with his family on board. He had only filed and flown IFR a few times before he received his rating. When airborne, he contacted departure control. The controller responded: "New clearance. Ready to copy?" It wasn't a major change, but due to his inexperience, it overloaded him. He lost control of the airplane and it crashed, killing all on board. If ever an accident could have been prevented by more thorough training, this was it.

GPS en route navigation

En route IFR navigation by GPS (Global Positioning System) requires panel-mounted equipment certified by the FAA as meeting the complex requirements of TSO (Technical Standard Order) C129. In addition, each individual installation must pass an FAA inspection. Handheld GPS units are not approved for IFR.

Furthermore, "aircraft using GPS navigation under IFR must be equipped with an approved and operational alternate means of navigation appropriate to the flight," according to AIM 1-1-22, b 1 (b). For all practical purposes, an IFR fight proceeding by GPS must be backed up by VOR navigation. The FAA is considering transitioning to an all-GPS system of air navigation, but this is still many years away. The *Instrument Rating Practical Test Standards* lists no requirements for competence in en route navigation by GPS. So plan your flights for VOR navigation without regard to GPS, even though you may be conducting your training in an airplane equipped with an approved, panel-mounted GPS system.

The flight log

As a VFR pilot, you already have considerable experience planning cross-country flights and navigating with VORs. Your IFR planning is nothing more than an extension of what you have already learned, with more detail in some areas and greater emphasis on other points that are not so crucial to VFR flying. Over the years I developed a flight log form (Fig. 9-1) that covers all information needed for VFR and IFR cross-country flights. I find that if my private pilot students use this log form for the cross-country phase of their training, they make the transition from VFR to IFR flight planning very easily.

Please be my guest; copy the form and use it for your VFR and IFR flying. It gives you a logical, step-by-step way to work your way through all the important elements of flight planning. If you fill out the log properly, nothing will be overlooked.

During the flight itself, the form functions as a running log with places to enter time en route, ETA, actual time of arrival, ground speed between fixes, and fuel consumption. Equally important, it provides a quick reference for all the detailed information you need to conduct the flight without a lot of confusion and fumbling around. Cockpit organization is the key to a good IFR flight.

Let's go through the form step-by-step so you can see how useful it is when planning a flight. We'll be taking a hypothetical IFR cross-country from Westchester County Airport, at White Plains, N.Y., to Broome County Airport, at Binghamton, N.Y., on a typical actual IFR day in late November. I have chosen this example because I have found that this trip to Binghamton contains all the elements of a good training flight. The principles and techniques that apply here are valid everywhere else.

IFR flight planning occurs in two phases. In the first phase you decide your route of flight and fill in your flight log with all the information available before obtaining a weather briefing. In the second phase, get your weather briefing, then fill in the information affected by the weather, such as ground speed and the choice of an alternate airport.

Planning the route

The first step is plotting the route between your departure and your destination airports, including the appropriate *standard instrument*

FIX		INTERSEC	ROUTE	M.C.	DIST.	G/S	TIME ENR	TIME ARR	DATE
	NAME	NAME	VIA	TO	TO	EST	EST	EST	FLIGHT PLAN ROUTE
		IDENT							
IDENT.	FREQ	FREQ RADIAL	ALT	FROM	REM	ACT	ACT	ACT	
									ATC CLEARS N _____
									CLEARANCE LIMIT

AC ID	POSITION	TIME	ALT.	IFR/VFR	EST. NEXT FIX	NEXT FIX	PIREPS

Fig. 9-1. *Flight log form for planning and logging IFR cross-country flights.*

PILOT:

TAS:	IAS:	CALL SIGN	FREQ.	ASGND ALT
RPM:	MP:			
TAKEOFF: RWY ∣ WIND ∣ KTS ∣ ALT		ATIS		
		CLNC		
		GND		
Field Elev:		TWR		
ALTIMETER ERROR		DEP C		
LANDING:				
Field Elev: RWY ∣ WIND ∣ KTS ∣ ALT				
TYPE OF APPROACH:				
TIME ON:				
TIME OFF:				
TOTAL TIME ENROUTE:		ATIS		
		APP C		
TACH ON	TACH OFF	TWR		
		GND		

WINDS ALOFT

Sta.	A	Dir.	Vel	Temp	A	Dr	VI	T	A	D	V	T

VFR WX AT:

	TANK ON	TANK OFF		TANK ON	TANK OFF

departure (SID), if there is one. Open up the en route low-altitude chart, or charts, that cover the area—in this case, L-25. Notice which Victor airways make a logical route to the destination; V252 from Huguenot VOR goes directly to Binghamton VOR, so that is the best Victor airway for the en route portion of this flight (Fig. 9-2).

You could also get from Huguenot to Binghamton by departing Huguenot on V162 to intercept V126 to Lake Henry, then taking V149 to Binghamton. This would lengthen the trip unnecessarily so V252 is the better choice. However, on many IFR cross-country flights the best route to your destination might involve two or more Victor airways, and there is nothing wrong with combining several. Also, an airway segment might have several Victor airway numbers. The airway heading southwest from Huguenot on L-25 is numbered V205-252 above the line and 489 below. Which one would you list on your log and flight plan? Either one. It doesn't matter as far as ATC is concerned. Pick the numbered route that will carry you far-thest toward your destination.

Standard instrument departures (SIDs)

Note that there is no Victor airway from Westchester Airport to Huguenot. How do you get to Huguenot? Utilize the Westchester's published SID called "Westchester Nine" (Fig. 9-3). If no SID is avail-able at your departure airport, you would simply file "direct" to the first VOR on the airway you intend to use.

Your IFR flight planning should always include a study of the SIDs available at your departure airport because ATC will expect you to use them. SIDs (and arrival procedure STARs) are now published in the same NOS booklets or packets that contain instrument approach procedure charts. SIDs are also published by Jeppesen and are in-cluded as separate sheets along with Jeppesen's approach charts for an airport.

Preferred routes and TECs

Before you enter the planning results on the flight log, check two more items that often can be very helpful. They are Preferred IFR Routes and Tower En Route Control (TEC). (In the beginning of a student's instrument training, I don't emphasize preferred IFR routes and TECs because they can be confusing.) You should know about them and become familiar with them.

Preferred IFR routes are listed toward the back of the *Airport/Facility Directory* (A/FD)—the green book. As the name implies, preferred IFR routes have been established by the FAA to guide pilots in their flight planning and to minimize route changes during the flight. If a preferred route doesn't cover your entire flight, use whatever segments you can. You are more likely to get the route you file for and less likely to have the route amended in the air if you can use all or part of a preferred route.

TEC routes and airports are also listed toward the back of the A/FD. TEC makes it possible to fly from one approach control point to another without entering air route traffic control center (ARTCC) airspace. Designed primarily for instrument flights below 10,000 feet, TEC is usually quicker and simpler than routing through center airspace. But it's available only between the paired airports listed in the A/FD; if your departure airport is not paired with your destination airport, you cannot use TEC. If you can use a TEC route, enter the acronym TEC in the remarks section of the flight plan.

Browse through the preferred routes and the TEC sections of the A/FD. Examine how the information is set up and check to see if your usual departure point fits into either category or both categories. If so, these two special procedures will simplify your planning and save time.

Approach planning

After you have worked out the route to the destination, you should study the Standard Terminal Arrival Charts (STARs), if any apply, as well as the instrument approach procedure charts for the destination airport. STARs help you plan your transition from the en route phase to the approach phase.

Look for two things when you study STARs and approach charts. First you need to familiarize yourself with *all* the approaches available at the destination because you must be prepared for the specific approach that ATC assigns. In the beginning of your instrument training, the amount of detail packed on these small charts might be quite mystifying. For example, Binghamton Airport, the destination in our flight-planning example, has no less than five different instrument approach procedures!

No one expects you to be fully cognizant of all the approach procedures on your first few training flights. But you must examine all the

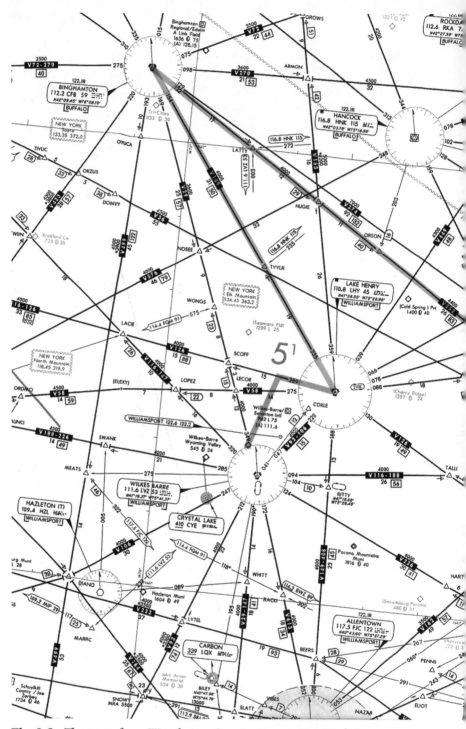

Fig. 9-2. *The route from Westchester County Airport, New York, to Binghamton, New York, as plotted on an en route low altitude chart (L-25). Also shown is the route to the alternate, Wilkes-Barre, Pennsylvania.*

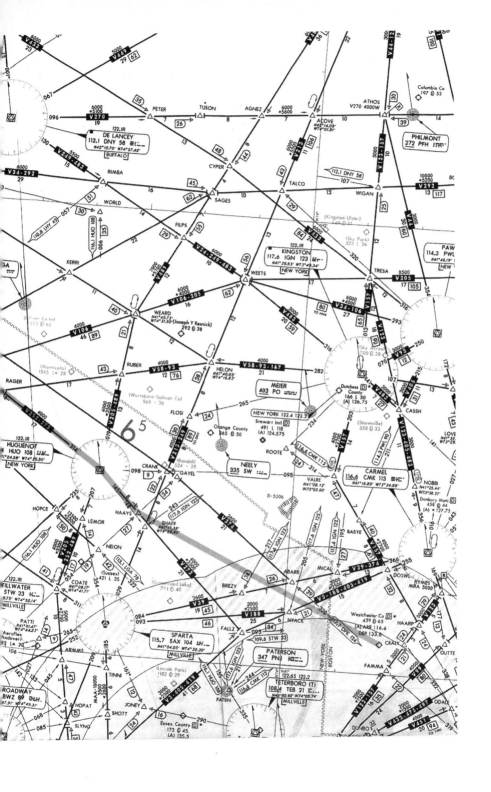

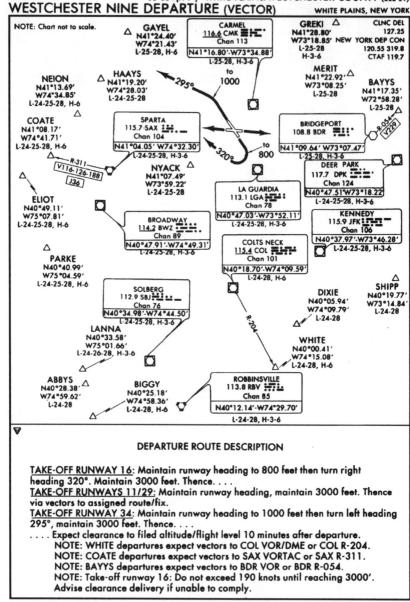

97254

SL-651(FAA)   WHITE PLAINS/WESTCHESTER COUNTY (HPN)

WESTCHESTER NINE DEPARTURE (VECTOR)   WHITE PLAINS, NEW YORK

DEPARTURE ROUTE DESCRIPTION

TAKE-OFF RUNWAY 16: Maintain runway heading to 800 feet then turn right
heading 320°. Maintain 3000 feet. Thence. . . .
TAKE-OFF RUNWAYS 11/29: Maintain runway heading, maintain 3000 feet. Thence
via vectors to assigned route/fix.
TAKE-OFF RUNWAY 34: Maintain runway heading to 1000 feet then turn left heading
295°, maintain 3000 feet. Thence. . . .
. . . . Expect clearance to filed altitude/flight level 10 minutes after departure.
    NOTE: WHITE departures expect vectors to COL VOR/DME or COL R-204.
    NOTE: COATE departures expect vectors to SAX VORTAC or SAX R-311.
    NOTE: BAYYS departures expect vectors to BDR VOR or BDR R-054.
    NOTE: Take-off runway 16: Do not exceed 190 knots until reaching 3000'.
    Advise clearance delivery if unable to comply.

WESTCHESTER NINE DEPARTURE (VECTOR)   WHITE PLAINS, NEW YORK
97254   WHITE PLAINS/WESTCHESTER COUNTY (HPN)

Fig. 9-3. *Westchester Nine Departure, the standard instrument
departure (SID) from Westchester County, N.Y., Airport, as depicted
on an NOS chart.*

approaches for your destination every time you file a flight plan. You will be amazed at how much information you will be able to absorb after you do this a few times.

Don't forget the departure airport; familiarize yourself with its instrument approaches in case a door pops open or a passenger becomes sick or for some other reason you have to return immediately after takeoff.

Airport diagrams

Study the airport diagrams first, just as you would on a VFR cross-country. Note the field elevation. Study the runway diagram and note the number of runways and their headings, lengths, and widths. Will you be able to use all runways? Do mountains or other obstructions affect an instrument approach? How high are the obstructions? Check over the taxiways because getting from your runway to your destination on the field can sometimes be the most complicated part of the trip. (One time, taxiing to the FBO at Montreal took longer than the flight from Ottawa!)

Initial approach fixes (IAFs)

Look at all the VORs, NDBs, marker beacons, intersections, and holding patterns. How will you get from the Victor airway you've been flying to the most likely instrument approach? Ask yourself which approach you will probably be assigned when you are handed off to approach control; consider the prevailing wind. Then pick out the VOR, NDB, or intersection on your inbound route that is closest to the final approach course. Chances are, this fix will be designated as an *initial approach fix* (IAF) on the approach chart.

For our hypothetical flight to Binghamton, the wind is generally northwesterly, so the most likely approach is ILS RWY 34 (Fig. 9-4). We could simply continue on V252 to Binghamton VOR and use that as our destination fix. But that would involve a lot of maneuvering to get back to the final approach course for Runway 34.

A better choice for a destination fix would be Latty intersection. Latty is on both V252 and the final approach course for Runway 34 and Latty appears both on the L-25 chart and the approach chart for ILS RWY 34. In most cases, approach control will give you radar vectors to the final approach course for the approach.

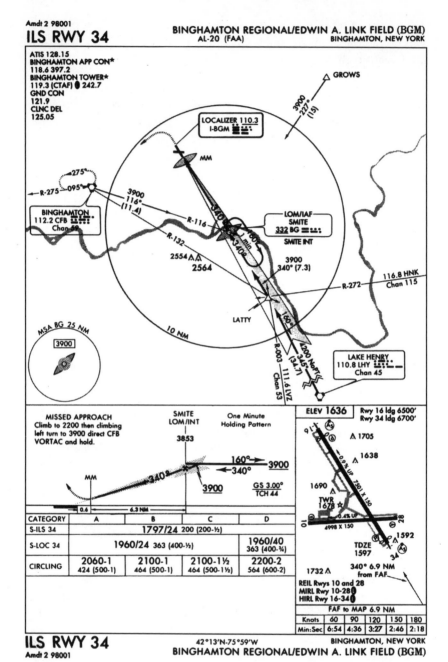

Fig. 9-4. *The ILS 34 approach to Binghamton. Note the location of LATTY intersection, the final fix on the IFR flight to Binghamton discussed in the text.*

But if there is no radar available, or you have lost radio communications, you will be expected to make the transition from the Victor airway to the final approach course on your own. Beware: your instructor might insist that you make the full approach! So you must plan ahead.

Now we are ready to fill out the flight log. In the "Flight Plan Route" section at the center top portion of your flight log form, write in the abbreviations that will spell out your planned route (Fig. 9-5), in this case—WESTCHESTER 9 SID, HAAYS INTERSECTION, V273 HUGUENOT, V252 LATTY INTERSECTION. This is also entered in Block 8 of the FAA's flight plan form for the "route of flight."

En route fixes

Now let's fill in more details of your flight. Look at the eight columns in the left one-third of the flight log (Fig. 9-6). No mysteries here, identical to a VFR flight with a few extra details. Spell out the full name of all VORs used for fixes along your route. You might not be familiar with the three-letter identifiers, at least not in the beginning, and the names of some VORs can be very confusing.

For example, who could possibly guess that the identifier HPN stands for Westchester County Airport in White Plains, New York? So to

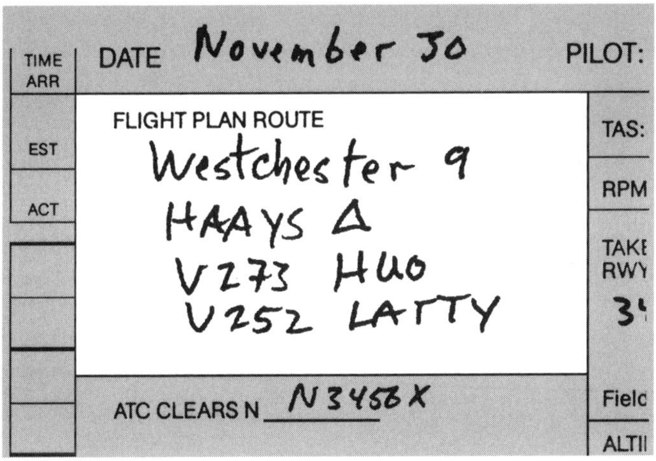

Fig. 9-5. *The flight log form has the flight-plan route block filled in for the flight to Binghamton.*

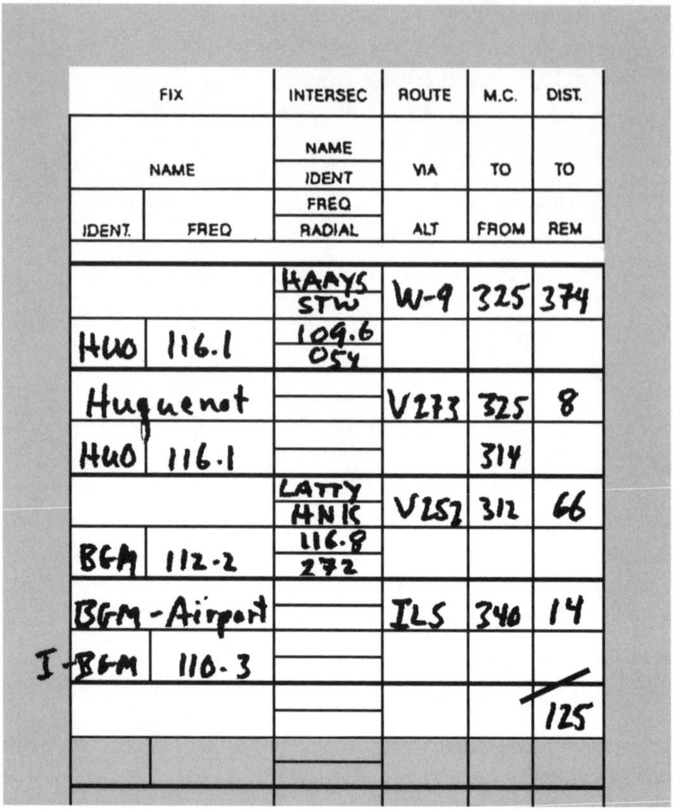

FIX		INTERSEC	ROUTE	M.C.	DIST.
NAME		NAME	VIA	TO	TO
		IDENT			
IDENT.	FREQ	FREQ	ALT	FROM	REM
		RADIAL			
		HAAYS STW	W-9	325	374
HUO	116.1	109.6 054			
Huguenot			V273	325	8
HUO	116.1			314	
		LATTY HNK	V252	312	66
BGA	112.2	116.8 272			
BGM-Airport			ILS	340	14
I-BGM	110.3				
					125

Fig. 9-6. *Fixes, magnetic courses, and mileages entered for the flight to Binghamton. Note: The initial heading of 3258 to HAAYS is an estimate because departure headings from the airport are radar vectors.*

avoid confusion, write out the station name as shown with its identifier and frequency in the boxes below.

If the navigational fix is an intersection, use the column under "intersec," enter the name of the intersection as shown and list the identifier, the frequency, and the radial identifying the intersection. Note that we have entered LATTY intersection here because of its usefulness in locating the final approach course and it is common to both the en route chart and ILS 34 at Binghamton.

The next three columns are for routes, altitudes, magnetic courses, and distances; fill in as shown. The altitude in the third column is what you shall request for the first leg of the flight. The ground speed, time en route, and time of arrival (time arr) columns will have to wait until the weather briefing provides wind information at the planned cruise altitude.

Actual ground speeds, actual times en route, and actual times of arrival will be entered while in flight. This versatile form not only provides a structure for flight planning, but also becomes a very handy log to monitor the flight's progress.

Communications frequencies

A few more details can be filled in before the weather briefing. Look up departure and destination airports in the A/FD and enter the frequencies as shown in the three columns at the far right of the form (Fig. 9-7). If the departure airport has an *automatic terminal information service* (ATIS), write this information in at the top of the first column.

The abbreviations for the other entries are: CLNC, clearance delivery; GND, ground control; TWR, tower; and DEPC, departure control. Use the blank boxes to list the en route frequencies as they are assigned.

The column on the far right is reserved for any changes of assigned altitudes. A typical change of en route frequencies might be: "Contact New York Center now, one two eight point five." When you make the frequency change and New York Center advises "Descend and maintain four," you simply write in the new altitude next to the frequency. The sequence of entries in the appropriate boxes would look like this: NY CTR 128.5 40 (for 4,000 feet). Once again the flight log form functions as a flight planning guide as well as a running log during your flight.

Toward the bottom of the radio frequency columns are spaces for arrival frequencies, including APP C for approach control.

While working on the right side of the form, fill in the indicated airspeed (IAS) you plan to fly (top of form) and the power setting, to maintain this airspeed. Let's assume that we're making this flight to Binghamton in a Cessna 182. We enter the revolutions per minute (RPM) and manifold pressure—2300 and 23"—that we estimate will maintain 125 knots (Fig. 9-8).

Harris

CALL SIGN	FREQ.	AS
ATIS	135.8	
CLNC	127.25	
GND	121.8	
TWR	119.7	
DEP C		
ATIS	128.15	
APP C	118.6	
TWR	119.3	
GND	121.9	

Fig. 9-7. *The sequence of frequencies for the flight to Binghamton.*

Field elevation

When still on the right side of the flight log form, enter the elevation of departure and destination airports (Fig. 9-9). Because you don't know the altitude of takeoff and landing runways you will be assigned, enter the field elevations.

In your VFR flying days, the field elevations you picked off your sectional charts were accurate enough for flying a good pattern and

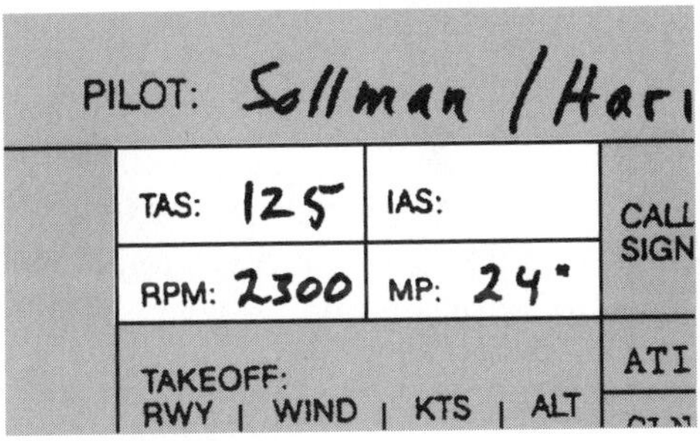

PILOT: *Sollman / Hari*

TAS: **125**	IAS:	CALL SIGN
RPM: **2300**	MP: **24°**	
TAKEOFF:		ATI
RWY | WIND | KTS | ALT		

Fig. 9-8. *Performance details for the flight to Binghamton.*

RPM: **2300**	MP: **24°**	SIGN
TAKEOFF:		ATIS
RWY | WIND | KTS | ALT		CLNC
34		GND
Field Elev: **439'**		TWR
ALTIMETER ERROR		DEP
LANDING:		
Field Elev: **1636'**		
RWY | WIND | KTS | ALT		
34		
TYPE OF APPROACH:		

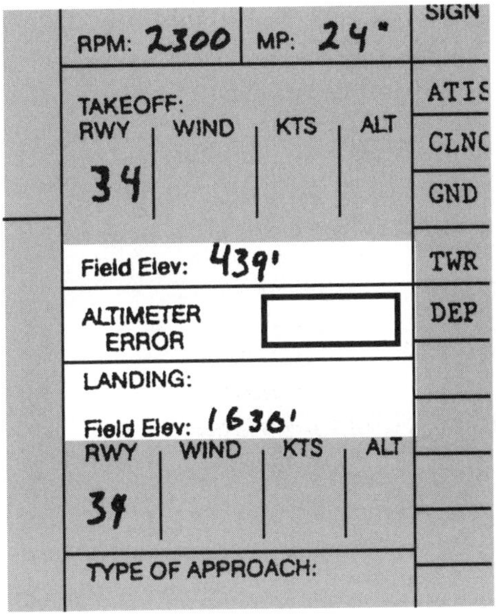

Fig. 9-9. *Field elevations for the flight to Binghamton.*

making a good landing. But field elevation is the highest point on an airport's usable runways, and not necessarily the most important elevation for an instrument approach to a selected runway. More precise field elevations are utilized.

Take a closer look at the approach chart for the ILS 34 approach at Binghamton (Fig. 9-4). On the airport diagram you will see "TDZE 1597" at the approach end of Runway 34; TDZE stands for *touch down zone elevation.* This is the precise field elevation for the landing area of this instrument approach and it is this TDZE of 1,597 feet on which the landing minimums for landing on Runway 34 are based, not the field elevation of 1630 feet. Keep in mind the fact that no matter which runway you end up using, the TDZE might be substantially different from the field elevation.

The same applies to the departure airport. You will need an equally precise elevation to set the altimeter accurately. At the ramp at Westchester County Airport, the spot elevation is 388 feet, which is 51 feet lower than the published field elevation (highest point) for Westchester!

If an altimeter had the maximum allowable instrument error of 70 feet, this could result in an error totaling 121 feet if you used Westchester's official field elevation. Most precision instrument approaches have a decision height 200 feet above the TDZE. If the aircraft altimeter were off by 121 feet, you could have serious problems on a precision approach.

If you are operating from an unfamiliar field and cannot find the elevation figure for the parking ramp, use the TDZE elevation nearest the run-up area for the takeoff runway.

Airport services

This completes the information you can enter on the flight log form prior to the weather briefing (Fig. 9-10).

But some unofficial planning items can make the difference between an easy, comfortable journey and a difficult trip. Just because the airport is listed in the A/FD does not mean it is adequate for you.

A lot of important information is not listed in the A/FD. Are there special noise abatement procedures? Will the FBO be open for parking, fuel, and other services when you arrive? Does the airport have more than one FBO? Is there convenient transportation to the office meeting, to a hotel, or to that golf course you've been wanting to play for so long?

Answer these questions before departure. The best way to obtain this information is from somebody on the ground at your destination who knows the area. Determine ahead of time which FBO you'll be using and give them a call. *AOPA'S Airport Directory* is an excellent current source of telephone numbers for FBOs. This annual guide also provides a wealth of information on transportation, lodging, and other services.

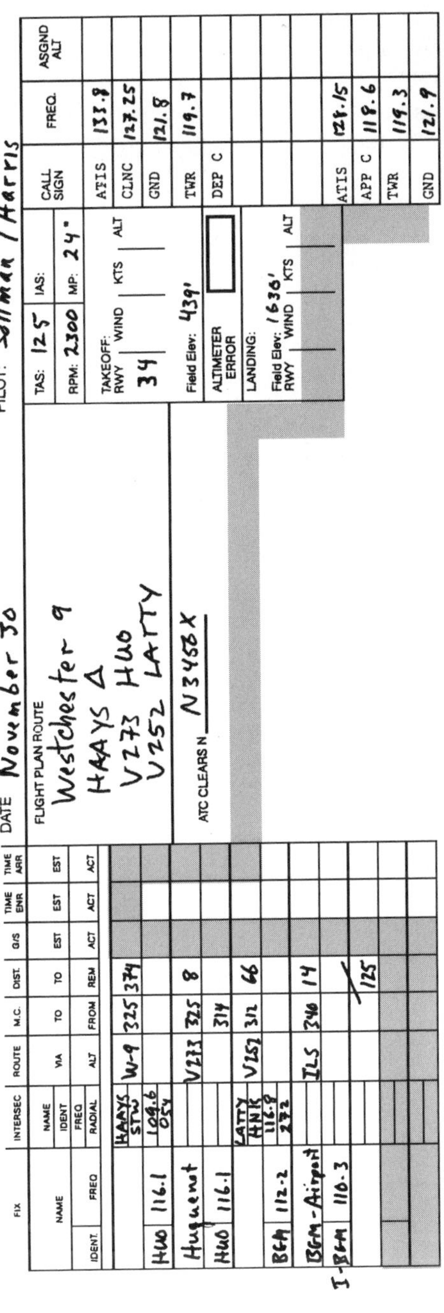

PILOT: Sollmer / Harris

DATE November 30

FLIGHT PLAN ROUTE
Westchester 9
HAAYS A
V273 HUO
V252 LATTY

ATC CLEARS N: N3458X

TAS: 125 IAS:
RPM: 2300 MP: 24"

TAKEOFF:
RWY 34 WIND ___ KTS ___ ALT

Field Elev: 439'

ALTIMETER ERROR

LANDING:
RWY ___ WIND ___ KTS ___ ALT
Field Elev: 1636'

CALL SIGN	FREQ	ASGND ALT
ATIS	135.8	
CLNC	127.25	
GND	121.8	
TWR	119.7	
DEP C		
ATIS	124.15	
APP C	118.6	
TWR	119.3	
GND	121.9	

FIX NAME / FREQ (IDENT)	INTERSEC NAME / IDENT / FREQ / RADIAL	ROUTE VIA / ALT	M.C. TO / FROM	DIST TO / REM	G/S EST / ACT	TIME ENR EST / ACT	TIME ARR EST / ACT
HUO 116.1	HAAYS STW 104.6 054	W-9	325	34			
Huguenot							
HUO 116.1		V273	325 / 314	8			
	LATTY HNK 116.9 272	V152	312	66			
BGM 112.2		ILS	340	14			
I-BGM 110.3	BGM–Airport			125			

Fig. 9-10. *Flight log for the flight to Binghamton with all the details that can be filled in prior to the weather briefing.*

All charts, approach plates, and other study materials excerpted from government and other publications contained in this textbook ARE NOT LEGAL FOR NAVIGATION. They are included for illustration and study purposes only! Use current legal charts only!

The materials used in this textbook are intended to prepare the instrument student with the most realistic and practical materials available. Even the purist who went out and bought all these charts would find that within two months they are all obsolete anyway—within a few weeks or a few months at most.

More than 60 years ago, the most common aeronautical chart was an automobile road map, which was readily available at any gas station at no charge. My, how times have changed.

10

Inoperative equipment

Minimum equipment lists and approved flight manuals

As pilots, we are very familiar with preflight inspections to determine airworthiness. We are also aware of the regulation that puts us completely to blame if we fly an airplane in an unairworthy condition. But what is "legal airworthiness"? When you move up to multi-engine airplanes, there are many more items you need to check, and therefore, many more items that could render the airplane not airworthy. While conducting an unscientific survey of pilots, I found that surprisingly few were sure about how to handle inoperative equipment or understood the term "legal airworthiness."

For instance, say that while performing a routine preflight inspection of an airplane, you discover the suction gauge that monitors the vacuum system is broken. Can you legally take off with the gauge inoperative? Is the airplane legally airworthy? The commonsense answer is yes. After all, the suction gauge has nothing to do with the operation of the engines, nor with the wings' ability to create lift.

Because it does not involve the structural integrity of the airplane, it should not affect airworthiness, right? Wrong! Actually, it does factor into legal airworthiness, but not in a way that most pilots would consider. As pilots, we are primarily concerned that the airplane will fly, the engines will run, and the airplane will stay in one piece. But legal airworthiness goes much further than all that.

Recently the FAA made a ruling that allows a pilot to fly a plane even if certain items on the airplane are inoperative. The ruling says that inoperative equipment may be flown "as long as the equipment is not essential to the safe operation of the aircraft." So who determines what is essential and what is not? Ordinarily, the answer to an airworthiness question lies within the judgment of the pilot, but in this case, several regulations must guide the pilot's judgment. There are presently two paths to follow: (1) developing a minimum equipment list and (2) FAR part 91.213.

Operations with a minimum equipment list

Minimum equipment lists (MEL) have been around since 1964 for air carriers, and since 1978 for Part 135 operators with multiengine airplanes. In December 1988 the FAA established the program for Part 91 operators to use MELs.

An MEL is simply a list of the equipment on a specific airplane and a statement as to whether the aircraft can be flown if a certain piece of equipment is inoperative. When an item is discovered to be inoperative, the pilot goes to the MEL, finds the entry for that item, and checks if the plane must be grounded until maintenance is performed. Once an MEL is in place, it is quick and easy to use. But getting it into place is the problem.

The MEL under Part 91 is approved for one individual airplane at a time. The process begins when the airplane is first manufactured. During the airplane's type certification process, the manufacturer submits what is known as a *proposed master minimum equipment list* (PMMEL) to a branch of the FAA called the *Flight Operations Evaluation Board* (FOEB). The FOEB is composed of FAA personnel who are avionics, airworthiness, operations, and aircraft certification specialists. The PMMEL is a working document that is used by the people at the FOEB.

Eventually, the FOEB determines the minimum operative instruments and equipment required for safe flight, and they transform the list into a master minimum equipment list (MMEL). The FOEB specialists make recommendations to the FAA's *Aircraft Evaluation Group* (AEG), who are responsible for the publication of the MMEL.

The approved MMEL is then passed on to the various *flight standards district offices* (FSDO) across the United States. The FSDOs have the capability to download MMELs of many different airplanes. In fact, the FAA has developed MMELs for most of the FAA type certificated aircraft in use today. Single-engine airplanes can get a generic MMEL. Multiengine airplanes require a specific MMEL, such as for a BE-76.

In order to have your own MEL for your own airplane, you must first obtain a copy of your airplane's master minimum equipment list, and the MMEL's preamble from the nearest FAA flight standards district office. An FAA inspector and the operator must meet with you to discuss operations using an MEL. After the inspector is convinced that you fully understand MEL operations, a *letter of authorization* (LOA) is issued (Fig. 10-1).

Now you must do some work. Using the MMEL, airplane flight manuals, and maintenance manuals, you need to revise the list to meet the individual needs of your airplane. Not all airplanes have carburetor air temp gauges, but for those that do, there must be instructions for how to proceed if this gauge is inoperative. To customize your list, you will use what are called "M and O procedures."

In the MEL, items are listed in columns. Beside each item is the letter "M" for maintenance or the letter "O" for operations. "M" indicates that a specific maintenance procedure must be accomplished before takeoff. This usually means a qualified maintenance person must do the work. "O" indicates that the procedure required can be accomplished by the pilot or flightcrew. The letter of authorization together with the MMEL, preamble to the MMEL, and the M and O procedures becomes your airplane's MEL. The process flow chart is shown in Fig. 10-2.

After all the MEL parts are together, the MEL is also considered a *supplemental type certificate* (STC) for that one specific airplane. The STC allows that airplane to be operated by the MEL conditions and not the Federal Aviation Regulations. In a sense, the STC becomes the federal regulations for that one airplane. The STC cannot be transferred. If you sell the airplane, the STC is void, and the new owner must start the process over again.

As you can see, getting an MEL has many steps and each can be time-consuming. However, once the MEL is in place, it can make life easier. For instance, say a pilot is at an airport far away from the home base and company maintenance shop. Before takeoff for a flight home, the pilot sees that the right fuel gauge is not working. According to FAR 91.213, the airplane would be grounded because 91.213 requires that each tank have an operating fuel quantity indicator. The pilot would have to get the gauge repaired before takeoff, delaying the flight and requiring out-of-town maintenance work, which could be expensive.

However, if the airplane has an MEL, the pilot might be able to bypass the FARs. I have an approved MEL for a multiengine airplane

Flight Standards District Office
Portland-Hillsboro Airport
3355 N.E. Cornell Road
Hillsboro, OR 97124

July 25, 1991

Mr. John Dough, President
John Dough Enterprises
Hangar 9, Suite 203
Portland-Hillsboro Airport
Hillsboro, OR 97124

Dear Mr. Dough:

This letter is issued under the provisions of FAR § 91.213(a)(2) of the Federal Aviation Regulations (FAR) and authorizes John Dough Enterprises only to operate Cessna Citation 500, N81149, Serial No. 12345, under the Master Minimum Equipment List (MMEL), using it as a Minimum Equipment List (MEL).

This letter of authorization and the MMEL constitute a Supplemental Type Certificate for the aircraft and must be carried aboard the aircraft as prescribed by FAR § 91.213(a)(2).

Operations must be conducted in accordance with MMEL. Operations and maintenance (O and M) procedures for the accomplishment of rendering items of equipment inoperative must be developed by the operator. Those procedures should be developed from guidance provided in the manufacturer's aircraft flight and/or maintenance manuals, manufacturer's recommendations, engineering specifications, and other appropriate sources. Such operations or maintenance procedures must be accomplished in accordance with the provisions and requirements of FAR Part 91, Part 145, or Part 43.

A means of recording discrepancies and corrective actions must be in the aircraft at all times and available to the pilot in command. Failure to perform O and M procedures in accordance with Part 91, Part 145 or Part 43 as appropriate, or to comply with the provisions of the MMEL, preamble, O and M procedures and other related documents, is contrary to FAR and invalidates this letter. All MMEL items that contain the statement "as required by FAR" must either state the FAR by part and section (e.g., 91.205) with the appropriate FAR carried aboard the aircraft, or the operational requirements/limitations required for dispatch must be clearly stated. When the MMEL is revised by the Flight Operations Evaluation Board (FOEB), John Dough Enterprises will be notified by post card of the revision. John Dough Enterprises must then obtain a copy of the revision from this Flight Standards District Office (FSDO), or the FSDO having jurisdiction, and incorporate any changes as soon as practicable including O and M procedures as required.

John Dough Enterprises must develop O and M procedures that correspond with those listed in the MMEL. John Dough Enterprises must also list the "as required by FAR" by specific FAR part and section, or state the operational requirements/limitations for aircraft dispatch. These items must be contained in a procedures document that is placed in the office of the first

Fig. 10-1. *Sample letter of authorization.*

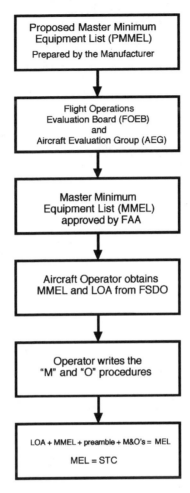

Fig. 10-2. *The minimum equipment list process.*

that allows the airplane to legally fly with one fuel gauge broken. In the MEL there is an O procedure listed by the fuel gauge entry, directing the pilot to top off the fuel tank in question and then fly less than two hours. The pilot in that situation could simply fill the tank and take off for home. If home was more than two hours away, a fuel stop would be needed. Either way, the airplane will get fixed at the home base and the flight is not delayed.

You can see that it would not take too many of these situations before the MEL will pay for itself.

MELs are revised by the pilot or operator, so they can be very practical and pilot-friendly once developed. Contact your local FSDO to get your airplane's MMELs and help throughout the process.

Operations without an MEL

Option two, the use of FAR 91.213, is available to any aircraft that does not have an MEL and falls into one of these categories:

1. Small rotorcraft
2. Nonturbine powered airplanes
3. Glider
4. Lighter-than-air aircraft

Light twins with reciprocating engines, therefore, are not required to have a minimum equipment list. If an approved MEL is not used, the pilot then determines airworthiness by using what I call the "four-step test," shown in Fig. 10-3. If the inoperative or missing instruments or equipment passes all four steps, you can fly the airplane if the equipment is taken out or placarded.

Let's look at each step of the process individually.

Step 1

Step 1, adopted from FAR 91.213(d)(2)(i), asks, "Is the inoperative equipment in question required by the VFR-day type certificate?" The VFR-day type certificate is the set of rules that were followed when the airplane's design was originally certificated to fly by the FAA. The problem is that most of today's general aviation airplanes were certified under some very old regulations. The airplane might be younger than the regulations, yet the airplane might still be "certificated" under the old regulations. In fact, these regulations are so old they are not Federal Aviation Regulations, but rather Civil Aviation Regulations (CARs).

The old CAR Part 3 still has jurisdiction over airplanes today when not specifically superseded by a newer FAR. When the FARs replaced the CARs, the CAR Part 3 became FAR Part 23. But Part 23 even today refers back to its regulation ancestor. So CAR Part 3 is still used to determine the VFR-day type certificate requirements.

Although you probably have never even seen CAR Part 3, I'll bet you know some of its rules. For instance, most pilots know that fuel quantity indicators are required for each fuel tank and that the only time that gauge must read accurately is *when the tank is empty*. This calibrated-to-empty rule is well known, but you won't be able to find it in the FAR. It is in CAR 3—CAR 3.672 to be exact. Do you know what regulation requires an airplane to have a compass correction card? It's

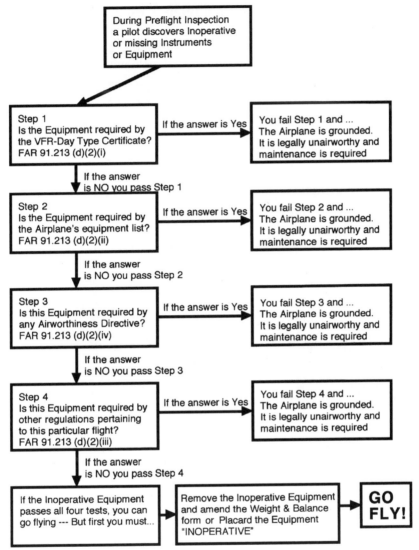

Fig. 10-3. *FAR 91.213. The four-step test.*

CAR Part 3.758. Are airplanes required to have a master switch? Yes, according to CAR 3.688.

There are several others like this that we have taken for granted. So where can these old rules be found? In a library, perhaps, or an FSDO office, or maybe no place. So here they are, the dusty, but still applicable, Civil Aviation Regulations:

CAR 3.655–Required Basic Equipment

The following table shows the basic equipment items required for *type and airworthiness certification of an airplane:*

(a) Flight and navigational instruments

  (1) Airspeed indicator

  (2) Altimeter

  (3) Magnetic direction indicator

(b) Powerplant instruments

  (1) For each engine or tank

    (i) Fuel quantity indicator

    (ii) Oil pressure indicator

    (iii) Oil temperature indicator

    (iv) Tachometer

  (2) For each engine or tank (required in reference section)

    (i) Cylinder head temperature indicator

    (ii) Fuel pressure indicator (if pump fed engines used)

    (iii) Manifold pressure indicator (if altitude engines used)

    (iv) Oil quantity indicator

(c) Electrical equipment (required in reference section)

  (1) Master switch arrangement

  (2) Adequate source(s) of electrical energy

ITEM NO	EQUIPMENT LIST DESCRIPTION
D67-A	Recorder, Engine Hour Meter
D82-S	Outside Air Temperature Gauge
D85-R	Tachometer Installation, Engine
D88-S	Indicator, Turn Coordinator
D88-O	Indicator, Turn & Bank
D91-S	Rate of Climb Indicator

Fig. 10-4. *Sample equipment list. (Not specific to a particular airplane.)*

(3) Electrical protective devices

(d) Miscellaneous equipment

(1) Approved safety belts for all occupants

(2) Airplane flight manual if required by 3.777

Now compare this list to today's FAR 91.205, the regulation that lists the instruments and equipment required to fly a United States registered airplane during the daytime in VFR conditions:

FAR 91.205 (b)

1. Airspeed indicator

2. Altimeter

3. Magnetic direction indicator

4. Tachometer for each engine

5. Oil pressure gauge (each engine)

6a. Temperature gauge (each liquid-cooled engine)

6b. Oil temperature gauge (each air-cooled engine)

7. Manifold pressure gauge (each altitude engine)

8. Fuel quantity gauge for each tank

9. Landing gear position indicator (retractable gear airplanes)

10. Flotation gear and flare gun (flights over water beyond glide-to-shore distance if operated for hire)

11. Safety belts

12. Front seat shoulder harness

13. Emergency locator transmitter

You can see that 91.205 (b) is the offspring of CAR 3.655. Although the lists are very similar, they are not identical. Using FAR 91.205 (b) in Step 1 of the four-step test can still leave the pilot vulnerable. The best way to determine exactly what your airplane requires is to obtain what is called a *type certificate data sheet* (TCDS) for your specific airplane. The TCDS is a document issued by the FAA that describes the aircraft's airworthiness requirements relating to a specific type, make, and model of aircraft. Although the TCDS can be obtained from the FSDO, ask your mechanics first; they probably already have one.

What books and manuals are required to be carried in the airplane? There is always a debate over operating limitations and where these limitations can be found. Different manufacturers publish pilot operating handbooks (POH), and others offer information manuals. Are any of these publications required for airworthiness? Once again, the old CAR 3.655 (d) helps answer the question.

CAR 3.655(d) Miscellaneous equipment

(2) Airplane Flight Manual if required by 3.777.

In regard to an Airplane Flight Manual, CAR 3.655 refers us to CAR 3.777. The old CAR 3.777 states:

> (a) An Airplane Flight Manual shall be furnished with each airplane, having a maximum certified weight of more than 6,000 pounds.

> (b) For airplanes having a maximum certified weight of 6,000 pounds or less, an Airplane Flight Manual is not required. Instead, the information prescribed in this part for inclusion in an Airplane Flight Manual shall be made available to the operator by the manufacturer in the form of clearly stated placards, markings, or manuals.

So while the old CARs were the only regulations to go by, an airplane weighing 6,000 pounds or less did not require an airplane flight manual. But in 1979 the newer FARs were in force, and the regulations changed to include airplanes that were less than 6,000 pounds. FAR Part 21 concerns the certification procedures for aircraft products and parts. It states:

> (a) With each airplane that was not type certificated with an Airplane Flight Manual and that has had no flight time prior to March 1, 1979, the holder of a type certificate shall make available to the owner at the time of delivery of the aircraft a current approved Airplane Flight Manual.

This means that airplanes heavier than 6,000 pounds always required an airplane flight manual (AFM), but after March 1, 1979, all airplanes, regardless of weight must have an approved AFM.

What then constitutes an "approved" airplane flight manual? FAR 23.1581 also spells out the AFM requirements. FAR 23.1581 says that the AFM must be kept in a binder so that pages can be inserted (no bound books) and that the AFM must be kept in a "suitable fixed container" that is readily accessible to the pilot. FAR 23.1581 goes on to say that the AFM must contain a table of contents and must include all that is required in FAR 23.1583, 23.1585, and 23.1589. As you might imagine, these regulations seem to go on forever, so here are the highlights:

The AFM must include information on the airplane's maximum weight, maximum landing weight, and maximum takeoff weights for each altitude, ambient air temperature, and required takeoff distances. There must be center of gravity limits, speeds, flight maneuvers, and kinds of flight operations (IFR, VFR day, VFR night). The AFM must have the airplane's serial number, not just tail number, on the title page, and the entire book must be "updatable."

The owner subscribes to a service whereby the manufacturer will send revision pages to the owner as revisions are made. When the owner receives a revision page, he or she removes the old page and replaces it with the new page. This is why the book must not be bound, but rather loose-leaf in a binder.

This requirement illustrates the difference between an airplane's POH and an approved AFM. The POH is a reprint of the AFM at the time of publication, but the POH is not updated. Any revision that comes out will not be replaced in a POH, and the POH will therefore become obsolete. The POH in most cases is still adequate as a reference book for study purposes and is definitely cheaper than an approved AFM.

So if you are flying an airplane that had no flight time before March 1, 1979, you must fly with an approved airplane flight manual on board or the airplane is to be grounded. I personally double-laminated all my airplane's AFM title pages so that the required serial number page could not accidentally tear out with wear and render the airplane unairworthy.

If any inoperative equipment is one of the items on the type certificate data sheet list or if the AFM is missing, then the airplane cannot legally leave the ground. If the inoperative equipment is not on this list and you have an approved AFM, then, congratulations, you passed the first test and can graduate to the second test.

Step 2

The second step comes from FAR 91.213 (d)(2)(ii) and asks, "Is the inoperative equipment specifically required by the manufacturer in the aircraft's equipment list?" To find this answer, you must go to the aircraft's operating handbook (Fig. 10-4) or other documents that contain the equipment list. After finding the list, you must understand the manufacturer's code.

Some manufacturers append a letter to a part number that signifies that a particular piece of equipment is required for flight on their airplane. For example, in part number R-001, the *R* stands for "required," and therefore, this equipment must be in good working order before flight.

Other letters are also used as codes. *S* can mean "standard." Equipment listed as standard means that the equipment comes with the airplane when it is built, but it can be broken and still legally fly. *A* can mean "additional." Just like options for a car, additional equipment must be ordered separately from the basic airplane. Additional equipment is usually not required for airworthiness, and therefore, an inoperative piece of additional equipment can be flown unless the additional equipment substitutes for a required piece of equipment. These codes can be found in the equipment list of the aiplane's handbook.

Other manufacturers code differently. In Fig. 10-5 the pressure gauge for instrument air has a minus (−) sign under the columns marked VFR day and VFR night. This means that the manufacturer does not require this gauge to be operating for VFR flight. However, note the "1" that appears under the IFR day and IFR night columns. This means that one operating instrument air gauge is required by the manufacturer before the airplane can be flown into IFR conditions.

In the same figure, note the heading "Engine indicating instruments." Here, two engine tachometers are required for any flight. Although listed under the manufacturer's requirements, tachometers for every engine are also a requirement of 91.205. (This is obviously a twin-engine airplane, since two tachometers are required.) There is no regulation or manufacturer requirement, however, for the next entry, "Exhaust gas temperature" indicator. This instrument has an asterisk (*) under every column to indicate that this piece of equipment is optional and can therefore be flown when broken.

Every pilot should become very familiar with the manufacturer-provided equipment list of the airplanes that they fly. The second inoperative equipment test can only be passed with assistance from the POH or FAA-approved AFM.

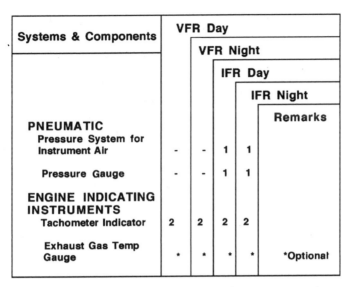

Systems & Components	VFR Day	VFR Night	IFR Day	IFR Night	Remarks
PNEUMATIC					
Pressure System for Instrument Air	-	-	1	1	
Pressure Gauge	-	-	1	1	
ENGINE INDICATING INSTRUMENTS					
Tachometer Indicator	2	2	2	2	
Exhaust Gas Temp Gauge	*	*	*	*	*Optional

Fig. 10-5. *Sample equipment list. (Not specific to a particular airplane.)*

Step 3

The third step comes from FAR 91.213(d)(2)(iv) and asks the question, "Is the inoperative equipment specifically required by any airworthiness directives?" An *airworthiness directive* (AD) is a recall. When parts of aircraft are found to be defective, they must be fixed or replaced. But how do aircraft owners find out about the problem?

Here is the process: A problem with an aircraft part is often determined by the National Transportation Safety Board (NTSB) through similar accident reports, by the FAA, or by the aircraft manufacturers themselves. When a part needs repair or replacement, the FAA is notified and a mailing is sent out to owners of that particular aircraft. The FAA knows who owns the airplanes because of the aircraft registration.

Once the owner gets the word of the AD, he or she must comply with the provisions of the AD. The AD might specify that the aircraft cannot be flown until a repair is made, or it might allow the repair to be made at the next scheduled inspection. The AD might even be recurring, requiring the owner to have a particular item inspected periodically for the life of the aircraft.

If any airworthiness directive states that a particular part or instrument must be operating in a particular airplane for flight, and if that part is in fact found to be inoperative, then the aircraft does not pass Step 3 and must be grounded.

If you are a renter pilot, you might not have ready access to information concerning ADs. The actual AD notice goes to the owner, but you, the pilot, are held responsible for flying only legally airworthy airplanes. If you get ramp-checked by an official of the FAA, how can you prove that the airplane you have flown is completely airworthy? You cannot pass the buck by saying, "I just rented this plane." The pilot is held accountable for the complete airworthiness.

Since a pilot could easily get hung by this situation, it is always best to rent from a dependable FBO that has a good maintenance track record. Spend some time with the chief mechanic and have the mechanic show you the existing ADs of the airplane you intend to rent. Ask about their AD tracking and research system. Educate yourself so that you will know what items and inspections might be required on a particular airplane. After doing some research if you discover that an inoperative piece of equipment is not specifically required by an airworthiness directive, then the equipment passes the third step.

Step 4

The fourth and final step depends on the particular conditions of the flight: "Will the weather conditions, the time of day, or the airspace to be flown in require specific equipment?" If the flight will be at night, the airplane will need instruments and equipment all operational that are listed in 91.205(c) for VFR night flight. If the flight will penetrate IFR flight conditions, even more instruments and equipment are required under 91.205(d) for IFR flight. Certain airspace types, namely airspace A, B, and C, require a Mode C transponder, so when you are flying into one of these airspace types, an altitude reporting transponder becomes required equipment.

Flights above certain altitudes and for specific duration require the availability of supplemental oxygen. For instance, if the flight was planned to fly above 15,000 feet in a nonpressurized airplane, supplemental oxygen for each person would also be required equipment.

The fourth test, listed under FAR 91.213(d)(2)(iii), says that "any other rule" applies. That means regulations 91.205, 91.207, 91.211, 91.215, and others must also be consulted before inoperative equipment can pass the final test.

Let's assume that all four tests are passed. Can the pilot now legally fly the airplane? No, not yet! FAR 91.213(d)(3) says that we are not finished. Before flight the inoperative equipment must:

1. Be removed from the aircraft, or deactivated under 91.213(d)(3)(i) or,

2. Be placarded with a sign that says "Inoperative" under 91.213(d)(ii) and,

3. The discrepancy is recorded in the maintenance records of the airplane under FAR 43.9.

Removing inoperative equipment from aircraft is usually not very practical. In addition, if you take equipment out of the airplane, the weight and balance form for the aircraft will become inaccurate, grounding the airplane anyway. The easiest method is to deactivate the equipment and place a sticker that says "Inoperative" or simply "Inop" over the equipment. Deactivation of electrical equipment can be accomplished by pulling the equipment's circuit breaker and placing a plastic collar around the breaker. This will prevent the breaker from inadvertently being pushed in.

To be completely legal, you might have to fly with a panel of "Inop" placards, which will not instill much confidence in your passengers. And, of course, you would expect the FAA to stipulate the type of sign, and they have not let us down. Advisory Circular 91-67 says the placard must be a label or decal with letters at least ⅛ inch high.

How long can you fly with an "Inop" placard on a piece of equipment? Not forever. FAR 91.405 (c), titled Maintenance Required, says that "the owner or operator of an aircraft shall have any inoperative instrument or item of equipment permitted to be inoperative by 91.213(d)(2) to be repaired, replaced, removed, or inspected at the next required inspection." The next required inspection is usually a 100-hour inspection or an annual inspection. This means items that pass the four-step test and then are properly placarded still have a deadline to be fixed or removed.

Applying the four-step test

The question posed at the beginning of the chapter asked whether you can legally take off with an inoperative suction gauge that monitors the vacuum system. Let's look at how we can apply the four-step test to answer this question.

Step 1 Is a suction gauge (or instrument air gauge) required for VFR flight during the day under part 91.205 and/or a TCDS? After looking down a sample TCDS list, we see that a suction gauge is not on the list of required items, so this gauge passes the first test.

Step 2 Is a suction gauge required by the manufacturer? This gauge is not required for day or night VFR. If the proposed flight will remain in VFR conditions, the second test is passed.

Step 3 Is the suction gauge required due to an airworthiness directive? You'll need to do some research to say for sure. If a visit to the maintenance technician's office did not reveal such an AD, the gauge passes the third test.

Step 4 Is this suction gauge required for the specific conditions of this flight? No regulation requires a suction gauge just because you fly to a certain type airspace or to a specific altitude. As long as this flight remains VFR, the gauge passes the fourth test.

The suction gauge made it all the way through without grounding the airplane. Now the suction gauge must be labeled as inoperative and the discrepancy must be recorded. Now you can take off with the assurance that flight even with this equipment broken is legal.

Ramp checks

The issue of flight with inoperative equipment is a complicated one. With all the paperwork required of an MEL and all the regulations involved with operating without an MEL, the pilot can get confused. But when you taxi in someday and are greeted by a person saying, "Good morning, I'm from the FAA and this is a ramp check," it will be a bad time to be confused. Study your airplane's AFM, the regulations, and apply the four-step test.

A last word on FAA ramp checks. The FAA performs these inspections on three occasions:

1. For normal surveillance

2. When an FAA inspector observes an unsafe operation

3. After an accident or incident

The inspector will ask the pilot for:

1. The pilot's, and any other required crew member's, Pilot Certificate and Medical Certificate.

2. The airplane's Airworthiness Certificate. The airplane's tail number and Airworthiness Certificate numbers must match, and the certificate must be original—no photocopies.

3. The airplane's Registration Certificate with matching numbers. If the airplane's ownership has changed within the past 120 days, there should be a pink temporary registration certificate. If 120 days have passed since the temporary certificate was signed, then the airplane is grounded until the permanent certificate arrives. If this certificate is late, call your local FSDO for assistance.

4. The radio station license (not required for flights in the United States).

5. POH, or an FAA-approved AFM, whichever is required.

6. The current and up-to-date weight and balance data sheet.

The FAA inspector who performs the ramp check must also have identification. He or she should present to the pilot an Aviation Safety Inspector's Credential, FAA Form 110A. This credential allows the holder access to aircraft, airports, and accident sites. In addition, the inspector might need an FAA form 8000-39 to gain access to secured areas of airports without escort. Both these credentials will have the inspector's photo on it.

The inspector will not board the aircraft without permission, but if permission is denied, you can bet that the inspector will write down the airplane's tail number and will then take action to ground that airplane and that pilot so that an inspection can be made at some point. The FAA's policy is not to do blanket ramp checks at air shows, fly-ins, wings weekend seminars, and so forth. The FAA inspector is also not supposed to cause a delay to a flight while doing the ramp check inspection.

It is also possible to get an evaluation from the FAA as to the complete airworthiness of your airplane without threat of enforcement by attending a PACE program. PACE is the Pilot and Aircraft Courtesy Evaluation where the FAA inspectors wear the white hat. During the PACE inspection, an FAA inspector will perform a courtesy ramp check and report to you any problems. This gives the pilot/operator the opportunity to become aware of any items that could have gotten them into trouble. After you complete the PACE program, the inspector gives you a T-shirt that says "I survived PACE"!

11

Weather, flight plans, and decision making

Weather and decision making

Some sources claim adverse weather contributes to as much as 40% of all general aviation accidents. By definition, instrument flying exposes pilots and their passengers to the threats of thunderstorms, turbulence, ice, and reduced visibility. Let's take a brief look at the weather decision-making process and how to get a good weather briefing.

Upon completion of the first phase of flight planning, you must determine if there is anything in the current or forecast weather that might hinder the flight. If anything, weather considerations are simpler for an IFR flight than for a VFR flight. In IFR flying, it's assumed that you'll be in the clouds all the way until breaking out of the overcast during the final approach.

In VFR flying, a good part of the weather analysis is devoted to figuring out where the clouds are, whether to go above, below, or around them, and whether you can safely land at the destination before everything begins closing in. One joy of an instrument rating is setting off on an IFR cross-country when VFR-only pilots are still back at the FBO agonizing over all those VFR weather decisions.

Your decision whether or not to make an IFR flight boils down to five major go/no-go decision factors:

- Thunderstorms
- Turbulence
- Icing

- Fog
- Departure and destination weather minimums

Thunderstorms

Thunderstorms are formed when moisture is combined with rising columns of unstable warm air. The rising warm air cools and the moisture condenses into droplets of rain. Heat is released during condensation, and this additional heat, in turn, increases the speed and power of the rising column of air.

As a thunderstorm begins to tower thousands of feet up into extremely cold temperatures, the rising column of warm air finally cools and a strong column of descending cold air begins to plunge toward the ground outside the core of rising warm air. Rain droplets carried to the top of the thunderstorm are frozen and begin to spill out of the top of the storm as hail.

This pattern is typical of the summertime thunderstorm buildups. But we don't want to fool ourselves with the idea that thunderstorms are just a summertime problem. Thunderstorms can happen in January most anywhere in the country. Anytime you have rising, unstable air and moisture in the atmosphere, thunderstorms can develop.

This is a simplified description of a very complex process; you should learn more about thunderstorms by studying weather-related publications. The FAA book *Aviation Weather* has a particularly good chapter on thunderstorms and good discussions about turbulence, icing, and fog.

What does all this mean to an instrument pilot? The point is that thunderstorms contain powerful columns of rising warm air surrounded by equally powerful columns of descending cold air.

The updraft in a mature thunderstorm cell might exceed 6,000 feet per minute (fpm). Structural limits of most general aviation aircraft might be exceeded when the aircraft passes through the shear between the updraft and downdraft. The vicinity of a thunderstorm must be avoided at all times. The odds are against you in a lightplane.

Hail

Other thunderstorm dangers are not so obvious. When hail spews out of the anvil shape at the top of a thunderstorm, it can come down in clear air as far as 8–10 miles from the storm. A good rule of thumb to avoid getting knocked out of the sky by a shaft of hail is to circumnavigate a towering thunderstorm by 20 miles or more.

A wide circumnavigation will also keep the aircraft outside the *gust front* that rings a mature thunderstorm. The gust front is an area of heavy turbulence caused by the descending currents of cold air reaching the ground and spreading outward. These turbulent, descending currents are called *downbursts*. The gusting cold air currents are a cause of *low-level wind shear* and can transform an otherwise routine instrument approach into a disaster or, at best, a hostile environment that you have to fight all the way.

Embedded thunderstorms

Of particular concern to the IFR pilot are embedded thunderstorms. As the name implies, embedded thunderstorms are hidden by the low-level clouds associated with frontal systems. As a front moves, its wedgelike leading edge forces columns of air aloft, creating columns of rising warm air. The moisture present in the clouds, plus the rising warm air, make for an ideal thunderstorm scenario. Surrounded by clouds, the embedded thunderstorm cannot be spotted visually, and you could penetrate one without warning.

The possibility that you might encounter an embedded thunderstorm can be predicted by meteorologists with fair accuracy. The National Weather Service (NWS) systematically measures air stability (the *lapse rate*), temperature, and humidity at various altitudes, and analyzes these variables along with other factors to develop forecasts.

Microbursts

Also associated with thunderstorms are microbursts—powerful downdrafts caused by descending cold air outside the column of ascending warm air in the core of a storm cell. As these cold downdrafts reach the surface, they produce sudden vertical and horizontal wind shears. The downdrafts can reach 6,000 feet per minute. Horizontal winds near the surface can reach 45 knots, resulting in a 90-knot wind shear from headwind to tailwind for a plane taking off or landing, as shown in Fig. 11-1.

Low-level wind shear detection systems are now in place at many airports. "The early detection of a wind shear/microburst event, and the subsequent warning(s) issued to an aircraft on approach or

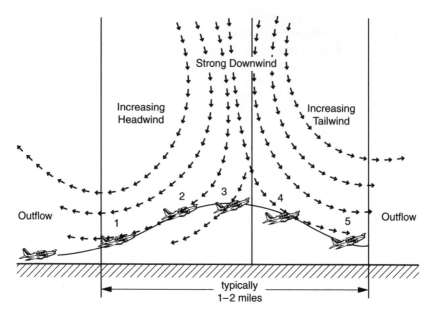

Fig. 11-1. *A microburst during takeoff. The quick shift from headwind to tailwind can reduce performance enough to risk an impact.*

departure, will alert the pilot/crew to...a situation that could become very dangerous!" says AIM.

While this information is of great value in deciding whether or not to delay a takeoff or a landing, it doesn't help much in planning a flight. AIRMETS and SIGMETS do contain advisories about low-level wind shear. But microbursts cannot be predicted as yet, and they are of such short duration that there is not much point in reporting them—they'll be gone by the time the METAR gets out. Avoid flying in areas of thunderstorm activity and you will stay clear of microbursts.

Thunderstorm forecasts and reports

If thunderstorms are forecast, this will always be stated in the terminal and area forecasts that are reported during a weather briefing.

Thunderstorms are also included in AIRMETs and SIGMETs (Airman's Meteorological Information and Significant Meteorological Information). These advisories are issued to amend area forecasts and to announce weather that is significant or potentially hazardous to flight. The advisories cover thunderstorms, turbulence, high winds, icing, low ceilings, and low visibilities.

AIRMETs concern weather of less severity than weather covered by SIGMETs. AIRMETs and SIGMETs will normally be part of all weather briefings. If not, ask for them. The news they contain—bad or good—is important to making your go/no-go decision.

Ground-based weather radar accurately pinpoints developing storms and tracks them as they grow in size and strength. Air traffic control radar, on the other hand, blocks out weather to a large extent to follow air traffic more precisely. It is important to understand this distinction because air traffic controllers cannot always provide deviation information to get around thunderstorms. Thunderstorms don't always show up clearly on traffic control radar.

Data from specialized weather radar locations is plotted on radar summary charts. The information from these charts can help you during preflight planning to plot a course that is clear of the thunderstorms.

With centralization of flight service stations (FSSs) and the increasing reliance on telephone briefings, many pilots obtain weather information without ever seeing the printed versions of radar summary charts, surface analyses, area and terminal forecasts, hourly weather reports, and other reports and charts provided by the National Weather Service.

Nevertheless, you must know what information these different reports and charts contain. Your knowledge should be sufficient so you can visualize what the weather briefer is reading to you over the telephone and you should know what to ask for if you need special information. There are also many questions about aviation weather services on the instrument written examination.

Purchase a copy of the FAA's book *Aviation Weather Services* and study the descriptions of forecasts, reports, and charts relevant to lightplane IFR flying.

Locate the FSS serving your area and visit it. The FSS personnel will show you the material they use for weather briefings and discuss it.

If thunderstorms are forecast for any portion of your flight—or they are developing under the watchful eye of a weather radar—choose a different route or destination or don't make that flight. With all the assistance available there is no excuse for flying into a thunderstorm, whether VFR or IFR. During rapidly increasing thunderstorm activity, a *severe weather avoidance plan* (SWAP) might be in effect to help all aircraft fly routes with more favorable weather conditions.

Turbulence

While you can always count on getting a rough ride in turbulence in and around thunderstorms, turbulence can also occur along fast moving weather fronts and around intense high- and low-pressure boundaries.

For the instrument pilot, avoiding turbulence takes on another dimension. It's not just a matter of avoiding severe and extreme turbulence that might cause structural damage to your aircraft. You must also consider the effect of light and moderate turbulence on personal performance as an instrument pilot.

It is much harder to hold headings and altitudes while flying by reference to instruments when the pilot is also buffeted by turbulence. Maximum attention and a lot of work by the pilot are required to maintain heading and altitude in IFR flight in turbulent conditions. Fatigue sets in much more quickly than on a smooth flight.

Fatigue is not just uncomfortable, it is a very real danger. As the AIM points out in the chapter "Medical Facts for Pilots": "Fatigue continues to be one of the most treacherous hazards to flight safety, as it might not be apparent to a pilot until serious errors are made."

Turbulence in clear air presents a problem, too. If you have to contend with turbulence over a long stretch during the VFR portion of an IFR flight, you will be fatigued, tense, and less confident when descending into the clouds to make an instrument approach. Tired, tense, and anxious is not the best way to commence a good instrument approach.

High winds

Turbulence is also associated with high winds. Expect light turbulence when winds of 15–25 knots are forecast or reported; moderate turbulence in winds of 25–50 knots. Base the go/no-go decision on whether or not you can handle winds of up to 50 knots and the associated turbulence. Above 50 knots lies the realm of severe and extreme turbulence. Turbulence in this range can cause structural damage to a lightplane, as well as slam you around unmercifully. Don't try it!

If there is a possibility of light or moderate turbulence during the flight—and particularly if it looks as if the turbulence might last for a long time—decide if you will be able to handle an instrument approach in lousy weather conditions and keep headings and altitudes under control. If there is any doubt, plan a shorter flight, or pick a destination for a VFR approach and landing, or wait for another day.

Terminal and area forecasts contain wind information; the area forecast specifically details turbulence. AIRMETs and SIGMETs detail high winds and turbulence.

A further aid is the surface weather analysis chart that you became so familiar with during your VFR training. If the isobars are tightly packed with a very small amount of space or no space between them, expect high velocity winds and turbulence.

A pilot weather report (PIREP) is even more conclusive evidence of turbulence. If the pilot of an aircraft similar to yours reports turbulence along your route, take it seriously. Here is someone who has just been through a troublesome or even hazardous experience and is telling it like it is—something no one in a weather station on the ground can ever do.

If a pilot reports turbulence along the route of flight, pick a different route or a different altitude or cancel the flight if you don't think you can handle it. While airborne, don't hesitate to ask the controller to solicit PIREPs on turbulence—or icing or any other conditions that might affect the safety of the flight. Likewise, do not hesitate to report any turbulence as soon as it is safe to do so—fly the airplane first in moderate or extreme turbulence. Other airmen and the controllers will be grateful for the PIREP.

Icing

The situation is different with icing conditions because it is almost impossible to forecast icing with certainty. But the conditions that produce icing are well known and if these conditions are present, assume that the airframe will pick up a load of ice—make other plans, either a different route of flight or alternate transportation.

Aircraft structural icing requires moisture and below-freezing temperatures. But ice might form on the aircraft when flying inside clouds that are above the freezing level. If the freezing level is at or near the ground, icing might occur immediately upon entering clouds after takeoff.

Consider this for a moment. Takeoff and departure are among the busiest and most intense moments of an IFR flight. You are making the transition from visual to instrument flight—sometimes very quickly if the departure ceiling and visibility are low. You are switching from tower to departure control and departure control might well have a clearance amendment. You might not be aware of it, but every human sense is alert to abnormal sounds from the engine, unusual control pressures, and unusual instrument readings—is this any time to be worrying about ice?

The most dangerous form of icing occurs when a mass of rainy, warm air overlies or overrides a cold air mass. As the warmer rain falls through the colder air mass below, the rain becomes cold enough to freeze upon impact with the surfaces of the aircraft. The cold droplets hit, splatter, and coat the surfaces they strike with layers of ice that can build up rapidly.

This rapid buildup can add hundreds of pounds of weight to a small aircraft very quickly. The ice destroys aerodynamic characteristics of wings, control surfaces, and propellers. In extreme cases, 2–3 inches of ice can form on the leading edge of an airfoil in fewer than five minutes. It takes only a half-inch of ice to reduce the lifting power of some aircraft by 50 percent. No doubt about it—icing can be lethal.

Freezing level

If a flight is going to be in the clouds, select an altitude that is beneath the freezing level and avoid areas where warm, moist air overlies colder air. Where do you get information on the freezing level and icing? The area forecast is the first place to look. The National Weather Service "sounds" the atmosphere for temperatures at hundreds of locations throughout the United States and plots freezing levels from these soundings.

Once again the most reliable information comes from PIREPs. If a pilot reports icing along the proposed route of flight, pick a different route or cancel the flight. The pilot isn't guessing or forecasting about this hazard; the pilot can actually see the ice building up on the aircraft. This is much more useful than a forecast, which can only predict conditions favorable for icing.

(If ATC issues a clearance—prior to the flight or while in flight—and you have determined that icing conditions are possible in that area prescribed by the clearance, do not accept the clearance, and explain why by citing, if necessary, the weather information that you obtained and should have written on the flight-planning form.)

Fog

If you are on the ground watching a wall of fog move in, the fog seems to have a sinister, menacing appearance. And with good reason. In a matter of minutes, fog can reduce visibility to zero. In a thick "pea soup" that might be encountered along the coast of New England, fog can be so dense that you literally cannot see where you are walking. This can be very bad news for a pilot, particularly if you are unprepared for it and the fog comes as a surprise.

The best preparation is to understand the conditions that produce fog and then avoid those conditions. Fortunately, this is not very complicated. Fog forms either when air is cooled to its dew point or when the dew point is raised by the presence of additional moisture. Raising the dew point is usually accomplished by the evaporation of water from falling precipitation or by the passage of a body of air over a wet surface.

Fog classifications are: radiation, advection, upslope, precipitation-induced, and ice. Develop a better understanding about fog by studying weather-related publications.

Temperature/dew point spread

Dew point is the temperature at which air is saturated and the water vapor begins to condense and produce visible moisture; dew point is expressed as a temperature. If the air temperature drops to the dew point temperature, fog forms. This is the most common situation. Along coastal areas, however, moist air moving in from the sea can raise the dew point and produce fog.

Because fog forms so low to the ground, there is nothing like it for closing an instrument runway or an entire airport. Fog can form rapidly, making it especially hazardous. You can start an instrument flight with sufficient ceiling and visibility at the destination to make a comfortable approach, then arrive and find a blanket of fog that makes it impossible to land.

Hourly aviation weather reports will always indicate the presence of fog. The hourly reports also contain temperature and dew point. The difference between the two numbers is the temperature-dew point spread, or simply "the spread" as it is commonly called.

When the spread is about 5°F, be alert for fog. For example, if the temperature is 48°F in the most recent report for the destination airport, and the dew point is 43°F, be prepared for fog on arrival. The possibility of fog also appears in terminal forecasts.

If the destination is socked in with fog, do not depart until the fog begins to lift. Likewise, if the spread is narrowing, according to reports, and drops to 5°F or less, do not depart until the situation begins to improve.

Departure and destination weather minimums

Fog is only one of many reasons for low ceilings and visibilities. Other factors are rain, snow, haze, smoke, low-pressure areas, and frontal systems. Ask yourself whether or not you will be able to land upon arrival and determine—according to regulations—if you need an alternate airport and what the alternate might be. And there is a third item relating to minimums that many pilots overlook: Is there sufficient ceiling and visibility at the *departure* airport to return for an immediate landing, if necessary?

Let's take this third item first. Basically, under FAR 91, you can take off in any weather—including zero-zero—if you're not carrying passengers or cargo for hire. See FAR 91.175 (f). Read the regulation closely and notice that takeoff minimums apply to FAR Part 121, 125, 127, 129, and 135 operations only. Part 91 operations are not covered by this regulation.

A takeoff in conditions approaching zero-zero is certainly not very smart. What do you do if a door pops open or the engine doesn't sound right? Where can you land?

The departure alternate

Two answers to this question should always be part of flight planning. First, do not depart on an instrument flight if the ceiling and visibility at the departure airport are too low for a safe return in an emergency, possibly requiring an instrument approach.

Or second, if the ceiling and visibility at departure are too low to return, select a departure alternate—a nearby airport with adequate ceiling and visibility for a safe arrival. Establish a route to that field and study the ATC frequencies and details of the approach as if the airport were the final destination. In other words, be completely prepared to land at this departure alternate if necessary.

If departing from an airport that has no instrument approach, select the second solution. Pick out the nearest field with an instrument approach plus a ceiling and visibility that are greater than your "personal minimums" (which are subsequently described in this chapter) and be prepared to make that approach. It's certainly much smarter to have an "out" or alternative when taking off under IFR conditions, no matter what type of airport you are departing from.

Destination minimums

After analyzing the departure situation, you must next anticipate the weather at the destination airport. Now let's take a closer look at the meaning of minimums. You are familiar with ceilings and visibilities from VFR training; in IFR flying, they take on additional importance.

The *Aeronautical Information Manual* (AIM), published by the FAA, defines ceiling and visibility in the glossary. Take time to browse through all definitions in the AIM glossary because it contains a gold mine of information relating to IFR operations, and each nugget is as clear and concise as anything you will ever find.

Ceiling is the height above the earth's surface of the lowest layer of clouds or obscuring phenomena that is reported as broken, overcast, or obscuration, and not classified as thin or partial.

Visibility

The AIM "Pilot/Controller Glossary" lists and defines several visibility classifications in increasing order of precision.

Ground visibility. Prevailing horizontal visibility near the earth's surface as reported by the National Weather Service or an accredited observer.

Prevailing visibility. The greatest horizontal visibility equaled or exceeded throughout at least half the horizon circle, which need not necessarily be continuous.

Runway visibility value (RVV). The visibility determined for a particular runway by a transmissometer. RVV is used in lieu of prevailing visibility in determining minimums for a particular runway.

Runway visual range (RVR). An instrumently derived value...that represents the horizontal distance a pilot will see down the runway from the approach end...RVR, in contrast to prevailing or RVV, is based on what a pilot in a moving aircraft should see looking down the runway. RVR is used in lieu of RVV and/or prevailing visibility in determining minimums for a particular runway. Which definition is used when flight planning?

Use the most precise visibility measurement available. Weather reports for airports with ILS approaches will have a ground visibility and an RVV or RVR for the runway in use. Visibilities at airports with ADF, VOR, or other nonprecision approaches will usually be given in the less precise prevailing visibility.

Importance of the visibility minimum

Why does the FAA go to such great lengths to define visibility?

The answer is quite clear: "No pilot operating an aircraft...may land that aircraft when the flight visibility is less than the visibility prescribed in the standard instrument approach procedure being used." (FAR 91.175 (d). In other words, you cannot land if the visibility is below minimums.

Furthermore, according to another section of the same FAR, you cannot descend below a minimum altitude if the visibility prescribed for the approach is less than minimums. There is a good reason for this because it is possible to be clear of clouds upon reaching altitude minimums, but you might be unable to see far enough down the landing runway to make a safe approach because of fog, rain, snow, or other runway-obscuring condition.

Understand from the beginning that *visibility*, not ceiling, determines whether or not you can initiate an instrument approach.

Let's continue planning the flight to Binghamton, New York. Which ceiling and visibility minimums apply to this flight? Turn first to the approach chart for ILS RWY 34 to Binghamton and look at the landing minimums section (Fig. 11-2).

Approach categories

Notice categories A, B, C, and D. These are "aircraft approach categories" that are based upon actual final approach speeds. Category A is for approach speeds from 0 (for helicopters) to 90 knots. Unless you are training for an instrument rating in a Learjet or something equally exotic, category A will be used on all training flights. (For further information on aircraft approach categories, refer to the first page of any set of NOS approach charts.)

Because you plan to make the ILS 34 approach, use the minimums listed for the straight-in approach (S-ILS 34): 1797/24. Numbers after the minimums have specific information:

- 200: The height above touchdown (HAT) at the ceiling minimum
- 200-½: The ceiling and RVR converted to prevailing visibility in statute miles

The ceiling and visibility minimums for planning this flight are the third set of numbers: 200 feet and one-half mile. Now check the other Binghamton approach charts to determine if any have lower minimums; shifting winds might change arrival runways. There are five instrument approaches to Binghamton. No minimums are lower than those for ILS 34, although the minimums for ILS 16 are close at 300 feet and one-half mile.

To make a go/no-go decision about this flight, look for a ceiling of no less than 200 feet and a visibility of no less than one-half mile. If ceiling and visibility are forecast to be lower than this at the esti-

CATEGORY	A	B	C	D
S-ILS 34		1797/24 200 (200-½)		
S-LOC 34		1960/24 363 (400-½)		1960/40 363 (400-¾)
CIRCLING	2060-1 424 (500-1)	2100-1 464 (500-1)	2100-1½ 464 (500-1½)	2200-2 564 (600-2)

Fig. 11-2. *Landing minimums for Binghamton ILS RWY 34 approach, as shown on the NOS approach chart.*

mated time of arrival (ETA), you will probably not be able to land. It is perfectly legal to file IFR and fly to Binghamton and attempt an approach when the weather is forecast to be below minimums (air carrier pilots are not allowed to do this), but what's the point of making the trip if you can't land?

Well, you might go to Binghamton and try an approach just to see if the weather had improved enough to land. You might get lucky! On the other hand, the weather could be much worse upon arrival and a landing might be impossible. This can happen on even the best-planned flights, especially in winter when weather systems pick up speed and sweep across the country much faster than in the summer.

You must pick an alternate airport for a safe landing if you can't get into Binghamton. This is not just a good idea, it's required by FAR 91.169. You might have to pick an alternate even if Binghamton is forecast to be VFR upon arrival. Many people are fooled by this because they think, "The airport is going to be VFR when I get there so I don't need an alternate." That's not true. To file to an airport IFR without listing an alternate, you need twice the ceiling required for VFR. Review FAR 91.169 for a thorough understanding of its impact.

The "one, two, three rule"

If, for one hour before ETA through one hour after ETA, the ceiling and visibility are forecast to be 2,000 feet and 3 miles, according to FAR 91.169 (b), a pilot must select an alternate airport and include the selection on the flight plan. That is the "one, two, three rule": *one* hour, *two* thousand feet, and *three* miles.

Another rule of thumb to help remember the alternate airport requirements is: You must have VFR conditions plus 1,000 feet for ETA ± 1 hour or you must file an alternate.

Selecting an alternate

How do you pick an alternate airport? Obviously, you can't pick an alternate that is so far away you'll run out of fuel before arrival. And there's not much point in picking an alternate where the weather is so bad it prevents landing.

The FARs are grounded in good common sense on these two points. First, FAR 91.167 says you must carry enough fuel on an IFR flight to:

- Complete the flight to the first airport of intended landing
- Fly from that airport to the alternate, if one is required, and
- Fly after that (the alternate airport) for 45 minutes at normal cruising speed

You can list an airport as an alternate only if the ceiling and visibility forecast for the alternate at your time of arrival will be at or above 600 feet and 2 miles (if the airport has a precision approach), or 800 feet and 2 miles (if it has only an ADF, VOR, or other nonprecision approach).

There might be a catch to this, however. Some airports might not be authorized for use as alternates, while others available might have *higher* minimum requirements than 600/2 and 800/2 because of local conditions such as hills, towers, or radio towers. How can you find this out?

Turn to the "E" section of your set NOS approach charts and find the listing of "IFR Alternate Minimums." An airport with minimums that deviate from the standard 600/2 and 800/2 will be listed in this section (Fig. 11-3). Note that an airport such as Farmingdale Republic is not allowed for use as an alternate when the control tower is not in operation.

If the airport selected as an alternate is listed in this section, use the minimums in this section. If the choice of an alternate does not appear in this section, use the 600/2 and 800/2 minimums discussed above.

Personal minimums

Just as most airlines qualify their crews to fly certain minimums, many competent instrument pilots set higher minimums for themselves than the published minimums. This is very smart! A brand new instrument pilot might want to start out with a 1,000-foot ceiling and 3 miles visibility (VFR minimums) until gaining more experience and confidence. (It's kind of scary being up there in the clouds without your friendly flight instructor!) This can then be lowered, depending upon frequency of flights, until reaching the lowest minimums available.

 ALTERNATE MINS

98169

INSTRUMENT APPROACH PROCEDURE CHARTS

 IFR ALTERNATE MINIMUMS

(NOT APPLICABLE TO USA/USN/USAF)

Standard alternate minimums for non precision approaches are 800-2 (NDB, VOR, LOC, TACAN, LDA, VORTAC, VOR/DME or ASR); for precision approaches 600-2 (ILS or PAR). Airports within this geographical area that require alternate minimums other than standard or alternate minimums with restrictions are listed below. NA - means alternate minimums are not authorized due to unmonitored facility or absence of weather reporting service. Civil pilots see FAR 91. USA/USN/USAF pilots refer to appropriate regulations.

NAME	ALTERNATE MINIMUMS
ALBANY, NY	
ALBANY INTL	ILS Rwy 1[1]
	ILS Rwy 19[1]
	VOR/DME or GPS Rwy 1[2]
	VOR Rwy 1[3]
	VOR or GPS Rwy 19[2]
	VOR or GPS Rwy 28[2]

[1]ILS, Categories B, C, 700-2; Category D, 800-2¼. LOC, Category D, 800-2¼.
[2]Category D, 800-2½.
[3]Category C, 800-2¼; Category D, 800-2½.

ALLENTOWN, PA
LEHIGH VALLEY INTL .... ILS Rwy 13
ILS, Categories A,B,C, 700-2; Category D, 700-2¼. LOC, Category D, 800-2¼.

ALTOONA, PA
ALTOONA-BLAIR COUNTY .... ILS Rwy 20[1]
VOR or GPS-A[2]
[1]Categories A,B, 900-2;Category C, 900-2½; Category D, 1100-3.
[2]Category D, 1100-3.

BATAVIA, NY
GENESEE COUNTY .... VOR/DME or GPS-A
Category D, 800-2¼.

BRADFORD, PA
BRADFORD
REGIONAL .... VOR/DME or GPS Rwy 14
NA when BFD FSS closed.

CORTLAND, NY
CORTLAND COUNTY-
CHASE FIELD .... VOR or GPS-A
Categories A,B, 1100-2,Categories C,D, 1100-3.

NAME	ALTERNATE MINIMUMS
DUBOIS, PA	
DUBOIS-JEFFERSON COUNTY	ILS Rwy 25
LOC, NA.	

ELMIRA, NY
ELMIRA/CORNING REGIONAL .... ILS Rwy 6
ILS Rwy 24
NDB Rwy 24
Categories A,B, 1300-2; Categories C,D, 1300-3.
NA when control tower closed.

ERIE, PA
ERIE INTL .... ILS Rwy 6[1]
ILS Rwy 24[1]
NDB Rwy 6
NDB Rwy 24
RADAR-1
NA when control tower closed.
[1]ILS, 700-2.

FARMINGDALE, NY
REPUBLIC .... ILS Rwy 14
NDB or GPS Rwy 1
NA when control tower closed.

HARRISBURG, PA
CAPITAL CITY .... ILS Rwy 8
Categories A,B, 900-2; Categories C,D, 900-2¾.
NA when control tower closed.

HARRISBURG INTL .... ILS Rwy 13[1]
ILS Rwy 31[1]
VOR or GPS Rwy 31[2]
[1]ILS, Categories C,D, 700-2. LOC, NA.
[2]Categories A,B, 900-2, Category C, 900-2¾, Category D, 900-3.

 ALTERNATE MINS

98169

Fig. 11-3. *If IFR landing minimums are other than 600-2 for precision approaches and 800-2 for nonprecision approaches, they will be listed in this IFR Alternate Minimums section of the NOS instrument approach procedures booklet.*

If you don't fly very often and are barely meeting currency requirements, it might be prudent to raise personal limits to 500 and 2, for example. Many pilots make it a policy to fly instruments every three or four weeks. Many also sign up for instrument refresher instruction every three months if they feel a little rusty.

To maintain currency as an instrument pilot, regulations require 6 instrument approaches plus "holding procedures" and "intercepting and tracking courses through the use of navigation systems" each 6 months. Does legally current mean that you are competent to fly the published minimums? *Absolutely not.* Set personal minimums with which you are comfortable and with which you feel confident.

Weather factors reviewed

This brief review should help you focus on necessary elements of a weather briefing:

1. Are there any weather conditions that might make it difficult or impossible to complete the flight as planned: thunderstorms, turbulence, icing, or fog?

2. Are the ceilings and visibilities high enough at departure and destination airports for a safe IFR takeoff and for a safe (and legal) IFR approach and landing? Does the weather meet your personal minimums?

3. Is an alternate needed? If so, what alternate airports can you reach that meet the minimum requirements for ceiling and visibility?

Look again at the flight log form and notice in the lower right-hand corner the words "VFR WX AT" (Fig. 11-4). This item is not found in any FAA publications; it is based upon years of IFR flying and IFR instruction.

When the chips are down, where can you find VFR conditions? Where can you go for a safe VFR landing if all electrical power—radios, transponder, electrically powered instruments, pitot heat, lights—is lost? Where can you go if an instrument flight becomes horrendous?

Always learn where the nearest VFR conditions are to safely abort the IFR flight and land VFR. A weather briefer will provide this in-

VFR WX AT:

	TANK ON	TANK OFF		TANK ON	TANK OFF

Fig. 11-4. *Space for noting nearest VFR weather on flight log form.*

formation if requested. If there is no VFR weather at an airport within range, don't go.

"When in doubt, wait it out!"

Weather briefings and flight plans

It wasn't too long ago that getting the weather meant talking directly to a weather briefer, either by telephone or during a personal visit to a Flight Service Station (FSS). Pilots lucky enough to have an FSS located at their airport could go in and examine all the weather maps, forecasts, and reports, then talk to a weather briefer in person.

For most pilots this personal contact is a thing of the past. The FAA has transitioned to a centralized network of Automated Flight Service Stations (AFSSs) that provide much more service than has been available previously from the nonautomated stations. FAA-funded Direct User Access Terminal Service (DUATS) now allows computer users to get complete "official" aviation weather briefings directly, as well as file IFR flight plans. The bottom line, as always, is money. The FAA simply does not have the budget to operate the number of Flight Service Stations it once had, nor can it justify the cost of one-on-one personal briefings for everyone.

Computer weather services: the "big" picture

Now is the time when a computer is handy for the "big picture." If you are a member of Aircraft Owners and Pilots Association (AOPA) and have an Internet connection, call up <www.aopa.org> and check out the weather pages offered at this Web site. With color maps and up-to-the-minute area and terminal forecasts and reports and much more, the AOPA Web site is a superb resource. Place a bookmark or placemark at this location so you can go quickly to it whenever you wish.

If you are not a member of AOPA, another good weather Web site source is the Weather Channel at <www.weather.com>. This site presents free weather in greater detail than what you see on the Weather Channel's TV report—and you don't have to wait until what you want to see cycles around again after the commercials! A special feature allows you to customize the information by location. Another Web site worth exploring is AccuWeather at <www.accuwx.com>. This site provides free color weather maps with a five-hour time delay. If you take out a subscription to the service, you will get real-time information, plus access to 35,000 weather products. Try the "five hours free" offer and see if it might not be worthwhile for you to subscribe to this excellent service. The Internet offers many other aviation-related sites. So surf the 'net and place bookmarks or placemarks at those sites that best serve your interests. With a little practice you will be surprised at how quickly you can go right to what you want.

Many professional pilots make a habit of checking the "big picture" on the weather *every day*, regardless of flight schedules. For instrument students, this habit will improve your understanding of how weather phenomena and patterns develop and why. These are excellent teaming experiences, and much of the best information is free.

DUATS

On the morning of a flight, check local TV stations and Internet sites again to see how the weather has changed overnight. Cross-check with local newspaper weather reports and maps. Are you ready to make a "go/no-go" decision? If so, and you have a computer with a modem, your next best move is to get a complete aviation weather briefing by DUATS for your departure, destination, and route of flight.

DUATS can be accessed toll-free 24 hours a day by pilots in the 48 contiguous states with current medical certificates. DUATS provides alphanumeric preflight weather data, NOTAMS, and information on traffic delays, which can be printed out easily for later reference and study. The two DUATS providers also offer free aviation weather graphics.

DUATS will also file IFR flight plans; so it is truly a one-stop service. A record is kept of all briefings, thus DUATS provides solid evidence of compliance with FAR 91.103 (a)—which is not possible with any other computer weather services.

DUATS is provided free of charge by these two commercial services under contract to the FAA:

DTC (Data Transformation Corporation)
1-800-245-3828—Modem access to weather briefings
and filing flight plans
1-800-243-3828—Regular help line for customer service and
information

GTE (Contel)
1-800-767-9988—Modem access to weather briefings and
filing flight plans
1-800-345-3828—Regular help line for customer service and
information

TAFs and METARs

Printing out the coded weather from DUATS provides an additional benefit. The printouts are ideal study guides for mastering TAF and METAR codes. TAFs (Terminal Aerodrome Forecasts) are airport forecasts; METARs (Meteorology Aviation Routines) are hourly and special weather reports by location. If it has an "F," it's a forecast. If it has an "R," it's a report. That's the easy part.

Let's look at the actual METAR and TAF reports as supplied by DUATS for Washington National Airport (DCA) on a rainy winter day. There is a pattern to all those numbers and letters, and once you understand the pattern, you should have no difficulty reading the reports and forecasts. To understand the following METAR, let's break it into the basic blocks in which it is organized:

METAR KDCA 031551Z 33005KT 9SM -RA SCT046 BKN060 OVC080 08/02
A3023 RMK AO2 RAB41 PRESRR SLP235 P0000 T00780017

- METAR KDCA—Type of report and station location. In this case, a METAR reported from Washington National Airport (DCA). (The "K" is an international designator for the 48 contiguous states.)

- 031551Z—Date and time of report: day 3 of the month of the current month, 1551 Zulu time
- 33005KT—Wind from 330 degrees at 05 KTS
- 9SM—Visibility 9 statute miles
- -RA—Light rain
- SCT046—Scattered clouds at 4,600 feet
- BKN060—Broken clouds at 6,000 feet
- OVC080—Overcast clouds at 8,000 feet
- 08/02—Temperature/dew point in degrees Celsius
- A3023—Altimeter setting 30.23

Next comes the remarks section:

RMK AO2 RAB41 PRESRR SLP235 P0000 T00780017

- RMK A02—Remarks, Automated Observation from location type 2 that can discriminate between rain or snow. An Automated Observation with no rain/snow discrimination would be designated AO1
- RAB41—Rain began at 41 after the hour
- PRESRR—Pressure rising rapidly
- SLP235—Sea level pressure 1023.5 millibars
- P0000—Precipitation less than one hundredth of an inch in the last hour
- T00780017—Temperature 7.8° Celsius, dew point 1.7° Celsius in the last hour, not rounded off

The remarks section of METARs is intended to refine the data reported in the main sections and to provide additional information, such as the status of equipment. This is useful—sometimes critical—information, but it can be hard to decipher. Both AIM, Chapter 7, and *Aviation Weather Services*, Chapters 2 and 4, cover the fine points of METARs and TAFs in considerable detail. If you always make it a habit to look up the codes you don't understand, you will soon see that the same abbreviations and sequences are used over and over again.

Now let's look at the TAFS, the forecasts, for Washington National on the same rainy day covered by the METAR above. TAFs are simpler to decode than METARS; they use the same sequence of groups as METARS—wind, visibility, significant weather, clouds. A TAF usually covers a 24-hour period. As you work through the TAF below, you will see that it is composed of a series of simplified METARS:

TAF KDCA 031130Z 031212 VRB03KT P6SM SCT100 OVC200 TEMPO 12214 BKN100

FM1400 35007KT P6SM SCT070 OVC100

FM1800 01010KT P6SM SCT040 OVC080 PROB40 1822 -RA BKN040

FM2200 02013KT P6SM BKN040 TEMPO 2202 4SM -RA BKN025

FM0200 02014KT 4SM -RA OVC025

FM0600 03015G25KT 3SM RA BR OVC012

- TAF KDCA—Type of report and station location. In this case, a TAF forecast for Washington National Airport (DCA). (The "K" is an international designator for the 48 contiguous states.)
- 031130Z—Date and time the forecast is actually prepared: day 3 of the current month, 1130 Zulu time
- 031212—Date and time of the beginning of the forecast's validity: day 3 of the current month, 1212 Zulu time
- VRB03KT—Wind variable at 03 knots
- P6SM—Visibility greater than 6 statute miles. (The "P" stands for "plus.")
- SCT100—Scattered clouds at 10,000 feet
- OVC200—Overcast clouds at 20,000 feet
- TEMPO 1214 BKN100—Temporarily at 1214 Zulu: broken clouds at 10,000 feet
- FM1400 35007KT P6SM SCT0700 OVC1000—From 1400 Zulu, wind 350 at 07 knots, visibility greater than 6 statute miles, scattered clouds at 7,000 feet, overcast clouds at 10,000 feet
- FM1800 01010KT P6SM SCT040 OVC080 PROB40 1822 -RA BKN040—From 1800 Zulu, wind 010 at 10 knots, visibility greater then 6 statute miles, overcast clouds at 8,000 feet, probability 40% at 1822 Zulu: light rain, broken clouds at 4,000 feet
- FM2200 02013KT P6SM BKN040 TEMPO 2202 4SM -RA BKN025—From 2200 Zulu, wind 020 at 13 knots, visibility

greater than 6 statute miles, broken clouds at 4,000 feet, temporarily at 2202 Zulu: visibility 4 statute miles, light rain, broken clouds at 2,500 feet

- FM0200 02014KT 4SM -RA OVC025—From 0200 Zulu, wind 020 at 14 knots, visibility 4 statute miles, light rain, overcast clouds at 2,500 feet
- FM0600 03015G25KT 3SM RA BR OVC012—From 0600 Zulu, wind 030 at 15 knots gusting to 25 knots, visibility 3 statute miles, moderate rain and mist ("BR"), overcast clouds at 1,200 feet

There is much additional aviation weather and information available from DUATS beside METARs and TAFs. I suggest that you print out one full briefing so that you can improve your understanding of how Flight Service reports items as area forecasts (FAs), Winds and Temperatures Aloft forecasts (FDs), AIRMETS, SIGMETS, pilot reports (UAs), radar weather reports (SDs), NOTAMs, and ATC delays and advisories.

Forecast reliability

When planning an instrument flight always ask: How good are the weather forecasts?

Pilots should understand the limitations and capabilities of present-day weather forecasting. Don't be lulled into complacency by fancy weather graphics and four- and five-day forecasts! They don't always hold up!

Pilots who understand limitations of observations and forecasts usually make the most effective use of forecasts. The safe pilot continually views aviation with an open mind, understanding that weather is always changing and knowing that the older the forecast, the greater the chance that parts of it will be wrong. The weather-wise pilot looks upon a forecast as professional advice rather than an absolute surety. To have complete faith in weather forecasts is almost as bad as having no faith at all.

According to FAA summaries of recent forecast studies, pilots should consider:

- Up to 12 hours—and even beyond—a forecast of good weather (ceiling 3,000 feet or more, and visibility 3 miles or more) is more likely to be correct than a forecast of conditions below 1,000 feet or less than 1 mile.

- If poor weather is forecast to occur within 3–4 hours, the probability of occurrence is better than 80 percent.
- Forecasts of poor flying conditions during the first few hours of the forecast period are most reliable when there is a distinct weather system, such as a front or a trough; however, there is a general tendency to forecast too little bad weather in such circumstances.
- Weather associated with fast-moving cold fronts and squall lines is the most difficult to forecast accurately.
- Errors occur when attempts are made to forecast a *specific* time that bad weather will occur. Errors are made less frequently when forecasting that bad weather will occur during a *period* of time.
- Surface visibility is more difficult to forecast than ceiling height. Visibility in snow is the most difficult of all visibility forecasts.

Predictable changes

According to FAA studies, forecasters can predict the following at least 75 percent of the time:

- Passage of fast-moving cold fronts or squall lines within ± 2 hours, as much as 10 hours in advance.
- Passage of warm fronts or slow-moving cold fronts within ± 5 hours, up to 12 hours in advance.
- Rapid lowering of ceilings below 1,000 feet in prewarm front conditions within ± 200 feet and within ± 4 hours.
- Onset of a thunderstorm 1–2 hours in advance, providing radar is available.
- Time rain or snow will begin, within ± 5 hours.

Unpredictable changes

Forecasters cannot predict the following with an accuracy that satisfies present aviation operational requirements:

- Time freezing rain will begin
- Location and occurrence of severe or extreme turbulence
- Location and occurrence of heavy icing
- Location of the initial occurrence of a tornado

- Ceilings of 100 feet or zero before they exist
- Onset of a thunderstorm that has not formed
- Position of a hurricane center to closer than 80 miles for more than 24 hours in advance

1-800-WX-BRIEF

With these sobering thoughts from the FAA about forecast limitations, it is time to call 1-800-WX-BRIEF for a briefing for the flight we have planned from Westchester County to Binghamton, New York. You may get an AFSS that is not in your immediate vicinity, but don't worry. The automated system switches calls to the most available AFSS, and whomever you reach will provide all the information you need wherever you are calling from. Chances are that a live specialist will take your call. If not, you will get a recorded menu of services to choose from. The recorded menu will provide area briefings, hourly observations, forecasts, special announcements, instructions for filing flight plans, and many other items. If you hear the acronyms "PATWAS" or "TIBS," this means you are connected to a "Pilot's Automatic Telephone Weather Answering Service" or a "Telephone Information Briefing Service." Both will provide a menu of services from which you can select the briefing and filing items you want.

It is a typical IFR day in late November in the Northeast. TV weather reports and a newspaper weather map that morning show (Fig. 11-5) that a low-pressure system with plentiful rain has been moving northeastward up the Atlantic Coast. It has been raining off and on at Westchester County Airport and we can see from personal observation that the ceiling is low and visibility is reduced at the airport.

A go/no-go decision must be made. What weather elements should be considered in making the go/no-go decision? Recall the list:

- Thunderstorms
- Turbulence
- Icing
- Fog
- Current and forecast weather at departure and destination airports
- Availability of alternate airports

Today's High Temperatures and Precipitation

Fig. 11-5. *Newspaper weather map on day of flight discussed in text.*

- Nearest VFR
- Personal minimums

Now let's add two items:

- Forecast winds aloft
- NOTAMs

Winds aloft are necessary to determine the estimated time en route (ETE) for each leg of the flight and the total time (TT) en route, just as you did during VFR flight planning.

NOTAMs are a critical item on all flights and are important factors when making the IFR go/no-go decision. It doesn't take much effort to imagine what might happen at the end of a long, tiring IFR flight if you suddenly discovered that a key component of the best instrument approach was "out of service."

Faster service

Faster and better service is available by initially telling the briefer:

1. The N number, which immediately identifies you as a pilot, not just someone from the general public calling to find out what the weather is like.

2. The type of airplane; light single-engine, high performance multiengine, and jet airplanes present different briefing problems.

3. "Planning an IFR flight from (departure airport) to (destination airport)."

4. Estimated departure time (in Zulu time).

5. Whether or not you can go IFR (if you have not clarified the point in step 3 and VFR is an option). The briefer doesn't know anything when you call and needs to know whether to provide an IFR briefing or a VFR-only briefing.

The briefer will call up the information for the flight on a computer display and will proceed step-by-step through a briefing appropriate for the flight.

Transcribing the weather

A fancy form is unnecessary for a weather briefing. Simply list the categories of information in the proper sequence. The trouble with preprinted weather briefing forms is they leave too much room for unnecessary information and too little room for necessary information.

Below is the briefing sequence; items may be omitted if they are not factors in the proposed flight.

1. **Adverse conditions.** Significant meteorological and aeronautical information that might influence an alteration of the proposed flight: hazardous weather, runway closures, VOR and ILS outages, and the like.

2. **Synopsis.** A brief statement of the type, location, and movement of weather systems, such as fronts and high- and low-pressure areas, that might affect the proposed flight.

3. **Current conditions.** A summary of reported weather conditions applicable to the flight from METARs, PIREPs, and the like.

4. **En route forecasts.** Forecasts in logical order: departure, climbout, en route, descent.

5. **Destination forecasts.** The destination's expected weather plus significant changes before and after the estimated time of arrival (ETA).

6. **Winds aloft.** Forecast winds for the proposed route.

7. **NOTAMS.**

8. **ATC delays.** Any known ATC delays and flow control advisories that might affect the proposed flight.

A simple way to set this up is to list the eight items on the left-hand margin of a blank sheet of paper as a reminder (Fig. 11-6) and leave the rest of the page for notes. Photocopy a supply of lists with the sequence to save time during a weather briefing.

Weather shorthand

Copying a weather briefing verbatim is unnecessary but I strongly suggest recording the vital highlights. Figure 11-7 is a list of easy-to-use "shorthand" weather symbols and letters based upon coding used with TAFs and METARs. You'll remember what they mean when encountered on various textual weather reports and forecasts. Eventually a personal shorthand will develop. Don't hesitate to ask the briefer to read the weather slowly, especially in the beginning when the briefing form and the shorthand might be unfamiliar. Plan on having two or three blank forms at hand to write big and still get everything in.

November weather briefing

The following briefing is quoted from an actual briefing at Westchester County Airport for a flight to Binghamton in November. The briefer reads the adverse conditions from the computer screen, then proceeds in sequence through the other items on the briefing checklist.

1. ADVERSE CONDITIONS
2. SYNOPSIS
3. CURRENT CONDITIONS
4. EN ROUTE FORECASTS
5. DESTINATION FORECASTS
6. WINDS ALOFT
7. NOTAMS
8. ATC DELAYS

Fig. 11-6. *A quick and simple way to set up a page for jotting down a weather briefing.*

WEATHER PHENOMENA

BR—Mist
DS—Dust Storm
DU—Widespread Dust
DZ—Drizzle
FC—Funnel Cloud
+FC—Tornado/Water Spout
FG—Fog
FU—Smoke
GR—Hail
GS—Small Hail/Snow Pellets

HZ—Haze
IC—Ice Crystals

PE—Ice Pellets
PO—Dust/Sand Swirls
PY—Spray
RA—Rain
SA—Sand
SG—Snow Grains
SN—Snow
SQ—Squal
SS—Sand Storm
UP—Unknown Precip. (Automated Observations)
VA—Volcanic Ash

DESCRIPTORS

BC—Patches
BL—Blowing
DR—Low Drifting
FZ—Supercooled/Freezing

MI—Shallow
PR—Partial
SH—Showers
TS—Thunderstorm

CLOUD TYPES

CB—Cumulonimbus

TCU—Towering Cumulus

ABBREVIATIONS

AO1	Automated Observation without precipitation discriminator (rain/snow)
AO2	Automated Observation with precipitation discriminator (rain/snow)
AMD	Amended Forecast (TAF)
BECMG	Becoming (expected between 2-digit beginning hour and 2-digit ending hour)
BKN	Broken
CLR	Clear at or below 12,000 feet (AWOS/ASOS report)
COR	Correction to the observation
FEW	1 or 2 octas (eights) cloud coverage
FM	From (4-digit beginning time in hours and minutes)
LDG	Landing
M	In temperature field means "minus" or below zero
M	In RVR listing indicates visibility less than lowest reportable sensor value (e.g., M0600)
NO	Not available (e.g., SLPNO, RVRNO)
NSW	No Significant Weather
OVC	Overcast
P in RVR	Indicates visibility greater than highest reportable sensor value (e.g., P6000FT)
P6SM	Visibility greater than 6 SM (TAF only)

Fig. 11-7. *Weather briefing shorthand based on TAF/METAR codes.*

ABBREVIATIONS (*Cont.*)

PROB40	Probability 40 percent
R	Runway (used in RVR measurement)
RMK	Remark
RY/RWY	Runway
SCT	Scattered
SKC	Sky Clear
SLP	Sea Level Pressure (e.g., 1001.3 reported as 013)
SM	Statute mile(s)
SPECI	Special Report
TEMPO	Temporary changes expected (between 2-digit beginning hour and 2-digit ending hour)
TKOF	Takeoff
T01760158, 10142, 20012, and 401120084	In Remarks—examples of temperature information
V	Varies (wind direction and RVR)
VC	Vicinity
VRB	Variable wind direction when speed is less than or equal to 6 knots
VV	Vertical Visibility (Indefinite Ceiling)
WS	Wind Shear (In TAFs, low level and not associated with convective activity)

Fig. 11-7. (*Continued*)

Adverse conditions

"Covering your route of flight this morning we have SIGMET November Six for occasional severe icing in clouds and precipitation above ten thousand feet.

"We have AIRMET Oscar Four for occasional moderate rime icing from the freezing level to fourteen thousand feet and Oscar Seven for occasional moderate turbulence below ten thousand and moderate to severe low level winds. And here's a NOTAM—Binghamton radar is out of service.

"Westchester at zero niner hundred had wind calm, visibility one-quarter statute mile in light rain and fog, vertical visibility one hundred feet, temperature nine Celsius, dew point eight Celsius.

"Along your route there are scattered to broken clouds below one thousand feet; then two to three thousand broken and four to five thousand broken, variable overcast.

"Binghamton at zero niner hundred had winds from zero three zero at niner knots, visibility twelve statute miles, clouds scattered at six hundred feet, broken at two thousand feet, temperature six Celsius, dew point five Celsius.

"We have a pilot report from seven miles west of Kingston VOR from a BE thirty-three at five thousand feet reporting a base of scattered clouds at one thousand two hundred feet, broken at twelve thousand, visibility five to ten miles. Temperature plus eight, occasional light turbulence."

Forecasts

"Westchester prior to eleven hundred local is forecasting wind from three two zero at seven knots, visibility two statute miles in fog, overcast at seven hundred feet. Occasionally visibility four statute miles in light rain and fog, overcast at one thousand one hundred feet. After eleven hundred local you can expect wind from three one zero at ten knots, visibility greater than six statute miles, overcast at two thousand feet variable three thousand five hundred feet overcast.

"En route, Poughkeepsie is forecasting the same improving conditions.

"Binghamton between ten and twelve hundred local is forecasting wind from two six zero at niner knots, visibility greater than six statute miles, overcast at one thousand five hundred feet, scattered clouds at five thousand feet, with a forty percent probability at eleven-thirty of visibility two statute miles in fog, light rain showers and broken clouds at five hundred feet.

"After twelve hundred Binghamton is forecasting wind from two seven zero at twelve knots, visibility greater than six statute miles, one thousand two hundred scattered variable to broken, two thousand broken."

Winds aloft

"Winds at three, six, and niner thousand feet at Kennedy are: one eight zero at two four, one niner zero at two eight, and two zero zero at three two, with a temperature of plus three.

"At Wilkes-Barre at three, six, and niner thousand feet, the winds are two three zero at six, two zero zero at two seven, and two zero zero at four two, with a temperature of plus one.

"The freezing level at Kennedy is between ten and eleven thousand feet. At Binghamton the freezing level is nine to ten thousand feet."

Alternate

Binghamton is not forecast to be above the "VFR plus 1,000-foot for ETA ±1 hour rule of thumb for requiring an alternate because the ceiling is too low; an alternate airport is required.

"Wilkes-Barre looks good. They are currently reporting wind from two eight zero at thirteen knots, visibility twenty-five statute miles, broken clouds at twenty thousand feet, temperature nine Celsius, dew point five Celsius. After ten hundred and for the rest of the day, Allentown is forecasting wind from two six zero at ten, visibility greater than six statute miles, four thousand five hundred broken variable to scattered.

Not only is Wilkes-Barre a good alternate, but it's also a convenient, easily located airport. Wilkes-Barre is also a good candidate to list on the log as a nearby airport with VFR conditions. Now, all the information needed to make a go/no-go decision has been obtained and written down on the weather briefing form (Fig. 11-8).

Go or no-go?

What does all this mean? Is it go or no-go?

Return to the checklist of five go/no-go factors and ask whether any will prevent flight as planned:

- Thunderstorms
- Turbulence
- Icing
- Fog
- Departure and destination weather minimums

Thunderstorms. No mention of thunderstorms in SIGMETs, AIRMETs, reports, or forecasts. "Go."

Turbulence. The briefer mentioned an AIRMET for occasional moderate turbulence below 10,000 feet, a PIREP for occasional light turbulence in the vicinity of Kingston, and winds aloft of 32 and 42 knots at 9,000 feet. AIRMET Oscar Seven is a concern with moderate to severe low-level winds.

But turbulence reported in the AIRMET and the PIREP would not damage the airplane. The flight progresses into improving weather,

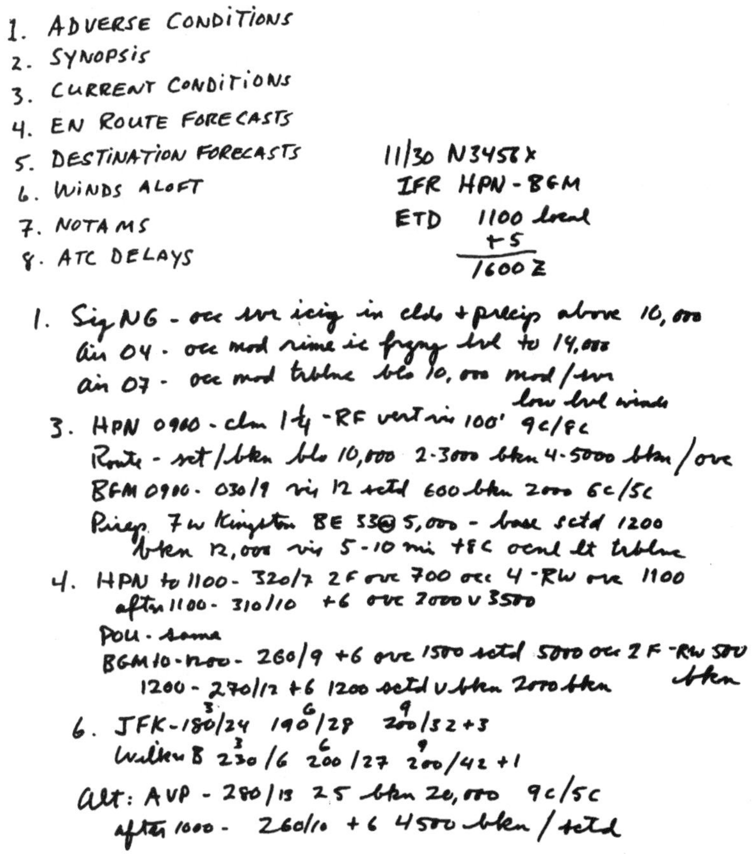

1. *ADVERSE CONDITIONS*
2. *SYNOPSIS*
3. *CURRENT CONDITIONS*
4. *EN ROUTE FORECASTS*
5. *DESTINATION FORECASTS*   11/30 N3458 X
6. *WINDS ALOFT*   IFR HPN - BGM
7. *NOTAMS*   ETD 1100 local
8. *ATC DELAYS*   +5
   ——————
   1600 Z

1. Sig N6 - occ svr icing in clds + precip above 10,000
   Air O4 - occ mod rime ic frzng lvl to 14,000
   Air O7 - occ mod trblnc blo 10,000 mod/svr
                                   low lvl winds
3. HPN 0900 - clm 1¼ -RF vert vis 100' 9c/8c
   Route - sct/bkn blo 10,000 2-3000 bkn 4-5000 bkn/ovc
   BGM 0900 - 030/9 vis 12 sctd 600 bkn 2000 6c/5c
   Precip 7 w Kingston BE 53@5,000 - base sctd 1200
        bkn 12,000 vis 5-10 mi +8c ocnl lt trblnc
4. HPN to 1100 - 320/7 2F ovc 700 occ 4 -RW ovc 1100
   aftn 1100 - 310/10 +6 ovc 2000 v 3500
   POU - same
   BGM 10-noo - 260/9 +6 ovc 1500 sctd 5000 occ 2 F -RW svv
        1200 - 270/12 +6 1200 sctd v bkn 2000 bkn   bkn
6. JFK - 180/24 190/28 200/52 +5
   Wilkes B 230/6 200/27 200/42 +1
   Alt: AVP - 280/13 25 bkn 20,000 9c/5c
        aftn 1000 - 260/10 +6 4500 bkn / sctd

Fig. 11-8. *Details of the weather for the IFR flight from Westchester County to Binghamton, as copied over the telephone.*

so any turbulence-induced fatigue should not be as important as it might be heading into deteriorating weather.

Surface winds at Westchester and Binghamton should range from calm to 13 knots and that's manageable. "Go."

Icing. Icing is mentioned in SIGMET November 6 and AIRMET Oscar 4. A westbound (even thousand) cruise altitude of 8,000 feet will stay below the freezing level. "Go."

Fog. Westchester reported dense fog at 0900, but it seems to be dissipating. Neither Binghamton nor Wilkes-Barre, the alternate, have fog. Go, but delay takeoff at Westchester until 1100 so the ceiling and visibility will be sufficient for an instrument approach if an emergency return is necessary.

Minimums. Departure minimums will be a "go" after 1100. Binghamton is forecast to be well above minimums when we arrive. "Go."

Practice this reasoning prior to each instrument flight and it will become easier to weigh the major factors in the go/no-go decision with solid evidence for the decision. Occasionally, the judgment calls will be too close for comfort, or the weather will unpredictably improve one hour and worsen the next hour. We have all seen days like that. For doubtful situations like these, apply the old pilot's rule of thumb: "If in doubt, wait it out."

In this case, it's a "go." Complete the flight log by entering estimated ground speeds, computing the time en route for each leg, and supplying additional details.

Estimated climbout time

Sometimes instrument students are perplexed about how to compute an accurate time en route to the first fix because that phase of flight involves takeoff, turns, and climbs. I teach students a simple and surprisingly accurate method to figure time to the first fix. Measure the distance from departure to the first fix, estimate winds from the forecast winds aloft reports, compute ground speed and time en route, and add one minute for each 2,000 feet of climb.

This will be very close to the actual time to reach the first fix. If an airplane is a slow climber, such as a Cessna 152, consider adding one minute for each 1,000 feet of climb.

Absolute precision on the first leg is not that important. ATC won't enforce the ±3-minute standard for arrival times because they might amend the clearance with additional turns and changes in altitude. But a reasonably accurate time en route to the first fix will ensure that the time en route for the entire flight will also be reasonably accurate.

Wind and ground speed

En route to Binghamton, the distance to the first fix, HAAYS intersection, is a shade farther than 37 nautical miles. Kennedy—the weather observation point nearest our departure airport—was reporting these winds:

180/24 at 3,000 feet

190/28 at 6,000 feet

200/32 at 9,000 feet

The winds at 6,000 feet—190/28—are a good compromise for the climbout to 9,000 feet. Allow 90 knots for the climb speed on a course of 325° and the wind of 190/28 will yield a ground speed of 108 knots. This ground speed computes to 21 minutes en route to HAAYS. A minute for each 2,000 feet of climb adds 4.5 minutes to reach 9,000 feet. Time en route to HAAYS is 21 + 4.5 = 25.5 minutes. Round that off to the nearest even number and enter 26 minutes for the first leg. Use Wilkes-Barre winds at 9,000 feet (200/42) for the remainder of the flight to Binghamton. Ground speed for the approach should be 90 knots.

Calculator options

Many pilots learned to figure wind problems on the E6B computer. This device is perfectly satisfactory for flight-planning calculations. The FAA has approved electronic calculators for use on tests, and there are several good ones that will handle all your flight planning calculations. Other pilots do many calculations mentally, using rules of thumb shown in Fig. 11-9. Select a system and stick to it.

Instructor note. Understand all calculation methods—E6B, electronic calculators, and rules of thumb—because you never know which system a student will prefer to use.

Enter the Kennedy (JFK) and Wilkes-Barre (AVP) winds in the lower right corner of the flight log as shown in Fig. 11-10. While working in this area fill in the VFR WX AT: section for reference, if necessary.

Compute and fill in the estimated ground speeds and times en route as in Fig. 11-11.

The estimated total time en route adds up to 1 hour and 8 minutes, which is written as 1 + 08.

Flight plan to alternate

Plan a diversion to the alternate, Wilkes-Barre. Weather might be improving, but always be prepared for diversion to an alternate airport. (If you forget to carry an umbrella on a threatening day, it is sure to rain.)

TIME TO CLIMB
1. Estimate distance to reaching altitude. (Include vector, reversal, or circle.)
2. Use filed true airspeed for time/distance calculation.
3. Add one minute for each 2,000 feet of climb.

GROUND SPEED	
Crosswind Angle	**Effect on True Airspeed**
0–15°	Full value of wind speed
30°	.9 wind speed
45°	.7 wind speed
60°	.5 wind speed
75°	.3 wind speed
90°	.1 wind speed

TIME TO FLY
Drop the last digit of the air speed (ground speed, if known); you will fly that many miles in 6 minutes.
Examples:
  Speed 150 - 15 miles every 6 minutes
   Miles to go - 7. Time to fly - 3 minutes.
   Miles to go - 20. Time to fly - 8 minutes.
  Speed 90 - 9 miles every 6 minutes.
   Miles to go - 5. Time to fly - 3 minutes.
   Miles to go - 35. Time to fly - 24 minutes.

	WIND CORRECTION ANGLE	
True Airspeed	**Crosswind Angle**	**Wind Corr. Angle**
100 - 120 - 140	90°	½ wind speed
	45°	⅓ wind speed
150 - 180 - 200	90°	⅓ wind speed
	45°	¼ wind speed

Fig. 11-9. *"Rules of thumb" calculations for time to climb, ground speed, time to fly, and wind correction angle.*

Plan for V149 to Lake Henry VOR (LHY), which is also the initial approach fix (IAF) for the Wilkes-Barre approach that will most likely be in use, ILS 22. Radar vectors to the ILS 22 final approach course are likely, but plan to execute the complete approach in case radio communication is lost or the approach control radar fails.

Identical methods for computing the times for climbout, en route, and descent reveal results shown in Fig. 11-12. Add the times en route from Westchester to Binghamton and thence to Wilkes-Barre

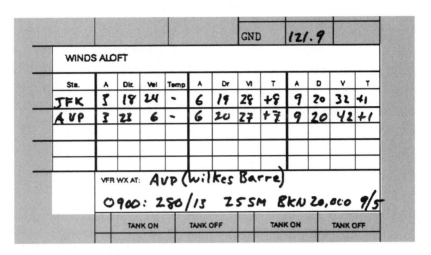

				GND			121.9					

WINDS ALOFT

Sta.	A	Dir.	Vel	Temp	A	Dir	Vel	T	A	D	V	T
JFK	3	18	24	-	6	19	28	+8	9	20	32	+1
AVP	3	23	6	-	6	20	27	+7	9	20	42	+1

VFR WX AT: **AVP (Wilkes Barre)**

O900: 280/13 25SM BKN 20,000 9/5

	TANK ON	TANK OFF		TANK ON	TANK OFF

Fig. 11-10. *Filled-in winds aloft forecast and nearest VFR weather blocks on flight log form for IFR flight to Binghamton.*

DIST.	G/S	TIME ENR	TIME ARR
TO	EST	EST	EST
REM	ACT	ACT	ACT
374	108	26	
8	144	4	
66	134	30	
14	109	8	
125		68	
		(1+08)	

Fig. 11-11. *Ground speed and estimated time calculations for IFR flight to Binghamton. Note: Actual ground speed and actual times of arrival (ACT) are filled in as the flight progresses.*

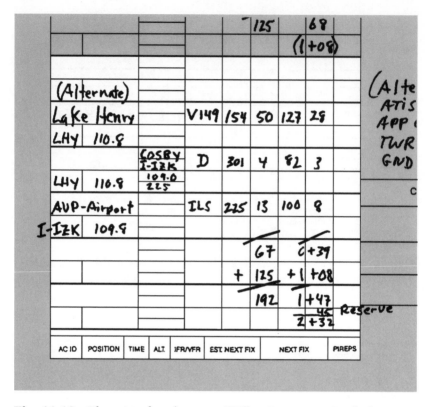

Fig. 11-12. *Planning for alternate, Wilkes-Barre, on IFR flight to Binghamton.*

plus the 45 minutes required by regulation (FAR 91.167) to obtain a total time for the flight, including alternate (plus 45 minutes), of 2 hours and 32 minutes, well within the fuel range of the aircraft. The completed flight log is shown in Fig. 11-13.

In-flight notations

Note the ample amount of space remaining for entering in-flight items. Locate where the following in-flight information goes:

- Actual ground speed and actual time en route
- Estimated and actual times of arrival (ETA and ATA)
- Position report sequence (lower left) if needed
- Space for clearances (center) and below that more space for alternate frequencies, ATIS, and clearance limits

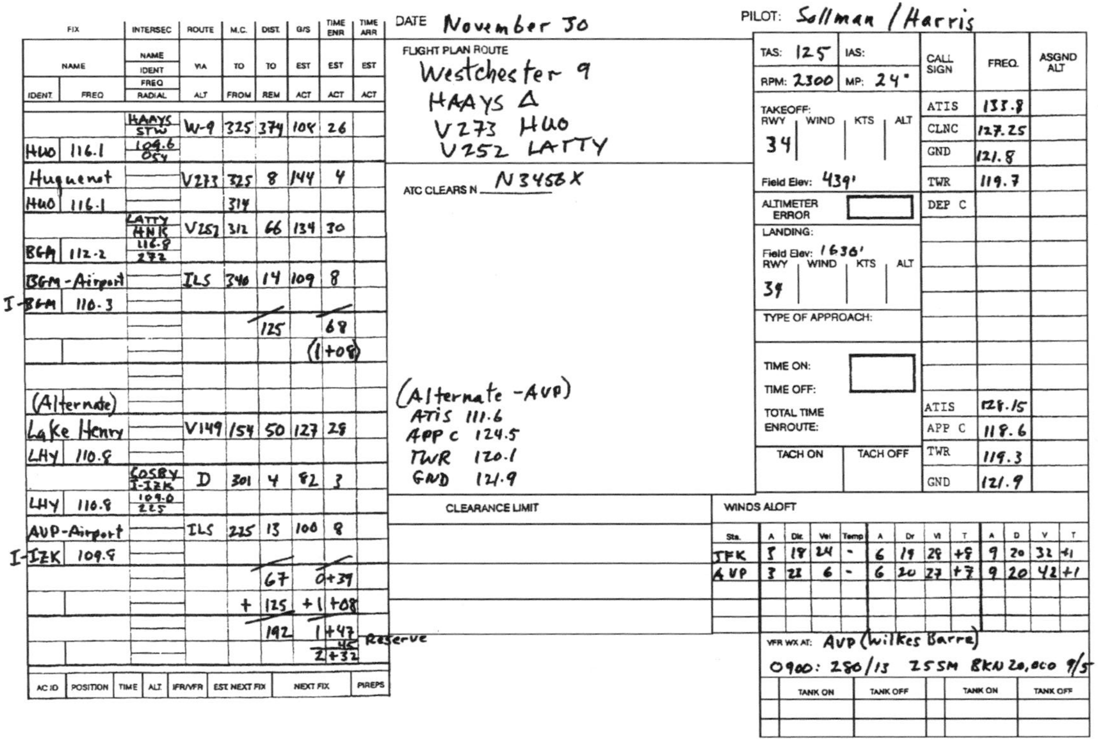

PILOT: *Sollman / Harris*

DATE *November 30*

FLIGHT PLAN ROUTE
Westchester 9
HAAYS △
V273 HUO
V252 LATTY

ATC CLEARS N *N3465X*

	TAS:	125	IAS:		
	RPM:	2300	MP:	24"	

TAKEOFF:
RWY | WIND | KTS | ALT
34

Field Elev: *431'*

ALTIMETER ERROR

LANDING:
Field Elev: *1636'*
RWY | WIND | KTS | ALT
34

TYPE OF APPROACH:

	CALL SIGN	FREQ.
ATIS		*133.8*
CLNC		*127.25*
GND		*121.8*
TWR		*119.7*
DEP C		

TIME ON:
TIME OFF:
TOTAL TIME
ENROUTE:

	CALL SIGN	FREQ.
ATIS		*127.15*
APP C		*118.6*
TWR		*119.3*
GND		*121.9*

(Alternate -AVP)
ATIS *111.6*
APP C *124.5*
TWR *120.1*
GND *121.9*

CLEARANCE LIMIT

TACH ON | TACH OFF

WINDS ALOFT

Sta.	A	Dir	Vel	Temp	A	Dir	Vel	Temp	A	Dir	Vel	Temp	A	Dir	Vel	Temp
IPT		3	19	24	-	6	19	24	+9	9	20	31	+1
AVP		3	21	6	-	6	20	23	+2	9	20	41	+1

VFR WX AT: *AVP (Wilkes Barre)*
0900: 250/13 25SM BKU 20,000 9/5

FIX		INTERSEC		ROUTE	M.C.	DIST	G/S	TIME ENR	TIME ARR
NAME	FREQ	NAME IDENT FREQ RADIAL	VIA ALT		TO FROM	TO REM	EST ACT	EST ACT	EST ACT
HUO 116.1		HAAYS STU 108.6 084	W-9 325 334	335	334	104 26			
Huguenot			V213	325	8	144	4		
HUO 116.1				314					
BGM 112.2		HNK 116.9 273	V151 317	66	134	30			
I-BGM 110.3		BGM-Airport	ILS 346	14	109	8			
				125			69 (1+08)		
(Alternate)									
Lake Henry			V149 154	50	123	29			
LHY 110.8							0+37		
LHY 110.8							1+08		
AVP-Airport			ILS 225	13	100	8			
I-IZK 109.9		COSBY I-IZK 109.9 215	D 301	4	92	3			
					147		1+08		
					192		1+47 2+31		NC RESERVE

| AC/D | POSITION | TIME | ALT | IFR/VFR | EST NEXT FIX | NEXT FIX | PIREPS |

233

Fig. 11-13. *The complete log for the IFR flight to Binghamton showing all details that can be filled in prior to obtaining a clearance.*

- Takeoff and landing runway information, and which approach might be expected
- Time off, time on, total time en route, and tachometer (or Hobbs) reading at the beginning and end of the flight
- Fuel management logs (lower right)

Filing the flight plan

The completed flight plan is shown in Fig. 11-14.

Use the equipment code in Fig. 11-15; add the code to the briefing card for later reference if a piece of equipment fails, or you are flying a different airplane.

Student note. Use the instructor's name, address, and telephone number because an instrument-rating student cannot legally file an IFR flight plan. The only exception is the instrument flight test.

When filing by telephone, stay on the line for a specialist to copy the flight plan rather than simply recording your message. Request information that has been updated since the previous briefing.

The FAA stated in *FAA Aviation News*: "Live briefers are a good source of unpublished Notices to Airmen concerning important data about airport or runway closures, military flight route training activity, obstructions to flight, outages and shutdowns, etc. The latest pilot reports of weather conditions aloft . . . are also more likely to be available from FSS specialists than from prerecorded messages."

Abbreviated briefings

There are certain "magic words" that I will introduce from time to time to simplify instrument procedures. One very handy pair is *abbreviated briefing*.

Request an abbreviated briefing for:
- Updated information, such as NOTAMS, to supplement recorded information.
- Updated specifics of a previous full-length briefing.
- One or two specific items. This would be the case, for example, when filing IFR on a "severe clear" day to a nearby

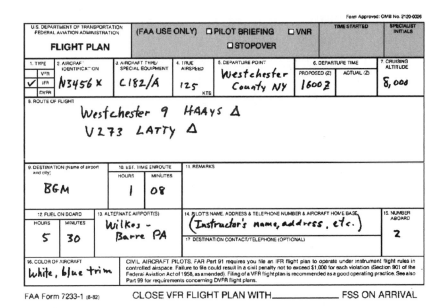

Fig. 11-14. *The completed flight plan for the IFR cross-country to Binghamton.*

/X	No transponder.
/T	Transponder with no altitude encoding capability.
/U	Transponder with altitude encoding capability.
/D	DME, but no transponder (per preceding).
/B	DME and transponder, but no altitude encoding capability.
/A	DME and transponder with altitude encoding capability.
/M	TACAN only, but no transponder.
/N	TACAN only and transponder, but with no altitude encoding capability.
/P	TACAN only and transponder with altitude encoding capability.
/I	RNAV and transponder with altitude encoding capability.
/C	RNAV and transponder, but with no altitude encoding capability.
/W	RNAV but no transponder.
/G	Global Positioning System (GPS)/Global Navigation Satellite System (GNSS) equipped aircraft with oceanic, en route, terminal, and GPS approach capability.

Fig. 11-15. *ATC equipment codes for use on flight plan.*

airport that you are familiar with. Request a terminal forecast to make sure VFR conditions will continue through the ETA.

Try abbreviated briefings a couple of times during training when weather conditions are favorable. This procedure can save a lot of time when used properly.

Outlook briefings

Another good pair of magic words is *outlook briefing*. Request this service for a short, live briefing to supplement other sources regarding a departure time more than six hours away.

Outlook briefings are particularly helpful when it's harder than usual to outguess the weather. There is no way to know when a stalled low pressure system might start moving unless a pilot has access to winds aloft charts, constant pressure charts, and other highly technical information. Even then, the charts might not be much help.

The one-call technique

Another method to speed up the weather briefing and filing process is combining the weather briefing and flight plan filing in one telephone call.

Completion of phase one flight planning—elements completed prior to calling flight service—will reveal all details to complete the flight plan form, except estimated time en route and alternate airport.

The weather briefing will quickly establish whether or not an alternate airport is necessary and the briefer will assist with the selection.

Estimated time en route to Binghamton is based upon a true airspeed (TAS) of 125 knots for the Cessna 182. It's a simple matter to bracket this airspeed in increments of 10 knots on either side and make a series of time-speed-distance computations.

Quick estimates

Before calling flight service, compute several figures for time en route in our example using 125 total nautical miles en route. Bracket the 125-knot TAS with headwinds of 30, 20, and 10 knots and tailwinds of 10, 20, and 30 knots to reveal:

125 @ 95 = 1 + 19 (30 knot headwind)
125 @ 105 = 1 + 12 (20 knot headwind)
125 @ 115 = 1 + 05 (10 knot headwind)
125 @ 125 = 1 + 00 (zero wind factor)
125 @ 135 = 0 + 56 (10 knot tailwind)
125 @ 145 = 0 + 52 (20 knot tailwind)
125 @ 155 = 0 + 38 (30 knot tailwind)

When the briefer gives the winds, simply pick the closest estimate of ground speed. A tailwind off the left quarter during this flight probably means a tailwind component of approximately 20 knots, based on the forecast winds for Wilkes-Barre. This produces a time en route of 52 minutes.

Add one minute for every 2,000 feet of climb to the planned altitude of 8,000 feet and the total time en route is 56 minutes. (Simplify the climb estimate by automatically adding 5 minutes for climbs up to 10,000 feet and 10 minutes for climbs above 10,000 feet.)

Total time en route

Return to the flight log and see the estimated 68 minutes for total time en route. Is the 56-minute quick estimate close enough for filing purposes? Certainly. Time en route establishes when ATC expects the approach to begin in the event of two-way communications failure.

Emergency procedures are subsequently explained in greater detail, but let's briefly consider lost communications. If two-way radio communication is lost, ATC will expect a pilot to carry out the lost communications procedure in FAR 91.185, which pertains to IFR operations during a two-way radio communications failure.

Fly to the destination following the last assigned routing and altitude. According to the FAR: "Begin descent from the en route altitude or flight level upon reaching the fix from which the approach begins [initial approach fix], but not before—

"The expect-approach-clearance time (if received);" or "If no expect-approach-clearance time has been received, at the estimated time of arrival, shown on the flight plan, as amended with ATC." ATC will expect an arrival time as filed. Enter a holding pattern if early. If late,

ATC will protect the approach airspace for 30 minutes, then ATC will initiate lost aircraft procedures.

The quick estimate indicates early arrival at Binghamton, which is perfectly all right. The approach would begin out of the holding pattern when the time en route expired.

If the estimated time en route changes either way by more than 5 minutes, notify ATC about a revised ETE or ETA. Not only is this good insurance to cover the possibility of two-way radio communication failure, but it will also help ATC sequence traffic efficiently.

Don't use the one-call procedure if concerned about any of the five go/no-go weather factors. Get a thorough briefing and study the impact of weather on the flight, then make a second call to update the weather information and file the flight plan.

"Cleared as filed"

ATC has its magic words too. If your flight planning has been thorough, you just might hear the welcome words "cleared as filed" when you receive your clearance.

Phase one of a thorough flight plan covers all the details prior to contacting flight service: route, destination, departures, approaches, frequencies, and the big picture of the weather.

Phase two was either a two-call, a one-call, or a recorded communication with flight service to obtain the detailed weather and other information necessary to make a go/no-go decision and to file the flight plan.

Follow this approach to planning an IFR flight from the very first day and the planning becomes easier, faster, and much more interesting. Following this procedure and filing an IFR flight plan for every instrument training flight—even if it's just out into the local area for practice in basic maneuvers—will provide a better chance of passing the instrument checkride.

Save all logs and flight plan forms. It was hard work to get all the information necessary for those flights, so save the information and use it again. Always check the information against the current charts for changes. Use routes and approaches that worked well on previous flights and especially remember any amended clearances received. Amended clearances might suggest better routes or fixes that can be incorporated into future flight planning.

12

Airplane, instrument, and equipment checks

Now that you've planned your flight, evaluated the weather, and filed your flight plan, you need to determine that the airplane is airworthy and obtain your instrument clearance.

You have completed the flight log, checked the weather, and found that the flight can be safely completed. The flight plan is filed and the pilot is prepared but what about the airplane? Is it ready for an IFR flight in actual instrument weather?

Several items on the preflight checklist require special attention for an IFR flight; see Fig. 12-1 for a basic list. Consider adding items for a specific airplane and equipment.

Fuel quantity

Some checks are obvious. Turn on the electrical power and check fuel gauges to make sure there is enough fuel to fly to the destination plus 45 minutes at normal cruising speed. If an alternate airport is required, verify enough fuel to the destination, then to the alternate plus 45 minutes. This item is first on the checklist so the pilot can continue the checklist while waiting to refuel.

It's always a good practice to start an instrument flight with full fuel tanks if weight permits, visually confirmed. Nothing is as useless as the "runway behind, the air above, and the fuel in the fuel truck."

During a round-robin trip, at each stop ask "Should I top off again?" If in doubt, top off. There is no reason to worry about running out of fuel in addition to all the other concerns.

PRETAXI CHECK

Before starting:
- ☐ Outside antennas—all secure
- ☐ Fuel—to destination, alternate, plus 45 minutes
- ☐ ATIS—copied
- ☐ Altimeter—set, error noted
- ☐ Airspeed, VSI—both on zero
- ☐ Magnetic compass—shows correct heading
- ☐ Clock—running and set correctly
- ☐ Pitot heat—working
- ☐ Lights—all working
- ☐ VOR—checked within 30 days
- ☐ Charts and logs—sequenced

After starting:
- ☐ Ammeter—checked
- ☐ Suction gauge—normal
- ☐ COM radios—all checked and set in sequence
- ☐ NAV radios—all checked and set for departure
- ☐ Marker beacon lights—checked
- ☐ Heading indicator—set
- ☐ Attitude Indicator—normal
- ☐ Alternate static source (if any)—working
- ☐ Clearance—copied
- ☐ Transponder—code set, checked, turned to standby

TAXI CHECK
- ☐ Heading indicator—responding normally to turns
- ☐ Attitude indicator—normal and stable
- ☐ Turn coordinator—responding normally to turns

PRETAKEOFF CHECK
- ☐ Approach charts—emergency return chart on top
- ☐ NAV radios—double-check set for departure
- ☐ COM radios—Departure control on #2
- ☐ Heading indicator—aligned with runway centerline

RUNWAY ITEMS—"STP"
- ☐ Strobes (or rotating beacon)—on
- ☐ Transponder—on
- ☐ Pitot heat—on

Fig. 12-1. *IFR checklist items.*

ATIS

While the electrical power is on, tune in the ATIS frequency if available at the departure airport and write down the current information. Set the altimeter to the current altimeter setting as reported by ATIS. If there is a difference of more than 75 feet between what the altimeter reads and the ramp elevation where the plane is parked, don't go. There is something seriously wrong with the altimeter.

Reference the altimeter with the ramp elevation where the plane is parked, not the general field elevation. Airports are seldom level. Look at the airport diagram on the approach chart and select the elevation nearest the parking ramp. As noted in the flight-planning chapter, there can be considerable difference between official airport elevation and ramp elevation.

Altimeter and airspeed errors

There is frequent confusion about what action to take with altimeter error. Use the given barometric pressure and note the error on the flight log in the appropriate spot. Fly the altitudes shown on the altimeter and disregard the error until the approach.

Add the error to published minimums, regardless of whether this error is plus or minus. For example, the White Plains decision height on the ILS 16 approach is 639 feet. If the altimeter error is 50 feet, add it to the minimum and use a decision height of 689 feet regardless of whether the error is plus 50 feet or minus 50 feet. Always add the difference to be on the safe side.

Check the airspeed indicator and the vertical speed indicator, which should indicate zero and zero. Most vertical speed indicators have an adjustment screw on the lower left corner of the instrument for calibration to zero.

If there is no adjustment screw, allow for the error when interpreting indications in flight. For example, if the instrument shows a 50 foot-per-minute climb on the ramp, that is level flight in the air.

Electrical equipment

Turn on and check the pitot heat and all outside lights, including landing and taxi lights, even if the flight will occur during daylight hours. Pitot heat is necessary if any part of the flight is in the clouds, especially when climbing through a cloud layer after takeoff. It's a very strange feeling to see airspeed drop toward stalling speed during an instrument climbout because ice is building up in the pitot tube!

Lights might be necessary for visual identification of the airplane by tower controllers during an approach in low visibility.

Many questions and doubts have arisen about the final item of the checklist. Before takeoff the *pitot heat must be* ON, whether the take-off is into severe clear, cloud, rain, or icing conditions ahead. This is the way I have been teaching other instructors for many years. In addition, if the plane has electric prop heat, it should also be turned on.

It is theorized that a B-727 crashed because the crew missed this item on the checklist, and recently it has surfaced as the most probable underlying cause of the crashes of a number of Malibu airplanes, which had shown no bad traits during flight testing and certification.

Proper pilot training and indoctrination could have prevented some of these accidents and loss of life. Remember the "law of primacy" that you were taught as a student flight instructor? If proper habit patterns had been taught from the beginning, some of these lives might have been saved. When we train pilots, we bear a heavy responsibility to "teach 'em right the first time." We have no way of knowing whether a student will wind up flying a B-747, Citation, Learjet, or P-210. We must train them right! Now!

Although it is not legal to fly in "known ice," anyone who has flown IFR for any length of time has encountered icing conditions where they were least expected—a complete surprise. Then ATC might be in a "bind" and unable to give us relief as fast as we would like, such as a lower or higher altitude to get us out of the ice. This is just the time that the pilot forgets to put on the pitot heat. The power drain is minimal for pitot heat—you can hardly see the ammeter needle move when the pitot heat switch is turned ON and OFF to check. If there is any doubt that the alternator/generator can't handle this added load, then this is not a real instrument airplane. I haven't seen an airplane in the last 10 years that couldn't handle the load, and similarly I haven't heard of many pitot heating elements burning out. So get with it and teach it right.

VOR checks

Verify that the record of VOR checks is current. The VOR check is probably the most overlooked or ignored regulation. There is nothing mysterious about the check nor is the regulation hard to understand. FAR 91.171 says that no person may use VORs on an IFR flight unless the system has been checked within the preceding 30 days and found to be within certain limits.

A record of VOR checks must be kept in the plane. If no log entry is found attesting to a VOR check within the preceding 30 days, it is illegal to make an instrument flight unless the VORs are checked before takeoff. In spite of the legal requirements, it doesn't make sense to depart on an instrument flight without knowing that the VORs are accurate and within limits.

If the receiver and indicator have been checked within the 30-day limit, everything is OK. If not, determine if a VOR check is possible before taking off. VOR checks will be explained later in this chapter. The point here is that checking the status of the VOR systems is an important item on the IFR preflight inspection.

Cockpit organization

The basic principle of cockpit organization is twofold:

- Organize charts and flight log to avoid fumbling for a vital piece of information
- Check and set up communication and navigation radios ahead of time according to the expected sequence of frequencies

One very real hazard associated with cockpit confusion is vertigo. Sudden head movement while searching for a chart might cause an attack of vertigo. It might be more severe when reaching back to pick up something from a flight case on the rear floor, then snapping around to an upright position.

How to sequence logs and charts

My favorite device for holding charts and logs is a standard 8½ × 11 clipboard. A folding flight desk that rests on the lap, or a kneeboard that straps to a thigh works well.

I place the flight log and flight plan on top. Underneath those I place the en route chart folded so I can see the route and airways or fixes that might be in any amended clearance. Approach charts for the departure airport are placed under the en route chart in case I have to return for an instrument approach shortly after takeoff. Next come approach charts for the destination airport with the expected approach on top. After these come all approaches for the alternate airport.

Radio preparation

Before starting an engine, determine what sequence of frequencies will be used for communications and navigation. Write the frequencies in sequence, starting from the top. A top-to-bottom sequence used consistently will reduce the chances of selecting the wrong radio at the wrong time.

In most cases, the departure communications sequence will be:

- ATIS
- Clearance delivery or pretaxi clearance
- Ground control
- Tower
- Departure control
- Air route traffic control center

Tune the ATIS frequency first on the number one communications radio. Move down and tune clearance delivery. (Clearance delivery is often handled by ground control. Some very large airports might also have a gate control frequency. Airline pilots monitor the gate control frequency for clearance to start engines, thereby saving fuel.)

Listen to transmissions on the first frequency, adjust the volume to a comfortable level, then adjust the squelch sensitivity until the "noise" begins; readjust the squelch until the "noise" stops.

Follow the same procedure with the second radio in the communications sequence.

Presetting frequencies

Some newer navcoms will store up to nine additional frequencies. I once preset nav frequencies to fly all the way from Palm Beach, Florida, to Myrtle Beach, South Carolina, without dialing in a new VOR frequency. The same features are also available on the com side. I frequently preprogram the ATIS, approach, tower, and ground control at the destination airport, which cuts down on the workload in the air.

Use the top-to-bottom approach to check navigation radios as well. I check the VOR receivers' accuracy on every flight. Look in the

"VOR Receiver Check" section in the back of the A/FD and see which checks are available at your home airport as well as other frequently used airports (Fig. 12-2). Set up the test frequencies before starting, as you did with the communications radios.

VOR checks with VOT

The best method of checking a VOR receiver and indicator is with a *VOR test facility* (VOT). The VOT is a special VOR ground facility that transmits only the 180° radial. Tune in the VOT frequency listed in the A/FD and dial in 180 with the *omni bearing selector* (OBS) knob. The needle should center in the TO position. An easy way to remember the setup and expected indication is "Cessna 182:" The C in Cessna is for center, one-eighty is the radial, and two is the direction (TO). The maximum permissible bearing error is ±4° with the VOT.

Unfortunately, only a limited number of airports in the United States have a VOT. Many are very busy, high-density hubs—such as

VOR RECEIVER CHECK
MASSACHUSETTS
VOR RECEIVER CHECK POINTS

Facility Name (Arpt Name)	Freq/Ident	Type Check Pt. Gnd. AB/ALT	Azimuth from Fac. Mag	Dist. from Fac. N.M.	Check Point Description
Gardner (Fitchburg Muni)	110.6/GDM	A/1500	102	13.0	Over intersection of rwys.
Gardner (Metropolitan)	110.6/GDM	A/2000	097	1.9	Over intersection of taxiway and rwy.
Gardner (Orange Muni)	110.6/GDM	A/1500	292	10	Over parachute jump circle.
Gardner (Worcester Regional)	110.6/GDM	A/2000	167	18.8	Over intersection of Rwys 11–29 and 15–33.
Lawrence (Plum Island)	112.5/LWM	A/1500	089	11.8	Over apch end Rwy 10.
Lawrence (Tew-Mac)	112.5/LWM	A/1500	224	9.9	Over apch end Rwy 21.
Marthas Vineyard (Marthas Vineyard)	114.5/MVY	G	216	0.7	On runup block for Rwy 06.
Nantucket (Nantucket Memorial)	116.2/ACK	G	242	1.9	On runup area at apch end Rwy 24.
Providence (Fall River Muni)	115.6/PVD	A/1500	097	15	Over intersection of rwys.
Putnam (Southbridge Muni)	117.4/PUT	A/1700	328	12	Over intersection of taxiway and rwy.

VOR RECEIVER CHECK
VOT TEST FACILITIES (VOT)

Facility Name (Airport Name)	Freq.	Type VOT Facility	Remarks
Laurence G. Hanscon	110.0	G	
Gen. Ed. Lawrence Logan Intl.	111.0	G	
Worcester Regional	108.2	G	

Fig. 12-2. *VOR receiver check information as published in the Airport/Facilities Directory.*

Kennedy, La Guardia, and Los Angeles—where instrument training flights are not always practical. If you don't have access to a VOT, the next best check is with a VOR located on the field or close enough to receive on the ground. Tune both VOR receivers to this station and note the indicated bearings TO that station. The maximum permissible difference between the two bearings is 4°.

Ground and airborne VOR checks

The A/FD also lists other ground and airborne checkpoints (Fig. 12-2). Again, see what checkpoints are available. One disadvantage to ground checkpoints is that they might be located on the opposite side of the airport. And because airborne checkpoints require a cross-check between a VOR radial and some point on the ground, such as the end of a runway, they won't be much good on an IFR flight in the clouds!

Another method of checking VOR receivers is by tuning the same VOR in flight and noting the bearings from the station. The maximum permissible variation between the two bearings is 4°. FAR 91.171, which covers VOR checks, also prescribes an airborne procedure for a single VOR receiver and for dual VOR receiver checks.

To legally meet the 30-day requirement, VOR receiver checks must be logged with date, type of check, place, bearing error, and signature of the person making the check.

Take advantage of all opportunities to make airborne checks. Log them properly and keep the log in the airplane to avoid getting stuck on the ground on a routine IFR day simply because the VOR receivers have not had a legal check within the last 30 days.

ILS check

If there is an ILS system at the departure airport, check out the ILS receivers after the VOR receiver check. Tune in the localizer frequency on the number one navigation radio, listen to the identification signal, and observe the needles for correct movement. The needles won't steady up in one position because the airplane is not aligned with the final approach course while at the parking ramp. If you hear a clear identification signal, the red warning flag on the instrument face disappears, and the needles are "alive," the receiver is operating properly.

Check the ILS identifier, then turn the receiver off or turn the volume down to the lowest level and leave the localizer frequency on the number one receiver. You will be ready to make an emergency return shortly after takeoff without fumbling around for the correct frequency. Likewise, set the number-two navigation receiver to the first en route VOR station and it is ready to proceed on course without cockpit confusion.

ADF check

Get a good check of the ADF by tuning the frequency of the outer compass locator at the departure airport and observe the swing of the needle. When the needle steadies it should point toward the locator. With a little trial and error you can soon establish the relative bearing from the ramp to the locator and use this to check the accuracy of the ADF. Use the same procedure if the departure airport has an NDB approach and no ILS.

If there is neither a compass locator nor an NDB near the airport, use a local AM radio station to check the ADF. Try to find a station near the airport that is shown on a sectional chart and plot the bearing to the station and see how well the ADF needle matches that bearing.

Transponder check

Finally, turn the transponder to the TEST position. If the indicator light comes on and blinks, the transponder has run an internal circuitry check and everything is functioning normally.

Be careful not to switch the transponder to the ON or ALT (mode C) positions because the transponder signal might be sensed by the traffic control radar antenna and fed into the computer to indicate that an IFR flight has commenced. Leave the transponder in the STANDBY position until cleared onto the runway for takeoff and include the transponder (Transponder ON) in the final list of runway items.

Tips to reduce cockpit confusion

The best ways to reduce cockpit confusion are:

- Organize logs and charts on a clipboard with the flight log on top and the other material in the proper sequence. Do this in the planning room before starting out to the airplane.

- Tune the communications and navigation radio frequencies to check before starting the engine. Then use a top-to-bottom sequence for checking out each piece of radio navigation equipment. If frequencies checked are different from departure frequencies, tune in the departure frequencies with the ILS of the departure airport on the number one navigation receiver. Then you can keep one step ahead throughout the flight.

All the checks above may be performed on battery power unless the temperature is below freezing. When it gets that cold, of course the battery has less power and must be conserved. In cold weather it is a good idea to start the engine first, while the battery is still fresh, then perform the radio checks.

Gyro instruments

Other instrument checks can only be made after starting the engine. Prior to taxiing, set the heading indicator to match the heading on the magnetic compass. Next, check the attitude indicator to ensure that it confirms level flight. (Pitch attitude cannot be set on the ground because there is no way of knowing when the airplane is sitting precisely in a level attitude.)

Any drift by either instrument during taxi indicates a malfunction that should be diagnosed and repaired prior to IFR flight.

13

Clearances and communications

Copying clearances seems to be a great stumbling block in the minds of many instrument pilots. There is no need for this to happen. If you start out copying clearances the right way, you will soon find that this is one of the easier elements of instrument flight. Because an IFR flight plan is filed for every training flight, you will have to copy IFR clearances from ATC—and read them back correctly—beginning with the first training flight. Based upon experience with hundreds of students, those with the best success in mastering clearance copying are those who are prepared to copy and read back clearances on every flight.

Practice clearances

These tips will soon have you copying clearances like an airline pilot. But practice, practice, practice is the best way to become competent in copying clearances, especially in the beginning when the jargon is unfamiliar and the controllers seem to set new records for fast talk!

I urge students to buy a multiband portable radio with aviation bands or a hand-held transceiver to practice copying clearances delivered to other flights. The transceiver is a better investment because it can also be used to communicate with ATC if communications are lost while on an IFR flight plan. Other handy uses for a transceiver are subsequently discussed. Read those clearances into a tape recorder then play the tape back and attempt to copy the clearances. It will be fun and it will develop a competence and confidence—within three or four lessons—that is unbelievable.

If you are unable to listen to actual clearances being delivered, ask an instructor to read simple clearances for practice.

Instructor note. Why not make up an audiotape that students can use? A cassette recorder and a few short sessions in a parked aircraft on a busy day at a large airport will produce a tape recording for students to use at home.

Clearance shorthand

It's impossible to copy a clearance in long hand and get it right. Every pilot develops a clearance "shorthand" that works well. Look over the list of simplified clearance shorthand symbols in Fig. 13-1 and practice using them. If a personal clearance shorthand is more comfortable and more in keeping with your personal taste, be my guest and use symbols from the list that are helpful. All that's necessary is an ability to read back the clearance promptly and accurately. If the flight has been planned correctly, there should be little need to refer back to the clearance, and never in an emergency!

On the first few instrument-training flights, tell ATC that you are an instrument student and ask the controller to "please" read the clearance slowly. When controllers realize that they are working with a student, they will read the clearance slower rather than read it over three or four times.

Be frank about expertise; ATC will cooperate in most cases. Controllers get in the habit of talking fast and they assume that any pilot can keep up with them. Conversely, an FSS specialist will quickly request slower speaking if a pilot talks too fast when filing a flight plan.

This advice is not just for students. When an experienced pilot is at an unfamiliar airport expecting an unfamiliar clearance, the pilot should request a slower delivery. ATC will usually cooperate.

Another tip on clearance copying: If the controller reads the clearance a little too fast and parts are missed, keep on copying and leave a blank spot for the missed information. Then read back what you have and ask for a repeat of any section that wasn't clear. Don't give up—keep on writing! The last thing clearance delivery wants to hear is "Please repeat everything after 'seven two Romeo cleared to...'"

Handling amended clearances

Now let's look at what's involved in copying clearances in the air. It's almost impossible to fly IFR anywhere these days without receiving

60	altitude—6,000'
A	airport
<	after passing
>	before reaching
C	ATC clears
X	cross
D	direct
EAC	expect approach clear-ance
EFC	expect further clearance
CAF	as filed
H	hold
H-W	hold west, etc.
M	maintain (altitude)
O	VOR or VORTAC
RL	report leaving
RP	report passing
RR	report reaching
RV	radar vectors
RY Hdg	runway heading
LT	left turn after takeoff
RT	right turn after takeoff
V	Victor airway
til	until further asking
↑	climb to
↓	descend to
→	intercept

Fig. 13-1. *Clearance copying shorthand.*

an amended clearance or two—and they always seem to come as a surprise. Be prepared for them and avoid surprises.

Always have a blank spot on the flight log or an extra blank piece of paper handy to reach instantly when the controller issues a clearance.

Take a pencil and start writing on that piece of paper on the left-hand edge while looking at the attitude indicator because it is the basic reference for control of the airplane. Keep the airplane straight and level.

Learn to copy clearances without looking at the paper by practicing on the ground. Try this experiment: Place a piece of paper on the clipboard, stare off into the distance at an object imagining it to be an attitude indicator, and write "I am a very good IFR pilot" several times, beginning from the left edge. It doesn't have to be perfect, and it's your writing so you'll be able to read it, even if it runs uphill or downhill a little. Try this a couple of times to see how easy it is.

Now go back and play some of those recorded clearances. Copy them also without looking. Airborne clearances seldom cover more than two or three items at a time. Develop a clearance shorthand and practice copying full clearances on the ground without looking. There should be no difficulty copying the shorter airborne clearances.

For example, let's say that an amended clearance while airborne said: "Cessna three four five six Xray cleared direct Haays intersection maintain eight, report passing six." All you would have to write, using the clearance shorthand, Fig. 13-1 is: "D HAAYS M 80 RP 60."

Obtaining clearances

After all this practice it is time to get on with the flight. An ATC clearance must be obtained. Complete the preflight and get the cockpit organized before worrying about the clearance, then decide whether or not to start the engine before calling for a clearance. On a cold day it might be wise to start up, then call clearance delivery; engine and instruments can warm up while waiting for the clearance.

On the other hand, the delay might be extensive and the engine might run for a long time prior to taxi. If renting the plane according to the time on the Hobbs meter, this is a needless expense. Instead, call for the clearance then wait for it without the engine running. Today's transistorized radios use very little power and you can listen for up to 15 minutes or so and still have plenty of starting power, unless the battery is weak or the temperature is very cold.

Avoid this dilemma entirely by using a hand-held transceiver to call clearance delivery. Again, the battery-powered transceiver is a backup radio in the event of a lost communications emergency.

When an IFR flight plan is filed, the information goes into an ATC computer. If the computer determines that the flight will not conflict with other traffic, clearance delivery might simply state the magic words: "Cleared as filed."

If there is a conflict, ATC will resolve the conflict by assigning a different altitude, a different route, or by clearance to a fix that is short of the destination. If cleared short of the destination, you will receive an amended clearance to the destination when the traffic conflict has been resolved. The flight plan normally remains in the computer until two hours after the proposed departure time. If the clearance is not requested by that time, the flight plan will be erased from the computer's memory unless an extension is requested.

Clearance on request

Here is a point that many people misunderstand. ATC will not ordinarily get your clearance from the computer until requested. I have seen quite a few students sit on the ramp with their engine running waiting in vain for some message from ATC. It is the IFR pilot's responsibility to inform clearance delivery or ground control when ready to copy the clearance.

The radio communication usually goes something like this:

> ***Pilot:*** *"Cessna three four five six Xray IFR Binghamton with information Romeo (the ATIS)."*

> ***ATC:*** *"Cessna three four five six Xray. Clearance on request."*

"Clearance on request" means that the clearance delivery controller does not have your clearance immediately available, or that there is a problem with it that must be resolved with ATC before it can be read to you. Sometimes, however, ATC will come right back with the full clearance; always have pencil in hand and be prepared to copy the clearance. In most cases there will be a delay between the time you call clearance delivery and when you actually receive the clearance.

Unacceptable clearances

If the clearance is unacceptable, read it back anyway, then explain. Valid reasons include:

- No survival gear for a long section that is over water
- Icing conditions
- Routing will add 50 minutes to the flight and exceed legal fuel reserves

The pilot in command is responsible for the safety of the flight; any compromises with safety are unacceptable.

When operating from an airport with a control tower, ATC will issue a clearance on either a clearance delivery frequency or on ground control. If operating from an airport not served by a control tower (or if the tower is closed) there are several ways to get the clearance.

Remote communications outlets (RCOs)

Check first to see if the uncontrolled departure airport has a remote communications outlet (RCO). The RCO transmitter and receiver antenna located on the airport is linked by landline to flight service or ATC. You can find RCOs listed in the A/FD in the communications section of the airport listing (Fig. 13-2) Some RCOs are also shown on sectional charts. If an RCO is located at an airport, request and copy a clearance and receive a "release" for takeoff from that RCO.

Check the RCO listings carefully. Sometimes an RCO will receive on 122.1 MHz and transmit on a VOR frequency. Numerous RCOs are found throughout the country at uncontrolled fields, as well as some controlled fields where towers do not operate 24 hours a day. Use an RCO to request and copy a clearance and receive a void time for takeoff.

Void time clearances

If there is no radio facility to issue an IFR clearance, it must be received by telephone. A clearance issued by telephone is called a *void time clearance* because ATC will always set a time limit after which the clearance is void. When you file IFR by telephone from an uncontrolled field, ask whom to call to pick up a clearance. Flight service will provide a telephone number, which might be that flight service station's number or an ATC telephone number.

Call back 10 minutes prior to the proposed departure time. Flight service or ATC will read the clearance over the telephone and you will read it back. Make sure everything is ready to take off immediately. If unable to take off within the time limit, call ATC back by telephone before the clearance is void and request a later "time window" for release and a new void time. The time window is the time between the release and the void time.

MASSACHUSETTS

NANTUCKET MEM (ACK) 3 SE UTC−5(−4DT) N41°15.18′ W70°03.61′   **NEW YORK**
48 B S4 **FUEL** 100LL, JET A ARFF Index A   **H−3I, L−25D**
RWY 06−24: H6303X150 (ASPH) S−75, D−170, DT−280 HIRL CL 0.3%up NE.   **IAP**
  RWY 06: MALSF. VASI(V4L)—GA 3.0°. Thld dsplcd 539′. **RWY 24:** SSALR. TDZL.
RWY 15−33: H3999X100 (ASPH) S−60, D−85, DT−155 MIRL
  RWY 15: REIL. Building. **RWY 33:** REIL. VASI(V4R)—GA 3.0°TCH 43′.
RWY 12−30: H3125X50 (ASPH) S−12
  RWY 12: Trees. **RWY 30:** Trees.
AIRPORT REMARKS: Attended continuously. Be aware of hi-speed military jet and heavy helicopter tfc vicinity of Otis
  ANGB. Deer and birds on and invof arpt. Rwy 12−30 VFR/Day use only aircraft under 12,500 lbs. Arpt has noise
  abatement procedures ctc Noise Officer 508−325−6136 for automated facsimile back information. PPR 2 hours
  for unscheduled air carrier ops with more than 30 passenger seat, call arpt manager 508−325−5300. When twr
  clsd ACTIVATE MALSF Rwy 06; SSALR Rwy 24; HIRL Rwy 06−24; MIRL Rwy 15−33 and twy lgts—CTAF. Rwy 24
  SSALR unmonitored when arpt unattended. Twy F prohibited to air carrier acft with more than 30 passenger
  seats. Fee for non-commercial acft parking over 2 hrs or over 6000 lbs. NOTE: See Land and Hold Short
  Operations Section.
WEATHER DATA SOURCES: ASOS (508) 325−6082. LAWRS.
COMMUNICATIONS: CTAF 118.3 **ATIS** 126.6 (508−228−5375) (1100−0200Z‡) Oct 1−May 14, (1100−0400Z‡) May
  15−Sept 30. **UNICOM** 122.95
  BRIDGEPORT FSS (BDR) TF 1−800−WX−BRIEF. NOTAM FILE ACK
  RCO 122.1R 116.2T (BRIDGEPORT FSS)
Ⓡ **CAPE APP/DEP CON** 126.1 (1100−0400Z‡) May 15−Sept 30, (1100−0300Z‡) Oct 1−May 14.
  BOSTON CENTER APP/DEP CON 128.75 (0400−1100Z‡) May 15−Sept 30, (0300−1100Z‡) Oct 1−May 14.
  TOWER 118.3 May 15−Sep 30 (1100−0300Z‡), Oct 1−May 14 (1130−0130Z‡).
  GND CON 121.7 **CLNC DEL** 128.25
AIRSPACE: CLASS D svc May 15−Sep 30 1100−0300Z‡, Oct 1−May 14 1130−0130Z‡ other times CLASS G.
RADIO AIDS TO NAVIGATION: NOTAM FILE ACK.
  (H) VOR/DME 116.2 ACK   Chan 109 N41°16.91′ W70°01.60′   236°2.3 NM to fld. 100/15W.
  WAIVS NDB (LOM) 248 AC   N41°18.68′ W69°59.21′   240° 4.8 NM to fld.
  NDB (HH−ABW) 194   TUK   N41°16.12′ W70°10.80′   115° 5.5 NM to fld.
  ILS/DME 109.1   I−ACK   Chan 28   Rwy 24.   LOM WAIVS NDB. ILS unmonitored when twr clsd.

NAUSET N41°41.51′ W69°59.39′   NOTAM FILE BDR.   **NEW YORK**
NDB (MHW) 279   CQX   at Chatham Muni. NDB unusable 220°−280° byd 20 NM.   **L−25D**

NEFOR N41°37.30′ W71°01.06′   NOTAM FILE EWB.
NDB (LOM) 274   EW   056°4.3 NM to New Bedford Regional.

NEW BEDFORD REGIONAL (EWB) 2 NW UTC−5(−4DT) N41°40.57′ W70°57.42′   **NEW YORK**
80 B S4 **FUEL** 80, 100LL, JET A OX 3, 4 LRA   **H−3I, L−25D, 28I**
RWY 14−32: H5000X150 (ASPH) S−33, D−48, DT−95 MIRL   **IAP**
  RWY 14: Bush. **RWY 32:** REIL. VASI(V4L)—GA 3.0°TCH 52′. Trees.
RWY 05−23: H4997X150 (ASPH) S−30, D−108, DT−195 HIRL
  RWY 05: MALSR. Trees. **RWY 23:** MALSR. VASI(V4L)—GA 3.0° TCH 51′. Thld dsplcd 413′. Trees.
AIRPORT REMARKS: Attended 1100−0500Z‡. Arpt CLOSED to touch and go ldg and training 0300−1000Z‡ daily. Birds
  and deer on and invof arpt. VASI Rwys 23 and 32 ops 24 hours. When twr clsd ACTIVATE HIRL Rwy 05−23, MIRL
  Rwy 14−32, MALSR Rwy 05 and Rwy 23, REIL Rwy 32—CTAF. Flight Notification Service (ADCUS) available.
  NOTE: See Land and Hold Short Operations Section.
WEATHER DATA SOURCES: ASOS 126.85 (1200−0300Z‡) LAWRS.
COMMUNICATIONS: CTAF 118.1 **ATIS** 126.85 508−994−6277. (1200−0300Z‡) **UNICOM** 122.95
  BRIDGEPORT FSS (BDR) TP 1−800−WX−BRIEF. NOTAM FILE EWB.
Ⓡ **PROVIDENCE APP/DEP CON** 128.7 (1100−0500Z‡)   **BOSTON CENTER APP/DEP CON** 124.85 (0500−1100Z‡)
  TOWER 118.1 (1200−0300Z‡) **GND CON** 121.9
AIRSPACE: CLASS D svc 1200−0300Z‡ other times CLASS G.
RADIO AIDS TO NAVIGATION: NOTAM FILE PVD.
  PROVIDENCE (H) VORTACW 115.6   PVD   Chan 103   N41°43.46′ W71°25.78′   112° 21.4 NM to fld. 50/14W.
    HIWAS.
  NEFOR NDB (LOM) 274   EW   N41°37.30′ W71°01.06′   056° 4.3 NM to fld.
  ILS/DME 109.7   I−EWB   Chan 34   Rwy 05.   LOM NEFOR NDB. (ILS unmonitored when twr clsd. BC
    unusable beyond 15° either side of LOC centerline and beyond 12 NM). Back course DME unusable beyond 12
    NM.

Fig. 13-2. *Typical listing in the* Airport/Facilities Directory *for a remote communications outlet (RCO).*

Taxi checks

Three important checklist items should be emphasized on an instrument flight.

Turn coordinator. How do the symbolic airplane and the ball move in a turn while taxiing? The airplane's "wings" tilt in the direction of the turn and the ball slides in the opposite direction to the outside of the turn. (On the older turn-and-slip indicator, the needle moves in the direction of the turn.)

The gyro of the turn needle is electrically powered. If the needle doesn't move, or moves erratically, it is either an instrument failure or an electrical failure. In either case, turn back. The airplane is not safe for an instrument flight.

Attitude indicator. The "wing" of the symbolic airplane should remain aligned with the horizon line. If the horizon display behind the symbolic wing pitches up or down or tilts beyond the slight movements seen during taxiing, it is an unreliable instrument. The gyros of the attitude indicator and the heading indicator are vacuum driven.

Cross-check the heading indicator. If it is drifting off heading, the vacuum system has probably failed. Check the suction gauge. If it is not within limits—usually 4.6–5.4" of mercury—turn back because the airplane is not safe for flight. If the suction gauge reads normally and the attitude indicator is not showing a normal display, turn back.

Heading indicator. A failure in the vacuum system will also affect the heading indicator. Headings shown on the heading indicator should change during a taxiing turn, then steady up to correctly match the magnetic compass on long, straight taxiways. Cross-check with the attitude indicator and suction gauge. If they show abnormal indications, it is a failure in the vacuum system. If the attitude indicator and suction gauge are normal and the heading indicator is erratic, it is an instrument failure in the heading indicator.

Rolling engine run-up

Normally those are the only checks made while the airplane is rolling. You do not want to get distracted while taxiing, particularly at night. However, there is an important exception to this rule when heavy commercial traffic is sharing the taxiways. The pilots of those big jets will certainly get upset if you come to a full stop and block the traffic for an engine run-up and takeoff checklist.

Do the run-up and takeoff checklist while taxiing. But first, discuss this technique with your instructor and practice it a few times with the instructor aboard when the taxi traffic is light.

Runway checks

Six items are on the instrument runway checklist:

1. **Correct Approach Chart.** The departure airport's approach in use in case you have to return shortly after takeoff.

2. **Nav Radios Set.** Nav 1 for approach in use in case you have to return quickly; nav 2 for first airborne fix.

3. **Correct Departure Control Frequency.** Be ready for a change to departure control quickly and without fumbling when tower hands you off.

4. **Lights On.** All ON for night operations, strobes and rotating beacon on for day IFR flights.

5. **Transponder** ALT **or** ON **and Set.** The correct code as given in the clearance.

6. **Pitot Heat** ON. Even on a clear day get in the habit of doing this to avoid inadvertently overlooking this important safeguard when flying into visible moisture.

7. **Check Heading Indicator Against Runway Heading.** Sometimes the heading indicator will drift off while taxiing. Reset if necessary.

Just before adding power for takeoff, do an "STP" check: strobes, transponder, pitot heat.

And finally, note time of takeoff just before adding power. Takeoff time is doubly important on an instrument flight for fuel calculations and for planning the arrival in the event of lost communications.

IFR communications

Departure is the pilot's busiest time on an IFR flight: controlling the plane through a wide variety of situations, from parking and taxi through takeoff and into an instrument climb; transitioning between flight using outside visual references and flight by instruments; trying to maintain an efficient climb with turns to comply with the departure clearance.

You must be listening for the airplane's N-number among many others that tower might be working and be ready to respond quickly and correctly when tower has an amended clearance or issues a frequency change.

Reply to every ATC communication promptly. This is especially important during readback of an amended clearance. Read it back immediately; if something puzzles you, figure it out later. ATC wants to hear from you right now. If a clearance is totally incomprehensible, say "stand by for readback" then call back and read back the entire clearance even if only the next section of the instructions is figured out. Research the rest of the clearance later.

Standard Phraseology

Use concise, standard phraseology and a professional tone of voice when working with ATC and the controllers will be inclined to assist as much as they possibly can. We've all heard transmissions like this: "This is Cessna November one two three four. I'm over (long pause), ah, Hartford. And I'm cruising at four thousand five hundred feet, departed from my home base at, ah (pause) ten o'clock on a VFR flight plan to New Jersey. Request permission to go through the New York Class B. Over."

Do you think ATC will show any enthusiasm for clearing this pilot through one of the country's busiest areas? Of course not. Time on the radio is very precious, especially around busy areas such as New York.

ATC is much more likely to grant a request if transmissions are brief and to the point. The more professional you sound, the more cooperation you will receive from ATC. This is especially true with IFR communications.

Who, who, where, what

Each time you switch to a new controller the initial call sequence consists of who, who, where, and what:

- Who you are calling
- Who you are (always give the full N number on the first call)

- Where you are located (east ramp, outer marker inbound, etc.)
- What your request or message is

Use the full registration to avoid a possibility of confusing similar numbers, such as 62876 and 67876.

As a general rule abbreviate the call sign after the controller begins doing so, and use the controller's abbreviation. This speeds things up considerably. There is no need to say "November" for flights within the United States. ("N" is the international designation for aircraft registered in the United States.)

For example, if the full identification is N3456X, the initial call would be "Cessna three four five six Xray." If the controller came back with "Five six Xray turn right to zero niner zero," use "five six Xray" until switched to another controller.

If ATIS is available, be sure to copy the ATIS information before calling clearance delivery or ground control for taxi clearance (or approach control for an arrival clearance). Give ground control the code for the ATIS and tell them that you are IFR. Ground control needs to know this to obtain a "release" from departure control or the air route traffic control center. You cannot take off on an IFR flight until the controlling authority for the airspace issues a release to enter that airspace on takeoff.

Calling ground control

The call to ground control might sound something like this:

"Westchester ground, Cessna three four five six Xray, terminal ramp, IFR Binghamton, information Delta. Ready to taxi."

Delta would be the code for the current ATIS information. It is changed every hour or sooner if there is a significant change in safety information. It is important, especially in IFR weather, to let the ground controller know that you have the latest information. If you don't, the ground controller will request verification. This results in extra transmissions that clutter up the frequency and wastes everybody's time.

Say it on the first transmission and avoid using the phrase "with the numbers" instead of the ATIS code. "With the numbers" tells the ground controller that you either didn't bother to obtain the ATIS information or that you got it and you could not remember the code.

You will remain on ground control until ready for takeoff, then ATC will switch you to tower. On the first call to the tower controller give the full call sign: "Westchester tower, Cessna three four five six Xray, ready for takeoff, IFR." Include "IFR" to alert the tower controller that the flight is IFR.

When switched to departure control after takeoff, the next call might be: "New York Departure, Cessna three four five six Xray, out of one thousand for three thousand [feet]."

Nearly every first contact with a new controller requires acknowledgment of altitude or the altitude passing through to an assigned altitude. This helps the controller verify that the actual altitude is the same as that reported by the Mode C transponder and shown on the controller's radarscope.

When you hear nothing further

When one facility switches you over to another facility, it's a *hand-off*. What if it comes time for a handoff—from tower to departure control, for example—and you receive no further instructions? Wait and hope for something to happen? Call someone? If so, who? And say what? This situation is not uncommon when the traffic is heavy. Sometimes the silence right after takeoff makes it feel like the tower has forgotten you.

If you have taken off and hear nothing further, continue climbing until at least 500 above the ground and established on departure heading. Then call the tower and ask: "Do you want four five Xray to go to departure control?" That will alert the tower to the fact that the handoff has not taken place. Don't just sit there and say "Well, they didn't ask me to go to departure control, so I'm not going to do it." As the pilot-in-command, get things straightened out.

Managing frequencies

Many experienced pilots keep a running list of the various frequencies used during a flight so that it is easy to return to the last frequency in case they or the controller makes a mistake in the next frequency. Another technique is to alternate between comm 1 and comm 2. Others like to work with only comm 1 while en route, leaving comm 2 tuned to the emergency frequency, 121.5 MHz, with the volume adjusted to a comfortable level.

If, for some reason, no one answers after switching to a newly assigned frequency, simply go back to the last assigned frequency, give the abbreviated call sign, and say "unable" with the name and frequency of the facility you couldn't contact. For example the transmission might be:

"Five six Xray unable Boston Center one three three point one."

ATC should respond to an "unable" message with further instructions.

Required reports

A major difference between VFR and IFR communications is that many IFR situations *require* reports to ATC. Pilots must make the following reports at all times, whether or not in radar contact with ATC:

- When leaving an assigned altitude.
- When changing altitude on a "VFR on top" clearance.
- When unable to climb or descend at a rate of at least 500 feet per minute.
- On commencing a missed approach. The report must include a request for clearance to make another approach attempt or to proceed to another airport.
- When average true airspeed at cruising altitude varies by more than 5 percent or 10 knots.
- Upon reaching a holding fix. The report must include time and altitude.
- When leaving a holding pattern.
- If, in controlled airspace, VOR, ADF, or ILS equipment malfunctions, or there is any impairment of air-to-ground communications (in case ATC is able to receive).
- Any information relating to the safety of the flight.
- Any weather conditions that have not been forecast or when encountering hazardous conditions that have been forecast.

When a flight is not in radar contact (Such as when ATC says "Radar contact lost" or "Radar service terminated" or while flying "VFR on top"), pilots must make these additional reports:

- En route position reports upon reaching all compulsory reporting points. These are indicated on en route charts by a solid triangle. Position reports include:
  - Position
  - Time
  - Altitude
  - ETA and name of next reporting point
  - Name only of next succeeding reporting point

For example: "Cessna three four five six Xray, Carmel, one five, six thousand, Kingston two zero, Albany."

- When leaving a final approach fix inbound on final approach.
- When it becomes apparent that a previously submitted estimate is in error in excess of 3 minutes.

Canceling IFR

Although you will prepare and file IFR flight plans for all training flights in this syllabus, you will probably terminate the IFR portion of many flights to conduct training exercises under VFR. This is a simple procedure. If you are able to carry out the rest of the flight in VFR conditions and wish to do so, simply tell ATC, "Cancel my IFR flight plan." Your request will be granted immediately—ATC will be glad to have one less airplane to control.

You, rather than ATC, will then become responsible for maintaining safe separation from other aircraft. As you work on the training exercises under the hood, the person in the other seat must act as safety pilot and must be qualified to do so.

There is one more thing to think about when canceling IFR. What kind of airspace are you in? You know you were in controlled airspace of some kind when you were operating IFR under ATC control. But what was it—B, C, D, or E airspace?

Each of these classifications limits the conduct of VFR flights in some respects, so be prepared to have ATC remain in contact with you and issue binding instructions if necessary. As part of your planning routine, make it a practice to determine what type of airspace you will encounter along your route of flight if you cancel IFR, as well as what restrictions apply. Consult AIM for details about VFR operations in each type of airspace.

Familiarity with the ATC system through actually using it builds confidence and competence and it is great practice!

Radio contact lost

Anticipate a handoff when you have been with one controller for quite some time and the controller requests an ident. Occasionally you might lose radio contact when flying at low altitudes or in areas of high terrain, or a long distance from the transmitter site. Call for a radio check; if there is no response after two or three attempts, try to reestablish communications through another nearby frequency listed on the IFR charts, such as flight service, approach control, or even a tower. As a last resort use the emergency frequency 121.5 MHz.

Always consider phraseology before transmitting and then speak in the most professional manner possible. Always be brief and to the point. Listen for a break in other transmissions and key the microphone immediately when a break occurs. Quickly give aircraft N-number and a concise message on the first call. That will save precious radio time for everyone on that frequency.

Section 3

Instrument charts and procedures

14

En route charts

Flying instrument departure, en route, and arrival procedures means reading and interpreting instrument charts. Most new instrument pilots in the United States learn using the U.S. National Oceanic Service (NOS) government charts. The government publishes low-altitude (below 18,000 feet mean sea level) and high-altitude (flight level 180 and above) en route charts, instrument approach "plates," Standard Instrument Departure (SID) and Standard Terminal Arrival (STAR) procedures, and a host of other instrument and visual navigational products.

Commercial providers also have instrument charts for sale. Most notably there's Jeppesen. Jeppesen takes the government data and presents it in a format that many feel to be superior and easier to use than the NOS charts. Jeppesen products are used worldwide and are the charts of choice for most airlines and many professional pilots.

Other chart providers have entered the market. Howie Keefe's Air Chart Systems reproduces the NOS charts in handy bound volumes. They eliminate a lot of the paperwork of filing chart updates by not sending customers a completely new set of charts every update cycle, but instead publishing "pen-and-ink changes" to the bound volumes. Pilots are responsible for looking through the change notices and writing changes on the charts themselves before flight.

With the rapid advance of computer capability and the promise of future "paperless" cockpits where chart data is called up on a computer screen as needed, it's possible that other chart vendors and formats may enter the picture. Regardless of which source a pilot uses in obtaining instrument navigational products, however, they all use the same government data as a primary source, and it's easy for most pilots to "jump" back and forth from using one product to another without much problem. With that in mind, let's look at the available instrument flying charts, using the NOS format.

Let's turn our attention now to the en route charts. We'll only be working with Jeppesen charts; National Ocean Service (NOS) charts will be easy to transition to when necessary. The low-altitude (LO) charts in the continental United States and Canada are effective up to but not including 18,000 feet MSL. If you intend to fly higher, use either the high (HI) or high/low (H/L) charts.

Look at the index map on the cover of the chart (Fig. 14-1), and note the gray squares and rectangles. These are found around some very large cities where the airway/facility congestion is so bad that a lot of information has to be omitted. To solve this problem, Jeppesen has larger-scale area charts that should be used whenever you are flying into or out of these specific areas. The area charts are referenced on the en route charts by a heavy dashed line with the identifier and location name.

Many of the symbols that are found on the area charts are shown on the en route charts in reduced scale so that pilots who are overflying the area can do so without having to open an area chart. But remember, if you are going to land or take off within that area, you need to use the area chart. Many important intersections and other features are shown on the area charts.

The LO and H/L Jeppesen charts have a unique feature that Jeppesen calls a Zigdex. You know how much of a problem it is to fold and unfold large charts in the confined space of small cockpits. This is even more difficult when you are trying to control a plane under IFR conditions. On the back of the chart (Fig. 14-2), note that the upper edge has been trimmed off at an angle so that when it is folded up, it zigzags back and forth like the switchbacks of a mountain trail. These folds divide the face of the chart into three sections. Each fold is designated by a major city found within that specific section. To find a specific area on the chart, it is only necessary to open it up at the correct fold. For example, if you're going to be operating around Green Bay, you would open chart 19 at fold 2. From then on, all you have to do is open and close the folds like the pages in a book, which is a lot easier than trying to open and refold the entire chart.

Most elevations on the en route chart are in feet above mean sea level (MSL), although some are in flight levels, in which case they will be noted as such.

◀20 US (LO)   JEPPESEN   US (LO)19▶
1 INCH = 20 NM   1 INCH = 20 NM
UNITED STATES
LOW ALTITUDE ENROUTE CHARTS

MEETS FAA REQUIREMENTS FOR AERONAUTICAL CHARTS
Within the continental U.S., the airways shown on these charts are effective up to but not including 18,000' MSL. At 14,500' MSL and above, all airspace is controlled Class E airspace. Shaded blue areas shown on the face of enroute charts designate airspace that is not controlled (Class G) below 14,500' MSL. The Jet Route structure is superimposed to show its relationship with the low altitude airways. For operational detail for Jet Routes at or above 18,000' MSL, use US(HI) charts.

REVISION DATA

CHART US(LO)19 2 DEC 94 Stevens Point, Wisc VORTAC station declination changed. Red Wing, Minn NDB freq changed. Mason City, Iowa VOR holding desig. Plugs Int formation bearing changed (Gopher, Minn R-333). Wetly Int renamed Apnel (Wiarton, Ont R-316).

CHART US(LO)20 2 DEC 94 Manchester, NY VORTAC shutdown. Picaa Int renamed Buker (Berlin, NH R-178). Alert Area CY(A)-611(T) revoked (NE of Montreal, Que VOR).

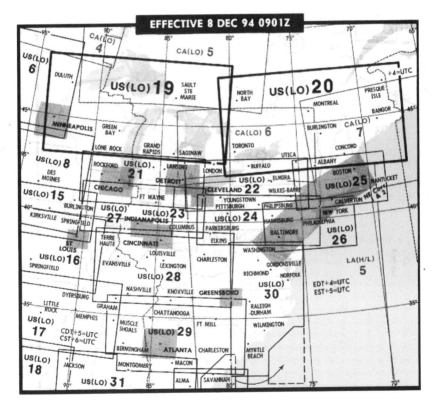

Fig. 14-1 *Reproduced with permission of Jeppesen Sanderson, Inc. Not for use in navigation.*

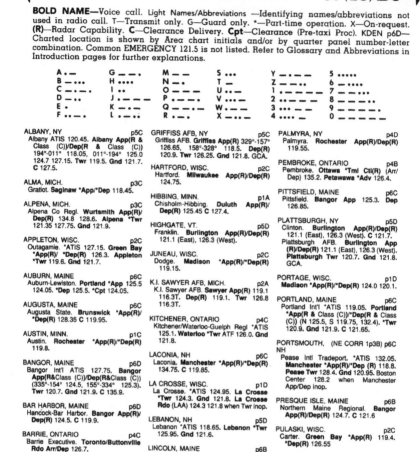

Fig. 14-2 *Reproduced with permission of Jeppesen Sanderson, Inc. Not for use in navigation.*

En route communications are shown in two places on the chart: on the face of the chart and on the end folds (Fig. 14-2) for convenience when planning a flight. Except on the high-altitude charts, terminal communications are also shown in the tabulations on the end folds.

The lower part of the facing page of the chart (Fig. 14-3) also describes the restricted airspace found on the chart. A circle explaining

AIRSPACE RESTRICTED AREAS
LEGEND

CY-Canada
(Canadian Alert Areas additionally coded as (A) Acrobatic (H) Hang Gliding
(P) Parachute Dropping (S) Soaring (T) Training)

22 APR 94
R-4201A
GND-23000
0800-1600 LT
TUE-SAT
O/T BY NOTAM
MINNEAPOLIS ARTCC

R-4201B
GND-9000
0001 LT SAT-
2359 LT SUN
O/T BY NOTAM
MINNEAPOLIS ARTCC

R-4202
GND-9000
NOTAM
MINNEAPOLIS ARTCC

R-5201
GND-23000
CONT 4/1-9/30
GND-20000
0600-1800 LT
10/1-3/31
& BY NOTAM
BOSTON ARTCC

R-5207
GND-2000
0730-1815 LT
MON-FRI

R-6501A
GND-4000
0700-2300 LT
MON-FRI
0000-2359 LT
SAT-SUN
O/T BY NOTAM
BURLINGTON APP

R-6501B
4000-13600
INTERMITTENT
BURLINGTON APP

R-6901A
GND-20000
CONTINUOUS
5/1-9/30
0800-2200 LT
MON-THUR
0800 LT FRI-
2200 LT SUN
O/T BY NOTAM
10/1-4/30
MINNEAPOLIS ARTCC

R-6901B
GND-20000
NOTAM
MINNEAPOLIS ARTCC

CY(R)-508
GND-1400
DAYLIGHT

CY(A)-520(T)
GND-4000
DAYLIGHT

CY(R)-604
GND-1500

CY(R)-613
GND-800

CY(R)-614
GND-1300

CY(A)-622(F)
GND-2000
DAYLIGHT
MON-FRI
4/1-11/30
NOTAM

CY(R)-624
GND-1300
DAYLIGHT
BELL HELICOPTER/
TEXTRON

CHIPPEWA MOA
7000-FL 180
TUE-SAT
0930-1130 &
1330-1530 EST
O/T BY NOTAM
MINNEAPOLIS ARTCC
EXCLUDES FL 180

FALLS 1/2 MOAs
500 AGL-FL 180
1200-2000 LT TUE
0800-1600 LT
WED-SAT
O/T BY NOTAM
MINNEAPOLIS ARTCC
EXCLUDES FL 180

VOLK EAST MOA
8000'-FL180
1200-2000 LT TUE
0800-1600 LT
WED-SAT
O/T BY NOTAM
CHICAGO ARTCC
EXLUDES FL 180

VOLK SOUTH MOA
500 AGL-FL 180
1200-2000 LT TUE
0800-1600 LT
WED-SAT
O/T BY NOTAM
CHICAGO ARTCC
EXCLUDES FL 180

VOLK WEST MOA
500 AGL-FL 180
1200-2000 LT TUE
0800-1600 LT
WED-SAT
O/T BY NOTAM
MINNEAPOLIS ARTCC
EXCLUDES FL 180 &
R-6904A/B when active

CRUISING ALTITUDES

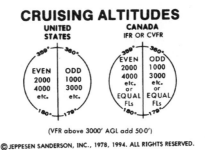

UNITED STATES

CANADA IFR OR CVFR

EVEN 2000 4000 etc. | ODD 1000 3000 etc.

EVEN 2000 4000 etc. or EQUAL FLs | ODD 1000 3000 etc. or EQUAL FLs

(VFR above 3000' AGL add 50·0')

Fig. 14-3 *Reproduced with permission of Jeppesen Sanderson, Inc. Not for use in navigation.*

the hemispherical rule altitudes is on the bottom of the page, with a note below it to add 500 feet for VFR altitudes.

Road maps in the sky

The charts are not all drawn to the same scale, so check the scale printed along the upper or side margin of the chart face before measuring anything on it. Jeppesen manufactures navigation chart plotters that have a compass rose as well as all of the various scales marked on them. These plotters come in very handy when measuring distances and bearings.

Let's take a look at some of the symbols to be found on the en route chart. Radio identification and communications can be seen in Fig. 14-4.

All navaid facilities that form an airway or a route component are enclosed in shadow boxes. If they are not collocated, they will be so noted in parentheses below the box. The name of the facility, its frequency, three-letter identification, and Morse code identifier will be found inside the box, and if it has DME capability that is frequency-paired, a small D will be included preceding the frequency.

There are still quite a few TACAN stations operating throughout the country. Some of these have frequency pairing, which means that the VOR frequency is also the DME frequency. In this case, the frequency is shown with the three-letter ID. If it is not frequency-paired, the ghost VOR frequency will be shown in parentheses below the TACAN information.

To facilitate aircraft using area navigation such as *loran-C, inertial navigation system* (INS), and *GPS*, the H/L and HI altitude charts also show the geographical coordinates below the code ID for all navaids forming airways and routes.

When landing aids perform en route functions, they are identified by round-cornered boxes. As in the navaid shadow box, the frequency is provided in the box, as well as the DME if it is available and frequency-paired. These facilities include LOC, SDF, LDA, MLS, and KRM. Certain facilities have not been previously defined in this book; Jeppesen defines them as:

> *Simplified Directional Facility (SDF)—A navaid used for nonprecision instrument approaches. The final approach course is similar to that of an ILS localizer except that the SDF course may be offset from the runway, generally not more than 3 degrees, and the course may be wider than the localizer, resulting in a lower degree of accuracy.*
>
> *Microwave Landing System (MLS)—An instrument landing system operating in the microwave spectrum which provides lateral and vectored guidance to aircraft having compatible avionics equipment.*
>
> *KRM—A system which provides three-axis position indication, similar to ILS, used in eastern Europe.*

Road maps in the sky   273

ENROUTE CHART LEGEND

NAVAID IDENTIFICATION

```
┌─STOUT─┐
│(R) 14.1 STO│
│··· - ───│
└───────┘
```

Navaid identification is given in shadow box when navaid is airway or route component, with frequency, identifier, and Morse Code. DME capability is indicated by a small "D" preceding the VOR frequency at frequency paired navaids. VOR and VOR-TAC navaid operational ranges are identified (when known) within the navaid box except on USA and Canada charts. (T) represents Terminal; (L) represents Low Altitude; and (H) represents High Altitude.

Heavier shadow boxes are gradually replacing existing shadow boxes. There is no difference in meaning.

```
┌─           ─┐
│             │
└─           ─┘
```

On HIGH/LOW altitude enroute charts, geographical coordinates (latitude and longitude) are shown for navaids forming high or all altitude airways and routes. On Area charts, geographical coordinates are shown when navaid is airway or route component.

```
┌─KADENA─┐
│ D 12.0 KAD │
│N26 22.4 E127 48.0│
│  335 KD  │
│N26 20.0 E127 44.8│
└────────┘
```

Some L/MF navaids are combined in the shadow box even though they are not part of the airway/route structure, except on US and CA charts. They are used for course guidance over lengthy route segments when airway/track is designated into a VOR.

```
┌─BENBECULA─┐
│ D 14.4 BEN │
│ ··· - ─── │
└──────────┘
(Not Co-located)
```

When VOR and TAC/DME antennas are not co-located, a notation "Not Co-located" is shown below the navaid box.

MOODY
113.3 VAD
TAC-80

KENNEY
254 ENY

TAPTHONG
POINT
```
┌─         ─┐
│(T) 15.5 TH │
│ ─  ···· │
└─         ─┘
```

LIPTON
TAC-88 LPT
(114.1)

GRAND VIEW
D**115.4 GND**

GREAT BARRINGTON
MASS 739
395 GBR

```
─────LOC─────
( 108.7 IMBS )
 ·· ─── ···· ···
```

LAYTON
─── ───

(Domestic)
BATHURST
DME CH-19
BTH

Off-airway navaids are unboxed on Low and High/Low charts. TACAN/DME channel is shown when VOR navaid has frequency paired DME capability. When an L/MF navaid performs an enroute function, the Morse Code of its identification letters are shown. (Off-airway VORs are boxed except on US and CA charts.)

When TACAN or DME are not frequency paired with the VOR, the TACAN is identified separately. The "Ghost" VOR frequency, shown in parentheses, enables civilian tuning of DME facility.

The navaid frequency and identification are located below the location name of the airport when the navaid name, location name, and airport name are the same.

LOC, SDF, LDA, MLS, and KRM navaids are identified by a round cornered box when they perform an enroute function. Frequency identification and Morse Code are provided. DME is included when navaid and DME are frequency paired.

Fan Marker name and code.

```
                                    ┐
```
Australia Domestic DME. Operates on 200 MHz and requires airborne receiver specific to this system.
```
                                    ┘
```

COMMUNICATIONS

RADIO FREQUENCIES
Frequencies for radio communications are included above NAVAID names, the NAVAID. These frequencies are also shown at other remoted locations. in the 120 MHz range, are shown with the numbers "12" omitted; 122.2 is etc. HF and LF frequencies are not abbreviated.

when voice is available through Radio Frequencies, which are shown as 2.2, 122.35 as 2.35,

```
  2.2-2.45-5680
┌─RIVER─┐
│ D 14.6 RIV │
│ ··· · ──·· │
└──────┘
```
River Radio transmits on 114.6 and transmits and receives on 122.2, 122.45 MHz and HF frequency 5680. SSB indicated single side band not available.

```
  2.1G-RIV
┌─CANYON─┐
│ 113.9 CNY │
└────────┘
```
River Radio (RIV) guards (receives) on 122.1 and transmits through Canyon VOR on 113.9.

```
  2.6-RIV
┌─────────┐
│ DIAMOND │
└─────────┘
    ⊙
```
River Radio transmits and receives on 122.6 located at Diamond. Small circle enclosing dot denotes remote communication site.

```
   2.2-2.4
┌─TAPEATS─┐
│ D 12.2 TPT │
└─────────┘
  HIWAS
MIA WX-* 2.0
┌─MIAMI─┐
│ D 115.9 MIA │
│N25 57.8 W080 27.6│
└──────┘
```
Tapeats Radio transmits and receives on 122.2 and 122.4. Telephone symbol indicates additional communications in communications panel listed under Tapeats.

HIWAS — Hazardous Inflight Weather Advisory Service. Broadcasts SIGMETS, AIRMETS and PIREPS continuously over VOR frequency.

```
  2.3-PTM
  2.6-PTM
┌─PHANTOM─┐
│ 364 PTM │
└─────────┘
```
River Radio transmits and receives at Phantom on 122.3. Additonally, Phantom Radio transmits and receives on 122.6.

```
  (RIVER FSS)
┌─LAVA─┐
│ D 115.3 LVA │
│ ·─·· ···─ ·─ │
└─────┘
```
River Radio transmits through Lava VOR on 115.3, but is not capable of receiving transmissions through the VOR site.

```
2.2-2.6-3.6(AAS)
GRAND ARIZ
   1285
```

```
  3.6 AAS
NORTHSIDE
   390
```

```
U-2.8 MF/10 NM
NORTHSIDE
   390
```

```
3.6 ATF MOOSE
NORTHSIDE
   390
```

Grand Radio is located at the airport and transmits and receives on 122.2 and 122.6. Additionally, Grand Radio provides AAS (Airport Advisory Service) on 123.6.

Terminal Radio frequencies and service may be included over airport or location name. Radio call is included when different than airport or location name. Mandatory Frequencies (MF), Aerodrome Traffic Frequencies (ATF) or UNICOM (U) frequencies include contact distance when other than the standard 5 nm.

Fig. 14-4 *Reproduced with permission of Jeppesen Sanderson, Inc. Not for use in navigation.*

Flight service station (FSS) VHF frequencies in the United States and Canada will be found above the navaid names. Frequencies shown are usually all in the 120-MHz range and are shown with the first two digits (12) omitted; 3.6 denotes 123.6 MHz.

Figure 14-4 has five samples of how FSS River Radio operates. In the first example, you will see that River VOR is on frequency 114.6, with the identification RIV. It has DME and will transmit and receive on 122.2 and 122.45 MHz. It also transmits on 114.6 MHz. In addition it has two-way high-frequency communications capability on 5680.

The flight route runs across Canyon VOR on 113.9 MHz with the identifier CNY. Note that it does not have DME because there is no small D before the frequency. Also note that River Radio remotes through it, transmitting on the VOR frequency and receiving on 122.1 MHz. The letter G following the frequency 122.1 indicates that it guards (listens to) that frequency but cannot transmit over it. The letter RIV following the G lets you know that you should use River Radio as the calling facility.

Sometimes you will find a *remote communications outlet* (RCO), here shown as diamond. This is not a VOR; hence, there is no VOR frequency, identifier, or Morse-code symbology, but you can see above the box that River Radio transmits and receives on 122.6 MHz, and the location of the antenna is shown by the dot enclosed in the circle.

You might run across an L/MF facility such as Phantom (364 MHz PTM) that is used by both River Radio on 122.3 MHz and Phantom Radio on 122.6 MHz.

Finally, selected VORs enable an FSS to transmit on the VOR frequency but do not have any receiving capability. Such a situation is illustrated at LAVA (115.3 MHz LVA).

The telephone symbol tells you that the facility has communications capabilities other than those shown on the charts. These will be found in the communications panel under the name of the facility. Transcribed weather broadcasts are indicated by the letters TWEB in parentheses over the name of the facility (Fig. 14-5). An asterisk (*) indicates that a facility's operation or service is not continuous.

As far as restricted airspace is concerned, it is important to recognize that the words and figures found near the restricted airspace block

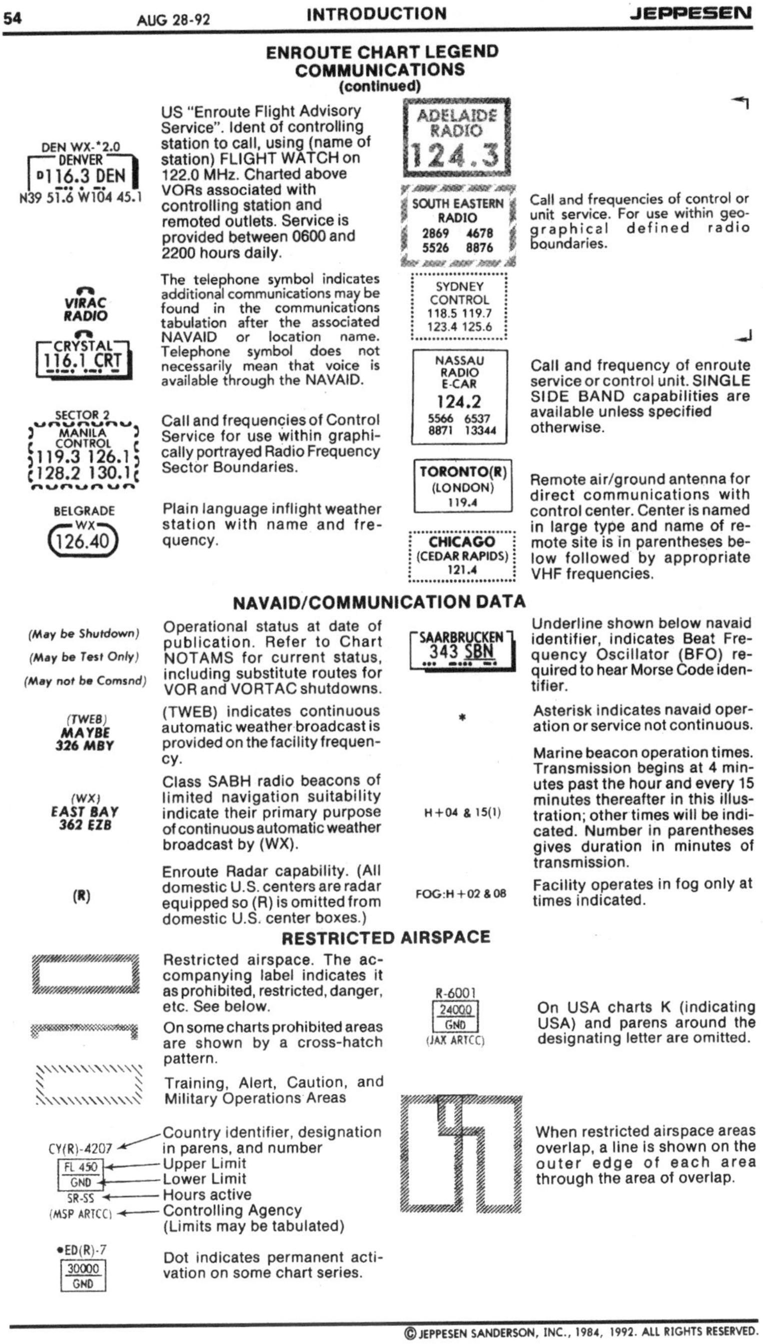

54   AUG 28-92   **INTRODUCTION**   **JEPPESEN**

ENROUTE CHART LEGEND
COMMUNICATIONS
(continued)

DEN WX-*2.0
DENVER
◘116.3 DEN
N39 51.6 W104 45.1

US "Enroute Flight Advisory Service". Ident of controlling station to call, using (name of station) FLIGHT WATCH on 122.0 MHz. Charted above VORs associated with controlling station and remoted outlets. Service is provided between 0600 and 2200 hours daily.

ADELAIDE
RADIO
124.3

⌐

SOUTH EASTERN
RADIO
2869   4678
5526   8876

Call and frequencies of control or unit service. For use within geographical defined radio boundaries.

VIRAC
RADIO
CRYSTAL
116.1 CRT

The telephone symbol indicates additional communications may be found in the communications tabulation after the associated NAVAID or location name. Telephone symbol does not necessarily mean that voice is available through the NAVAID.

SYDNEY
CONTROL
118.5 119.7
123.4 125.6

⌐

NASSAU
RADIO
E-CAR
124.2
5566   6537
8871   13344

Call and frequency of enroute service or control unit. SINGLE SIDE BAND capabilities are available unless specified otherwise.

SECTOR 2
MANILA
CONTROL
119.3 126.1
128.2 130.1

Call and frequencies of Control Service for use within graphically portrayed Radio Frequency Sector Boundaries.

BELGRADE
WX
126.40

Plain language inflight weather station with name and frequency.

TORONTO(R)
(LONDON)
119.4

CHICAGO
(CEDAR RAPIDS)
121.4

Remote air/ground antenna for direct communications with control center. Center is named in large type and name of remote site is in parentheses below followed by appropriate VHF frequencies.

NAVAID/COMMUNICATION DATA

(May be Shutdown)
(May be Test Only)
(May not be Comsnd)

Operational status at date of publication. Refer to Chart NOTAMS for current status, including substitute routes for VOR and VORTAC shutdowns.

SAARBRUCKEN
343 SBN

Underline shown below navaid identifier, indicates Beat Frequency Oscillator (BFO) required to hear Morse Code identifier.

(TWEB)
MAYBE
326 MBY

(TWEB) indicates continuous automatic weather broadcast is provided on the facility frequency.

*

Asterisk indicates navaid operation or service not continuous.

(WX)
EAST BAY
362 EZB

Class SABH radio beacons of limited navigation suitability indicate their primary purpose of continuous automatic weather broadcast by (WX).

H+04 & 15(1)

Marine beacon operation times. Transmission begins at 4 minutes past the hour and every 15 minutes thereafter in this illustration; other times will be indicated. Number in parentheses gives duration in minutes of transmission.

(R)

Enroute Radar capability. (All domestic U.S. centers are radar equipped so (R) is omitted from domestic U.S. center boxes.)

FOG:H+02 & 08

Facility operates in fog only at times indicated.

RESTRICTED AIRSPACE

Restricted airspace. The accompanying label indicates it as prohibited, restricted, danger, etc. See below.

On some charts prohibited areas are shown by a cross-hatch pattern.

R-6001
24000
GND
(JAX ARTCC)

On USA charts K (indicating USA) and parens around the designating letter are omitted.

Training, Alert, Caution, and Military Operations Areas

CY(R)-4207 — Country identifier, designation in parens, and number
FL 450 — Upper Limit
GND — Lower Limit
SR-SS — Hours active
(MSP ARTCC) — Controlling Agency (Limits may be tabulated)

When restricted airspace areas overlap, a line is shown on the outer edge of each area through the area of overlap.

●ED(R)-7
30000
GND

Dot indicates permanent activation on some chart series.

Fig. 14-5 *Reproduced with permission of Jeppesen Sanderson, Inc. Not for use in navigation.*

have a given order that will include the designation, both the upper and lower limits, the hours of operation, and the controlling agency. Many of them will have infrequent hours of operation and will list NOTAMs rather than days and times; check the current NOTAMs to see if these areas will be activated during the time you plan to be near the area.

Airports are depicted similar to the way they are on an airway or aeronautical chart; specific symbols can be seen in Fig. 14-6. An airport that does not have a Jeppesen approach chart is designated by printing the airport name in both uppercase and lowercase letters. The elevation shown below the name is in feet above mean sea level.

If an airport has a Jeppesen approach chart, the name of the location, by which it is indexed, is printed all in uppercase letters. If the airport name is different from the location name, the airport name will be printed in small type below the location using both uppercase and lowercase letters. This is useful information if you know the airport name and its approximate location but not the actual listed name. In this instance, look around the general area until you locate the airport name and note the chart designation name above it.

Now, look about halfway down the left side of the page (Fig. 14-6). Here you can see that there are four different groups of symbols for reporting points listed as FIXES. These are for compulsory reporting points, on-request reporting points, low-altitude reporting points, and low-altitude noncompulsory reporting points.

With so much radar available that has excellent reliability, fewer and fewer compulsory reporting points are on the airways. Most compulsory points are at the VORs, and even these are becoming fewer in number. Unused information is usually forgotten; examine Fig. 14-6 to refresh your memory.

If a holding pattern depicted on the chart is a DME holding pattern, the first number shown will be the DME distance of the fix in relation to the DME facility upon which the holding pattern is predicated. The second number is the DME distance at the outbound limit. In the example shown, the fix is at 31 DME, the holding pattern is shown as utilizing nonstandard (left) turns, and you would turn inbound again at 39 DME.

If the length of the holding pattern will be other than standard, the time (in minutes) will be printed in white numerals inside a dark diamond.

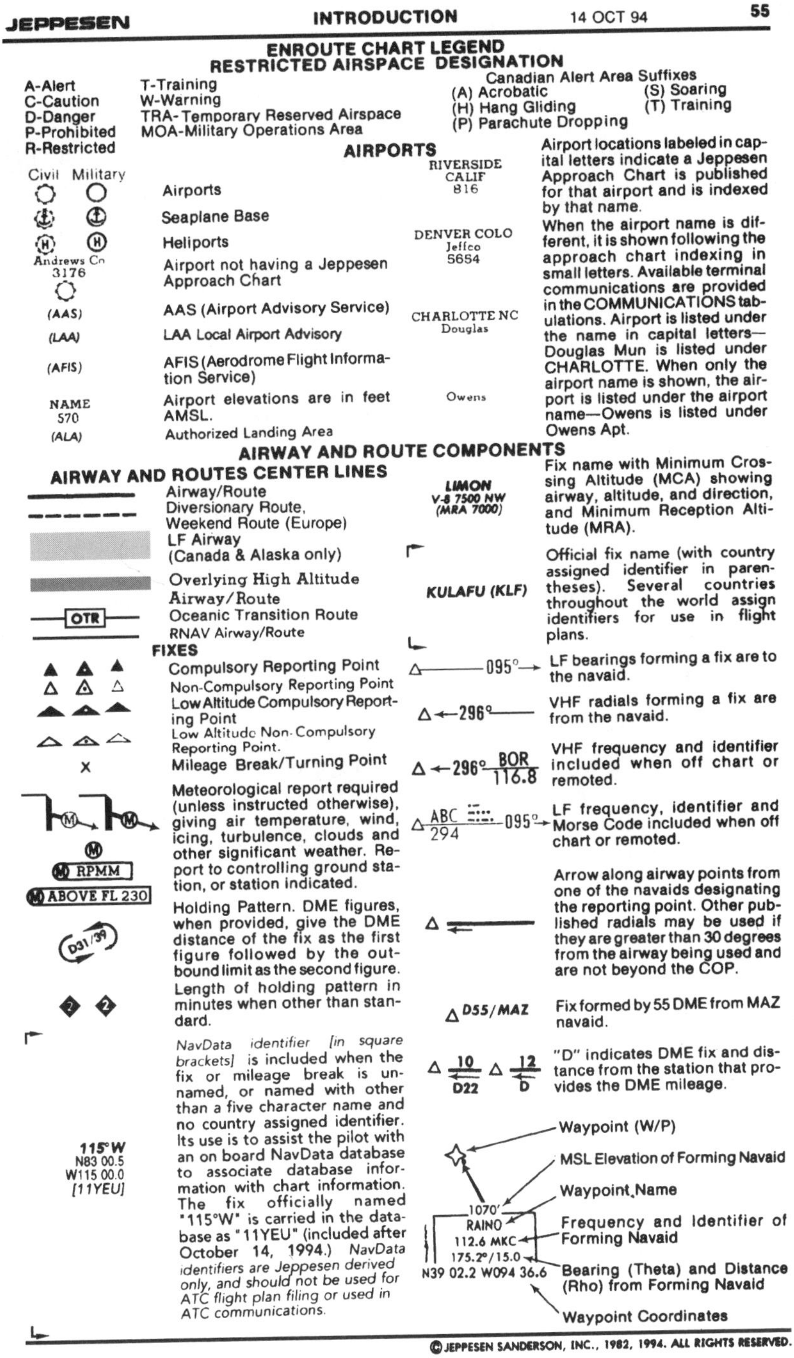

JEPPESEN   INTRODUCTION   14 OCT 94   **55**

ENROUTE CHART LEGEND
RESTRICTED AIRSPACE DESIGNATION

A-Alert
C-Caution
D-Danger
P-Prohibited
R-Restricted

T-Training
W-Warning
TRA- Temporary Reserved Airspace
MOA-Military Operations Area

Canadian Alert Area Suffixes
(A) Acrobatic   (S) Soaring
(H) Hang Gliding   (T) Training
(P) Parachute Dropping

AIRPORTS

Civil	Military		
○	○	Airports	
⊕	⊕	Seaplane Base	
Ⓗ	Ⓗ	Heliports	

Andrews Cn
3176

○ Airport not having a Jeppesen
   Approach Chart

(AAS) AAS (Airport Advisory Service)

(LAA) LAA Local Airport Advisory

(AFIS) AFIS (Aerodrome Flight Informa-
        tion Service)

NAME
570   Airport elevations are in feet
      AMSL.

(ALA) Authorized Landing Area

RIVERSIDE
CALIF
816

DENVER COLO
Jeffco
5684

CHARLOTTE NC
Douglas

Owens

Airport locations labeled in capital letters indicate a Jeppesen
Approach Chart is published
for that airport and is indexed
by that name.

When the airport name is different, it is shown following the
approach chart indexing in
small letters. Available terminal
communications are provided
in the COMMUNICATIONS tabulations. Airport is listed under
the name in capital letters—
Douglas Mun is listed under
CHARLOTTE. When only the
airport name is shown, the airport is listed under the airport
name—Owens is listed under
Owens Apt.

AIRWAY AND ROUTE COMPONENTS
AIRWAY AND ROUTES CENTER LINES

━━━━━ Airway/Route
─ ─ ─ ─ Diversionary Route,
         Weekend Route (Europe)
▬▬▬▬ LF Airway
         (Canada & Alaska only)
████ Overlying High Altitude
         Airway/Route
─[OTR]─ Oceanic Transition Route
━━━━━ RNAV Airway/Route

FIXES

▲ ▲ ▲ Compulsory Reporting Point
△ △ △ Non-Compulsory Reporting Point
▲▲ ▲▲ Low Altitude Compulsory Report-
         ing Point
△△ △△ Low Altitude Non- Compulsory
         Reporting Point
✕ Mileage Break/Turning Point

Ⓜ Ⓜ Meteorological report required
(unless instructed otherwise),
giving air temperature, wind,
icing, turbulence, clouds and
Ⓜ other significant weather. Report to controlling ground sta-
Ⓜ RPMM tion, or station indicated.

Ⓜ ABOVE FL 230

(D31/39) Holding Pattern. DME figures,
when provided, give the DME
distance of the fix as the first
figure followed by the outbound limit as the second figure.

◆ ◆ Length of holding pattern in
minutes when other than standard.

┌ NavData identifier [in square
brackets] is included when the
fix or mileage break is unnamed, or named with other
than a five character name and
no country assigned identifier.
Its use is to assist the pilot with
an on board NavData database
to associate database information with chart information.
The fix officially named
"115°W" is carried in the database as "11YEU" (included after
October 14, 1994.) NavData
identifiers are Jeppesen derived
only, and should not be used for
ATC flight plan filing or used in
ATC communications.

115°W
N83 00.5
W115 00.0
[11YEU]

LIMON
V-8 7500 NW
(MRA 7000)

KULAFU (KLF)

△────095°→

△←296°─

△←296° BOR
        116.8

ABC ⋯
294 ─── 095°→

△ ⇌

△ D55/MAZ

△ 10 △ 12
D22   D

Fix name with Minimum Crossing Altitude (MCA) showing
airway, altitude, and direction,
and Minimum Reception Altitude (MRA).

Official fix name (with country
assigned identifier in parentheses). Several countries
throughout the world assign
identifiers for use in flight
plans.

LF bearings forming a fix are to
the navaid.

VHF radials forming a fix are
from the navaid.

VHF frequency and identifier
included when off chart or
remoted.

LF frequency, identifier and
Morse Code included when off
chart or remoted.

Arrow along airway points from
one of the navaids designating
the reporting point. Other published radials may be used if
they are greater than 30 degrees
from the airway being used and
are not beyond the COP.

Fix formed by 55 DME from MAZ
navaid.

"D" indicates DME fix and distance from the station that provides the DME mileage.

Waypoint (W/P)

MSL Elevation of Forming Navaid

Waypoint,Name

1070'
RAINO
112.6 MKC
175.2°/15.0
N39 02.2 W094 36.6

Frequency and Identifier of
Forming Navaid

Bearing (Theta) and Distance
(Rho) from Forming Navaid

Waypoint Coordinates

Fig. 14-6 *Reproduced with permission of Jeppesen Sanderson, Inc. Not for use in navigation.*

Minimum altitudes, fixes, and reporting points

Although many fixes show a *minimum reception altitude* (MRA), there are many parts of the country in which you can fly all your life and never encounter a *minimum crossing altitude* (MCA). The pilot/controller glossary defines MCA:

> *The lowest altitude at certain fixes at which an aircraft must cross when proceeding in the direction of a higher minimum enroute IFR altitude (MEA).*

FAR 91.177, which deals with minimum altitudes for IFR operations states:

> *(b) Climb. Climb to a higher minimum IFR altitude shall begin immediately after passing the point beyond which that minimum altitude applies, except that when ground obstructions intervene, the point beyond which the higher minimum applies shall be crossed at or above the applicable MCA.*

You can see that the reasoning for an MCA is that you will find higher obstructions somewhere beyond the fix (and usually quite close to it), and you will want to be above them before you get into that territory.

LIMON intersection's MCA is explained in Fig. 14-6. The symbol will first show the airway on which the MCA is applicable, the altitude of the MCA, and the direction of travel for which it is applicable. Figure 14-7 shows what LIMON might look like on a chart. Let's say that V-8 at LIMON is an NW/SE airway and you are eastbound on V-5, an E/W airway, planning to turn SE on V-8. In this case, the MCA would not apply to you.

If you were either westbound on V-5 and planning to turn to the northwest or you were northwestbound on V-8, the MCA would be in effect and would require crossing the fix (LIMON) at or above the MCA.

Many fixes are formed by a VOR nonairway radial that crosses an airway; take a look at the symbol about midway down the right-hand column of Fig. 14-6. In this case, the airway, which has an on-request reporting point on it, could be oriented in any direction and is not illustrated. The 296-degree radial from the BOR VOR (on VHF frequency 116.8) is not part of any airway, but it helps to identify the

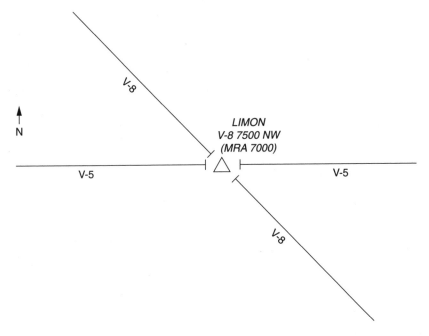

Fig. 14-7

on-request reporting point. The VOR itself is off to the right of the on-request reporting point; once again, directions shown in regard to VORs are the radials FROM rather than the bearings TO the facility. Directions with regard to LF facilities, such as NDBs, are printed in green on the chart (as opposed to blue for VHF aids) and are bearings to the station.

If the fix is formed by a VOR radial or an LF bearing that is being re-moted or is off the chart, the symbol will include not only the radial or bearing, but also the identifier of the VOR or LF facility above the line and the frequency below the line. You can see both of these il-lustrated in the case of BOR as explained above and by the LF facil-ity ABC shown just below it.

Two on-request reporting-point symbols are shown a little farther down in the right-hand column of Fig. 14-6. The numbers are shown above the airway running through these points that indicate segment distances. The distance of the segment between the two on-request reporting points is 10 miles. The distance between the on-request re-porting point and the next fix to the right (off the chart) is 12 miles. When you look below the course line, you see two more symbols.

The one between the fixes is "D22" with a small arrow pointing to the left. This tells you that the left-hand fix (to which the arrow points) is 22 miles from the DME facility. The symbol on the right, a D with an arrow, tells you that the DME distance (from the DME facility) to the right-hand reporting point is the same as the segment distance printed above the course line, in this case, 12 miles.

Note the square star in the last illustration in Fig. 14-6. This is used to illustrate a *waypoint* (W/P) which is used in RNAV systems. These will be used more and more as GPS comes into play. The rest of the illustration indicates a waypoint that is formed in relation to a navigation facility such as a VOR. As we move more and more into GPS waypoints, all we will see are the latitude/longitude coordinates.

At the top of the left-hand column in Fig. 14-8, you can see that the airway and route designations are always shown in reverse (negative) type. About halfway down the left-hand column, take a look at some of the various ways to depict minimum en route altitudes (MEAs). MEA is defined as follows:

MINIMUM ENROUTE IFR ALTITUDE—The lowest published altitude between radio fixes which assures acceptable navigational signal coverage and meets obstacle clearance requirements between those fixes. The MEA prescribed for a Federal airway or segment thereof, area navigation low or high route, or other direct route applies to the entire width of the airway, segment, or route.

In the United States, an MEA is depicted as either an altitude or flight level, and if a flight level, it will be designated with the letters FL preceding the numerals. At times, the MEA will be higher in one direction than in the other, in which case the altitudes will be shown with appropriate directional arrows.

A *minimum obstruction clearance altitude* (MOCA) is depicted with a T following the altitude, while a route *minimum off-route altitude* (MORA) is designated with a lowercase letter "a" following the altitude. What is the difference between the two? Here is MOCA from the pilot/controller glossary:

MINIMUM OBSTRUCTION CLEARANCE ALTITUDE—The lowest published altitude in effect between radio fixes on VOR

Minimum altitudes, fixes, and reporting points 281

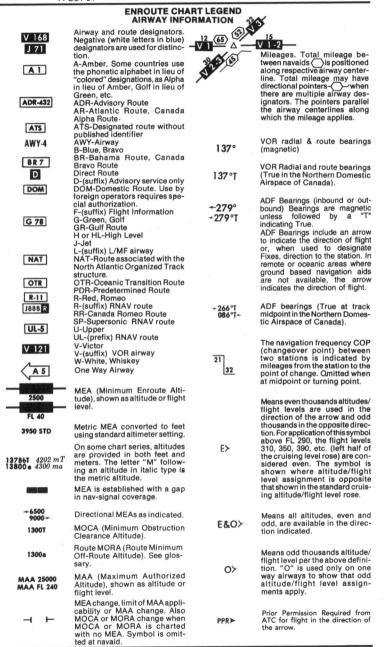

ENROUTE CHART LEGEND
AIRWAY INFORMATION

V 168 / J 71 — Airway and route designators. Negative (white letters in blue) designators are used for distinction.

A 1 — A-Amber. Some countries use the phonetic alphabet in lieu of "colored" designations, as Alpha in lieu of Amber, Golf in lieu of Green, etc.

ADR-432 — ADR-Advisory Route AR-Atlantic Route, Canada Alpha Route.

ATS — ATS-Designated route without published identifier

AWY-4 — AWY-Airway B-Blue, Bravo

BR 7 — BR-Bahama Route, Canada Bravo Route

D — Direct Route D-(suffix) Advisory service only

DOM — DOM-Domestic Route. Use by foreign operators requires special authorization. F-(suffix) Flight Information

G 78 — G-Green, Golf GR-Gulf Route H or HL-High Level J-Jet L-(suffix) L/MF airway

NAT — NAT-Route associated with the North Atlantic Organized Track structure.

OTR — OTR-Oceanic Transition Route PDR-Predetermined Route R-Red, Romeo

R-11 / J888 R — R-(suffix) RNAV route RR-Canada Romeo Route SP-Supersonic RNAV route

UL-5 — U-Upper UL-(prefix) RNAV route

V 121 — V-Victor V-(suffix) VOR airway W-White, Whiskey

A 5 — One Way Airway

Mileages. Total mileage between navaids ◯ is positioned along respective airway centerline. Total mileage may have directional pointers ◯ when there are multiple airway designators. The pointers parallel the airway centerlines along which the mileage applies.

137° — VOR radial & route bearings (magnetic)

137°T — VOR Radial and route bearings (True in the Northern Domestic Airspace of Canada).

←279° / ←279°T — ADF Bearings (inbound or outbound) Bearings are magnetic unless followed by a "T" indicating True. ADF Bearings include an arrow to indicate the direction of flight or, when used to designate Fixes, direction to the station. In remote or oceanic areas where ground based navigation aids are not available, the arrow indicates the direction of flight.

←266°T / 086°T← — ADF bearings (True at track midpoint in the Northern Domestic Airspace of Canada).

21 / 32 — The navigation frequency COP (changeover point) between two stations is indicated by mileages from the station to the point of change. Omitted when at midpoint or turning point.

2500 / FL 40 — MEA (Minimum Enroute Altitude), shown as altitude or flight level.

3950 STD — Metric MEA converted to feet using standard altimeter setting.

13786T 4202 mT / 13800a 4300 ma — On some chart series, altitudes are provided in both feet and meters. The letter "M" following an altitude in italic type is the metric altitude.

■■■ — MEA is established with a gap in nav-signal coverage.

←6500 / 9000← — Directional MEAs as indicated.

1300T — MOCA (Minimum Obstruction Clearance Altitude).

1300a — Route MORA (Route Minimum Off-Route Altitude). See glossary.

MAA 25000 / MAA FL 240 — MAA (Maximum Authorized Altitude), shown as altitude or flight level.

⊣ ⊢ — MEA change, limit of MAA applicability or MAA change. Also MOCA or MORA change when MOCA or MORA is charted with no MEA. Symbol is omitted at navaid.

E> — Means even thousands altitudes/flight levels are used in the direction of the arrow and odd thousands in the opposite direction. For application of this symbol above FL 290, the flight levels 310, 350, 390, etc. (left half of the cruising level rose) are considered even. The symbol is shown where altitude/flight level assignment is opposite that shown in the standard cruising altitude/flight level rose.

E&O> — Means all altitudes, even and odd, are available in the direction indicated.

O> — Means odd thousands altitude/flight level per the above definition. "O" is used only on one way airways to show that odd altitude/flight level assignments apply.

PPR▶ — Prior Permission Required from ATC for flight in the direction of the arrow.

Fig. 14-8 *Reproduced with permission of Jeppesen Sanderson, Inc. Not for use in navigation.*

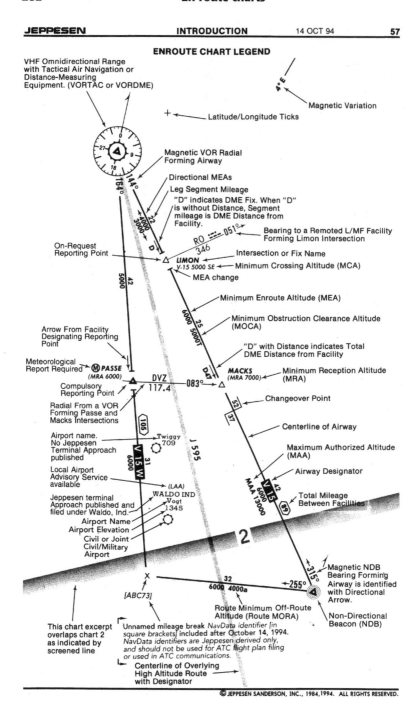

Fig. 14-9 *Reproduced with permission of Jeppesen Sanderson, Inc. Not for use in navigation.*

*airways, off-airway routes, or route segments which meets
obstacle clearance requirements for the entire route segment
and which assures acceptable navigational signal coverage
only within 25 statute (22 nautical) miles of a VOR.*

This means that if the route segment is more than 50 statute miles
long and you opt for the MOCA, you might not have proper VOR reception for part of the flight. So, in effect, the MOCA is almost a last-
resort altitude.

**MORA is strictly a Jeppesen altitude. You won't find it in the
ATC pilot/controller glossary.** Jeppesen's definition is:

*MINIMUM OFF-ROUTE ALTITUDE (MORA)—This is an alti-
tude derived by Jeppesen. The [route] MORA provides reference point clearance within 10 nm of the route centerline
(regardless of route width) and end fixes. A grid MORA altitude provides reference point clearance within the section
outlined by latitude and longitude lines. MORA values clear
all reference points by 1,000 feet in areas where the highest
reference points are 5,000 feet MSL or lower. MORA values
clear all reference points by 2,000 feet in areas where the
highest reference points are 5,000 feet MSL or higher. When a
MORA is shown along a route as "unknown" or within a grid
as "unsurveyed," a MORA is not shown due to incomplete or
insufficient information.*

A broken section in the route indicates there is either a change in the
MEA, MOCA, or MORA. It is there primarily to show a change in
MEA. This symbol will be omitted at a facility.

Other symbols

The total mileage between facilities is shown inside a six-sided
polygon that is positioned along and parallel to the centerline of the
airway. When two or more airways share the same course, the polygon has directional pointers indicating the airway to which the
mileage applies.

If the *changeover point* (COP) for the navigational radio is other
than at the midpoint of the leg or at the turning point on the route,
it will be shown by a Z-like symbol crossing the airway. (See Fig.
14-8: sixth item from the top right.) The numeral on each side of

this symbol shows the distance from the respective facility, at which point you should change the navigational radio's frequency. By doing so, you will be assured of adequate navigational reception along the entire route segment.

Take a look at the upper portion of the left-hand column in Fig. 14-10. This area shows the airway navaid/reporting point bypass illustrations. These illustrations depict how routes are illustrated when they pass through facilities or intersections that are not required for that specific route. Locate the VOR symbol (third from the top); when a facility is used for a particular route, the route centerline ends at the outer circle of the facility symbol. It usually ends just short of the symbol, leaving enough room to show the outbound radial (as in the case of V-15). If the facility is not required, the airway centerline will be extended through the center of the symbol and no radial will be shown (as in the case of V-76). If a report is not required for one of the routes that pass through a reporting point, that route will be shown passing directly through the reporting point symbol, or the route will be shown as a semicircle so as not to fill in the center of an open symbol.

Fitting all the pieces together

Until now, we've been looking at individual symbols. Let's put them all together to see how they might look as part of a chart; examine Fig. 14-9.

At the top you will see latitude (horizontal) and longitude (vertical) tick marks along the chart edges. To help further, these ticks are shown as small crosses (plus marks) along the face of the charts indicating each half degree of latitude or longitude.

Look down the airway marked V-15W. First off, you can see that it's 42 miles to PASSE, which is a compulsory reporting point when you are not in radar contact. PASSE requires a meteorological report as designated by the M within the circle preceding the name.

PASSE is formed by the intersection of V-15W and the 083-degree radial of the DVZ VOR, which has a frequency of 117.4 MHz and which lies somewhere off the left of the chart.

From the VOR at the top of the page to PASSE, the MEA is 5,000 feet, as you can see by the numerals below the airway centerline. The

58	14 OCT 94	INTRODUCTION	**JEPPESEN**

ENROUTE CHART LEGEND

AIRWAY NAVAID/REPORTING POINT BY-PASS

When an airway passes over or turns at a navaid or reporting point, but the navaid is not to be utilized for course guidance and/or no report is required, the airway centerline passes around the symbol. In cases where a by-pass symbol cannot be used, an explanatory note is included.

Airway J-26 does not utilize the navaid or reporting point.

Airway J-14 turns at the navaid or reporting point but does not utilize them. A mileage break "X" is included to further indicate a turn point.

Airway V-76 does not utilize the navaid. A note indicating the proper use of the navaid is included.

V-76 Disregards navaid.

Airway V-76 does not utilize the Int. A note indicating the proper use of the Int is included.

V-76 Disregards Int.

BOUNDARIES

```
::::::::::::::::::::::::::::
```
ADIZ, DEWIZ and CADIZ

```
.........................
. . . . . . . . . .
```
FIR, UIR, ARTCC or OCA boundary.

```
- ——— - - ———- -
```
International boundary.

```
⊢ ⊢ ⊢ ⊢ ⊢ ⊢
```
Time zone boundary.

```
    QNH
-o-o-o-o-o-o-o-o-o
    QNE
```
QNH/QNE-boundaries.

ALTITUDE LIMITS AND TYPES OF CONTROL

$\frac{4000}{\text{CTR, ATZ, TIZ}}$	CTR-Control Zone ATZ-Aerodrome Traffic Zone TIZ-Traffic Information Zone
$\frac{\text{FL 360}}{\text{UTA}}$	UTA-Upper Control Area
$\frac{\text{FL 70}}{4000}$ TMA	TMA-Terminal Control Area OCTA-Oceanic Control Area

CONTROLLED AIRSPACE

Controlled airspace shown in white.
Uncontrolled airspace shown as a tint.

Controlled airway/route.

Uncontrolled airway or advisory route.

Control Area boundary within controlled airspace (CTA, TMA).

U.S. Class B airspace. Waffle screen shows lateral limits.

Radio Frequency Sector Boundary.

Radio boundaries of .control or service unit.

Boundaries within TMAs or CTAs defining different altitude limits and/or sectorizations.

U.S. special VFR weather minimums for fixed wing aircraft are not authorized within the lateral boundaries of the surface areas of Class B, Class C, Class D, or Class E airspace designated for an airport.
Australia Mandatory Traffic Area. Traffic information is exchanged while operating to or from an airport without an operating control tower within the area.

Control Zone or Aerodrome Traffic Zone (controlled).

Aerodrome Traffic Zone (no control). Aircraft broadcast intentions on standard enroute frequency, and listen on same, when within such zones.
Japan Information Zone (no control) within which special VFR may be cleared by an air-ground station.

U.S. Class C airspace.

Canada Class C airspace.

ICAO AIRSPACE CLASSIFICATIONS

Airspace classification is designated by the letters (A) thru (G). Classification (A) represents the highest level of control and (G) represents uncontrolled airspace. The definitions of each classification is found in the Glossary portion of this section of the Enroute and Air Traffic Control section of this manual. The airspace classification letter is displayed in association with the airspace type and vertical limits.

┌─── TMA ───┐	┌── CTA ──
FL 145-165 **(A)**	FL 145-165 **(A)**
FL 50-145 **(D)**	──TMA──
	5000-FL 145 **(D)**
FL 45 /	──CTR──
CTR **(E)**	GND-5000 **(E)**

CTA(A)/UTA(A)
(UP TO FL 195**(D)**)

Fig. 14-10 *Reproduced with permission of Jeppesen Sanderson, Inc. Not for use in navigation.*

small arrow just prior to PASSE points from the facility that designates the reporting point.

The perpendicular stubs crossing the airway on either side of the fix at PASSE indicate that the MEA is different on each side. Looking to the southeast of PASSE, you will see that the MEA goes up to 6,000 feet MSL. Notice the absence of those stubs at MACKS (over on V-15) where the MEA is 6,000 feet MSL on both sides of the fix.

Moving along to the southeast side of PASSE (which has an MRA of 6,000 feet MSL, by the way), note that the distance from PASSE to the mileage break (indicated by the X) is 31 nautical miles. The 105 within the six-sided figure tells you that it's a total of 105 nautical miles between the facilities on V-15W. The mileage segments to the break, 42 and 31, only add up to 73 nautical miles. The remaining 32-mile segment extends ESE from the mileage break to an NDB station along the route. The mileage break in this instance also indicates a turn in the airway.

The alphanumerics within the square brackets pointing to that mileage break are Jeppesen NavData identifiers for those pilots using navigational receivers that rely on Jeppesen NavData update cards. More about them in the chapters on loran-C and GPS.

The shaded (screened) line running across the chart just above the mileage break symbol indicates that chart 2 overlaps this chart at that point.

As for the two airport symbols shown on the chart, you can determine that Jeppesen does not have any instrument approach procedure chart of Twiggy Airport because the airport name is in uppercase and lowercase. The field is 709 feet MSL. Jeppesen does provide an instrument approach chart for Vogt Municipal. Vogt is at 1,345 feet MSL, and the chart will be found listed under Waldo, Indiana.

The bearing to the NDB is the reciprocal of 255 degrees, which is 075 degrees. The MEA on this route segment is still 6,000 feet MSL, while Jeppesen has included a route MORA of 4,000 feet MSL. The beacon is a compulsory reporting point. The note by the NDB identifier tells us that the magnetic NDB bearing forming the airway is identified with directional arrows.

The NDB is not only part of V-15W, but also helps to form V-15. The bearing from it on V-15 northbound is 315 degrees. It is 42 nautical miles from the NDB to MACKS intersection, which has an MRA of 7,000 feet MSL. MACKS is formed by V-15 and the 083-degree radial of DVZ VOR. The changeover point from the NDB to the unnamed VOR is 37 nautical miles northwest of the NDB and 52 nautical miles southeast of the VOR. The total distance between these facilities can be found by adding these two changeover point mileages together, adding up the three route segments (22, 25, and 42), or simply by reading the overall distance shown inside the six-sided figure (89).

You can also find a *maximum authorized altitude* (MAA) symbol on this illustration. It's beneath the 6,000-foot MEA below the V-15 airway identifier. The MAA (defined in the glossary of this book) on this airway is 13,000 feet.

Moving up now, past the MACKS on-request reporting-point symbol, you will notice an arrow pointing from the VOR to the fix symbol with D47 below it. This indicates that you can determine the fix by the DME distance from the facility to the fix, which is 47 nautical miles. You can also add the two segment distances of 22 and 25 nautical miles to arrive at the same figure.

Proceeding northwest from MACKS, note that the MEA of 6,000 feet MSL is followed by 5000T, which is the MOCA between MACKS and LIMON. The stubs at LIMON indicate an MEA change similar to that at PASSE. Not only is 5,000 feet MSL the MOCA between LIMON and MACKS, but it is also the MCA at LIMON when heading southeast, as indicated by the wording V-15 5000 SE found under the same LIMON.

Northwest of LIMON, you will see another arrow from the VOR with the letter D below it, indicating that the DME distance from the VOR to LIMON is the same as the segment distance of 22 miles. LIMON is formed by the airway V-15 (the 144-degree radial of the VOR) and the 051-degree bearing to a remote L/MF facility that uses the identifier RO and the frequency 346 MHz. The short line with the small arrow following the bearing indicates that the L/MF facility is either remoted or off the right side of the chart.

Directional MEAs exist northwest of LIMON: 3,000 feet MSL heading toward the VOR and 4,000 feet MSL heading away from the VOR toward LIMON.

Refer to Fig. 14-10 to continue the discussion of en route charts.

A time zone boundary is indicated by a line of small Ts. This is useful information because the time zones wander around the chart, and you'd hate to arrive somewhere thinking it was 4:30 p.m. only to find that you were in the wrong time zone and everything closed up half an hour ago because it's really 5:30.

The lateral limits of the Class B airspace are shown as a waffled pattern of light blue/gray.

Look at the upper left-hand column in Fig. 14-11; a small grid shows shorelines and latitude/longitude lines. Notice the numbers 62, 75 (with a plus/minus sign), and 28 superimposed on the grids. This is Jeppesen's way of indicating the MORA for each section of the grid. In this case, the left grid section has a MORA of 6,200 feet MSL, the right section has a MORA of 2,800 feet MSL, while the center section has a MORA that is believed not to exceed 7,500 feet MSL, as depicted by the plus/minus sign (±).

This last part is significant: There are times when Jeppesen receives terrain height information where the values are given to them as plus or minus so many feet. In cases like this, Jeppesen will take the highest value and include it in the MORA with the plus/minus figure (±) to indicate that it is not an exact altitude. Because Jeppesen has used the highest elevation provided, they are confident that the terrain does not enter the 1,000- or 2,000-foot clearance space that the MORA is trying to achieve.

Look at the border information given at the bottom of the left-hand column. A name in the margin is the "next airway facility to which the total mileage is given."

Among the examples illustrated, you will see both "18 to WIND" and "32 to DADE," with respective MEAs. Outside the margin, you will see CHEJU and TAIPEI to indicate the next airway facility. The identifier and frequency of CHEJU indicate that the VOR designates an "on-chart reporting point, changeover point, or course change."

Note the next-to-last item on the right. Many times there is too much information to put on the chart; at least it's not possible to put it directly alongside the symbol to which it is referring. Other times the same information might apply to more than one area on the same chart. In these cases, it will be indicated by a number or letter in reverse print, called *ball flags*. When you see one or more of these symbols, you will find the information in a nearby box.

JEPPESEN   **INTRODUCTION**   AUG 28-92   **59**

ENROUTE CHART LEGEND

ORIENTATION

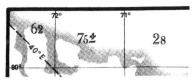

Grid shown at the intersection of units of latitude and longitude or by complete line.

Magnetic variation isogonic lines are indicated at the edge of the chart or are extended fully across the chart in a continuous dashed line.

Shorelines and large inland lakes are shown.

Grid Minimum Off-Route Altitude (Grid MORA) in hundreds of feet provides reference point clearance within the section outlined by latitude and longitude lines. Grid MORA values followed by a ± denote doubtful accuracy, but are believed to provide sufficient reference point clearance.

BORDER INFORMATION

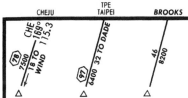

This area overlapped by chart indicated.

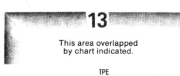

To Notes: Name outside the neatline is the next airway navaid to which the total mileage is given. Navaid identification is shown on all charts except the US(LO) and Canada/Alaska chart series. Reporting point name is shown when it is the airway termination.

To Notes: Name inside the neatline is the first reporting point outside the chart coverage to which the mileage and MEA are shown..

Airway lead information: The frequency and identifier of an off-chart navaid are shown when the navaid designates an on-chart reporting point, changeover point or course change.

MISCELLANEOUS

Outline indicates coverage of separate area chart. Information within this outline for terminal operation, may be skeletonized.

The area chart should be referred to if departure or destination airport is within this boundary to insure pertinent information is available.

On Enroute chart coverage diagrams, shaded symbol denotes Area chart coverage. Area chart name is included with shaded symbol.

Outline indicates an area covered elsewhere on the same or adjoining chart in enlarged scale. Information within this outline may be skeletonized.

Ball Flags: Number or letter symbol used to index information not shown at the point of applicability, but carried in a like-identified note within the same panel.

Reference number for INS Coordinates. These coordinates are tabulated elsewhere on the chart and identified in a like manner.

Fig. 14-11 *Reproduced with permission of Jeppesen Sanderson, Inc. Not for use in navigation.*

U.S. SERIES 800 AND 900 DESIGNATED RNAV ROUTES

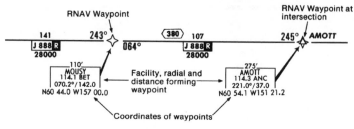

AUSTRALIA AND CANADA T RNAV ROUTES

Fig. 14-12 *Reproduced with permission of Jeppesen Sanderson, Inc. Not for use in navigation.*

Notice the small blue box with DENVER inside it on the right side of Fig. 14-11. This symbol indicates that an area chart is available for that area.

Finalizing this chapter, the square-shaped stars in Fig. 14-12 once again illustrate how Jeppesen presents RNAV waypoints, specifically GPS waypoints.

15

Departure and arrival procedures

Although it sounds like this chapter is going to be covering a lot of ground, it is actually going to just hit the highlights of a lot of subject matter because most of the symbology has been discussed.

Area charts ease clutter

Recall that area charts are large-scale charts for high-density areas. By using them, pilots arriving or departing from airports within these areas will have all the information they need without cluttering up the en route chart.

The lower section of Fig. 15-1 shows extra symbols that are found on the area charts. There is a communications section for the major airports shown on each chart. Also, major airports are illustrated by airport symbols that show runways. If the departure and arrival routes are different, the departure route is indicated by a solid line with an arrow, while the approach route uses a dashed line with an arrow. If the arrival and departure routings are the same, a heavy solid line is used without the arrow.

The area chart acts as a form of transition chart between the en route and the approach charts. If the terrain rises above 4,000 feet above the main airport on the chart, this information may be shown by depicting the terrain. When shown, it will be in one of two forms.

Prior to June 24, 1994, *area minimum altitudes* (AMAs) are something like MORAs, except they are not depicted in latitude/longitude grids. Instead, a general layout of the terrain is depicted (Fig. 15-2). The different heights of the terrain are shown by gradient tints *of green.* The AMA is shown within the different tinted

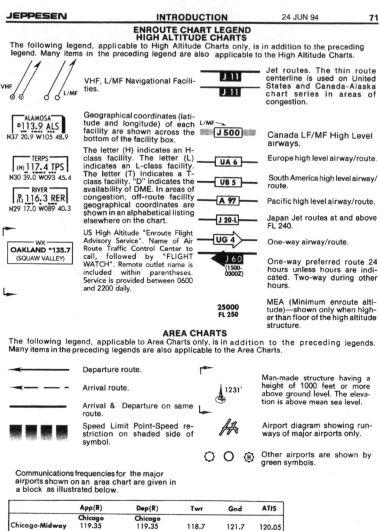

Fig. 15-1. *Reproduced with permission of Jeppesen Sanderson, Inc. Not for use in navigation.*

ENROUTE CHART LEGEND
AREA CHARTS (Continued)

TERRAIN INFORMATION

Terrain information may be depicted on area charts when terrain within the area chart coverage rises more than 4000 feet above the main airport. This information is portrayed using one of the two following methods:

1) Prior to June 24, 1994, terrain information was depicted as Area Minimum Altitude (AMA) envelopes printed in green.

2) After June 24, 1994 AMAs will gradually be replaced with generalized contour lines, values and gradient tints printed in brown.

AREA MINIMUM ALTITUDE (AMA) ENVELOPES
(Prior to June 24, 1994)

AMA envelopes portray an exaggerated layout of terrain when compared to the detail of contour lines. Terrain rise between two charted envelope lines is indicated by gradient tints.

The area between each envelope line includes an AMA figure which represents the terrain high point/man-made structure clearance altitude for the envelope area. AMA values clears all terrain and man-made structures by 1000 feet in areas where the highest terrain and man-made structures are 5000 feet MSL or lower. AMA values clear all terrain and man-made structures by 2000 feet in areas where the highest terrain and man-made structures are 5001 feet MSL or higher.

DME arcs and radials are included for relating position to AMA envelope.

NOTE: An MEA may be lower than an AMA because of locally lower terrain beneath an airway.

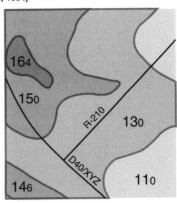

GENERALIZED TERRAIN CONTOURS
(After June 24, 1994)

Generalized terrain contour lines and contour values are depicted. Gradient tints indicate the elevation change between contour intervals. Contour lines, values and tints are printed in brown. Within the highest contour interval some, but not all, terrain high points may be included along with their elevation above mean sea level.

THE TERRAIN CONTOUR INFORMATION DEPICTED DOES NOT ASSURE CLEARANCE ABOVE OR AROUND TERRAIN OR MAN-MADE STRUCTURES. THERE MAY BE HIGHER UNCHARTED TERRAIN OR MAN-MADE STRUCTURES WITHIN THE SAME VICINITY. TERRAIN CONTOUR INFORMATION IS USEFUL FOR ORIENTATION AND GENERAL VISUALIZATION OF TERRAIN. IT DOES NOT REPLACE THE MINIMUM ALTITUDES DICTATED BY THE AIRWAY AND AIR ROUTE STRUCTURE. Furthermore, the absence of terrain contour information does not ensure the absence of terrain or structures.

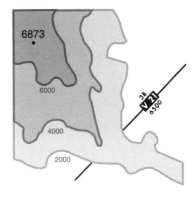

Fig. 15-2. *Reproduced with permission of Jeppesen Sanderson, Inc. Not for use in navigation.*

sections, or envelopes, the same way that the grid MORA is shown on the en route charts, with the larger of the numerals depicting thousands of feet and the smaller numerals depicting hundreds of feet.

As in the MORA altitudes, values of 5,000 feet or less allow for a 1,000-foot obstruction clearance, while values above 5,000 feet will clear obstructions within that area by 2,000 feet. Sometimes the AMA value will be higher than the MEA for a specific area because the MEA is only for the terrain beneath the airway itself.

On charts printed after June 24, 1994, the terrain will be shown in brown tints, and the heights (in mean sea level) will be printed along the terrain contour lines. The highest terrain in the specific contour area may be indicated by a dot and black print, but, and this is very important, these contour areas are for general information only. Man-made obstacles are not indicated, nor are all of the terrain heights. Read what Jeppesen says about it in the note alongside the symbols for the generalized terrain contours. These figures are for general information only.

As a rule, these terrain features will only be found on charts where some of the terrain might be higher than 4,000 feet above the area surface.

DME circles are drawn around certain major airports, with specific radials drawn on them. These are for reference points in relation to the AMA envelopes. Figure15-3, the San Francisco Area Chart 10-1, shows this information.

Three major airports are depicted on this chart: Oakland, San Francisco, and San Jose. More than a dozen other fields are shown. Some of those within the SFO Class B airspace include Navy Moffett, Palo Alto, San Carlos, and Hayward.

That circle of green dots around SFO can be explained as the symbol used to designate United States Class B, C, D, or E airspace within which special VFR weather minimums for fixed-wing aircraft are not authorized.

Runways at the major airports are depicted on the *area* chart, as are the ILS approaches at those airports on the chart that have an ILS available. Study the chart during preflight or during en route cruise because by the time you get there, it will be too late.

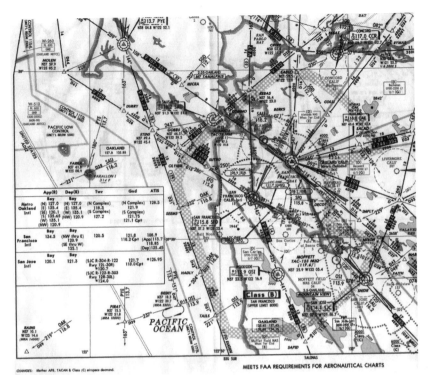

Fig. 15-3 *Reproduced with permission of Jeppesen Sanderson, Inc. Not for use in navigation. Reduced image for reproduction.*© JEPPESEN

The Class B trap

There is a trap in the Class B procedures for unwary IFR pilots. Refer to Fig. 15-4, the San Francisco (B) Chart 10-1A.

If you're like me, you probably don't like Class B terminal areas. They're too confusing the way they're designed, and if you're operating VFR, you're spending more time looking at the chart to be sure that you're within the proper airspace than you are looking outside for traffic. For that reason alone I'd rather file IFR. Here's where the trap comes in.

Each area of Class B has two figures in it, one above and one below a line. The figure above the line is the top of the Class B airspace for that sector. In other words, if you're flying above 8,000 feet, you're legally above the San Francisco Class B area and have nothing to worry about.

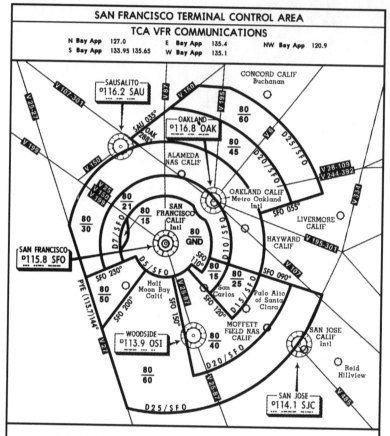

FLIGHT PROCEDURES

IFR Flights-Aircraft within the TCA are required to operate in accordance with current IFR procedures.

VFR Flights-

a. Arriving aircraft, or aircraft desiring to transit the TCA, should contact Bay Approach Control on the frequency depicted. Pilots on initial contact should state their position, direction of flight and destination. If holding of VFR aircraft is required, the holding point will be specified by ATC and will be a prominent geographical fix, landmark or VOR radials.

b. Aircraft departing the primary airport prior to taxiing are requested to advise the San Francisco clearance delivery position of the intended route of flight and altitude. Aircraft departing from other than San Francisco International Airport whose route of flight would penetrate the TCA should give this information to ATC on appropriate frequencies.

c. Aircraft desiring to transit the TCA may obtain an ATC clearance to transit the TCA when traffic conditions permit and will be handled on an equitable "first-come, first-served basis," providing the requirements of FAR 91 are met.

Fig. 15-4 *Reproduced with permission of Jeppesen Sanderson, Inc. Not for use in navigation.*

It's the lower number that's the most confusing. This is the floor of Class B for that sector. If you're below that floor, you're also legally clear; however, if you're operating IFR, and have been cleared for a visual approach, watch out. You're not allowed to descend below the floor of the Class B airspace on a visual approach. That rule is there to protect the VFR aircraft that might be flying around just under the floor. There is no floor once you're inside the traffic area or thereabouts. As you can see in the illustration, the floor is the ground all around SFO, for a distance of from 5 to 7 DME.

Then beware that if you accept a visual approach, and there are scattered clouds in the area, you won't be able to duck below them to remain visual if that move will take you below the floor of the Class B sector that you're in. That loophole gives you more reason to be careful before accepting a visual approach. Other than that warning, the Class B chart is fairly straightforward.

SIDs, STARs, and profile descents

Before talking about approach charts, let's take a look at some of the specific symbols on the SID and STAR charts. Figure 15-5 illustrates the heading information that will be found on these charts. As a rule, you will find the departure frequency for a SID in a box at the upper left of the heading. If there is more than one frequency sector involved in the area that the chart covers, the different frequencies will be shown on the plan view of the chart, separated by sector boundary symbols.

There is a note telling us that a full explanation of transition altitudes and transition levels can be found on the Jeppesen introductory page.

The charts are identified both by index numbers: 10-2 with suffixes for STARs and profile descents; 10-3 with suffixes for SIDs; and by the actual type of chart denoted in reverse print (white print on black space).

The bottom half of Fig. 15-5 shows various route identifications along with computer codes that are used. It also shows two different types of SIDs. These are the pilot nav SIDs, which means the pilot shall fly the entire SID, including the transition, from the information provided on the SID itself. There is also the vector SID, which indicates that the pilot shall be provided with radar vectors to an assigned route or fix depicted on the SID itself.

JEPPESEN   **INTRODUCTION**   JUN 8-90   **81**

SID, STAR, AND PROFILE DESCENT LEGEND

The following legend is applicable to Standard Instrument Departure (SID), Departure, Standard Terminal Arrival (STAR), Profile Descents, and Arrival Charts. Refer to the Chart Glossary for more complete definition of terms.

These charts are graphic illustrations of the procedures prescribed by the governing authority. A text description is provided, in addition to the graphic, when it is furnished by the governing authority. In some areas text information is required to perform a SID or STAR procedure. *Not all items apply in all areas.*

All charts meet FAA requirements for aeronautical charts. All altitudes shown on SIDs, STARs, and Profile Descent charts are MSL, unless otherwise specified.

COMMUNICATIONS AND ALTIMETER SETTING DATA

Departure Control frequencies are included with SIDs. The frequencies are listed in the heading of the chart or when frequency sectors are specified they may be displayed in the planview of the chart.

EAST SECTOR

HEADING | TERPS Departure (R) **126.9** | PLANVIEW

sector boundary ➝   symbol

TERPS DEPARTURE CONTROL 126.9

The ATIS frequency is provided on STARs in the heading of the chart.

ATIS **120.3**

The Transition Level and Transition Altitude are listed below the Communications. For a complete explanation of Transition Level and Transition Altitude see Introduction page 103.

TRANS LEVEL: FL 140
TRANS ALT: 13000'

CHART IDENTIFICATION

STARS/PROFILE DESCENTS   **SIDS**

(10-2) (10-2A), etc.   Index number   (10-3) (10-3A), etc.   Index number

[10-2]   [10-3]   Special chart issued to special coverages only. Contains modified information for your company.

STAR   Standard Terminal Arrival   SID   Standard Instrument Departure

ARRIVAL   Arrival Procedure   DEPARTURE   Departure Procedure

PROFILE DESCENT   Profile Descent

ROUTE IDENTIFICATION
TYPICAL EXAMPLES USING COMPUTER LANGUAGE
STARS   **SIDS**

MOORPARK FOUR ARRIVAL (FIM.MOOR4)   MILIS (ROCKI1.MILIS)
Arrival Name   Arrival Code   Transition Name   Transition Code

PILOT NAV SIDS

FRESNO (FAT.MOOR4)
Transition Name   Transition Code

Departure Name

ROCKI ONE DEPARTURE (ROCKI1.ROCKI) (PILOT NAV)
Departure Code   Primary Navigation is by pilot, not radar

PROFILE DESCENTS

Transition origination
FACILITY/FIX Code

—FILLMORE—
⟨ᴸ⟩ **112.5 FIM**
N34 21.4 W118 52.8

CIVET
D52/ILAX
D52/LAX
N34 02.1 W117 23.3

VECTOR SIDS

DENVER FIVE DEPARTURE (DEN5.DEN) (VECTOR)
SID where ATC provides radar navigational guidance to an assigned route or to a fix depicted on the SID. Vector SIDs indicate the fix or route to which the pilot will be vectored.

TYPICAL EXAMPLES NOT USING COMPUTER LANGUAGE
STARS   **SIDS**

ALPHA ARRIVAL (RWY 10)
Arrival Name   Specified runway to be used

RUNWAY 13 ARRIVAL

INDIA DEPARTURE
Departure Name

RUNWAY 13 DEPARTURE

PROFILE DESCENTS

RWYS 8L, 9R, 27L ← Identification shown in chart heading.

Fig. 15-5. *Reproduced with permission of Jeppesen Sanderson, Inc. Not for use in navigation.*

This does not mean that you won't receive radar vectors when performing a pilot nav SID because you can receive radar vectors at anytime throughout your flight.

Certain SIDs, STARs, and profile descents do not incorporate computer codes, examples of which are on the bottom of Fig. 15-5.

Notice in Fig. 15-6 that many of the symbols are the same as those found on en route charts. The routes are shown as solid lines with arrows, while transition tracks use a dashed line. For the transition portion, the name and computer identification of the route are found above the track while the MEA and segment mileage are below it.

If a SID includes a heading, it will be shown with the letters "hdg" following the heading to be flown. Visual flight tracks are depicted by a series of small arrows, while radar vectoring tracks will be illustrated by a series of short arrowheads.

Figure 15-7 shows various crossing restrictions and course guidance information. *Profile descent* tracks are shown using heavy solid lines with arrows. (A profile descent is an uninterrupted descent from cruising altitude to initial approach.) Some of the fixes require airspeed and/or altitude crossing restrictions that might include minimum and/or maximum crossing altitudes.

Near the bottom of Fig. 15-7 is a SID climb gradient/climb rate table. If a SID does not have a published minimum climb gradient, you can assume that there are no terrain or obstacle problems. In this event, a minimum climb gradient of 200 feet per nautical mile should be flown to the MEA or assigned altitude, unless any climb reductions are necessary to comply with published restrictions. During this climb, the aircraft is expected to cross the departure end of the runway at 35 feet AGL. You must also climb to 400 feet above the airport elevation before initiating any turn, unless one is specified in the procedures.

The standard FAA obstacle clearance slope is based on 152 feet per nautical mile. Climbing at a rate of 200 feet per nautical mile will assure a minimum of 48 feet of obstacle clearance per nautical mile of flight. "That ain't much," so be sure to keep climbing at a good, steady rate.

If, however, there are obstacles or terrain that will penetrate the 40-to-1 slope ratio that results from a climb of 152 feet per nautical mile, you will have a steeper climb gradient shown on the chart.

SID, STAR, AND PROFILE DESCENT LEGEND
GRAPHIC
(Charts are not drawn at a specific scale)

RADIO SYMBOLS

VORTAC/VORDME

VOR (VHF Omnidirectional Range)

TACAN (Tactical Air Navigation) or DME (Distance Measuring Equipment)

NDB (Nondirectional Radio Beacon)

LOC, LDA, or SDF Front Course

LOC Back Course

Locator with Outer Marker (LOM)

Outer or Middle Marker (OM) (MM)

RADIO IDENTIFICATION

┌─DENVER─┐
(D)(H)116.3 DEN
N39 51.6 W104 45.1

┌─PRACHINBURI─┐
201 PB
N14 06.0 E101 22.0

LOC
108.7 IMBS

LOC (BACK CRS)
089° 109.7 IMEX
(FRONT CRS 269°)

Navaid identification is given in shadow box with frequency, identifier, Morse Code and latitude & longitude coordinates. DME capability is indicated by a small "D" preceding the VOR frequency at frequency paired navaids. VOR and VORTAC facility operational ranges are identified (when known) within the navaid box. (T) represents Terminal; (L) represents Low Altitude; and (H) represents High Altitude.

Localizer navaids are identified by a round cornered box. Frequency identification and Morse Code are provided. DME is included when navaid and DME are frequency paired. Localizer back course facility boxes include front course bearing for HSI setting.

VERTICAL NOISE ABATEMENT PROCEDURES

RWY	VNAP
07, 15	A
25, 33	A or B

Vertical Noise Abatement Procedures (VNAP). For explanation of procedures, see Air Traffic Control section.

RESTRICTED AIRSPACE

PROHIBITED, RESTRICTED, DANGER AREAS
Prohibited, Restricted & Danger Areas are charted when referenced in SID or STAR source, plus any Prohibited Area within five (5) nautical miles of route centerline or primary airport.

R-2713 ◄── Designation (Type of area can be determined by P-Prohibited, R-Restricted, D-Danger.)
UNL ◄── Upper Limit
GND ◄── Lower Limit
(0800-2200 LT MON-SAT) ◄── Hours active
(IND ARTCC) ◄── Controlling Agency

ROUTE PORTRAYAL

SID/STAR Track

BOLES ──Transition name
Transition track
12000 ──Minimum enroute altitude (MEA)
25 ──Segment mileage

BOLES
(REX.BOLES3)
12000
25

On charts dated on or after Jul 26, 1985 the Transition name will include the route identification code, when assigned.

DF 11 ──► SID or STAR label of a particular route in some coverage areas

3.0

Cross at TL+10 and descend to 3000'

Crossing altitude instructions, Transition Level plus 1000'

Radar vectoring

Johns 25

Visual flight track

Flight Track segment flown with heading only.

150° hdg

Fig. 15-6. *Reproduced with permission of Jeppesen Sanderson, Inc. Not for use in navigation.*

SID, STAR AND PROFILE DESCENT LEGEND
GRAPHIC (Continued)
ROUTE PORTRAYAL (Continued)

115° *JAMES* (ATOMS I.JAMES) **45** 117°
JAMES
8000
60
65

—MAXSON—
D (H) 17.0 MXS
N36 42.0 W118 08.0

GILER
N36 18.0 W117 16.3

△ ←—297°— *JNS*
116.0

Changeover point (COP) on transition between
MXS VOR and Giler Int, JNS VOR is used for
track guidance at and after COP to Giler Int.

← TANGO HOTEL
[THOTL] △ ⇌—093°→ ——60/7000—————→ R273° (Ⓐ)

—UNDER—
D (H) 116.1 UDR
N30 38.3 W122 01.0

Inbound course on   VOR Radial
outbound VOR Radial.

⌐ *NavData identifier [in square brackets]* is included for STAR origination or SID termination fix
when the fix is unnamed, or named with other than a five character name and no country
assigned identifier. Its use is to assist the pilot with an on board NavData database to
associate database information with chart information. The fix officially named "TANGO
HOTEL" is carried in the database as "THOTL". (Included after October 14, 1994) *NavData
identifiers are Jeppesen derived only, and should not be used for ATC flight plan filing or used in
⌊ ATC communications.*

———————→ Profile Descent Track

MSA is provided when specified by the
controlling authority.

△ *DPK 060°* ×   N36 15.0 W119 17.2
114.7   At or below
8000'

TROZE
D15/DEN

Radar
vectoring

4000' | 2400'
090°— | —270°
3800' | 3600'

*MSA
TPS VOR*

Crossing altitude
restriction

Metering fix
(Beginning of
Profile Descent.
May also be radio
navaid.) ⌐

Cross at 16000' at 250 Kt
Descend and maintain 12000'
Vector to final

Inbound course on
outbound VOR radial (R-213°)

BYSON

D36/DEN
N39 22.3 W105 26.0

△ ←—033°

Crossing altitudes &
speed restrictions

Cross at or below FL 230
Cross at or above FL 190
Cross at 250 Kt
Descend and maintain 16000'

San Diago Intl-
Lindbergh

SID CLIMB GRADIENT/CLIMB RATE TABLE
This SID requires a minimum climb gradient
of 330' per nm to 9000'.
Climb gradient converted to climb rate in feet
per minute at specified ground speeds. ⌐

←—Required climb
gradient

⌐Arrival/departure
airport, highlighted with
circular screen after
⌊November, 1993.

Gnd speed–Kts	75	100	150	200	250	300
330' per nm	413	550	825	1100	1375	1650

Climb gradient

"MILITARY" notation
indicates military
source used for this
procedure.

LOST COMMUNICATIONS PROCEDURE ONLY
If not in contact with Departure Control one
minute after take-off:
Rwy 1: Climb straight ahead to 4000', climbing
right turn, thence intercept and proceed via
George R-039 to Mikes Int, then via (transition)
or (assigned route).

Symbol identifies the LOST
COMMUNICATIONS PROCEDURE
to be flown when communications
are lost with ATC after take-off.

MILITARY

Fig. 15-7. *Reproduced with permission of Jeppesen Sanderson, Inc.
Not for use in navigation.*

Many times Jeppesen will publish a climb gradient/climb rate table to help calculate the necessary rate of climb for an aircraft. This is similar to the descent/time-conversion tables published at the bottom of the approach plates that are explained in a subsequent chapter.

All you need to know is the actual *ground speed* in knots. Once you have that, just read the required rate of climb from the chart, interpolating as necessary. Using the chart in Fig. 15-7, for example, with a ground speed of 120 knots, you would need a rate of climb of approximately 660 feet per minute to result in a climb gradient of 330 feet per nautical mile. At 240 knots ground speed, the rate of climb would have to be 1,320 feet per minute.

This is an important item to check because as mentioned before there will be times when the aircraft won't be able to meet the requirement. And there will be times (wind shear or other factors) when you won't know until you are airborne that you can't make the necessary climb gradient. In such a case, you will have to let ATC know immediately so it can vector you to a safe area.

Finally, at the very bottom of Fig. 15-7 you will find an example (self-explanatory) of a lost communications block.

Looking at a profile descent

Figure 15-8 is the RUNWAY 24/25 PROFILE DESCENT for Los Angeles International Airport, and as you can see from the index number it is only one of a series for this airport.

The profile descent is a transitional routing that will take the aircraft from the en route phase of flight and join it to a visual or instrument approach by allowing the aircraft to descend and position itself at a steady rate of descent with the least amount of communications.

This procedure starts at the FILLMORE VOR, and there is one inherent trap in the profile depiction that has caught a number of high-time airline pilots. You are to cross FIM at an altitude between FL 190 and 15,000 feet. From the VOR, you will fly the 148-degree radial to SADDE intersection where you will turn left to join the Santa Monica SMO VOR 261-degree radial inbound.

En route to SADDE, you will be expected to cross the SYMON fix, which is 12 DME from FIL, at between 15,000 and 12,000 feet and at an airspeed of 280 knots (or as assigned by ATC). You will then be

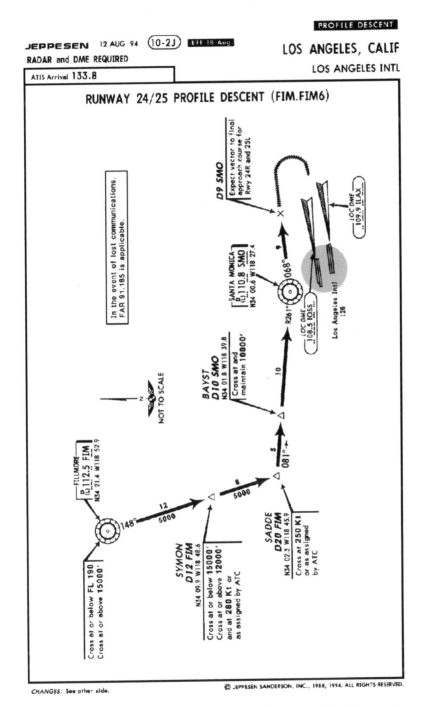

Fig. 15-8. *Reproduced with permission of Jeppesen Sanderson, Inc. Not for use in navigation.*

expected to slow down further to 250 knots (or as assigned by ATC) by SADDE.

From SADDE, you will only have 5 miles to descend to 10,000 feet at BAYST, which is 10 DME from SMO. At SMO, turn left to track on the SMO 068-degree radial and at 9 DME you should expect vectors to a final approach.

After beginning the profile descent, if ATC revises any part of the route or altitude, ATC will cancel the rest of the profile descent, and all routes, altitudes, and speeds will be provided by ATC. If ATC revises the speed, however, such a revision will void only the charted speed restrictions, and you will still be obliged to comply with the charted routes and altitudes. If you are unable to comply with the required routes and altitudes because of a speed change by ATC, you must notify the controller that you can no longer comply.

The trap is the minimum en route altitude figures shown between FIL and SADDE. These are for your information and emergency use only. You must comply with the altitudes shown at the fixes. If you go below the lower limit, or bust the upper limit, you will be subject to a violation.

It is important to note that a clearance for a profile descent is not a clearance for an instrument approach procedure. The last altitude shown on the profile descent is the final altitude unless cleared lower by ATC or cleared for a specific instrument approach procedure. In Fig. 15-8, descent below 10,000 feet is prohibited until cleared for a lower altitude.

Charted visual approaches

Many high traffic airports are beginning to use charted visual approaches, which are designed to provide an orderly flow of traffic, like SIDs, STARs, profile descents, and instrument approach procedures.

The charted visual approach headings are similar to IAP headings discussed in the next chapter. Two are included here to show what they look like and also to show you that you can get into trouble if you don't pay attention.

Take a look at Figs. 15-9 and 15-10. You will see that there are a number of similarities. Both require radar. Both begin at a point over the SMO VOR and track on the 068-degree radial. Both require

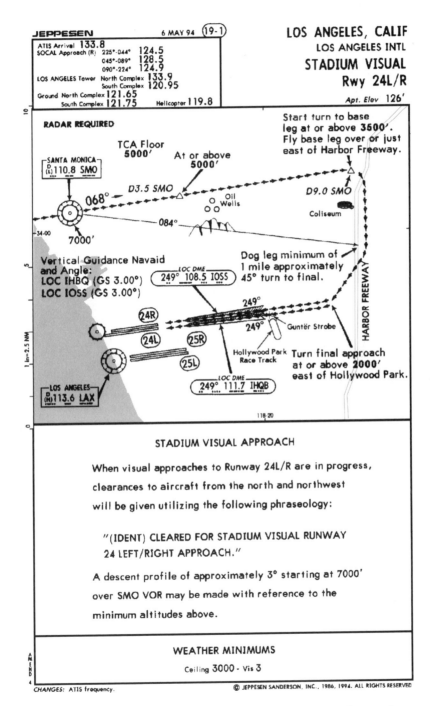

STADIUM VISUAL APPROACH

When visual approaches to Runway 24L/R are in progress,
clearances to aircraft from the north and northwest
will be given utilizing the following phraseology:

"(IDENT) CLEARED FOR STADIUM VISUAL RUNWAY
24 LEFT/RIGHT APPROACH."

A descent profile of approximately 3° starting at 7000'
over SMO VOR may be made with reference to the
minimum altitudes above.

WEATHER MINIMUMS

Ceiling 3000 - Vis 3

CHANGES: ATIS frequency.

Fig. 15-9. *Reproduced with permission of Jeppesen Sanderson, Inc.
Not for use in navigation.*

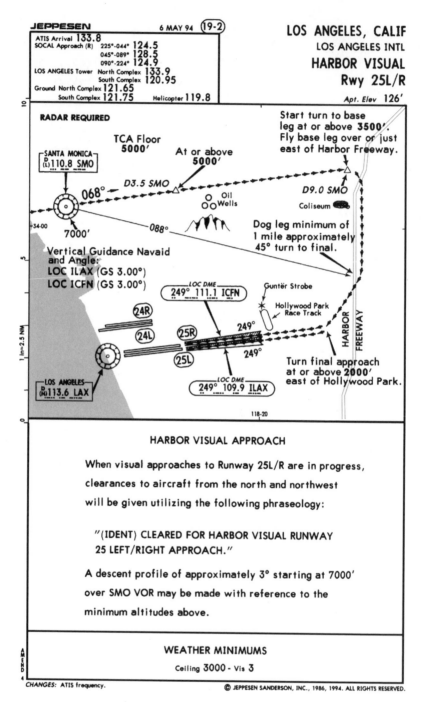

Fig. 15-10. *Reproduced with permission of Jeppesen Sanderson, Inc. Not for use in navigation.*

crossing the 3.5 DME fix at or above 5,000 feet because the floor of the Class B airspace in that area is 5,000 feet. From there, descent is allowed to 3,500 feet by the 9.0 DME, then execute a turn to the right to pass east of the Coliseum and the Harbor Freeway. Both procedures require a dogleg with a minimum of a 1-mile, 45-degree turn to final, and that turn has to be at or above 2,000 feet east of Hollywood Park.

Look at the bottom of each chart page to see that both require the same ceiling and visibility of 3,000 feet and 3 miles. Just above the weather requirements you will see the note telling you that both can be performed using a 3-degree descent profile if started at 7,000 feet over the SMO VOR.

Here's where the similarities end and where you have to be careful. The STADIUM VISUAL approach is to Runway 24L or 24R, while the HARBOR VISUAL approach is to Runway 25L or 25R.

On the Stadium Visual, you shall use the 084-degree radial of SMO as a lead-in radial, pass to the *north* of the Hollywood Park Race Track, and use one of the Runway 24 ILSs for vertical guidance and alignment angle. The LOC for 24R is 108.5 MHz, using an identifier of IOSS, while the LOC for 24L is 111.7 MHz with an identifier of IHQB. Both have a descent path of 3 degrees.

The Harbor Visual uses the SMO 088-degree radial as a lead-in radial, passes *south* of the Hollywood Park Race Track, and uses *one of* the Runway 25 ILSs. Runway 25R is on a frequency of 111.1 and has an identifier of ICFN, while Runway 25R uses a frequency of 109.9 with an identifier of ILAX. Like the Stadium Visual, both use a descent path of 3 degrees, and all four runways have the same inbound course of 249 degrees.

It is easy to make a mistake and use the improper chart, especially after a long day of flying on instruments, then landing during heavy haze into the setting sun, *and* with a lot of traffic around you.

Pay attention to the clearance you're given. Make sure you identify the chart itself, set the proper LOC frequency in the radio, and *make sure you positively identify the localizer.*

16

Approach chart: plan view

Airports with a published instrument approach might have one approach or perhaps 10 or more. Each approach is depicted individually, with very few exceptions, such as an NDB or LOC approach that might be combined with an ILS when both use the same routing.

The first approach to the airport is printed on the first page of the series; the plan view of the airport, as well as airport data and the takeoff and alternate minimums, are usually printed on the reverse side of that first page. The remainder of the approach chart pages for that airport will normally be printed on both sides of the page.

Figure 16-1 shows the general format for the approach chart and the airport chart. At many major airports the airport charts, taxiway charts, gate positions, and INS coordinates are printed on charts that precede the 11-1 chart. In this case, charts are numbered 10-9, 10-9A, 10-9B, and the like. This has been done because some of these airport layouts have become so very complex that even veteran pilots get lost without detailed information that won't fit on one page.

Each approach chart is a storehouse of information. Take the time to familiarize yourself with the chart legend so you can derive the necessary information. Review this information periodically. Remember, these approach charts are designed to lead you down the rosy approach path to the airport runways. Oftentimes they will allow you to thread the needle between mountains, antenna farms, or high-rise buildings when forward visibility is down to bare minimums.

Then, if you get to the missed approach point and still can't see the runway, the charts will lead you safely through the missed approach, but only if you understand and follow the approach and missed approach exactly. Figure 16-2 shows what you can learn from the chart's heading.

Approach charts are graphic illustrations of instrument approach procedures prescribed by the governing authority. All charts meet FAA requirements for aeronautical charts. The following legend pages briefly explain symbology used on approach charts throughout the world. *Not all items apply to all locations.* The approach chart is divided into specific areas of information as illustrated below.

FORMATS

The first approach procedure published for an airport has the procedure chart published on the front side with the airport chart on the back side. On major airports, the airport chart may preceed the first approach procedure. These locations will have expanded airport information that may occupy more than one side. When an airport has more than one published approach procedure, they are shown front and back on additional sheets. Blank pages will indicate "INTENTIONALLY LEFT BLANK".

APPROACH PROCEDURE CHART FORMAT	AIRPORT CHART FORMAT
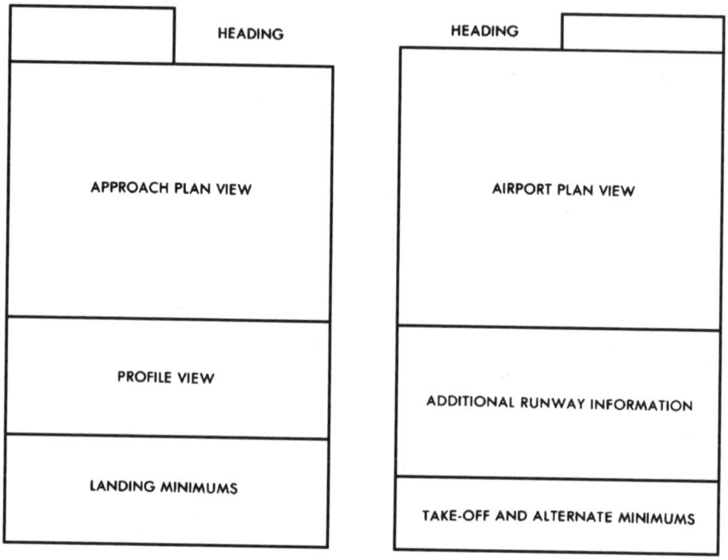 HEADING	HEADING
APPROACH PLAN VIEW	AIRPORT PLAN VIEW
PROFILE VIEW	ADDITIONAL RUNWAY INFORMATION
LANDING MINIMUMS	TAKE-OFF AND ALTERNATE MINIMUMS

Fig. 16-1 *Reproduced with permission of Jeppesen Sanderson, Inc. Not for use in navigation.*

APPROACH CHART LEGEND

HEADING

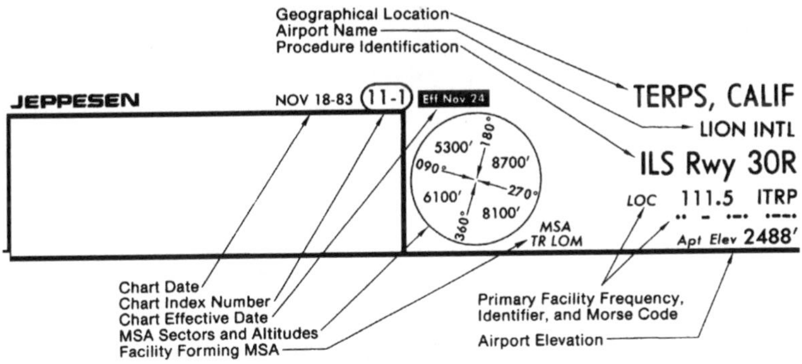

Geographical Location
Airport Name
Procedure Identification

JEPPESEN　　　NOV 18-83 ⓵⓵-1　Eff Nov 24　　　TERPS, CALIF

LION INTL

5300'
090°　　8700'
6100'　　270°
8100'

ILS Rwy 30R

LOC　111.5　ITRP
·· — ··· ·—·

MSA
TR LOM

Apt Elev 2488'

Chart Date
Chart Index Number
Chart Effective Date
MSA Sectors and Altitudes
Facility Forming MSA

Primary Facility Frequency,
Identifier, and Morse Code

Airport Elevation

The geographical name used is generally the major city served by the civil airport or installation name if a military airport. A hyphen before the airport name is used when the location name is part of the airport name. The charts are arranged alphabetically by the geographical location served.

NOTE:　U.S. Airway Manual: The civil approach charts covering the United States are arranged alphabetically by state. Within each state, the charts are arranged alphabetically by the name of the city served.

For each location, the charts are sequenced by the chart index number. This index number will appear as shown below:

First Digit:　represents the airport number and is an arbitrary assignment.

Second Digit:　represents the chart type as shown below:

0-area, SID, etc.	5-RESERVED
1-ILS, MLS, LOC,	6-NDB
LDA, SDF, KRM	7-DF
2-PAR	8-ASR
3-VOR	9-RNAV, vicinity chart,
4-TACAN	Visual Arrival or Visual Departure chart.

Third Digit:　represents the filing order of charts of the same type.

⌐Oval outlines of chart index numbers represent:

◯　Standard chart issued to Airway Manual subscribers.

◖　Special chart issued to special coverages only.

◖M　Standard chart that uses metric system units of measure.

In this numerical system—both procedure and airport—there will be gaps in the filing sequence because of deletions, expected expansion, selected distribution and tailoring for specific subscribers. Two procedures may be combined. Numbering, in this case, will be for the lowest number of the pair. ILS and NDB is a typical combination indexed as 11-1, 21-1, etc.

All chart dates are Friday dates. This chart date is not to be confused with the effective date. The effective date is charted when a chart is issued prior to the changes being effective. Charts under USA jurisdiction with an effective date are effective at 0901Z of that date.

Procedure identification is given below the airport name. This identification is per the applicable authoritative source (e.g. VOR-1, NDB (ADF) Rwy 16, etc.). The use of an alphabetical suffix indicates a procedure does not meet criteria for straight-in landing minimums (e.g. VOR-A, VOR-B, LOC (BACK CRS)-A, etc.).

MSA provides 1000 feet of obstruction clearance within the circle (or sector) within 25 nautical miles of the facility identified just to the lower right of the circle. If the protected distance is other than 25 nautical miles, the effective radius is stated beside the identifier of the central facility. The MSA value is supplied by the controlling authority.

Fig. 16-2 *Reproduced with permission of Jeppesen Sanderson, Inc. Not for use in navigation.*

First of all, check the calendar date against the chart *effective* date. You don't want to use a chart before it becomes effective. Why would you have a new chart before an effective date? If a very important change is scheduled in a procedure, the cartographers will issue the charts early enough to assure that all subscribers receive their copies in time to note the change. Additionally, all revisions are Friday dated, and changes effective after the printing date will be shown with an effective date noted. If the chart does not have an effective date shown, it is current and effective when received.

Creating the charts

Instrument procedures are developed under the United States Standard Terminal Procedures (TERPs). There are no regulations regarding the appearance of approach charts, so it is up to the cartographers to determine appearance. This is why the NOS and Jeppesen formats are so different.

Information that must be provided to the users comes through three sources: letters of transmittal from the *Federal Register,* the daily *National Data Flight Digest,* and NOTAMs. It is up to the cartographers to study these publications to assure that all charts are up-to-date.

The letters of transmittal include either a form 8260-3 for precision approaches or 8260-5 for nonprecision approaches. These forms print out all the necessary information for each specific approach. The cartographers then use this printed information to create a chart.

Howie Keefe's update concept

Many hours are spent by pilots to keep their manual revisions up-to-date. Not only that, because the changes to the Jeppesen charts come out weekly, a pilot on an extended flight might end up having old approach plates for her return flight. Most of the information is available about seven weeks before the changes take place.

Howie Keefe has done a lot of flying: long-time pilot, ex-Navy pilot, national air racer (modified P-51 *Miss America*), and holder of the transcontinental piston speed record (6 hours and 21 minutes from Los Angeles to Washington, D.C., set in *Miss America* in 1972). Tired of having out-of-date charts, tired of spending many hours revising manuals, and realizing that over 97 percent of the changes can be accomplished by pen and ink changes to the existing charts, Howie started the Air Chart Company in Venice, California.

He sends his revision notices, called the *Universal-Enroute Chart Update* and the *Universal-Approach Chart Update*, to his subscribers six weeks before they become current. No other changes can be made for an additional six weeks, so the user can now take an extended trip and know that the charts will be up-to-date. NOS and Jeppesen users can avail themselves of this service, which is really handy for professional people who don't have the time to make numerous manual revisions.

The approach chart updates are all in written, cryptic form on both sides of one sheet of paper. The subscriber checks her route to see what areas she will be flying through, then looks on the list to see if any of the changes will affect her. If they do, she merely makes a pen-and-ink correction to that specific chart.

Figure 16-3 shows the changes made to the Orlando Executive chart for the LOC BC RWY 25. The amended number was changed from 18 to 19, and the new date was inked in; a new fix was added called VITTU, which can be identified as the 9 DME or by radar; the minimum altitude for VITTU is 1,600 feet; the minimum altitude for the MARYB fix was changed to 1,500 feet; and the final approach fix was moved from MARYB to the old ORL 4 DME/radar fix that was named BRICE, and its minimum altitude was changed to 1,100 feet. In addition, the ceiling requirements were lowered to 480 feet for the straight-in approaches.

This encompassed many more changes than the average revision, which is usually no more than a radio frequency, yet, as you can see, these changes were handled neatly with a pen.

Examples from the approach chart updates are seen in Figs. 16-4 and 16-5. These revisions cover all changes made from March 1, 1989, through January 25, 1990. The important changes are listed more than once.

For example, under the SUMMARY OF MAJOR CHANGES in Fig. 16-4, look under GA (for Georgia) and see where it says that the VIDALIA (VDI) NDB was renamed ONYUN (UON) and moved.

This is expanded in Fig. 16-5 to tell you that the NDB has been moved 5 miles east to 32°13.4' north, and 082°17.9' west. The symbol following that information indicates the end of that item.

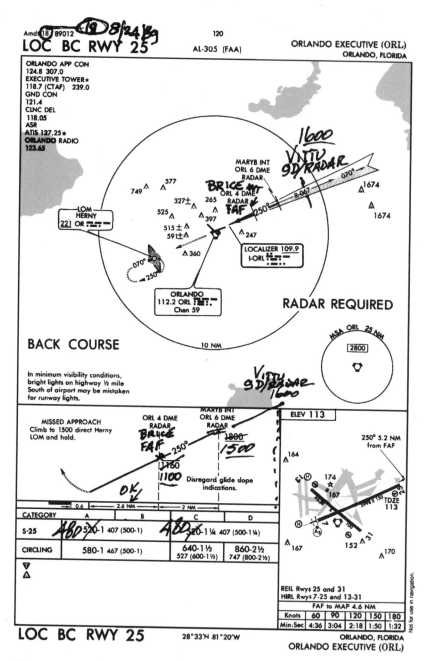

LOC BC RWY 25

ORLANDO EXECUTIVE (ORL)
ORLANDO, FLORIDA

AL-305 (FAA)

BACK COURSE

RADAR REQUIRED

In minimum visibility conditions, bright lights on highway ½ mile South of airport may be mistaken for runway lights.

MISSED APPROACH
Climb to 1500 direct Herny LOM and hold.

ELEV 113

Disregard glide slope indications.

CATEGORY	A	B	C	D
S-25	520-1 407 (500-1)		520-1¼ 407 (500-1¼)	
CIRCLING	580-1 467 (500-1)		640-1½ 527 (600-1½)	860-2½ 747 (800-2½)

REIL Rwys 25 and 31
HIRL Rwys 7-25 and 13-31

FAF to MAP 4.6 NM

Knots	60	90	120	150	180
Min:Sec	4:36	3:04	2:18	1:50	1:32

LOC BC RWY 25

28°33'N 81°20'W

ORLANDO, FLORIDA
ORLANDO EXECUTIVE (ORL)

Fig. 13-3

```
:: :: SUMMARY OF MAJOR CHANGES :: ::
```

AL-EVERGREEN-name>Middleton Field
AZ-PHOENIX(1/11/90)new TCA to replace ARSA
AR-BLYTHEVILLE-Hicks(HKA)NDB idnet>IUI
 -EL DORADO-new LOM(EL-418k)
 -TEXARKANA-new ATIS,frequency 120.2
CA-CAMARILLO-new control tower
 -DELANO-new AWOS,frequency 119.55
 -FRESNO-Air Terminal-tower freq 118.2
 -SACRAMENTO-Exec-rwy 2 loc freq>109.7
 -SACRAMENTO-Metro-loc freq 16R>110.3
 -SAN JOSE(SJC)VOR moved 2m west
CO-TELLURIDE-new CONES(ETL)VOR
CT-GROTON/NEW LONDON-gnd ctl>121.65
FL-CECIL(NZC)VOR ident>VQQ & freq>117.9
 -JACKSONVILLE(JVC)VOR>CRAIG(CRG)
 -LAKE CITY-tower freq 119.2
GA-ALBANY(ABY)VOR>PECAN(PZD)
 -ALBANY-name>SW Georgia Regional
 -SWAINSBORO(SBO)VOR is cancelled
 -TIFTON-TMA NDB>TIFTO LOM,freq>409k
 -TOCCOA(TOC)VOR>FOOTHILLS(ODF)
 -VIDALIA(VDI)NDB>ONYUN(UQN) & moved
ID-MCCALL(MYL)VOR(1/11/90)>DONNELLY(DNJ)
IL-CHICAGO-OHare-tower freq>126.9
IN-KOKOMO-new AWOS,frequency 109.8
IA-FT DODGE-new AWOS 124.75
KS-JUNCTION CITY-name>Freeman Field
 -MANHATTAN-new tower frequency 118.55
 -OLATHE-Executive-LOC 17 freq>111.1
 -SALINA(SLN)VOR moved 2.6m north
 -SALINA(SLN)VOR frequency>117.1
 -TOPEKA-Forbes-grnd control>121.7
LA-MONROE(MLU)VOR add 3° to radials
MA-BOSTON-twr freq rwy 4L/22R 128.8
 -BEVERLY-tower frequency>134.75
 -MARTHAS VINEYARD(MVY)VOR fre>114.5
MI-DETROIT-Willow Run(YIP)VOR cancelled
 -DETROIT-new VOR(DXO-113.4)
 -WEST BRANCH(BXC)VOR freq>113.2
MN-LITCHFIELD-new airport 2½m south of
    old airport.
 -MINNEAPOLIS-Anoka-tower closed
MS-GREENWOOD-new private control tower
 -JACKSON-Thompson name>Jackson Intl
 -JACKSON-Hawkins-tower freq>120.0
MO-FORNEY(TBN)VOR add 3° to radials
 -KANSAS CITY-Richards-new ATIS 127.2
 -POPLAR BLUFF-name>Poplar Bluff Muni

MT-KALISPELL(FCA)VOR add 2° to radials
NH-LEBANON-tower frequency>125.95
NJ-ATLANTIC CITY-Bader-control twr closed
NY-NEW YORK-JFK-rwy 31L freq>111.35
NC-CHARLOTTE-new TCA to replace ARSA
 -KINSTON-name>Kinston Regional
 -NEW BERN-name>Craven County Regnl
OH-MEDINA-Freedom Field>Medina MUNI
OR-CORVALLIS(1/11/90)new AWOS,freq 135.17
 -CORVALLIS(CVO)VOR(6/29)frequency>115.
 -HOOD(HDR)VOR(6/1)>PORTLAND(PDX)
 -KLAMATH FALLS(LMT)VOR add 2°+radials
 -KLAMATH FALLS-name>Klamath Falls Intl
 -LAGRANDE-new AWOS,freq 135.07
PA-BEAVER FALLS-new ATIS 118.35
SC-TOCCOA(TOC)VOR>FOOTHILLS(ODF)
TN-NASHVILLE-Metro-LOC 2R freq>110.75
 -MEMPHIS-new TCA to replace ARSA
TX-GREENVILLE-tower freq & CTAF>118.65
 -LUBBOCK-tower frequency 118.8>120.5
 -SWEETWATER-name>Avenger Field
UT-SALT LAKE CITY-new TCA to replace ARSA
VA-SHAWNEE(EEY)VOR(7/27)is cancelled
 -WINCHESTER-TZ LOM frequency>364k
WA-BELLINGHAM-gnd ctl 121.85>127.4
WI-KENOSHA-EN LOM frequency>389k
 -SHEBOYGAN-HE LOM frequency>338k
WY-DUNOIR(DNW)VOR add 2° to radials
 -FT BRIDGER-new AWOS,freq 118.8
 -FT BRIDGER(FBR)VOR add 2° to radials
 -JACKSON(JAC)VOR add 2° to radials
 -MUDDY MTN(MDM)VOR ident>DDY

* * TAKE-OFF & DEPARTURES * *

Check with FSS for airports in this
update for possible changes in Take-off
or Departures. In addition the airports
listed below have changes.

AL-TROY	LA-MANY
CA-OAKLAND	-RUSTON
-RAMONA	MS-LAUREL
-TRUCKEE	NE-FREMONT
-VACAVILLE	-MINDEN
CO-FT COLLINS	NY-MIDDLETOWN
-TRINIDAD	-PLATTSBURGH
ID-MTN HOME	OR-NORTH BEND
IL-CANTON	-THE DALLES
-CHICAGO-Midway	RI-PROVIDENCE
IN-SOUTH BEND	SC-FLORENCE
IA-DES MOINES	VA-BRIDGEWATER
KY-SPRINGFIELD	WV-WHEELING

Fig. 16-4 *Air Chart Systems. Not for use in navigation.*

The next item explains that VIDALIA approach control frequency
has been changed from 132.5 to 132.3 MHz. The next item notes the
effective dates of the NDB change.

Keefe includes other useful information in block form such as the
NOTAM CAUTION in the upper right of Fig. 16-5. This information
doesn't necessarily concern itself with chart changes.

Fig. 16-5 *Air Chart Systems. Not for use in navigation.*

The en route chart update is also handled on one large piece of paper. In addition to the cryptic symbols, a map of the United States is included. Stylized symbols appear on the map to speed changes.

Figure 16-6 is the northwestern United States. Simple changes are printed on the map, such as the Fresno Air Terminal Tower frequency being changed to 118.1 MHz; this is amplified in the other listings.

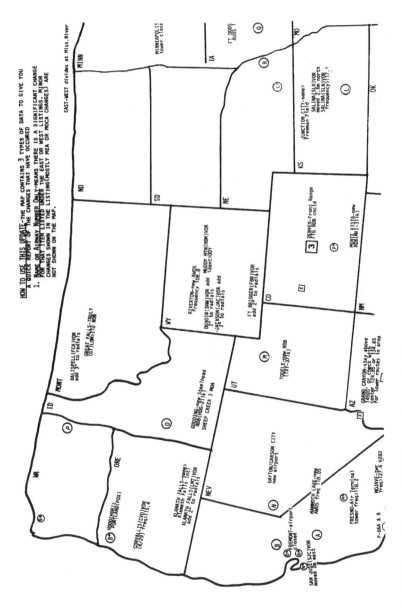

Fig. 16-6 *Air Chart Systems. Not for use in navigation.*

317

Sometimes there will be a major change in a routing. Note the boxed numeral 3 at Denver. This number refers to another section of the chart, which is labeled VIS-AIDS.

VIS-AIDS have been prepared to IFR specifications to be used with a low altitude en route chart. You can see this change at Denver in Fig. 16-7. In addition to the change itself, Keefe includes a synopsis of why the change was made.

Keefe also provides similar services for VFR pilots. With the increasing use of loran and GPS routings, Keefe has a very helpful bound volume called the *LORAN/GPS Navigator Atlas,* which among other things includes the lat/long of all fixes in the United States. It is also an excellent aid for the right-seat passenger, and is highly recommended for RNAV pilots as well as for long cross-country pilots who want the utmost in reference material.

Contact Air Chart for more information. The address and phone number are in the resources section at the back of this book.

Filing the charts

Returning to the Jepp charts, the number circled to the right of the chart issue date (Fig. 16-2) is the chart index number. The first digit represents the airport. If there are two or more airports in the same

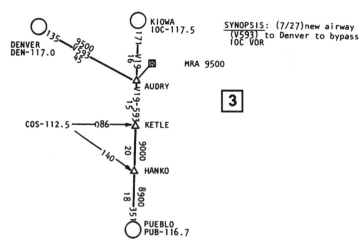

Fig. 16-7

vicinity, the chartmakers will make one number 1, another number 2, and so on. These numbers are purely discretionary.

The second digit will indicate the type of approach or the type of chart. The third digit establishes a filing order for charts of the same type at the same airport. For example, many airports will have more than one ILS or one VOR approach. In the case of the VOR approaches, they will be numbered 13-1, 13-2, 13-3, 13-4, and the like.

In the case of ILS approaches, the indexing is further broken down; 11-1 would be a CAT I approach to a specific runway, and 11-1A would be the CAT II and CAT IIIA approaches to the same runway.

The charts that have a zero (0) for the second digit are for SIDs, STARs, area charts, and the like. In these cases, the third digit is used to designate the type of chart and the suffix will be the filing index. For example, the area chart is a 10- 1, the Class B terminal chart is a 10-1A, STARs are 10-2 (suffixed by A, B, C), SIDs are 10-3 (also suffixed by letters), noise abatement charts are 10-4, taxi charts are 10-5, and the newest charts of all, the airport charts and their attendant pages, are 10-9 (again with letter suffixes).

Let's say that you are looking at the approaches to John Q Airport and note an approach numbered 13-2 but can't find a 13-1. Well, don't panic because there might not be a 13-1. Perhaps the procedure was abandoned, or perhaps ATC is planning on developing a 13-1 in the future. Jeppesen issues a check sheet every six months to verify that the manuals are up-to-date.

The best way to double-check that the charts are up-to-date is to save the last check sheet and the revision notes from that last check sheet to the present time. The check sheet is a listing of all of the pages in the manuals. Check off any page you might be missing. When you get a new check sheet, keep it, and throw away all the old check sheets and revision notices. That way you always have a reference. If the check sheet doesn't list a 13-1 for that airport, chances are there isn't one. If you are still in doubt, go through the revision notices to see if one was issued in the meantime. Most of the time Jeppesen will leave a blank page in that spot anyway and insert a note that the page was intentionally left blank.

What if the check sheet does list a 13-1? Review the revision notices to see if you were to have destroyed it in the meantime. If the search indicates that you should have a 13-1 and you still can't find it, drop a note to Jeppesen and get a replacement by return mail.

Notice the big circle under the chart's effective date. This is the *minimum safe altitude* (MSA) circle.

The MSA is set up to provide a 1,000-foot obstacle clearance within 25 nautical miles from a specific navigational facility. This facility is not necessarily within the ATA. The facility being used is identified outside the circle on the lower right side. If, for any reason, the radius of the MSA circle is other than 25 nautical miles, it shall be so noted.

If the circle is broken up into different sectors—defined in this case with inbound magnetic bearings so you can identify the proper sector directly from the compass—the altitudes then become known as *minimum sector altitudes.* They are designed for obstruction clearance only and will not always provide navigational reception. In fact, Jeppesen's glossary states that ". . . this altitude is for EMERGENCY USE ONLY and does not necessarily guarantee NAVAID reception." (Emphasis is Jeppesen's.) To reemphasize, another important thing to remember is that because the center of the circle is located at a navigational facility that the approach procedure is predicated upon, the MSA is not necessarily a 25-nautical-mile radius around the *airport.*

The last column to the right in the heading lists the geographical location of the airport and under that, the airport name. The charts are filed alphabetically by geographical location.

The procedure identification is listed below the airport name, and here is another area where caution must prevail. Certain airports have many approaches and it becomes necessary to identify each of them individually. But if you are bouncing around in moderate to severe turbulence or if you are tired and in a hurry to get down on the ground, you might hear "cleared for a VOR DME . . . approach." The airport might have several VOR DME approaches, and you might in haste turn to the VOR DME 20 chart when you were cleared to the VOR DME-A approach. So it behooves you to be sure of the approach you have been cleared for; one way to help is to always read back the approach clearance for confirmation.

If the approach has a letter suffix, such as VOR-A, or VOR DME-A, or DME-B, it means that the approach will not meet the criteria for straight-in minimums and is therefore a circling approach.

The primary facility frequency, the identifier, and the Morse-code dots and dashes fall in below the procedure identification followed by other pertinent information such as "Ops not continuous." The airport elevation is shown in feet above sea level (MSL).

The upper left corner of the approach charts (Fig. 16-8) shows the arrival air/ground communications data, listed systematically in the proper sequence as used on approach: ATIS, approach, tower, and ground control frequencies.

For awhile, prior to October 1984, Jeppesen was only putting the approach control frequency on the first chart at a location. On all other charts for that location, a note would indicate the first chart for that frequency. This saved time when making revisions because many times every chart at a location had to be reissued when only a frequency was changed, but the hassle of having to leaf back and forth in the approach charts prompted the change back to the original method of putting the frequencies on each chart.

Transition altitudes, where aircraft operating at flight levels change back to altitudes, and other pertinent altimeter setting information are shown in the lower part of the communications box.

Jeppesen uses a three-letter code that is found in the communications section. G means that the frequency is guarded (listened to) only, T means that the ATC only uses that frequency to transmit, and X means that the frequency is available upon request.

The bird's-eye view

Jeppesen defines the plan view at the top of Fig. 16-9. In effect, the plan view is a bird's-eye view of the approach with the flight path drawn in (appearing to be what you would *like* to see on a tracing paper after you get out of a simulator).

The note at the top of the page tells you that most approach plan views are drawn at a scale of 1 inch = 5 nautical miles. If they are drawn at a different scale, that is noted.

The plan view offers a tremendous amount of information, and you must understand each and every symbol to fly in a safe, professional manner. Many of the symbols require little or no thought, but some of them could use a little more explanation. For example, notice the

322   **Approach chart: plan view**

APPROACH CHART LEGEND
HEADING (continued)

COMMUNICATION AND ALTIMETER SETTING DATA

Communications for "arrivals" are given in normal sequence of use as shown below. See Airport Chart Legend, Introduction page 116, for other communications.

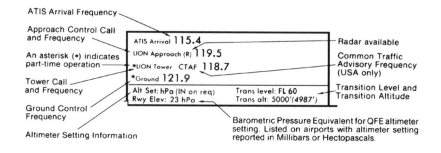

Transition level and transition altitude are listed on the bottom line of the communications and altimeter setting data box. Transition level and transition altitude are provided for all areas outside the 48 conterminous United States, Alaska and Canada.

Trans level: FL 60   The transition level (QNE) is the lowest level of flight using standard altimeter setting (29.92 inches of mercury or 760 millimeters of mercury or 1013.2 millibars or 1013.2 hectopascals.)

Trans alt: 5000'(4987')   The transition altitude (QNH) is the altitude at and below which local pressure setting must be used.

Altimeter setting units are listed on the bottom line of communications data box.

Barometric Pressure Equivalent in millibars or hectopascals enables aircraft operators who use QFE altimeter setting for landing to establish the QFE altimeter setting by subtracting the hectopascal or millibar equivalent from the reported QNH altimeter setting. The value shown is the barometric pressure equivalent for the height reference datum for straight in landing. The height reference datum will be the runway threshold elevation (Rwy), airport elevation (Apt), or the runway touchdown elevation (TDZ), as applicable.

Letter designations behind a frequency indicate operation as follows:

G-guards only
T-transmits only
X-on request

Bearings defining frequency sectors are clockwise outbound (e.g., 270° to 090° would be north of the airport.)

Fig. 16-8 *Reproduced with permission of Jeppesen Sanderson, Inc. Not for use in navigation.*

APPROACH CHART LEGEND
APPROACH PLAN VIEW

The plan view is a graphic picture of the approach, usually presented at a scale of 1 in = 5 NM. Plan views at scales other than 1 in = 5 NM are noted. Latitude and longitude are shown in 10 minute increments on the plan view neatline. Symbols used in the plan view are shown below.

NAVAIDS

NDB (Non-Directional Radio Beacon)

VOR (VHF Omni-Directional Range)

TACAN (Tactical Air Navigation facility) or DME (Distance Measuring Equipment)

VORTAC or VOR/DME

ILS, LOC, LDA, SDF, MLS or KRM Front Course

LOC Back Course

OFFSET LOC

Offset Localizer

Markers with or without locator, NDB, or Intersection. The triangle or circle in a marker or NDB symbol represents co-located intersection.

TERPS
^{D}115.4 TRP

THORNTON
281 TOT

TERPS
$^D_{(H)}$115.4 TRP

Navaid facility boxes include facility name, identifier, Morse code, and frequency. The shadow indicates the primary facility upon which the approach is predicated. In VORTAC and VORDME facility boxes the letter "D" indicates DME capability.

VOR, VORTAC and VORDME class is indicated by a letter "T" (Terminal), "L" (Low Altitude), or "H" (High Altitude) when available.

SAARBRUCKEN
343 SBN

Underline shown below navaid identifier, indicates Beat Frequency Oscillator (BFO) required to hear Morse Code identifier.

(OP NOT CONT)
or *

Indicates part-time operation.

TAC-112
CUSTARD
^{D}(116.5) CUS

TACAN facility box with "Ghost" VOR frequency for civil tuning of TACAN - only facilities to receive DME information.

Domestic DME
BIBOOHRA
CH-4 BIB

Australia Domestic DME Operates on 200 MHz and requires airborne receiver specific to this system.

NAVAIDS (continued)

ILS DME
257° 110.3 IDEN

ILS, LOC, LDA, or SDF facility box. It includes inbound magnetic course, frequency, identifier, and Morse code.

LOC (BACK CRS)
077° 110.3 IDEN
(FRONT CRS 257°)

Localizer Back Course facility box. Front course included for HSI setting.

MLS DME
035°Ch 516 MTRP
^{D}(109.7)

MLS facility box including inbound magnetic final approach course, MLS channel, identifier with Morse code and VHF "Ghost" frequency for manually tuning DME.

BEARINGS

106°⤙   Magnetic course
106°T⤙   True course

VOR cross radials and NDB bearings forming a position fix are "from" a VOR and "to" an NDB.

Morse code ident is charted on VOR radial/NDB bearing when forming facility is outside of planview.

Fig. 16-9 *Reproduced with permission of Jeppesen Sanderson, Inc. Not for use in navigation.*

symbol for an offset localizer, you will have to make some sort of turn to the landing runway when you get down to minimums, and because of this, the minimums will be higher than they would be for a straight-in ILS.

Notes on the page will provide such information as how many degrees the localizer is offset from the runway heading, as well as how far from the threshold the approach course crosses the runway centerline. These offset localizers are usually LDAs:

> *LOCALIZER TYPE DIRECTIONAL AID—A navaid, used for nonprecision instrument approaches with utility and accuracy comparable to a localizer but which is not a part of a complete ILS and is not aligned with the runway.*

An example of the notes mentioned above can be seen on the Honolulu, Hawaii, LDA DME Runway 26L approach (Fig. 16-10):

"Use IEPC LDA DME when on LOC course. Localizer course offset from landing runway by 45°. Final approach course crosses runway centerline 8100' from threshold."

Even though you have to make a 45-degree turn to get lined up with the runway, you have more than a mile and a half to accomplish it. The minimum visibility for this approach is 2 miles, so even though the course change is over 30 degrees, the 2-mile visibility provides enough time to line up with the centerline so the approach is considered a straight-in.

Referring back to Fig. 16-9, under bearing symbols, notice that bearings are magnetic courses unless the letter T is after the course direction, in which case it will be a true course. You will find true courses in the Far North.

Looking a little farther down, note that the symbol for EWD in the example also shows the Morse-code identifier and the explanation for this information: "charted on the VOR radial/NDB radial when forming facility is outside of plan view." You saw the same thing on the en route charts.

Take a look at the upper left-hand column of Fig. 16-11. Position fixes are depicted by filled and open triangles, like en route charts. If there are alternate means of identifying a fix, these means will be specified, such as in the example, where the fix can be either the

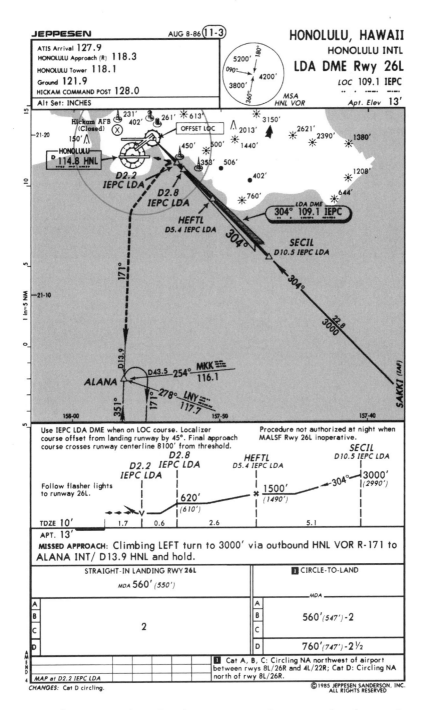

Fig. 16-10 *Reproduced with permission of Jeppesen Sanderson, Inc. Not for use in navigation.*

JEPPESEN   **INTRODUCTION**   24 JUN 94   **105**

APPROACH CHART LEGEND
APPROACH PLAN VIEW (continued)

POSITION FIXES

Position fixes are portrayed by a triangle. △ ▲

DME value will be portrayed as D10.0. When fix and co-located navaid name are the same, only the navaid name is displayed.

Allowable substitutions for identifying a fix are noted in the planview. At the pilot's request, where ATC can provide the service, ASR may be substituted for the OM. In addition, PAR may be substituted for OM and MM.

APPROACH TRANSITIONS

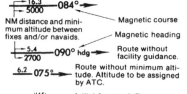

NM distance and minimum altitude between fixes and/or navaids.

Magnetic course

Magnetic heading

Route without facility guidance.

Route without minimum altitude. Altitude to be assigned by ATC.

(IAF)   Initial Approach Fix

(IF)   Intermediate Approach Fix

NoPT   No procedure turn, Race track pattern or any other type of course reversal procedure required or authorized without ATC clearance.

●   Flag notes -see applicable reference notes elsewhere on the plan view.

Cross at FL 110 and descend to 3000'

WAKER   Crossing altitude and descent instructions.

TERPS VOR   Approach transition inset. (Dog leg route, with off-chart turn). Also provided when route originates at an off-chart intersection designated only for approach use—such fixes are not charted on enroute and area charts.

NOT TO SCALE

JOHNS

APPROACH TRANSITIONS (continued)

NoPT Arrival Sector via Airway

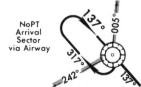

NoPT arrival sectors depict an area of approach transition routing to an approach fix. No procedure turn, Race Track Pattern or any type course reversal is required nor authorized without ATC clearance when an arrival course is within the charted sector and on an established airway radial to the fix.

Approach transition track, distance, and altitude from a defined fix is illustrated below.

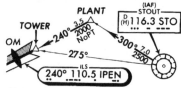

Note that the routes from STO to Plant to Tower are approach transitions, whereas the STO R-275° is not an approach transition. The STO R-275° has a small arrowhead and is a cross radial forming Tower. The STO R-300° has a large and small arrowhead indicating both an approach transition and a cross radial forming Plant. Plant and Tower are also formed by the IPEN localizer course.

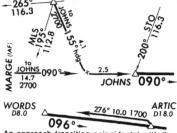

An approach transition coincidental with the approach procedure flight track is charted offset from the flight track for clarity.

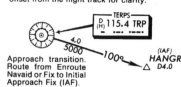

Approach transition. Route from Enroute Navaid or Fix to Initial Approach Fix (IAF).

Fig. 16-11 *Reproduced with permission of Jeppesen Sanderson, Inc. Not for use in navigation.*

outer marker or 6.2 DME. The note here also reveals that if you so request, you can substitute ASR for the outer marker. Due to the increased precision, you can ask for PAR to substitute for both the outer and the middle markers.

Transitioning for the approach

Moving on to the *approach transition* (formerly called the terminal route) information. In the second illustration down the column, the nautical mileage is shown above the course line, the minimum altitude is shown below the line, and the magnetic course is depicted on the line. Sometimes the route is merely a heading (rather than a bearing to or from a facility), in which case "hdg" will follow the magnetic course symbol.

The upper illustration on the right-hand column shows a NoPT (no procedure turn) sector. The NoPT sector is indicated by radials from the VOR. These radials are shaded on the NoPT side. If arriving on an airway within this sector, you are not allowed to make any type of course reversal without permission from ATC.

The NoPT is one of the most misunderstood symbols on the charts. Some pilots will say this means that you are not required to make a procedure turn, while others will say that procedure turns are not authorized. I have heard pilots argue both points, yet they are both half-right. The procedure turn is not required, nor is it authorized without ATC clearance. These procedures have been set up to increase the flow of traffic. Airspace is getting too congested to have aircraft wandering around procedure turns that take 2 minutes (minimum) to complete or as much as 5 minutes for the published procedure.

The procedure turn is merely designed to allow you to reverse direction so you will be on the inbound course. It also provides time to get down to the initial approach altitude.

The NoPT is found on DME arc approaches, straight-in approaches, and various approach transitions depicted on the charts, which also show procedure turns from overhead approaches. This is to help differentiate the various initial approaches and let you use the same charts if the terminal radar fails.

With more and more aircraft being compressed into the airspace, more and more routes seem to crop up on en route and approach charts. At times, these routings result in a mess of lines crisscrossing the charts, intermixed with numbers and symbols. It is important to understand such subtle things as the differences in the widths of the lines, for example, or in the position of the numbers in relation to the lines so that you will have a full understanding of what the charts are trying to explain.

The approach transitions are tracks that are defined by courses, minimum altitudes, and fixes that are designed to bring you from the *initial approach fix* (IAF) to the *final approach fix* (FAF). The mileage figures shown on these transitions are only between the fixes shown on the transition. They are not the mileage to the airport.

Again, this is obvious while sitting in an easy chair under a no-stress condition, but it is possible to misread or misunderstand it while you're bouncing around in a rainstorm with flickering panel lights at night, especially if it is at the end of a 14- or 16-hour day.

On the bottom of the left-hand column, you can see how Jepp insets a dogleg route with an off-chart turn. It is important to note that these insets are not to scale; also, the dogleg turn fixes are not charted on en route and area charts.

Look at the triangular-shaped transition halfway down the right-hand column and notice a route from Stout (STO) to TOWER. The route goes via PLANT intersection. Transition routes are depicted in heavy lines; they are the routes you must fly. You are not authorized to fly the 275-degree course from STO to TOWER because this light line with a small arrow only shows a cross-radial that helps to form the fix. You are only assured obstacle clearance when you are on the published route. The heavy line from STO to PLANT ends in a light line that indicates both an approach transition and a cross-radial.

In the next diagram below, you see two IAFs leading to an approach. HANDS is located by two cross-radials, one from STO and the other from MLS. Both facilities are off the chart as indicated by the complete identification of them. There is a dogleg 4.1 miles southeast of HANDS. From the dogleg, it is another 2.5 miles to JOHNS, which is identified by the 090-degree course and the 200-degree cross-radial from STO.

The second IAF is off the chart to the left. The IAF in this case is known as MARGE, and it is 14.7 miles to JOHNS as noted below the transition course line. There is also a minimum altitude of 2,700 feet shown on this section of the transition.

Because the approach procedure flight track is portrayed by a bold line, it would be difficult to show the approach transition segment when the segment coincides with the approach procedure track. The answer is to draw the approach transition alongside the approach-path line. The line signifying the approach transition will jut out from a fix and terminate in a heavy arrow pointing parallel to the approach path, as in the WORDS-ARTIC example in Fig. 16-11.

The upper left-hand column of Fig. 16-12 indicates how the approach transitions are depicted using DME arcs. The DME distance is shown alongside the arc itself, in this case the 18 DME. The minimum altitude is shown just inside the arc.

The next illustration down shows how lead radials are depicted as advisory points of when to begin the turn inbound.

The rest of the way in

An interesting example of an approach procedure flight track is shown near the bottom of the left-hand column. It shows a holding pattern designed to lose altitude over an NDB, after which the outbound leg of the approach procedure is a 105-degree magnetic course followed by a left turn to a 270-degree magnetic course inbound. As the holding pattern is not really part of the approach procedure flight track, it is depicted with a fine line rather than a bold one. Once inbound, the series of small arrows indicate that the remainder of the approach is a *visual* flight path.

The right-hand column shows procedure turns and course reversals. Four are illustrated here. The upper one is a schematic of a procedure that might be called the "generic" procedure turn. Remember that this is just a schematic and need not be followed exactly. The same is true of the 80/260 turn that is shown. All you are required to do in these two cases is to make the reversal on the same side of the approach course as depicted and to stay within the protected airspace at or above the minimum altitude.

The next two course reversals, the teardrop and the holding pattern, however, must be followed as published. This is an important point to remember.

Approach chart: plan view

330

APPROACH CHART LEGEND
APPROACH PLAN VIEW (continued)

APPROACH TRANSITIONS (continued)

Approach transitions via DME arcs are illustrated below with distance from facility, direction of flight, start and termination points of the arc. DME arc approach transitions may be started from any airway or authorized direct route which intercepts the arc. DME arc altitude is maintained until established on approach course.

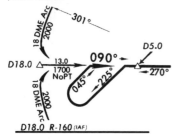

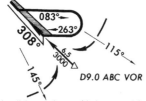

Lead radials may be provided as an advisory point for turning to the approach course.

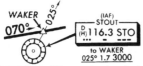

Approach transitions may be described under the originating navaid with course, distance, altitude, and terminating point.

APPROACH PROCEDURE FLIGHT TRACK

The approach procedure flight track is portrayed by a bold line. This track begins in the plan view at the same location where the profile begins.

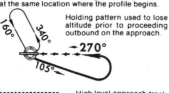

Holding pattern used to lose altitude prior to proceeding outbound on the approach.

················· High level approach track
➤➤➤➤➤➤➤ Visual flight track

PROCEDURE TURNS- COURSE REVERSALS

Schematic portrayal of procedure turn

45°/180° turn

80°/260° turn

Tear drop or Base turn. When course reversal is required, it must be flown as charted.

Holding pattern or Racetrack pattern. When course reversal is required, it must be flown as charted.

When a procedure turn, Racetrack pattern, Teardrop or Base turn is not portrayed, they are not authorized.

ALTITUDES

2300'	All altitudes in the plan view are "MINIMUM" altitudes unless specifically labeled otherwise. Altitudes are above mean sea level in feet. May be abbreviated "MIM."
MANDATORY 2400'	Mandatory altitudes are labeled "MANDATORY" and mean at the fix or glide slope intercept.
MAXIMUM 1900'	Maximum altitudes are labeled "MAXIMUM." May be abbreviated "MAX."
RECOMMENDED 2000'	Recommended altitudes are labeled "RECOMMENDED."

MISSED APPROACH

- - - - ➤ Initial maneuvering course for missed approach. Details of the missed approach are specified below the profile diagram.

Missed approach fix inset.

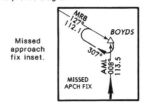

Fig. 16-12 *Reproduced with permission of Jeppesen Sanderson, Inc. Not for use in navigation.*

Odds and ends

Moving on to Fig. 16-13, as far as the missed approach is concerned, about all that is pictured on the chart is a heavy dashed line with a heavy arrowhead that shows the initial maneuvering course. The remainder of the procedure is written out in text form and can be found below the approach profile diagram.

Remember that if forced to execute a missed approach out of a circling approach, the initial climbing turn must be made toward the landing runway, then continue the turn until established on the missed approach course. The missed approach fix inset, shown near the top of the left-hand column, is depicted on charts when the missed approach fix is at a location outside the chart boundaries.

As a VFR pilot, you are conditioned to the color-coded sectional charts on which you can visualize what the en route terrain will look like just by looking at the topographical color on the chart. As you transition from VFR to IFR flying, you have to reevaluate the information available on the charts. As an IFR pilot, you will require even more information from charts, but color coding it would make the other information too hard to find and hard to read. Although the airport chart shows more ground detail, about all you will get from the approach plate plan view is the type of airport, large masses of water, restricted areas, specific obstructions that might encroach upon your airspace, and a 5-statute-mile-radius circle around the airport. This circle has been omitted on charts published after October 1, 1993.

Carefully study orientation details and reference points. In the first place, not all of the obstructions will be charted.

Prior to August 1988, many of the obstructions were indicated. Now, reference points lower than 400 feet above the airport elevation are not depicted, and none of the reference points shown can be relied upon to provide terrain or obstruction clearance because other higher obstacles might be in the same area.

"Why is that?" you ask. Well, it's to avoid clutter on the chart for one thing, and for another you aren't supposed to descend that low anyway, so the obstacles shouldn't really concern you. Remember, as long as you are on the prescribed approach course at or above the minimum allowable altitude, you won't have any problems with obstructions.

APPROACH CHART LEGEND
APPROACH PLAN VIEW (continued)

HOLDING PATTERN

Holding pattern not part of the approach procedure. DME figures, when provided, give the DME distance of the fix as the first figure followed by the outbound limit as the second figure. 3000 indicates the minimum holding altitude, (MHA).

Length of holding pattern in minutes when other than standard.

Holding patterns are generally not charted to scale.

Indicates procedure for leaving the holding pattern.

AIRPORTS

IFR airports in the area and VFR airports underlying the final approach are depicted.

    Airport to which the approach is designed

◯   Nearby Military airport

◌   Nearby Civil or joint use Military airport

Ⓗ Ⓗ   Heliport

⊕   Civil Seaplane Base

⊕   Military Seaplane Base

✪   Airport with light beacon

Ⓧ   Abandoned or closed airport

✈   An airport reference circle, 5 statute miles in radius, centered on the airport. Omitted after 1 OCT 93.

AIRSPACE

R-2402   Restricted airspace (Refer to the enroute chart for limitations.)

PROHIBITED AREA SC(P)-23

ORIENTATION DETAILS

River

Lake or large water area

★   Aeronautical Light/Beacon

TERRAIN HIGH POINTS AND MAN-MADE STRUCTURES

1. Some, but not all, terrain high points and man-made structures are depicted, along with their elevation above mean sea level. THIS INFORMATION DOES NOT ASSURE CLEARANCE ABOVE OR AROUND THE TERRAIN OR MAN-MADE STRUCTURES

TERRAIN HIGH POINTS AND MAN-MADE STRUCTURES (continued)

AND MUST NOT BE RELIED ON FOR DESCENT BELOW THE MINIMUM ALTITUDES DICTATED BY THE APPROACH PROCEDURE. Generally, terrain high points and man-made structures less than 400 feet above the airport elevation are not depicted.

2. Symbols for terrain high points and man-made structures:

✳   Natural terrain (peak, knoll, hill, etc.) Used prior to August 12, 1988.

•   Unidentified natural terrain or man-made. Used prior to August 12, 1988.

•   Natural terrain (peak, knoll, hill, etc.) Used after August 12, 1988.

⚒⚒⚒⚒   Man-made (tower, stack, tank, building, church)

∧   Unidentified man-made structure

4460′   Mean Sea Level elevation at top of TERRAIN HIGH POINT/MAN-MADE STRUCTURE.

±   Denotes unsurveyed accuracy

▼   Arrow indicates only the highest of portrayed TERRAIN HIGH POINTS AND MAN-MADE STRUCTURES in the charted planview. Higher terrain or man-made structures may exist which have not been portrayed.

⌐**GENERALIZED TERRAIN CONTOURS**

1. Generalized terrain contour information may be depicted when terrain within the approach chart planview exceeds 4000 feet above the airport elevation, or when terrain within 6 nautical miles of the Airport Reference Point (ARP) rises to a least 2000 feet above the airport elevation. THIS INFORMATION DOES NOT ASSURE CLEARANCE ABOVE OR AROUND THE TERRAIN AND MUST NOT BE RELIED ON FOR DESCENT BELOW THE MINIMUM ALTITUDES DICTATED BY THE APPROACH PROCEDURE. Furthermore, the absence of terrain contour information does not endure the absence of terrain or structures.

2. Terrain features are depicted using one of the two following methods:

a) Prior to June 24, 1994, terrain information was depicted as screened contour lines with contour values.

b) After June 24, 1994, screened contour lines will gradually be replaced with generalized contour lines, values and gradient tints printed in brown. Gradient tints indicate the elevation change between contour intervals.

L

Fig. 16-13 *Reproduced with permission of Jeppesen Sanderson, Inc. Not for use in navigation.*

17

Approach
chart: profile view

Officially, the profile view of the approach chart is described as the "vertical cross-section of the plan view." It is like watching the aircraft descend while standing off to the side of the approach path. The plan view depicts the flight path over the ground; the profile view shows the aircraft's altitude above the surface and is, therefore, a very crucial tool for the instrument pilot.

Three types of nonprecision approach profiles are shown in Fig. 17-1. Above them is a table of recommended altitudes/heights at various DME fixes. This table is designed to allow for a constant rate of descent; however, the altitudes/heights are only recommended figures, and the minimums depicted in the profile view still apply.

Regarding the tables, they do not always read from left to right as most other tables do. They are orientated to read in the same direction as the profile is depicted.

NONPRECISION APPROACHES

Examine the uppermost approach depicted on Fig. 17-1, which is the standard nonprecision approach that includes an overhead approach, an outbound leg, a procedure turn, and inbound leg. You begin by passing over the VOR and proceeding outbound on the 280-degree radial. Notice a group of numbers to the left of the profile. These numbers represent the procedure turn and define the turn's limits. In this case, the procedure turn must be completed within 10 nautical miles of the VOR, the fix that the turn is predicated upon. Remember this: The protected airspace begins at the fix

334   Approach chart: profile view

APPROACH CHART LEGEND
PROFILE VIEW

The top of the profile view on certain *non-precision* approaches contains a table of *recommended* altitudes/heights at various DME fixes to allow a constant rate of descent. The altitudes/heights are *recommended* only; minimum altitudes in the profile view apply. The table is sequenced in the same direction as the profile is portrayed.

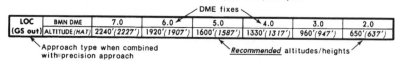

LOC	BMN DME	7.0	6.0	5.0	4.0	3.0	2.0
(GS out)	ALTITUDE (HAT)	2240'(2227')	1920'(1907')	1600'(1587')	1330'(1317')	960'(947')	650'(637')

Approach type when combined with precision approach — Recommended altitudes/heights

Notes pertaining to conditional use of the procedure are shown at the top of the profile. The note "Pilot controlled lighting" indicates that pilot activation is required as specified on the airport chart under Additional Runway Information.

The profile view schematically (not to scale) portrays the approach procedure flight track as a vertical cross section of the plan view.

NON-PRECISION APPROACH PROFILE (LOC, VOR, VORTAC, NDB, etc.)

M symbol representing the non-precision missed approach point (MAP), as shown below, is used on charts dated on or after 5 FEB 93. This symbol is omitted when more than one non-precision approach track is depicted.

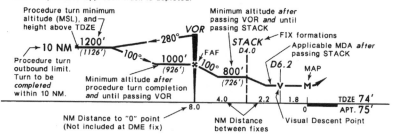

NON-PRECISION APPROACH PROFILE (LOC, VOR, VORTAC, NDB, etc.)
with constant rate of descent

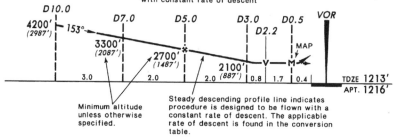

NON-PRECISION APPROACH PROFILE (VISUAL APPROACH)

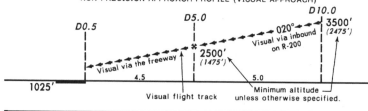

Fig. 17-1 *Reproduced with permission of Jeppesen Sanderson, Inc. Not for use in navigation.*

where the outbound track begins, not necessarily from the final approach fix, unless they are collocated. Also, the turn *must be completed* within the distance specified.

The large number above the line (1200') is the minimum altitude for the procedure turn, and the smaller number in parentheses below the line (1126') is the height in feet above the TDZE (touchdown zone elevation), the runway end, or the airport.

The most complete definition of the TDZE is in the pilot/controller glossary:

> *TOUCHDOWN ZONE ELEVATION—The highest elevation in the first 3,000 feet of the landing surface. TDZE is indicated on the instrument approach procedure chart when straight-in landing minimums are authorized.*

For more on the TDZE, look ahead and refer to Fig. 17-4, about midway down from the top left of the figure and notice that the number shown in parentheses (1200') is the:

> *Height in feet above airport, runway end, or TDZ elevation. Height is measured from airport elevation unless TDZE or runway end elevation is noted at the airport symbol.*

The airport symbol referred to is the heavy horizontal bar on the ground reference line (which is the next symbol down from the top left of Fig. 17-4, as well as that shown in the right side of the two top illustrations in Fig. 17-1). Going back to the nonprecision approach mentioned above, you will see that after completing the procedure turn and inbound on the 280-degree radial (equivalent to a magnetic course of 100 degrees, or 100 degrees TO on the VOR), you are allowed to descend a little lower, in this case down to 1,000 feet MSL (926 feet above the TDZE), until reaching the FAF.

Inbound from the FAF

The nonprecision FAF is identified by a Maltese cross symbol on the profile view. The FAF is described in the chart glossary:

> *FINAL APPROACH FIX (FAF)—The fix from which the final approach (IFR) to an airport is executed and which identifies the beginning of the final approach segment. It is desig-*

nated in the profile view of Jeppesen Terminal charts by the Maltese Cross symbol for nonprecision approaches and by the glide slope/path intercept point on precision approaches. The glide slope/path symbol starts at the FAF. When ATC directs a lower-than-published Glide slope/path Intercept Altitude, it is the resultant actual point of the glide slope/path intercept.

Inbound past the FAF (in this case the VOR), the course to the runway will be the 100-degree radial of the VOR (100 degrees FROM on the OBI). The numeral 8.0 under the VOR symbol and below the ground reference line indicates the distance, in nautical miles, between that point and the 0 point (in this case, the runway threshold). The zero point can be shown under middle markers, inner markers, or other points as well. It is not shown when a DME fix is at the same location. As an example, there is no mileage shown under the ground reference line at STACK. Indicated by D4.0, STACK is 4.0 miles from the VOR. The VOR in this case is also the FAF.

On the other hand, look at the center illustration on Fig. 16-2, and you will see underneath the letters LOM (outer compass locator) a DME distance of 5.8 from MTN, while the 4.7 is below the ground reference line.

Similar data is printed for the LMM (inner compass locator). The reason for this is that the DME to the markers is measured from a geographical point other than the 0 point, whereas the numerals beneath the ground reference line show the distance to the 0 point at the runway threshold.

Continuing with Fig. 17-1, after you pass the FAF inbound, use a step-down descent, descending first to 800 feet MSL until past STACK, which is a fix 4.0 DME along the 100-degree radial on the final approach course. Once past STACK, you can descend to the MDA.

Visualizing the VDP

Above the ground-reference line, notice 4.0 between the FAF and STACK, 2.2 between STACK and the V symbol, and 1.8 between the V symbol and the runway threshold. These are the segment distances between these points, all of which add up to the 8.0 miles from the FAF to the 0 point.

The V symbol is the visual descent point (VDP) that is described in the chart glossary:

VISUAL DESCENT POINT/VDP—A defined point on the final approach course of a nonprecision straight-in approach procedure from which normal descent from the MDA to the runway touchdown point may be commenced, provided the approach threshold of that runway, or approach lights, or other markings identifiable with the approach end of that runway are clearly visible to the pilot.

In other words, you are not to go below the MDA until you are past the VDP. When flying VFR, it's also a good idea to stay at or above the MDA until the VPD; this will give you a mental picture of what the runway environment will look like under real-life instrument conditions.

To save the trouble of having to add up all of the segment distances, the distance from the applicable facility, in this case D6.2, is included above the descent profile line.

In the center illustration of Fig. 17-1, notice a nonprecision profile using a steady rate of descent. You can find the rate of descent in the conversion table (which is discussed later), but the minimum altitudes shown at the various points of the approach still apply. In this example, the VOR is on the airport surface, and the missed approach point is 0.4 mile from the runway threshold. The small arrow pointing toward the runway indicates that the rest of the approach (from the MAP in this case) must be flown visually.

At times, you will find a nonprecision profile utilizing a visual approach as illustrated at the bottom of Fig.17-1, but this is self-explanatory.

A new symbol "**M**" has been appearing on the nonprecision approach profiles since February 1993. This indicates the missed approach point. It will be omitted if there is more than one nonprecision approach depicted on the chart.

Precision approaches cut a fine line

Precision approaches feature additional symbols. The first difference is between the procedure turn information and the FAF. The descent profile has been split into two different lines, one solid and one dashed.

Many airports will use locater beacons in conjunction with the outer marker. Sometimes this beacon will be incorporated into an NDB approach and the dashed line will signify the descent profile for the NDB approach. At other times, the glide slope might malfunction—either the airborne equipment or the ground equipment—in which case the ILS will revert to an LOC approach (minimums will be affected) and again the dashed line will signify the profile flight path.

Why do you see two different approach altitudes (1800' and 1400') inbound after the procedure turn? Because the entry to each type of approach is based on different thinking. Without glide path or DME information, you are never really sure where you are along the ground track, unless of course you are crossing one of the fixes. So, you want to get down as low as possible as soon as possible to reach the MDA before arriving at the MAP. On the other hand, when you do have the glide path information, you'll usually want to have the aircraft stabilized on the glide slope as soon as possible, certainly prior to crossing the LOM inbound.

The nonprecision FAF, as indicated by the Maltese cross, is the LOM in this illustration; however, the FAF for precision approaches is always the published glide slope intercept point, which in this case is reached prior to the LOM (unless ATC has authorized a lower-than-normal glide slope intercept altitude). The inbound ILS course of 043 degrees is illustrated inside the localizer symbol, and the glide slope altitude of 1,689 feet (1,615 feet above TDZE) at the LOM is under the two-dash Morse code identifier of the LOM.

To illustrate these points further, realize that from the glide path intercept point, which is slightly before crossing the FAF, to the DH, an aircraft on an ILS approach will continue down a steady descent path; however, an aircraft flying the localizer or NDB approach will descend to 1,400 feet MSL before crossing the FAF. Then it will descend to the MDA, level off, and maintain the MDA until it reaches the MAP.

Again, note the distances from fixes to the 0 point listed below the ground-reference line. The 4.2 above the line is the distance between the LOM and the MM.

Recall the last sentence in the earlier definition of the final approach fix. A lower glide slope/path intercept point can be used when directed by ATC. The lower diagram on Fig. 17-2 illustrates this very well. The step-down altitudes are shown for the nonprecision ap-

Precision approaches cut a fine line

APPROACH CHART LEGEND
PROFILE VIEW (continued)

PRECISION APPROACH PROFILE [ILS with LOC (GS out), or with NDB Approach]

⌐M symbol representing the non-precision missed approach point (MAP), as shown below, is used on charts dated on or after 5 FEB 93. This symbol is omitted when more than one non-precision ⌊approach track is depicted.

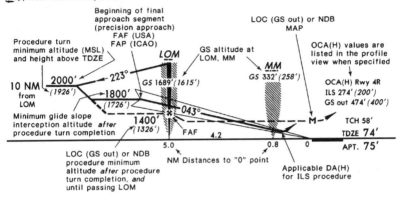

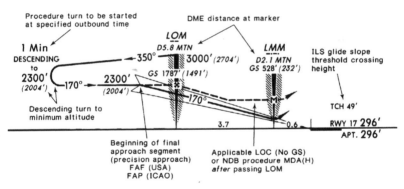

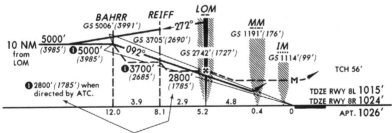

When ATC directs the lower noted altitude: For precision approaches, the altitude becomes the minimum glide slope intercept altitude and the resultant actual point of glide slope intercept becomes the FAF (USA).

Fig. 17-2 *Reproduced with permission of Jeppesen Sanderson, Inc. Not for use in navigation.*

proach on the chart, but the note states that the lower 2,800-foot altitude can be used as the glide slope/path intercept point when directed by ATC. In this case, the final approach point for the precision approach will still begin at the glide slope/path intercept point, not at the nonprecision FAF point at the LOM as indicated by the Maltese Cross symbol. You must always have a definite point at which to begin the final approach segment; without the glide slope/path indication in the cockpit, you will have to wait until the LOM can be identified.

DH, MAP, and other acronyms

Look at the differences in altitudes and geographical positions between the DH and the MAP. In the first place, understand that the DH is the altitude you will be at when you arrive at the MAP on an ILS approach, *if* you are properly on the glide slope.

On the other hand, on a nonprecision approach, when you reach the MDA, you are allowed to fly out the *time* to the MAP before initiating the missed approach procedure. In this case, descent as low as an ILS is not allowed due to a lack of precision either in altitude or azimuth or both. After all, if you are timing the approach, the actual ground position can vary slightly from the charted MAP, and if you don't get down to the MDA before reaching the MAP, you will still have to break off the approach if you don't have the runway environment in sight. The higher minimum altitude provides adequate obstruction clearance, compensating for the lack of precision.

Other notes might be placed in the profile view. One found in this illustration is the *threshold crossing height* (TCH), which in this case is 58 feet. You will only find this on precision approaches as it signifies the height at which the glide slope crosses the threshold. On some charts, its placement might lead you to believe that it refers to the nonprecision approach path that it appears to be attached to, but remember that it is the threshold crossing height *for the glide slope.*

If the *obstruction clearance altitude* (OCA) is lower than the charted DH or MDA, it will be listed as shown in the illustration. In this case, it is depicted (even though it is the same as the DH) because it is definitely lower than the MDA.

The center profile of Fig. 17-2 shows a descending turn in combination with the procedure turn. The procedure turn is to be started after flying outbound for a specific time (in this case, 1 minute). It also shows an LMM rather than just the MM shown in the upper illustration.

CAT II and CAT IIIA

Moving on to the top of Fig. 17-3, we find the granddaddies of all instrument approaches. You might fly your entire career and never get into equipment that will allow you to fly down to these minimums—as low as zero/zero on a CAT IIIA. Still, you should be aware of how they are portrayed.

In the first place, these are straight-in approaches in which neither procedure turns, base turns, nor any other turn is authorized. The aircraft has to be specially equipped and certified, and you have to be certified as well. The beginning of the approach is just like any other ILS, but notice where the descent ends. That's right, the runway. It flies you right to the ground.

The letters RA appear twice, once at the inner marker (IM) not used in normal ILS approaches, and again at a point between the MM and the IM. The numbers shown after the RA are for the radio altimeter height above the ground at those points. Those points are at 150 and 100 feet above touchdown or height above touchdown (HAT).

If that's the case, why don't the radio altimeters read 150 and 100 at these points? That's because radio altimeters read directly off the ground, and the ground at any given distance from the runway is very seldom the same elevation as the runway itself. In the first instance, the radio altimeter would read 199 feet, so the ground is 49 feet below the touchdown elevation. In the second instance the radio altimeter would read 113 feet, so the ground there is 13 feet lower than the touchdown zone.

Get an uncluttered look at some of the profile symbols by looking at the rest of the symbols on Fig. 17-3. Here is the VDP, the FAF, the visual flight track path, and others.

A recent new symbol is DA(H). Although this is really an ICAO symbol, Jeppesen is using it on its charts so it might as well be discussed. The DA is the decision altitude referenced to MSL (the DH is the decision height referenced to the touchdown elevation). Another new symbol is MDA(H), with the same mean-sea-level explanation.

342 Approach chart: profile view

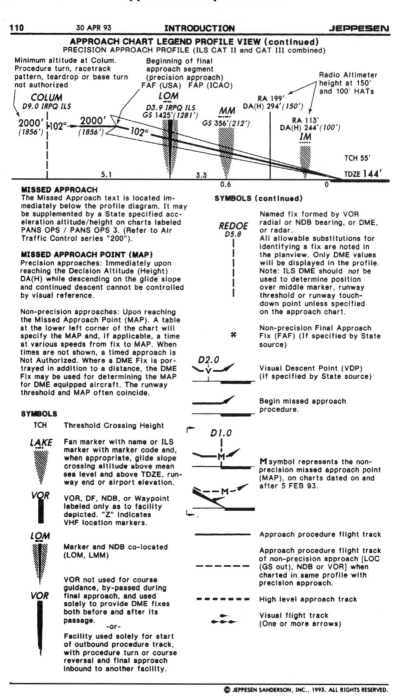

APPROACH CHART LEGEND PROFILE VIEW (continued)
PRECISION APPROACH PROFILE (ILS CAT II and CAT III combined)

Minimum altitude at Colum. Procedure turn, racetrack pattern, teardrop or base turn not authorized

Beginning of final approach segment (precision approach) FAF (USA) FAP (ICAO)

Radio Altimeter height at 150' and 100' HATs

COLUM
D9.0 IRPQ ILS

LOM
D3.9 IRPQ ILS
GS 1425'(1281')

MM

RA 199'
DA(H) 294'(150')

2000'
(1856') ├102°

2000'
(1856') ╲ 102°

GS 356'(212')

RA 113'
DA(H) 244'(100')
IM

TCH 55'

5.1   3.3   0.6   0   TDZE 144'

MISSED APPROACH
The Missed Approach text is located immediately below the profile diagram. It may be supplemented by a State specified acceleration altitude/height on charts labeled PANS OPS / PANS OPS 3. (Refer to Air Traffic Control series "200").

MISSED APPROACH POINT (MAP)
Precision approaches: Immediately upon reaching the Decision Altitude (Height) DA(H) while descending on the glide slope and continued descent cannot be controlled by visual reference.

Non-precision approaches: Upon reaching the Missed Approach Point (MAP). A table at the lower left corner of the chart will specify the MAP and, if applicable, a time at various speeds from fix to MAP. When times are not shown, a timed approach is Not Authorized. Where a DME Fix is portrayed in addition to a distance, the DME Fix may be used for determining the MAP for DME equipped aircraft. The runway threshold and MAP often coincide.

SYMBOLS

TCH   Threshold Crossing Height

LAKE   Fan marker with name or ILS marker with marker code and, when appropriate, glide slope crossing altitude above mean sea level and above TDZE, runway end or airport elevation.

VOR   VOR, DF, NDB, or Waypoint labeled only as to facility depicted. "Z" indicates VHF location markers.

LOM   Marker and NDB co-located (LOM, LMM)

VOR   VOR not used for course guidance, by-passed during final approach, and used solely to provide DME fixes both before and after its passage.
-or-
Facility used solely for start of outbound procedure track, with procedure turn or course reversal and final approach inbound to another facility.

SYMBOLS (continued)

REDOE
D5.8   Named fix formed by VOR radial or NDB bearing, or DME, or radar. All allowable substitutions for identifying a fix are noted in the planview. Only DME values will be displayed in the profile. Note: ILS DME should *not* be used to determine position over middle marker, runway threshold or runway touchdown point unless specified on the approach chart.

✳   Non-precision Final Approach Fix (FAF) (If specified by State source)

D2.0   Visual Descent Point (VDP) (if specified by State source)

Begin missed approach procedure.

D1.0   ─M─   M symbol represents the non-precision missed approach point (MAP), on charts dated on and after 5 FEB 93.

Approach procedure flight track

Approach procedure flight track of non-precision approach [LOC (GS out), NDB or VOR] when charted in same profile with precision approach.

High level approach track

Visual flight track (One or more arrows)

© JEPPESEN SANDERSON, INC., 1993. ALL RIGHTS RESERVED.

Fig. 17-3 *Reproduced with permission of Jeppesen Sanderson, Inc. Not for use in navigation.*

CATII and CATIIIA 343

APPROACH CHART LEGEND
PROFILE VIEW (continued)

SYMBOLS (continued)

2300' — All altitudes in the profile view are "MINIMUM" altitudes unless specifically labeled otherwise. Altitudes are above mean sea level in feet. May be abbreviated "MIM".

MANDATORY 2400' — Mandatory altitudes are labeled "MANDATORY" and mean at the fix or glide slope intercept.

MAXIMUM 1900' — Maximum altitudes are labeled "MAXIMUM". May be abbreviated "MAX".

OCL Rwy 04R 274' (200') — Obstruction Clearance Limit

OCA(H) Rwy 26 720' (263') — Obstruction Clearance Altitude (Height)

RECOMMENDED 2000' — Recommend altitudes are labeled "RECOMMENDED".

(1200') — Height in feet above airport, runway end, or TDZ elevation. Height is measured from airport elevation unless TDZE or runway end elevation is noted at the airport symbol.

TDZE **74'** / APT. **75'** — Touchdown Zone Elevation. (Runway End or Threshold Elevation when labeled RWY). / Official Airport Elevation

10 NM 1200'/(1126') — Procedure turn minimum altitude (MSL) / Height above TDZE, runway end, runway threshold, or airport.

Procedure turn outbound limit. When the outbound procedure track is depicted in the profile view, the turn limit is from the fix where the outbound track begins. The turn must be carried out within the specified distance.

Combined procedure turn (course reversals) and NoPT procedure flight tracks

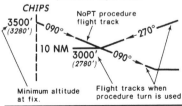

CHIPS

3500' (3280') 090°

10 NM 3000' (2780') 090° 270°

NoPT procedure flight track

Minimum altitude at fix. / Flight tracks when procedure turn is used

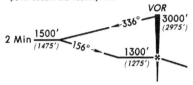

1 Min 080°→ ←260° 2000' (1900') — Racetrack used in lieu of procedure turn with holding limit, outbound and inbound bearings, and minimum altitude.

For a racetrack and holding in lieu of procedure turn, the outbound track corresponds to the plan view depiction beginning at a point abeam the facility/fix.

VOR

2 Min 1500' (1475') 336° **3000' (2975')** 156° **1300' (1275')**

Procedure based on 120 KT TAS. — When airspeeds are indicated in profile note, higher airspeeds require shortened times to assure remaining in the protected area.

Radar required. — Radar vectoring is required when it is the only approved method for providing a procedure entry and/or for identifying a terminal fix.

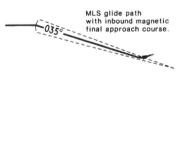

Glide Slope with inbound magnetic course of Localizer. —120°

Glide Slope, Glide path Intercept is the Final Approach Fix (FAF USA), Final Approach Point (FAP ICAO) for precision approaches. The glide slope symbol starts at the FAF/FAP.

035° — MLS glide path with inbound magnetic final approach course.

Fig. 17-4 *Reproduced with permission of Jeppesen Sanderson, Inc. Not for use in navigation.*

Missed approach

Recall the approach chart format; the heading information was at the top of the page, followed by the approach plan view, under that is the profile view, and finally, at the bottom of the page are the landing minimums.

The missed approach procedures are written out below the profile view just before the landing minimums. The missed approach is a very crucial area of the approach. Here you are at a minimum altitude, tooling along in the dark (so to speak), with the aircraft all dirtied up and ready to land. If you can't see the airport when you get to the MAP, it's time to get back up in the ozone where you will be safe. But in order to clean up, climb, and miss all of the buildings, trees, poles, mountains, and the like, the missed approach must be executed in an orderly fashion.

When reviewing the approach plates prior to filing the flight plan, you also reviewed the missed approach procedure. When there are two pilots aboard, it's a good idea for the one not flying to read the missed approach procedure aloud after passing the FAF inbound.

When there is only one pilot on board, I find it useful to copy the missed approach procedure in advance on a small file card or on one of those pages from the telephone answering pad and tape it to the glare shield where I can see it without having to take my hands off the controls. (Naturally, when taping the card up it has to be placed where it won't block the instruments from view.) I put the initial turn and altitude information in large capital letters so they'll catch my eye.

Remember, when the time comes to execute the missed approach, you will be pretty busy applying power, retracting the gear and flaps, transitioning from a descent to a climb regime, turning, resetting navigational radios, and reporting to the proper control facility, so you won't have time to scan a page in the manual for the procedure *and you definitely don't want to rely on your memory*.

Figure 17-4 presents a few more symbols that might be of interest. The bottom illustration in the left-hand column shows how a procedure turn and a NoPT flight path would be symbolized at the same facility.

You have already reviewed the procedure turn symbol. The upper right-hand symbol shows a holding pattern used in lieu of a procedure turn. Here notice the inbound and outbound bearings as well as the time of the inbound leg and the minimum altitude. Below this symbol is another one that shows a holding pattern and racetrack in lieu of a procedure turn that is depicted somewhat differently. Here you have the outbound track shown in the descending line that would begin abeam the VOR, and the inbound track showing how you would descend from the holding pattern when cleared for an approach.

Note the words "Procedure based on 120 KT TAS" in the next example. This indicates that if you have a faster aircraft, you will have to reduce all times to remain within the protected airspace.

Landing minimums

Landing minimums are presented in a standard-format table depending on the type of approach. To simplify the understanding of these minimums, the only information shown in each table will be those items that specifically affect that particular approach.

Examine the USA FORMAT found at the top of Fig. 17-5. The letters A, B, C, and D running down the left-hand border of the format signify the aircraft categories.

The table is broken up into various blocks. The DA(H) or MDA(H) will be shown in the upper part of the table, while the visibility requirements necessary to initiate the approach will be shown in the blocks below.

The format shown in Fig. 17-5 is just an illustration. The headings used on the actual charts might be somewhat different, depending upon which lights and auxiliary systems are available on a particular approach. Furthermore, the delineations of the various blocks will depend on many factors, like the airport itself, the surrounding terrain or aircraft performance, for instance.

The lowest minimums listed in any format will be found on the left side of the chart. Less-desirable minimums are shown in descending order to the right. In this case, note that all minimums to the left of the vertical line running from top to bottom are predicated upon straight-in approaches. The minimums to the right side of the line are for circling approaches.

Notice that the top two formats are for United States charts. The upper example is for charts published prior to October 15, 1992, while

APPROACH CHART LEGEND LANDING MINIMUMS (continued)

USA FORMAT - Prior to 15 October 1992 Effective date.

	STRAIGHT-IN LANDING RWY 36L						CIRCLE-TO-LAND	
	ILS				LOC (GS out)			
	DA(H) 212' (200')			DA(H) 262'(250')	MDA(H) 400' (388')		Max Kts	MDA(H)
	FULL	TDZ or CL out	ALS out	MM out		ALS out		
A	RVR 18 or ½	RVR 24 or ½	RVR 40 or ¾	RVR 24 or ½	RVR 24 or ½	RVR 50 or 1	90	560' (533')-1
B							120	
C							140	560' (533')-1½
D				RVR 40 or ¾	RVR 40 or ¾	RVR 60 or 1¼	165	580' (553')-2

USA FORMAT - Effective 15 October 1992 and all succeeding revisions.

	STRAIGHT-IN LANDING RWY 36L					CIRCLE-TO-LAND	
	ILS			LOC (GS out)			
	DA(H) 212' (200')			MDA(H) 400' (388')		Max Kts	MDA(H)
	FULL	TDZ or CL out	ALS out		ALS out		
A	RVR 18 or ½	RVR 24 or ½	RVR 40 or ¾	RVR 24 or ½	RVR 50 or 1	90	560' (533')-1
B						120	
C						140	560' (533')-1½
D				RVR 40 or ¾	RVR 60 or 1¼	165	580' (553')-2

WORLD-WIDE FORMAT

	STRAIGHT-IN LANDING RWY 36L					CIRCLE-TO-LAND	
	ILS			LOC (GS out)			
	DA(H) 212' (200')			MDA(H) 400' (388')		Max Kts	MDA(H)
	FULL	TDZ or CL out	ALS out		ALS out		
A	RVR 550m VIS 800m	RVR 720m VIS 800m	1200m	RVR 720m VIS 800m	RVR 1500m VIS 1600m	100	560' (533') -1600m
B						135	
C						180	630' (603') -2800m
D			1200m		RVR 1800m VIS 2000m	205	730' (703') -3600m

SIDESTEP INOPERATIVE COMPONENTS
For a runway identified as sidestep, such as SIDESTEP RWY 24L:
Inoperative light components shown in Rwy 24L column are those for the lights installed on *Rwy 24L, not* the lights for Rwy 24R.

CIRCLE-TO-LAND
Starting with charts dated July 28, 1989, *maximum aircraft speeds* for circling are shown in lieu of Aircraft Approach Categories. The maximum indicated airspeeds are shown in knots (kilometers per hour on Metric Edition charts).

U.S. STANDARD FOR TERMINAL INSTRUMENT APPROACH PROCEDURES (TERPS)

	CIRCLE-TO-LAND
Max Kts	MDA(H)
90	560' (533')-1
120	
140	560' (533')-1½
165	580' (553')-2

NEW INTERNATIONAL CIVIL AVIATION ORGANIZATION (ICAO) FLIGHT PROCEDURES

	CIRCLE-TO-LAND
Max Kts	MDA(H)
100	560' (533') -1600m
135	
180	630' (603') -2800m
205	730' (703') -3600m

Known deviations to the above speeds are charted. For the few countries that have not published maximum circling speeds, aircraft approach categories A,B,C and D will continue to be shown.
Aircraft Approach Categories in the straight-in minimum column can be read across the chart from left to right for referencing the circle-to-land information.
The fact that straight-in-minimums are not published does not preclude the pilot from landing straight-in, using published circling minimums, if he has the straight-in runway in sight in sufficient time to make a normal approach for landing. Under such conditions, and when Air Traffic Control has cleared him for landing on that runway, he is not expected to circle even though straight-in minimums are not published. If he desires to circle, he should advise ATC.

Fig. 17-5 *Reproduced with permission of Jeppesen Sanderson, Inc. Not for use in navigation.*

the next illustration down is for charts published after October 15, 1992.

Since October 15, 1992, none of the approaches in the United States require higher minimums for the loss of the MM.

Whenever you see a horizontal line, it will tell you that visibility minimums are different above and below the line. These differences are not always the same for each category in each situation, so very carefully check the landing minimums for the approach charts. This statement cannot be repeated too often.

Circling minimums

To the right of the vertical line are the landing minimums for circling approaches. You will find speeds listed on the left side of the column. These are the same V_{SO} speeds as you will find for the categories, but they simplify planning for you.

Recall that if an aircraft had to circle at a speed higher than what was listed for its category, the aircraft would occupy the next higher category for a circling approach. These speeds simplify that process.

That will not happen in most general aviation aircraft, but as you progress in flying you will probably move up to a high performance business aircraft or an airliner. Many of these will maneuver at a flap setting lower than the landing flaps, termed *maneuvering* or *approach flaps*. The V_{S1} will increase with these different flap settings, and may very well throw you into the higher approach category. It's a lot easier to read the numbers than it is to have to convert to letters.

The worldwide formats are also shown in Fig. 17-5. The biggest differences are that the worldwide charts show the distances in meters or kilometers, rather than in feet or miles. If meters are used, the full value is shown as well as the letter m; if only a number is shown, it is feet or miles. Also, the circling speeds are slightly higher.

Conversion tables are useful for a couple of reasons, but to give them more meaning, let's look at the examples shown in Fig. 17-6.

The obvious time to use this table is when you are flying an approach that uses time/speed to determine the MAP. It's also a good backup for any approach in which it can be utilized, especially non-precision approaches. On precision approaches it gives a fairly good

348　　　　　　　　Approach chart: profile view

APPROACH CHART LEGEND
LANDING MINIMUMS (continued)

CONVERSION TABLE

At the bottom of the approach chart page, there is a conversion table as shown below.

Gnd speed-Kts	70	90	100	120	140	160
GS　　2.50°	315	405	450	541	631	721
LOM to MAP 5.0	4:17	3:20	3:00	2:30	2:09	1:53

Gnd speed-Kts	70	90	100	120	140	160
VOR to MAP 3.9	3:21	2:36	2:20	1:57	1:40	1:28

The speed table relates aircraft approach speeds to the rate of descent for the ILS glide slope (descent in feet per minute). For non-precision approaches it relates speed to the distance shown from the final approach fix (FAF) or other specified fix to the missed approach point (MAP).

Some missed approach points are calculated on a time/speed basis after completion of the procedure turn inbound on final approach. The absence to a time/speed table means the MAP cannot be determined by time and a timed approach is Not Authorized.

Gnd speed-Kts	70	90	100	120	140	160
Descent rate D7.0 to D3.0	466	600	667	800	934	1067
MAP at D1.5						

Non-precision approaches designed to be flown at a constant rate of descent have a rate of descent provided in the conversion table. The conversion table specifies a rate of descent that allows arrival at minimum altitudes shown in the profile view. The descent rate is a recommended rate only. Minimum altitudes shown in the profile view apply.

Gnd speed-Kts		70	90	100	120	140	160
Rwy 5, 23, PAR GS 2.50°		315	405	450	541	631	721
Rwy 30 PAR GS　　2.55°		322	413	459	551	643	735

On PAR charts:
Speed table with rates of descent on PAR glide slope is provided.

Gnd speed-Kts	70	90	100	120	140	160
Descent Gradient 5.9%	418	538	597	717	836	956
MAP at VOR						

When provided by the State, a non-precision descent gradient is provided with a descent table in feet per minute.

Gnd speed-Kts		70	90	100	120	140	160
ILS GS 3.00° or							
LOC Descent Gradient 5.2%	377	484	538	644	753	861
MAP at MM							

For combined ILS and non-precision approaches, only one descent table is provided when the ILS glide slope angle and the descent gradient are coincidental.

Gnd speed-Kts		70	90	100	120	140	160
Glide path Angle 3.00°	377	485	539	647	755	863
FAF to MAP	5.1	4:22	3:24	3:04	2:33	2:11	1:55

On MLS charts the Glide path angle authorized for the procedure and rate of descent table is provided.

Amendment number of a procedure. An amendment number increase generally indicates a procedure change. →

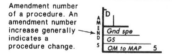

"MILITARY" notation, shown here on charts dated on and after JUN 8-90 indicates military source used for the procedure. →

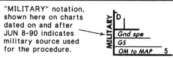

ON CHARTS DATED ON OR AFTER DEC 16-88 (NOT APPLICABLE TO USA AND CANADA)

⌐ "PANS-OPS" margin notation indicates that the State has specified that the instrument approach procedure complies with the ICAO Procedures for Air Navigation Services-Aircraft Operations (PANS-OPS) Document 8168, Volume II, 1st or 2nd Edition. Aircraft handling speeds for these procedures are shown on Introduction Page 2 under "AIRCRAFT APPROACH CATEGORY (ICAO)". Known deviations to these handling speeds are charted.

"PANS-OPS 3" further indicates that holding speeds to be used are those specified in Document 8168, Volume II, Third Edition.

"PANS-OPS 4" further indicates that the acceleration segment criteria have been deleted, as formerly published in Document 8168, Volume II, Third Edition.

Jeppesen International Air Traffic Control ("200" Series) pages provide an extract of the latest PANS-OPS Document 8168, Volume I. They highlight the major differences of Document 8168, Volume I and the earlier version, concerning holding speeds. Holding speed tables for both the earlier revision, and the later Edition 3 and 4, of PANS-OPS are included in these pages.

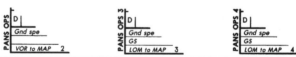

Fig. 17-6 *Reproduced with permission of Jeppesen Sanderson, Inc. Not for use in navigation.*

estimate of the rate of descent you'll need to stay on the glide slope. If you maintain this rate of descent, yet find yourself deviating from the glide slope, sometimes this will be the first indication that a wind shear exists.

The conversion table at the top left of Fig. 17-6 is for a precision approach. The upper line is ground speed in knots. This means that you will have to convert from statute miles per hour, unless the airspeed indicator is calibrated in knots, and then add or subtract the known or estimated headwind or tailwind components. Naturally, you will have to interpolate, but that should be no major problem. The second line shows the glide slope angle, in this case it is 2.50 degrees, as well as what the rate of descent should be, based upon the estimated ground speed, to stay on the glide slope.

The bottom line gives the amount of time (in minutes and seconds) from the nonprecision FAF to the MAP, useful for timed localizer approaches when the glide slope is out of service. In the example, it shows that the distance from the LOM to the MAP is 2.6 nautical miles, and time ranges from 2 minutes and 14 seconds to just 59 seconds, depending on the ground speed.

In the nonprecision approach table on the right, the glide slope line is omitted. The distance from the VOR to the MAP is 3.9 miles. The rest is self-explanatory.

If the time/speed table is missing, it means that you are unable to determine the MAP by time. In such a case, a timed approach is not authorized.

18

Airport charts

Airport layout charts are normally found on the reverse side of the first approach chart for the airport. As mentioned in Chapter 15 though, some of the airports are becoming so large and complex that it is necessary to produce even larger charts of the airport, the taxiways, and the gate locations at the terminals in some instances.

The airport format is made up of four sections just like the approach chart. The upper portion is the heading, the next one down is the airport plan view, below that you will see the runway data, and finally, at the bottom of the page, you find the takeoff and alternate airport minimums.

Examine a sample heading in Fig. 18-1. The location and airport name in this example are on the left side of the heading. The latitude and longitude are listed directly under the airport name; these coordinates are shown for the airport reference point (ARP), when one is charted at the airport; otherwise, the source of the coordinates will be shown. In the example you will also see that the airport is 7.7 miles from the TRP VOR.

The ICAO or United States airport identifier is printed in bold type just to the left of the chart index number.

The communications block is divided into two sections; the left side pertains primarily to frequencies used on the ground; the right side is for frequencies used in flight.

The frequencies are listed top to bottom in the order used during departure: ATIS, clearance delivery, ground, and tower. The departure control frequencies and the VOR test frequency, if applicable, are found on the right side of the block.

APPROACH CHART LEGEND
AIRPORT CHART FORMAT

The airport chart appears on the back side of the first approach chart. It contains information pertaining to the airport, air/ground communications, take-off minimums, alternate
minimums, and departure procedures. At major terminals, the airport chart may be expanded
and indexed separately to provide detailed information pertaining to taxiways, ramp or terminal parking areas, aircraft parking spot coordinates, start-up procedures, and low visibility
procedures.

HEADING

Geographic name, airport name, latitude and longitude, elevation, magnetic variation. location
identifier, index number, revision date and communications are given at the top of the page as
illustrated below. All communications for departing the airport are listed in order of use. The
designated Common Traffic Advisory Frequency (CTAF) is shown for U.S. public airports without control tower or where the tower is part-time. UNICOM, when available, is charted when
other local communication capabilities are not available.

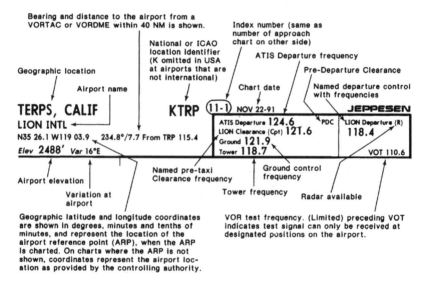

Fig. 18-1 *Reproduced with permission of Jeppesen Sanderson, Inc.
Not for use in navigation.*

New symbols

Some airport plan symbols will be new to you. To help understand
them, examine Fig. 18-2 and Fig. 18-3. As on the approach chart, the
symbols on the airport plan are all in white, black, and shades of
gray. This reduces the amount of detail that can be shown; therefore,
many symbols are used. Basically, paved runways are black, un-
paved runways are white, and taxiways and ramps are gray. Perma-
nently closed taxiways are shown by gray Xs.

New symbols

353

APPROACH CHART LEGEND
AIRPORT PLAN VIEW

SYMBOLS

Physical feature symbols used on the airport chart are illustrated below.

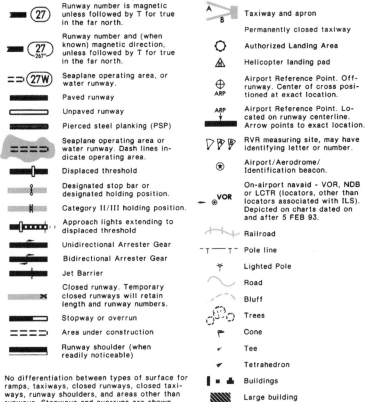

(27)	Runway number is magnetic unless followed by T for true in the far north.
(27) 267°	Runway number and (when known) magnetic direction, unless followed by T for true in the far north.
(27W)	Seaplane operating area, or water runway.
	Paved runway
	Unpaved runway
	Pierced steel planking (PSP)
	Seaplane operating area or water runway. Dash lines indicate operating area.
	Displaced threshold
	Designated stop bar or designated holding position.
	Category II/III holding position.
	Approach lights extending to displaced threshold
	Unidirectional Arrester Gear
	Bidirectional Arrester Gear
	Jet Barrier
	Closed runway. Temporary closed runways will retain length and runway numbers.
	Stopway or overrun
	Area under construction
	Runway shoulder (when readily noticeable)

A B	Taxiway and apron
	Permanently closed taxiway
○	Authorized Landing Area
⚠	Helicopter landing pad
⊕ ARP	Airport Reference Point. Off-runway. Center of cross positioned at exact location.
ARP	Airport Reference Point. Located on runway centerline. Arrow points to exact location.
▽ ▽ ▽	RVR measuring site, may have identifying letter or number.
✪	Airport/Aerodrome/Identification beacon.
⊙VOR	On-airport navaid - VOR, NDB or LCTR (locators, other than locators associated with ILS). Depicted on charts dated on and after 5 FEB 93.
	Railroad
-T—T-	Pole line
⍟	Lighted Pole
∿	Road
	Bluff
	Trees
☞	Cone
↰	Tee
↰	Tetrahedron
▮ ▪ ▲	Buildings
▨	Large building

No differentiation between types of surface for ramps, taxiways, closed runways, closed taxiways, runway shoulders, and areas other than runways. Stopways and overruns are shown regardless of surface, with the length, when known. Stopway and overrun lengths are not included in runway lengths.

ADDITIONAL INFORMATION

Runway end elevations are shown on the airport diagram if source is available.

Approach lights and beacons are the only lighting symbolized on the airport diagram. Approach lights are normally shown to scale in a recognizable form. For approach light symbols see page 121.

A representative selection of reference points known to Jeppesen is depicted. The elevation of reference points depicted is above mean sea level (MSL).

Latitude and longitude ticks at tenths of a minute interval are charted around most planview neatlines.

Feet 0 1000 2000 3000 4000 5000
Meters 0 500 1000 1500
Bar Scale

Fig. 18-2 *Reproduced with permission of Jeppesen Sanderson, Inc. Not for use in navigation.*

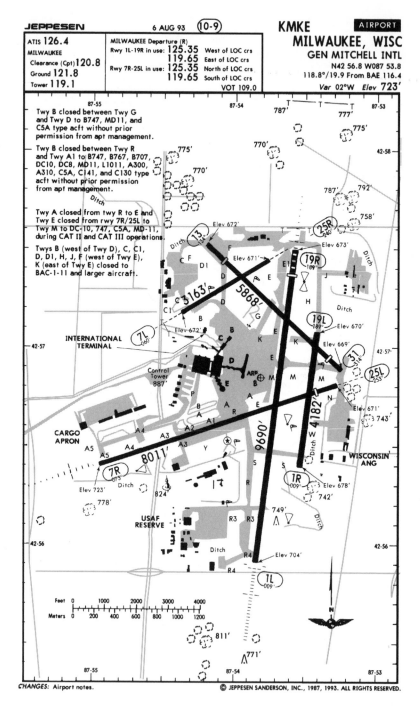

Fig. 18-3 *Reproduced with permission of Jeppesen Sanderson, Inc. Not for use in navigation.*

The runway numbers are all magnetic unless they are followed by a "T" to indicate that they are true headings. True headings are only found in the Far North. Most runways have the magnetic heading shown in smaller numbers below the runway number as seen in the second example down in the left-hand column of Fig. 18-2. A "W" will follow the runway number if it is a seaplane-operating area or a water runway.

Let's begin examining the Milwaukee airport chart by glancing at the communications block in Fig. 18-3 to see how KMKE delegates its departure control frequencies, depending on which runway is in use, and the departure route after takeoff.

The airport as seen from the air

On the airport plan view, notice that Runways 25L, 31, 19R, and 13 all have displaced landing thresholds as indicated by the white crossbar placed across the runway. The *airport reference point* (ARP), where the latitude and longitude coordinates are measured, can be seen just to the west of the midpoint of Runway 1L/19R.

The legend in the right-hand column of Fig. 18-2 explains that if the ARP is located off the runway, a circle with a cross inside it will be placed so that the center of the cross is on the ARP. The letters ARP will be below the circle. If the ARP is on a runway, the position will be indicated by an arrow from the letters ARP, and the circle and cross symbol will be omitted.

Every runway at Milwaukee has the elevation of both ends marked on the chart. If this information is omitted, the information is not available. Taxiways are identified alphabetically: G, M, DD.

Additional runway information

The only lights shown on the airport plan view are approach lights and beacons, and according to Jeppesen, while the approach lights are "normally shown to scale in a recognizable form," the easiest way to determine what type of system(s) is (are) installed is to read the information in the additional runway information table that is often located below the airport plan view.

In Milwaukee's case, however, this table appears on the reverse side of the plan view (Fig. 18-4). Some airports show this table on the reverse

JEPPESEN 6 AUG 93 (10-9A)

AIRPORT

MILWAUKEE, WISC
GEN MITCHELL INTL

GENERAL

Birds in vicinity of airport.
Low-level wind shear alert system.

ADDITIONAL RUNWAY INFORMATION

RWY		LANDING BEYOND Threshold	LANDING BEYOND Glide Slope	USABLE LENGTHS Threshold to Intersecting Runway	TAKE-OFF	WIDTH
1R ❶ 19L	MIRL			13/31 3450'	Turbojet NA	150'
❶ Closed 2200-0600 LT except to lightweight single engine aircraft.						
1L ❷	HIRL CL ALSF-I TDZ ❸PAPI-R RVR		8439'	7R/25L 4500' 13/31 7200'		200'
19R	HIRL CL MALSR ❸PAPI-R RVR	8915'	7861'	7R/25L 3700'		
❷ Runway grooved. ❸ (angle 3.00°)						
7R ❹	HIRL SSALR ❺PAPI-L RVR		6837'	1L/19R 5450' 1R/19L 6500'		150'
25L	HIRL REIL VASI-L	7339'				
❹ Runway grooved. ❺ (angle 3.00°)						
7L ❻	MIRL VASI-L (angle 3.1°)					100'
25R	MIRL VASI-L (angle 3.5°)					
❻ Closed to all jet aircraft and aircraft over 12,500 lbs. Closed 2200-0600 LT except to lightweight single engine aircraft.						
13 ❼	MIRL VASI-L	5137'		1L/19R 2150' 1R/19L 3550'		150'
31	MIRL VASI-R	5344'		7L/25R 3650'		
❼ Closed to turbojet aircraft without prior permission from airport manager. Closed 2200-0600 LT except to lightweight single engine aircraft.						

TAKE-OFF

	Rwy 1L			Rwys 13,19R 19L,25R,25L,31		Rwy 7R			Rwys 1R, 7L
	CL & RCLM any RVR out, other two req.	Adequate Vis Ref	STD	Adequate Vis Ref	STD	With Mim climb of 299'/NM to 1100' Adequate Vis Ref	With Mim climb of 299'/NM to 1100' STD	Other	
1 & 2 Eng	TDZ RVR 6 Mid RVR 6	RVR 16 or 1/4	[1] RVR 50 or 1	RVR 16 or 1/4	[2] RVR 50 or 1	RVR 16 or 1/4	[3] RVR 50 or 1	300-1	300-1
3 & 4 Eng	Rollout RVR 6		[1] RVR 24 or 1/2		RVR 24 or 1/2		RVR 24 or 1/2		

[1] FAR 135: Rwy 1L, RVR 18.
[2] FAR 135: Rwy 19R, RVR 24.
[3] FAR 135: RVR 24.

FOR FILING AS ALTERNATE

	Precision	Non-Precision	NDB Rwy 7R
A B C D	600-2	800-2	NA

AMEND 4

CHANGES: Usable lengths, take-off minimums.

Fig. 18-4 *Reproduced with permission of Jeppesen Sanderson, Inc. Not for use in navigation.*

side of the plan view, Milwaukee uses the 10-9A series (Fig. 18-4). For a complete listing of the abbreviations and an example of an additional runway information table, see Figs. 18-5 and 18-6.

This example shows the wealth of information that can be derived from this one small block. The definitions of the abbreviations are quite straightforward, but there are a few that might benefit from discussion. CL, for example, means centerline lights. If shown by itself, it means that these lights are arranged in what is considered to be a standard configuration. Centerline lights are white from the approach end of the runway until 3,000 feet from the departure end; from that point until 1,000 feet from the departure end, you will see alternating red and white lights; all of the centerline lights in the final 1,000 feet are red. These are a tremendous help during low-visibility conditions. Some airports use all-white centerline lights, and in that case, the symbol will be CL (white). If the centerline lights are in a nonstandard configuration, and the configuration is not known, the symbol is CL (non-std).

If the nonstandard configuration is known, it will be specified in the symbol. Look at the example shown in Fig. 18-5 to see CL (50W, 20R & W, 20R), which means that the first 5,000 feet from the approach end are white, the next 2,000 feet are red and white, and the last 2,000 feet are red.

Remember that this refers only to centerline lights, directly in the center of the runway, not the runway edge lights. Some people get the two confused. It is also important to fix the configuration of the lights in your mind. It would be very embarrassing to be rolling out at a high rate of speed, expecting red and white alternating lights and then red lights at the end of the runway, only to be on a runway that has all white lights.

A recent change in the abbreviations is the meaning of RL in the upper left. In the past it simply meant runway edge lights. Now it means that they are of low intensity or that the intensity has not been specified by the airport. High- and medium-intensity lights have always been identified by the addition of the letters MI or HI in front of the RL.

What you see ain't always what you get

Look again at the additional runway information table on Fig. 18-5, and find the columns listing the usable lengths for the runways.

APPROACH CHART LEGEND
ADDITIONAL RUNWAY INFORMATION (continued)

RWY	ADDITIONAL RUNWAY INFORMATION					USABLE LENGTHS LANDING BEYOND		TAKE-OFF	WIDTH
						Threshold	Glide Slope		
4R	HIRL	CL	ALSF-I	TDZ	grooved	RVR				150'
22L	HIRL	CL			grooved	RVR		6641'		
4L	HIRL	CL	HIALS	SFL					NA	
22R										150'
7	RL	VASI (angle 2.4°, TCH 10')								200'
25										
13	HIRL	CL	VASI	LDIN			11, 972'			
31	HIRL	CL	SSALR	VASI (non-std)	HST-H		11, 252'			150'

RUNWAY LIGHTS-ABBREVIATIONS

RL.........Low Intensity Runway Lights or intensity not specified.

HIRL......High Intensity Runway Edge Lights

Runway edge lights are white, except on instrument runways amber replaces white on the last 2000' or half of the runway length, whichever is less.

MIRL......Medium Intensity Runway Edge Lights

REIL......Runway End Identifier Lights (threshold strobe)

TDZ.......Touchdown Zone Lights

HST-H... High Speed Taxiway turn-off with green centerline lights. H indicates taxiway identification.

CL.........Standard Centerline Light configuration white lights then alternating red & white lights between 3000' and 1000' from runway end and red lights for the last 1000'.
-or-
Exact configuration is not known. Known non-standard configurations are stated as listed below

CL (white).......all lights are white full length of runway.

CL (non-std)....non-standard, configuration unknown

CL(50W, 20R & W, 20R)...non-standard, configuration known...first 5000' white lights; next 2000' alternating red & white lights; last 2000' red lights.

APPROACH LIGHTS-ABBREVIATIONS

ALS.......Approach Light System. Color of lights, if known to be other than white, is included.

HIALS....High Intensity Approach Light System

MIALS...Medium Intensity Approach Light System

SFL........Sequenced Flashing Lights

F..........Condenser-Discharge Sequential Flashing Lights/Sequenced Flashing Lights

ALSF-I...Approach Light System with Sequenced Flashing Lights

ALSF-II..Approach Light System with Sequenced Flashing Lights and Red Side Row Lights the last 1000'. May be operated as SSALR during favorable weather conditions.

SSALF....Simplified Short Approach Light System with Sequenced Flashing Lights

SALSF....Short Approach Light System with Sequenced Flashing Lights

MALSF...Medium Intensity Approach Light System with Sequenced Flashing Lights

RAI...... Runway Alignment Indicator

RAIL......Runway Alignment Indicator Lights (Sequenced Flashing Lights which are installed only in combination with other light systems)

RLLS......Runway Lead-in Lighting System

SSALR... Simplified Short Approach Light System with Runway Alignment Indicator Lights

MALSR...Medium Intensity Approach Light System with Runway Alignment Indicator Lights

SALS..... Short Approach Light System

SSALS... Simplified Short Approach Light System

MALS.....Medium Intensity Approach Light System

LDIN......Sequenced Flashing Lead-in Lights

ODALS...Omni-Directional Approach Light System

VASI......Visual Approach Slope Indicator (L or R indicates left or right side of runway only)

AVASI... Abbreviated Visual Approach Slope Indicator (L or R indicates left or right side of runway only)

VASI (3 bar)....Visual Approach Slope Indicator for high cockpit aircraft (L or R indicates left or right side of runway only)

T-VASI.. Tee Visual Approach Slope Indicator

AT-VASI.Abbreviated Tee Visual Approach Slope Indicator (L or R indicates left or right side of runway only)

VASI (non-std)..Visual Approach Slope Indicator when known to be non-standard

Fig. 18-5 *Reproduced with permission of Jeppesen Sanderson, Inc. Not for use in navigation.*

APPROACH CHART LEGEND
ADDITIONAL RUNWAY INFORMATION (continued)

APPROACH LIGHTS-ABBREVIATIONS (continued)

VASI......VASI/AVASI/NON-STD angles are shown when known to be less than 2.5° or more than 3.0°. T-VASI/ AT-VASI angles are shown at all times. VASI (3 bar) descent angles are shown when other than upwind angle 3.25°, downwind angle 3.00°.

PAPI......Precision Approach Path Indicator (L or R indicates left or right side of runway only)

PLASI....Pulsating Visual Approach Slope Indicator, normally a single light unit projecting two colors. (L or R indicates left or right side of runway only)

TRCV.....Tri-Color Visual Approach Slope Indicator, normally a single light unit projecting three colors. (L or R indicates left or right side of runway only)

TCH.......Threshold Crossing Height. Height of the effective visual glide path over the threshold.

MEHT.....Minimum Eye Height over Threshold. Lowest height over the threshold of the visual on glide path indication.

MEHT or TCH is shown (when known) when less than 60′ for the upwind bar of a VASI (3 bar) system or less than 25′ for all other systems including PAPI.

Fig. 18-6 *Reproduced with permission of Jeppesen Sanderson, Inc. Not for use in navigation.*

These are set up for three conditions: landing beyond the threshold, landing beyond the glide slope, and takeoff.

You will not see anything in the landing-beyond-threshold column unless the runway has a displaced threshold. Then, the figure shown is the effective length of the runway from the displaced threshold to the departure (rollout) end of the runway.

Just because there is a displaced runway for landing, that does not mean that the displaced part cannot be used for takeoff. There are times when there will be a takeoff displacement as well. This usually has to do with noise abatement or jet blast considerations. If no signs have been posted on the runway itself, or no notes are listed in the additional runway information columns, then all of the runway is available for takeoff.

For the usable length, landing beyond the glide slope, the figure is taken from a point abeam the glide slope transmitter to the departure (rollout) end of the runway. For a runway using PAR, it is from the point of the theoretical glide path/runway intersection to the departure end of the runway.

Glance at Fig. 18-5 and examine the data for Runway 1L/19R in the additional runway information table. Although the runway is shown on the plan view as being 9,690 feet long, the displaced threshold on 19R reduces its effective length to 8,915 feet when landing beyond the threshold. Wouldn't it be embarrassing if you figured the weight and balance for landing on a 9,690-foot runway by just taking a cursory glance at the plan view without looking at the additional information?

There is no reduction in usable length for landing beyond the threshold in the other direction (Runway 1L); therefore, that box is left blank. Because glide slope angles are set up to effect touchdown somewhere between 750 and 1,500 feet from the threshold, notice that for landing beyond the glide slope, Runway 1L has a usable length of 8,439 feet, and Runway 19R has a usable length of 7,861 feet. Runway 25L is restricted to 7,339 feet when landing beyond the threshold, and Runways 13 and 31 are restricted to 5,137 and 5,344 feet, respectively, when landing beyond the threshold.

Although takeoff usable lengths are not listed for any of the runways at Milwaukee, had they been, they would have been measured from the point the takeoff roll commences, to the end of the pavement that is usable for takeoff.

The note "Turbojet NA" in that column indicates that Runway 1R is not available for jet departures. At times, there are also other restrictions for a runway. These are indicated by small numbers in reverse print in black circles (1, 2, 3, and the like, appearing in white). Milwaukee is a good example of this. The reasoning behind some of the restrictions, besides sounding confusing at first, is hard to understand at times.

Tax your brains a bit and look at the restrictions concerning Runway 13/31; see if you can figure out why some of them came about.

Takeoff minimums

The last section of the airport chart is the takeoff and alternate minimums table. This block is explained quite well in Fig. 18-7. The block might be broken down into two main sections: takeoff minimums and alternate minimums. Or the takeoff minimum block might appear above the alternate minimum block, as in Milwaukee's case (Fig. 18-4), depending on the amount of information printed.

Notice that the takeoff minimums apply only to Part 121, 123, 125, 129, and 135 operators, but Part 91 operators would be wise to heed them.

APPROACH CHART LEGEND
TAKE-OFF AND ALTERNATE MINIMUMS (continued)

USA FORMAT

The title TAKE-OFF & IFR DEPARTURE PROCEDURE is used to indicate that both take-off minimums and IFR departure procedures are specified. In such cases, refer to the note IFR DEPARTURE PROCEDURE to the left and immediately below the minimum columns for the procedure.

"Adequate Vis Ref" is shown as a reminder that at least one of the following visual aids must be available. The Touchdown Zone RVR report, if available, is controlling. The Mid RVR report may be substituted for the Touchdown Zone RVR report if the Touchdown Zone RVR is not available.
(1) Operative high intensity runway lights (HIRL).
(2) Operative runway centerline lights (CL).
(3) Runway centerline marking (RCLM).
(4) In circumstances when none of the above visual aids are available, visibility or RVV ¼ statute mile may still be used, provided other runway markings or runway lighting provide pilots *with adequate visual reference* to continuously identify the take-off surface and maintain directional control throughout the take-off run.
("Forward Vis Ref", in lieu of "Adequate Vis Ref", is used on charts dated prior to July 28, 1989.)

STD denotes standard take-off minimums for FAR 121, 123, 125, 129 and 135 operators. Standard is RVR 50 or 1 for 1 & 2 Eng. RVR 24 or ½ for 3 & 4 Eng.

The IFR Departure for runways 29L/R require (when the weather is below 1000' ceiling-7 miles) a climb to 1800' MSL on runway heading before initiating a turn.

Applicable to FAR 121 and 129 operators. Applicable to FAR 135 operators of large aircraft and small transport category aircraft.

Operative Touchdown Zone and Rollout RVR reporting systems serving the runway to be used, both of which are controlling, or three RVR reporting systems serving the runway to be used, all of which are controlling. However, if one of the three RVR reporting systems has failed, a take-off is authorized provided the remaining two RVR values are at or above the appropriate take-off minimums.

To be eligible for the minimum shown in the columns below, a climb gradient of at least 290'/NM is required until reaching 1000' MSL.
If unable to meet climb requirement, 300' ceiling-1 mile apply.

Restrictions in this column, if any, apply to all operators.

Approaches with electronic glide slope.

LOC, VOR, etc. approaches.

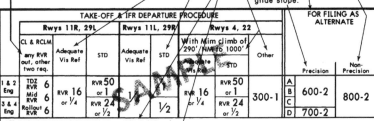

TAKE-OFF & IFR DEPARTURE PROCEDURE							FOR FILING AS ALTERNATE	
	Rwys 11R, 29L		Rwys 11L, 29R		Rwys 4, 22			
CL & RCLM any RVR out, other two req.	Adequate Vis Ref	STD	Adequate Vis Ref	STD	With Mim climb of 290'/NM to 1000' Adequate Vis Ref	STD Other	Precision	Non-Precision
1 & 2 Eng	TDZ RVR 6 Mid RVR 6	RVR 16 or ¼	RVR 50 or 1	RVR 16 or ¼	RVR 50 or 1	300-1	A 600-2 B C	800-2
3 & 4 Eng	Rollout RVR 6		RVR 24 or ½	½	RVR 24 or ½		C D 700-2	

IFR DEPARTURE PROCEDURE: Rwys 29L & 29R, when weather is below 1000-7 northbound departures (296° clockwise 116°) climb rwy heading to 1800' before turning.

Figures shown with RVR (runway visual range) represent readings in hundreds of feet. The figures without the RVR prefix represent visibility in statute miles or fractions thereof. For example: RVR 50 or 1 means 5000 feet RVR or one statute mile visibility; RVR 24 or ½ means 2400 feet RVR or one-half statute mile visibility.

Individual runway columns are shown whenever minimums are not the same for all runways The best opportunity runway is shown at the far left. Within each runway column, all conditions are specified, and minimums are positioned in ascending order, left to right. Columns are not established solely to identify runways with and without RVR when all other conditions are the same.

Altitudes listed in climb gradient requirements or for IFR departure procedures are above Mean Sea Level (MSL). Ceiling specified for Take-off minimums or Alternate minimums are heights Above Airport Level (AAL).

Fig. 18-7 *Reproduced with permission of Jeppesen Sanderson, Inc. Not for use in navigation.*

Study Fig. 18-7 very carefully. There is a lot of good information included. Runway 4 information is presented in three columns. The first two columns are headed by the words "With Mim climb of 290'/NM to 1000'." The first column under that specifies the visibility required under the definition of "adequate Vis Ref," which is defined in the upper left-hand side of the page. The next column is for "STD" visibility requirements, and although the standard takeoff visibilities are just that, standard, they are printed out in case someone were to forget. It would appear that there are no transmissometers installed on Runway 11L, 22, and 29R because the standard visibilities are only printed in terms of miles, not RVR feet.

Getting back to the breakdown for Runway 4, the other column is the visibility requirements for aircraft that cannot meet the climb criteria specified of 290'/NM to 1000'. The minimum climb criteria would certainly be indicative of obstructions above the normal obstruction clearance plane.

Reexamine Milwaukee and note how many options are available, depending on the runway and the equipment in use.

Here is another very important point. Some day you will run across instances where the takeoff minimums are higher than many of the landing minimums, due to obstructions; therefore, read the notes on the applicable airport charts very carefully.

As mentioned, these takeoff minimums are not for Part 91 operators, but it would certainly be wise to abide by them.

Alternate minimums

The alternate minimums block is very straightforward; for one example, see Fig. 18-8. This column is divided into two sections. The section on the left is further divided into three sections. The furthest to the left lists the minimums for the precision approach. The next column indicates the minimums required for using localizer or NDB approaches to Runway 2 or the localizer DME back course to Runway 20. The next column indicates the minimums required to use the NDB approach to Runway 20. All of these columns require that the control zone be effective.

You can use the minimums for the VOR 20 or VOR DME 20 approaches provided you have approved weather service. What is

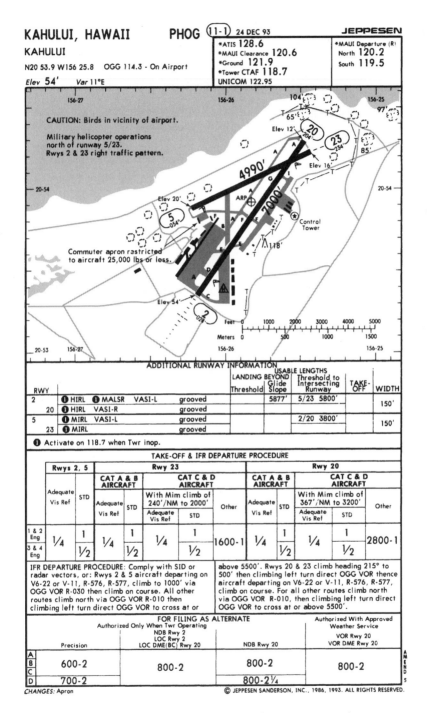

Fig. 18-8 *Reproduced with permission of Jeppesen Sanderson, Inc. Not for use in navigation.*

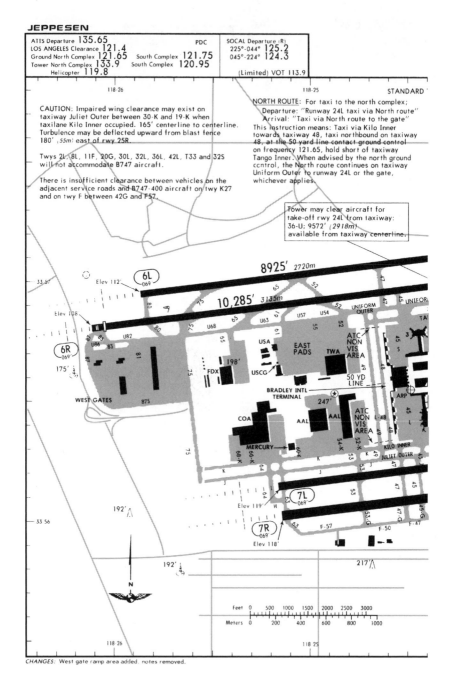

JEPPESEN

ATIS Departure 135.65		PDC	SOCAL Departure (R)
LOS ANGELES Clearance 121.4			225°-044° 125.2
Ground North Complex 121.65	South Complex 121.75		045°-224° 124.3
Tower North Complex 133.9	South Complex 120.95		
Helicopter 119.8			(Limited) VOT 113.9

118-26   118-25   STANDARD

CAUTION: Impaired wing clearance may exist on
taxiway Juliet Outer between 30-K and 19-K when
taxilane Kilo Inner occupied. 165' centerline to centerline.
Turbulence may be deflected upward from blast fence
180' (55m) east of rwy 25R.

Twys 2L, 8L, 11F, 20G, 30L, 32L, 36L, 42L, T33 and 32S
will not accommodate B747 aircraft.

There is insufficient clearance between vehicles on the
adjacent service roads and B747-400 aircraft on twy K27
and on twy F between 42G and F57.

NORTH ROUTE: For taxi to the north complex;
 Departure: "Runway 24L taxi via North route"
 Arrival: "Taxi via North route to the gate"
This instruction means: Taxi via Kilo Inner
towards taxiway 48, taxi northbound on taxiway
48, at the 50 yard line contact ground control
on frequency 121.65, hold short of taxiway
Tango Inner. When advised by the north ground
control, the North route continues on taxiway
Uniform Outer to runway 24L or the gate,
whichever applies.

Tower may clear aircraft for
take-off rwy 24L from taxiway:
36-U; 9572' (2918m)
available from taxiway centerline.

Fig. 18-9 *Reproduced with permission of Jeppesen Sanderson, Inc.
Not for use in navigation.*

30 DEC 94 (10-9) **KLAX**   `AIRPORT`
LOS ANGELES, CALIF
LOS ANGELES INTL
049.8°/1.3 from LAX 113.6   N33 56.6 W118 24.5

Var 14°E   Elev **126'**

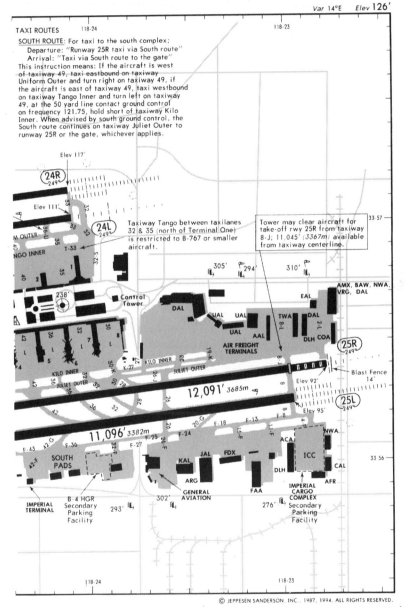

TAXI ROUTES   118-24

SOUTH ROUTE: For taxi to the south complex;
   Departure: "Runway 25R taxi via South route"
   Arrival: "Taxi via South route to the gate"
This instruction means: If the aircraft is west
of taxiway 49, taxi eastbound on taxiway
Uniform Outer and turn right on taxiway 49, if
the aircraft is east of taxiway 49, taxi westbound
on taxiway Tango Inner and turn left on taxiway
49, at the 50 yard line contact ground control
on frequency 121.75, hold short of taxiway Kilo
Inner. When advised by south ground control, the
South route continues on taxiway Juliet Outer to
runway 25R or the gate, whichever applies.

118-23

Elev 117'

24R -249°

Elev 111'

24L -249°

M OUTER

NGO INNER

T-33

Taxiway Tango between taxilanes
32 & 35 (north of Terminal One)
is restricted to B-767 or smaller
aircraft.

Tower may clear aircraft for
take-off rwy 25R from taxiway
8-J; 11,045' (3367m) available
from taxiway centerline.

33-57

305'   294'   310'

238'

Control
Tower

DAL

UAL UAL

UAL   AAL

AMX, BAW, NWA,
VRG, DAL

EAL

TWA   DAL

DLH COA

25R -249°

AIR FREIGHT
TERMINALS

KILO INNER
JULIET OUTER

KILO INNER

JULIET OUTER

Blast Fence
14'

Elev 92'

12,091' 3685m

Elev 95'

25L -249°

11,096' 3382m

F-18   F-13   F-8

F-43

F-36

F-27

F-25   F-24

NWA

ACA

SOUTH
PADS

KAL JAL   FDX

ICC

CAL

33-56

ARG

DLH

AFR

IMPERIAL
TERMINAL

B-4 HGR
Secondary
Parking
Facility

293'

302' GENERAL
AVIATION

FAA

IMPERIAL
CARGO
COMPLEX

276' Secondary
Parking
Facility

118-24

118-23

approved weather service? The weather bureau will suffice when it is operating, but when it is closed, you will be out of luck unless you happen to fly for a commuter or an air carrier that has its own certificated weather people.

Expanded airport charts

I talked earlier about the new expanded airport charts that are coming out for some of the major airports. Take a look at those that have been produced for Los Angeles International (Figs. 18-9 through 18-11).

Figure 18-9, in its original form, is a fold-out chart to show the complexity of LAX. There are cargo buildings, terminals, and outbuildings scattered all over the field, and the taxiways are quite complex. Another interesting thing about LAX is that there are three different tower frequencies shown in the communications block.

As at Milwaukee, the additional runway information section as well as the takeoff and landing minimum blocks are shown on the new series 10-9A chart, which need not be illustrated again.

The terminal area (gates and taxiways), Imperial Cargo Complex, and the West Pads with their attendant taxiways are complex enough that two additional charts (Figs. 18-10 and 18-11) have been created exclusively to illustrate those specific areas of the airport.

These two pages also list the latitude and longitude coordinates of every gate for the benefit of those pilots who are flying aircraft equipped with any form of area navigation equipment.

You can see that as terminals and airspace become more and more congested, the charts change to keep pace. Keep up with these rapid changes by periodic reviews.

Hopefully, by now you have learned the importance of studying all of the information available to you on Jeppesen charts, as well as in the explanatory pages. In the next chapter we'll take a look at the NOS charts to see how they compare.

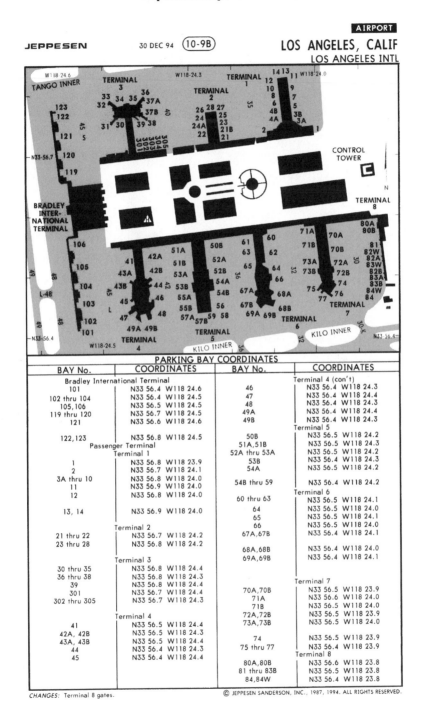

PARKING BAY COORDINATES

BAY No.	COORDINATES	BAY No.	COORDINATES
Bradley International Terminal			**Terminal 4 (con't)**
101	N33 56.4 W118 24.6	46	N33 56.4 W118 24.3
102 thru 104	N33 56.4 W118 24.5	47	N33 56.4 W118 24.4
105, 106	N33 56.5 W118 24.5	48	N33 56.4 W118 24.3
119 thru 120	N33 56.7 W118 24.5	49A	N33 56.4 W118 24.4
121	N33 56.6 W118 24.6	49B	N33 56.4 W118 24.3
			Terminal 5
122, 123	N33 56.8 W118 24.5	50B	N33 56.5 W118 24.2
Passenger Terminal		51A, 51B	N33 56.5 W118 24.3
Terminal 1		52A thru 53A	N33 56.5 W118 24.2
1	N33 56.8 W118 23.9	53B	N33 56.4 W118 24.3
2	N33 56.7 W118 24.1	54A	N33 56.5 W118 24.2
3A thru 10	N33 56.8 W118 24.0		
11	N33 56.9 W118 24.0	54B thru 59	N33 56.4 W118 24.2
12	N33 56.8 W118 24.0	**Terminal 6**	
		60 thru 63	N33 56.5 W118 24.1
13, 14	N33 56.9 W118 24.0	64	N33 56.5 W118 24.0
		65	N33 56.5 W118 24.1
Terminal 2		66	N33 56.5 W118 24.0
21 thru 22	N33 56.7 W118 24.2	67A, 67B	N33 56.4 W118 24.1
23 thru 28	N33 56.8 W118 24.2		
		68A, 68B	N33 56.4 W118 24.0
Terminal 3		69A, 69B	N33 56.4 W118 24.1
30 thru 35	N33 56.8 W118 24.4		
36 thru 38	N33 56.8 W118 24.3		
39	N33 56.8 W118 24.4	**Terminal 7**	
301	N33 56.7 W118 24.4	70A, 70B	N33 56.5 W118 23.9
302 thru 305	N33 56.7 W118 24.3	71A	N33 56.6 W118 24.0
		71B	N33 56.5 W118 24.0
Terminal 4		72A, 72B	N33 56.5 W118 23.9
41	N33 56.5 W118 24.4	73A, 73B	N33 56.5 W118 24.0
42A, 42B	N33 56.5 W118 24.3		
43A, 43B	N33 56.5 W118 24.4	74	N33 56.5 W118 23.9
44	N33 56.4 W118 24.3	75 thru 77	N33 56.4 W118 23.9
45	N33 56.4 W118 24.4	**Terminal 8**	
		80A, 80B	N33 56.6 W118 23.8
		81 thru 83B	N33 56.5 W118 23.8
		84, 84W	N33 56.4 W118 23.8

CHANGES: Terminal 8 gates.

Fig. 18-10 *Reproduced with permission of Jeppesen Sanderson, Inc. Not for use in navigation.*

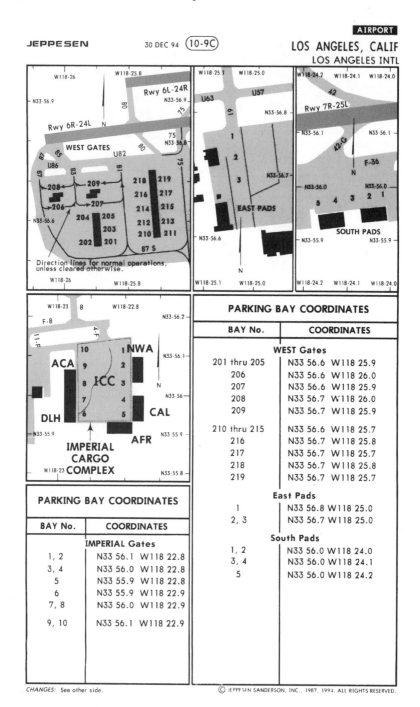

Fig. 18-11 *Reproduced with permission of Jeppesen Sanderson, Inc. Not for use in navigation.*

19

NOS approach charts

Recall that there are no hard and fast rules for cartographers to follow when producing approach plates, as long as all necessary information is presented. Many pilots think that the National Ocean Service (NOS) charts are confusing because they cram a lot of information on one sheet of paper. Others say the NOS charts are much better than the Jeppesen products because NOS charts give you all the information you need on one piece of paper without making you turn pages to discover items such as the airport diagram. This all boils down to personal choice, which will depend, to an extent, on what you used during instrument training.

The NOS charts come in volumes that are updated every 56 days with a change notice volume issued on the 28th day. This change notice is being replaced by NOTAMs. It is possible, though, to get them in either a spiral ring-type or loose-leaf style held together by metal clips or nonmagnetic plastic rings. These versions are available through the Air Chart Company in Venice, Calif., whose unique update service works equally well for Jeppesen charts.

General information and abbreviations for NOS charts appear near the front of the volume: Distances are in nautical miles (NM), except for the visibility, which will be in statute miles (SM); RVR will be in hundreds of feet; ceilings shown will be in feet above the airport elevation; and radials, bearings, headings, and courses are magnetic.

The approaches are arranged by airport name, and, if the airport has more than one, the approaches are listed with the NDB approach first, followed by ILS, LOC, LOC/DME, VOR, VOR/DME, and RNAV.

ILS by NOS

Figure 19-1 shows the ILS approach for Runway 9L at Opa Locka, Florida. The upper and lower left margins of the chart identify the specific approach (ILS RWY 9L) in bold letters. The upper and lower right margins identify the airport's name in uppercase letters and by its geographical location in smaller uppercase letters. The three-letter identifier follows the airport name. In the center of the lower margin (called the *trailer*), you will find the latitudinal and longitudinal coordinates.

The plan view comprises the upper two-thirds of the plate. The profile view is below the plan view on the left. The minimums information is below the profile view, followed by any notes. The airport diagram is located below the plan view on the right side of the plate (large airports will have a separate airport chart), including any appropriate notes, and the distance and time from the FAF to the MAP is under the airport diagram, when applicable.

Obviously, there is a great deal of information on one side of an approach plate, but some of it is difficult for an old or tired pair of eyes to read, and the pulp paper it's printed on doesn't make it any easier. If I were using these plates, I would make a magnifying glass part of my flight kit.

You should be aware of the large circle on the plan view. In Fig. 19-1, this circle depicts a 10-nautical-mile radius. It is not necessarily drawn around the airport, but rather around the final approach fix. This is just a reference circle, and although it is usually of 10-nm radius, it can vary. So check the radius when you look at the chart. The radius can be found along the edge of the circle. The airport might not even be within the circle, as in the case of Fig. 19-1, where this 10-mile ring just barely chops across some of the runways.

The MSA circle will be found somewhere within the plan view. Look at the bottom left portion of the plan view and see the MSA circle for Opa Locka. The data above the circle explains which facility the MSA is predicated on, in this case, the Miami (MIA) VOR, and the radius is 25 nautical miles. Note that in the northwest sector from a bearing to the VOR of 040 degrees (220-degree radial) clockwise through a bearing of 180 degrees (360-degree radial), the MSA is 1,500 feet MSL. The MSA for the rest of the circle is 2,900 feet MSL.

The radio frequencies you will use on approach are listed in the upper-left corner of the plan view. With the exception of the ATIS frequency, these are listed in the order that you will use them,

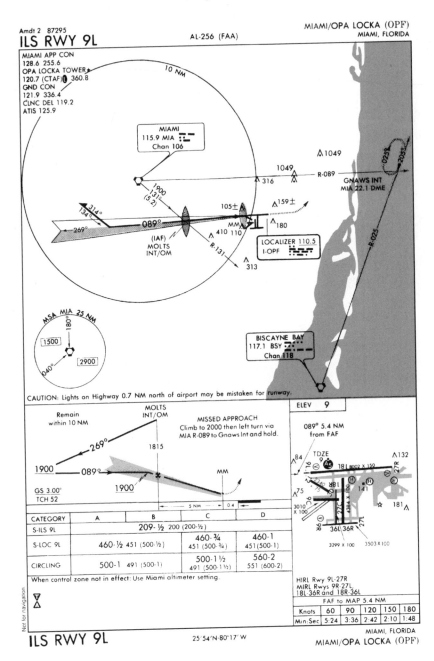

Fig. 19-1. *Not for use in navigation.*

beginning with approach control and continuing down through ground control. Any note concerning the airport, in this case the note about the highway lights, is printed near the bottom of the plan view.

Missed approach instructions are printed in the profile view, as are any procedure turn instructions. In the case of an ILS, both the glide slope angle and the threshold crossing height of the glide slope are indicated in the profile view section.

Notes concerning displaced thresholds and lighting are included with the airport diagram. On approaches other than ILS and LOC, a helpful arrow indicates the direction from which you will be approaching the airport.

Look at the minimums table of the approach plate. Notice the categories listed along the top and various approaches listed down the left side. In this case, the approach minimums are for straight-in approaches (ILS and LOC) to Runway 9L and for circling approaches to the airport.

A lot of page turning

The next section of the approach plate contains information that doesn't fit anywhere else. In the case of Opa Locka, you find a T and an A in reverse type set within black triangles. The T indicates that the takeoff minimums are not standard, or there are published departure procedures, or both. The A indicates that the alternate minimums are not standard. In either or both cases, you must turn to the list of tabulated data for airports in the front of the booklet.

There is only so much room on one 5 x 8 piece of paper; it's virtually impossible to include everything on one plate. As a result, NOS charts show nonstandard takeoff and landing minimums on pages other than the approach chart.

NOS explanatory pages are in the front of the book. Although the pages are laid out quite logically, they still require you to do a lot of page turning. The explanation of the approach categories is quite straightforward, especially regarding how you must change the category to a higher one if you are using a circling approach in which the aircraft will be maneuvering in a configuration that increases the stall speed.

G1
LEGEND
INSTRUMENT APPROACH PROCEDURES (CHARTS)

89348

IFR LANDING MINIMA

Landing minima are established for six aircraft approach categories (ABCDE and COPTER). In the absence of COPTER MINIMA, helicopters may use the CAT A minimums of other procedures. The standard format for portrayal of landing minima is as follows:

AIRCRAFT APPROACH CATEGORIES

Speeds are based on 1.3 times the stall speed in the landing configuration of maximum gross landing weight. An aircraft shall fit in only one category. If it is necessary to maneuver at speeds in excess of the upper limit of a speed range for a category, the minimums for the next higher category should be used. For example, an aircraft which falls in Category A, but is circling to land at a speed in excess of 91 knots, should use the approach Category B minimums when circling to land. See following category limits:

MANEUVERING TABLE

Approach Category	A	B	C	D	E
Speed (Knots)	0-90	91-120	121-140	141-165	Abv 165

RVR/Meteorological Visibility Comparable Values

The following table shall be used for converting RVR to meteorological visibility when RVR is not reported for the runway of intended operation. Adjustment of landing minima may be required – see Inoperative Components Table.

RVR (feet)	Visibility (statute miles)	RVR (feet)	Visibility (statute miles)
1600	¼	4000	¾
2000	⅜	4500	⅞
2400	½	5000	1
3200	⅝	6000	1¼

LANDING MINIMA FORMAT

In this example airport elevation is 1179, and runway touchdown zone elevation is 1152.

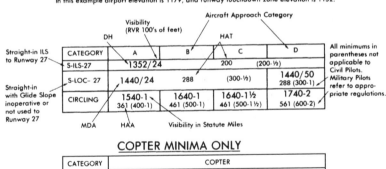

COPTER MINIMA ONLY

CATEGORY	COPTER
H-176°	680-½ 363 (400-½)

Copter Approach Direction   Height of MDA/DH
Above Landing Area (HAL)

No circling minimums are provided

Fig. 19-2. *Not for use in navigation.*

You will find the landing minima format at the bottom of Fig. 19-2 (as a tribute to NOS, I'll use the word minima for awhile!). It is different enough from the Jeppesen format to justify an explanation.

While Jeppesen puts the categories in a vertical column and breaks up the horizontal row to show the various minima for inoperative-components approaches, NOS charts list the categories horizontally, then show the various approaches in columns. In the case of the ILS approach, if the glide slope is inoperative, you only look down to the next line to find the minima for the LOC approach. Notice that in Fig. 19-2 the minima for the ILS and LOC are for straight-in approaches, signified by the S. If you use the ILS or LOC to reach the airport, but plan to circle to land, you will find the minima on the bottom line, labeled CIRCLING.

Unlike the Jeppesen charts, with the NOS versions, if you have inoperative components for the approach, you are obliged to turn to another page to see what changes must be made to the minima (Fig. 19-3). Although this format is fairly easy to understand, you would be hard pressed to find the correct page if you were already on the approach when something decided to quit. In fact, unless you had a copilot on board who could do the research, you would probably have to execute a missed approach until everything was straightened out. With the Jeppesen chart, you could glance at the bottom of the page and make the necessary adjustments to the MDA more efficiently. This is an important point, particularly when making an approach to an alternate, when low on fuel, or if landing at a busy field and a miss would put you way back in a long daisy chain.

Also noteworthy in Fig. 19-3 is the caution that reads:

> *This table may be amended by notes on the approach chart. Such notes apply only to the particular approach category(ies) as stated.*

Certain amendments are indicated in the block below the minima block (Fig. 19-4). The note's text increases the required visibility to RVR 5,000 for Category D aircraft conducting straight-in localizer approaches to Miami International's Runway 9L when the middle marker is inoperative. Once again, you have to carefully read everything on every chart.

Landing minimums published on instrument approach procedure charts are based upon full operation of all components and visual aids associated with the particular instrument approach chart being used. Higher minimums are required with inoperative components or visual aids as indicated below. If more than one component is inoperative, each minimum is raised to the highest minimum required by any single component that is inoperative. ILS glide slope inoperative minimums are published on instrument approach charts as localizer minimums. This table may be amended by notes on the approach chart. Such notes apply only to the particular approach category(ies) as stated. See legend page for description of components indicated below.

(1) ILS, MLS, and PAR

Inoperative Component or Aid	Approach Category	Increase DH	Increase Visibility
MM*	ABC	50 feet	None
MM*	D	50 feet	¼ mile
ALSF 1 & 2, MALSR, & SSALR	ABCD	None	¼ mile

*Not applicable to PAR, MLS, and Operators Authorized in their Operations Specifications.

(2) ILS with visibility minimum of 1,800 RVR.

MM*	ABC	50 feet	To 2400 RVR
MM*	D	50 feet	To 4000 RVR
ALSF 1 & 2, MALSR, & SSALR	ABCD	None	To 4000 RVR
TDZL, RCLS	ABCD	None	To 2400 RVR
RVR	ABCD	None	To ½ mile

*Not applicable to Operators Authorized in their Operations Specifications.

(3) VOR, VOR/DME, VORTAC, VOR (TAC), VOR/DME (TAC), LOC, LOC/DME, LDA, LDA/DME, SDF, SDF/DME, RNAV, and ASR

Inoperative Visual Aid	Approach Category	Increase MDA	Increase Visibility
ALSF 1 & 2, MALSR, & SSALR	ABCD	None	½ mile
SSALS, MALS & ODALS	ABC	None	¼ mile

(4) NDB

ALSF 1 & 2, MALSR, & SSALR	C	None	½ mile
	ABD	None	¼ mile
MALS, SSALS, ODALS	ABC	None	¼ mile

Fig. 19-3. Not for use in navigation.

Basic takeoff and alternate minima

The basic regulations that apply to minima are FAR 91.175(f) and 91.169(c). In effect, these regulations set up standard takeoff and alternate minimums. For takeoff, these require 1 statute mile visibility

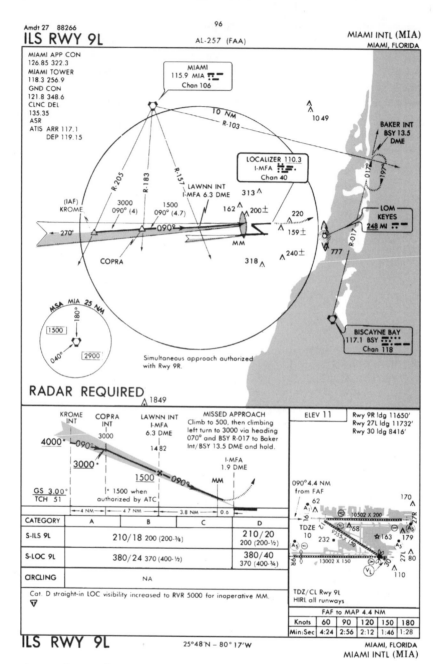

Fig. 19-4. *Not for use in navigation.*

for airplanes with one or two engines, and one-half statute mile visibility for those with more than two engines, except for Part 91 operators, as explained earlier.

Standard alternate minima require a 600-foot ceiling and 2-statute-mile visibility for precision (ILS and PAR) approaches and an 800-foot ceiling with 2-statute-mile visibility for nonprecision approaches. In the Jeppesen format, all of this information, including any nonstandard requirements, is listed under the takeoff and alternate minimum formats that are found on each airport chart.

The NOS charts present this information in a different manner. NOS lists takeoff and alternate minima in the front of the booklet. These pages (Figs. 19-5 through 19-8) spell out the standard minima as mentioned above and list each airport that has minima different from standard. To find out if the airport you are operating at has nonstandard minimums, you need only look at the note section of the individual approach chart.

There you will find a white T in an inverted black triangle if the takeoff minima are nonstandard and a white A in a black triangle if the alternate minima are nonstandard. Then turn to the appropriate page in the front of the book and check out the differences, or you can just look up the airports in these pages to begin with. In either case, it amounts to a lot of page turning or trying to memorize the standard minima, which leads to errors. The less you commit to memory, the better off you are.

As with all minima, the takeoff and alternate minima are based on obstacle clearance and aircraft performances. You will see, for example, that although there are no ceiling requirements in the standard takeoff minima, you might have them in nonstandard situations, such as seen in Figs. 19-5 through 19-8.

In the note section for Albert Whitted Airport at St. Petersburg, Florida (Fig.19-9):

When control zone not effective the following applies:
1. Use Tampa, FL altimeter setting.
2. Increase all MDAs 40 feet.
3. Alternate minimums not authorized.

The note goes on to say that you need ADF, DME, or RADAR for this approach. Looking below the table, you find the symbols that tell

88322

INSTRUMENT APPROACH PROCEDURES (CHARTS)

⚠ IFR ALTERNATE MINIMUMS
(NOT APPLICABLE TO USA/USN/USAF)

Standard alternate minimums for nonprecision approaches are 800-2 (NDB, VOR, LOC, TACAN, LDA, VORTAC, VOR/DME or ASR); for precision approaches 600-2 (ILS or PAR). Airports within this geographical area that require alternate minimums other than standard or alternate minimums with restrictions are listed below. NA - means alternate minimums are not authorized due to unmonitored facility or absence of weather reporting service. Civil pilots see FAR 91. USA/USN/USAF pilots refer to appropriate regulations.

NAME	ALTERNATE MINIMUMS	NAME	ALTERNATE MINIMUMS

ALBERT WHITTED — SEE ST. PETERSBURG, FL

ALEXANDER HAMILTON — SEE
CHRISTIANSTED, ST. CROIX, VI

CHARLOTTE AMALIE, ST. THOMAS, VI
   CYRIL E KING . . . . VOR-A
                               ILS Rwy 10
   1200-3

CHRISTIANSTED, ST. CROIX, VI
   ALEXANDER HAMILTON . . . . NDB Rwy 9, 1200-3†
                                ILS Rwy 9†
                            VOR Rwy 27*
   *Non-DME equipped aircraft 900-3.
   †NA when control tower closed.

CRAIG MUNI — SEE JACKSONVILLE, FL

CYRIL E KING — SEE CHARLOTTE AMALIE,
ST. THOMAS, VI

DAYTONA BEACH REGIONAL, FL
                               VOR Rwy 16
   Category D, 800-2 ¼

EUGENIO MARIA DE HOSTOS — SEE
MAYAGUEZ, PR

FORT LAUDERDALE, FL
   FORT LAUDERDALE EXECUTIVE . . . . .RNAV Rwy 8
                                NDB Rwy 8*
                               ILS Rwy 8†
   NA when control zone not in effect.
   *Category D, 800-2 ¼
   †ILS, Category D, 700-2

   FORT LAUDERDALE-HOLLYWOOD INTL
                                ILS Rwy 9L
                              ILS Rwy 27R
   ILS, 700-2

FORT MYERS, FL
   SOUTHWEST FLORIDA REGIONAL . . . NDB Rwy 6
                                ILS Rwy 6*
                              VOR Rwy 24
                              RADAR-1
   *ILS, Category E, 700-2 ¼ ; LOC, Category E,
   800-2 ¼
   NA when control tower closed.

GAINESVILLE REGIONAL, FL
                                 VOR-A
   Category C, 800-2 ¼ ; Category D, 800-2 ½ .

JACKSONVILLE, FL
   CRAIG MUNI . . . . VOR Rwy 13
   Category D, 800-2 ¼ .

KEY WEST INTL, FL
                                 NDB-A
   NA when control tower closed.

LAKELAND MUNI, FL
                                 NDB Rwy 5
                                 ILS Rwy 5
                              VOR Rwy 13
                            VOR Rwy 27†
   †Categories C, D, 800-2 ½
   NA when control zone not in effect.

LUIS MUNOZ MARIN INTL — SEE
SAN JUAN, PR

MAYAGUEZ, PR
   EUGENIO MARIA DE HOSTOS . . . . . VOR Rwy 9
   Categories A, B, C, 900-2 ½

SE-3

Fig. 19-5. *Not for use in navigation.*

89348

INSTRUMENT APPROACH PROCEDURES (CHARTS)

⚠️IFR ALTERNATE MINIMUMS
(NOT APPLICABLE TO USA/USN/USAF)

Standard alternate minimums for nonprecision approaches are 800-2 (NDB, VOR, LOC, TACAN, LDA, VORTAC, VOR/DME or ASR); for precision approaches 600-2 (ILS or PAR). Airports within this geographical area that require alternate minimums other than standard or alternate minimums with restrictions are listed below. NA - means alternate minimums are not authorized due to unmonitored facility or absence of weather reporting service. Civil pilots see FAR 91. USA/USN/USAF pilots refer to appropriate regulations.

NAME	ALTERNATE MINIMUMS	NAME	ALTERNATE MINIMUMS

ALBERT WHITTED — SEE ST. PETERSBURG, FL

ALEXANDER HAMILTON — SEE
CHRISTIANSTED, ST. CROIX, VI

CENTRAL FLORIDA REGIONAL – SEE
SANFORD, FL

CHARLOTTE AMALIE, ST. THOMAS, VI
CYRIL E KING . . . . VOR-A
ILS Rwy 10, 1200-3

CHRISTIANSTED, ST. CROIX, VI
ALEXANDER HAMILTON . . . . NDB Rwy 9, 1200-3†
ILS Rwy 9†
VOR Rwy 27*
*Non-DME equipped aircraft 900-3.
†NA when control tower closed.

CRAIG MUNI — SEE JACKSONVILLE, FL

CYRIL E KING — SEE CHARLOTTE AMALIE,
ST. THOMAS, VI

DAYTONA BEACH REGIONAL, FL
VOR Rwy 16
Category D, 800-2¼

EUGENIO MARIA DE HOSTOS — SEE
MAYAGUEZ, PR

FORT LAUDERDALE, FL
FORT LAUDERDALE EXECUTIVE . . . . .RNAV Rwy 8
NDB Rwy 8*
ILS Rwy 8†
NA when control zone not in effect.
*Category D, 800-2¼
†ILS, Category D, 700-2

FORT LAUDERDALE-HOLLYWOOD INTL
ILS Rwy 9L
ILS Rwy 27R
ILS, 700-2

FORT MYERS, FL
SOUTHWEST FLORIDA REGIONAL . . . NDB Rwy 6
ILS Rwy 6*
VOR Rwy 24
RADAR-1
*ILS, Category E, 700-2¼; LOC, Category E, 800-2¼
NA when control tower closed.

GAINESVILLE REGIONAL, FL
VOR-A
Category C, 800-2¼; Category D, 800-2½.

JACKSONVILLE, FL
CRAIG MUNI . . . . VOR Rwy 14*
ILS Rwy 32†
*Category D, 800-2¼.
†NA when control tower closed.

KEY WEST INTL, FL
NDB-A
NA when control tower closed.

LAKELAND REGIONAL, FL
NDB Rwy 5
ILS Rwy 5
VOR Rwy 9
VOR Rwy 13
VOR Rwy 27†
†Categories C, D, 800-2½
NA when control zone not in effect.

LUIS MUNOZ MARIN INTL — SEE
SAN JUAN, PR

MAYAGUEZ, PR
EUGENIO MARIA DE HOSTOS . . . . . VOR Rwy 9
Categories A, B, C, 900-2½

SE-3

Fig. 19-6. *Not for use in navigation.*

▼

▼

89348

INSTRUMENT APPROACH PROCEDURES (CHARTS)
▼IFR TAKE-OFF MINIMUMS AND DEPARTURE PROCEDURES
Civil Airports and Selected Military Airports
CIVIL USERS: FAR 91 prescribes take-off rules and establishes take-off minimums as follows:
(1) Aircraft having two engines or less – one statute mile. (2) Aircraft having more than two engines – one-half statute mile.
MILITARY USERS: Special IFR departure procedures, not published as Standard Instrument Departure (SIDs), and civil take-off minima are included below and are established to assist pilots in obstruction avoidance. Refer to appropriate service directives for take-off minimums.
Airports with IFR take-off minimums other than standard are listed below. Departure procedures and/or ceiling visibility minimums are established to assist pilots conducting IFR flight in avoiding obstructions during climb to the minimum enroute altitude. Take-off minimums and departures apply to all runways unless otherwise specified. Altitudes, unless otherwise indicated, are minimum altitudes in feet MSL.

NAME	TAKE-OFF MINIMUMS	NAME	TAKE-OFF MINIMUMS

ALBERT WHITTED — SEE ST. PETERSBURG, FL

ALEXANDER HAMILTON — SEE
CHRISTIANSTED, ST. CROIX, VI

BONIFAY, FL
TRI COUNTY
IFR DEPARTURE PROCEDURES: Rwys 1, 19, climb runway heading to 2000' before turning west.

BUNNELL, FL
FLAGLER COUNTY
IFR DEPARTURE PROCEDURE: Rwy 29, aircraft departing on a course between 270° clockwise to 360° climb on heading 270° to 1300' before proceeding on course.

CENTRAL FLORIDA REGIONAL – SEE
SANFORD, FL

CHARLOTTE AMALIE, ST. THOMAS, VI
CYRIL E KING . . . . Rwy 10, 400-1
IFR DEPARTURE PROCEDURES: Rwy 10, immediate climbing right turn to heading 120°. Continue climb to 2000 before turning north. Rwy 28, climb runway heading to 2000 before turning north.

CHRISTIANSTED, ST. CROIX, VI
ALEXANDER HAMILTON
Rwy 9, 1000-3 or standard with minimum climb of 300' per NM to 500.
IFR DEPARTURE PROCEDURE: Rwy 27, climb runway heading to 1100 before turning north. Rwy 9, for departures 250° clockwise thru 095°, climbing left turn to 1400 via heading 045° before proceeding on course.

CRAIG MUNI — SEE JACKSONVILLE, FL

CYRIL E KING — SEE CHARLOTTE AMALIE, ST. THOMAS, VI

DAYTONA BEACH REGIONAL, FL
Rwys 7R, 16, 25L/34, 300-1

DELAND MUNI-SIDNEY H. TAYLOR FIELD, FL
IFR DEPARTURE PROCEDURE: Rwys 5, 12, 30, 36, aircraft proceeding on a course between 120° CW to 260°, fly runway heading to 1800. Rwy 18, aircraft proceeding on a course between 120° CW to 260°, fly heading 120° to 1800. Rwy 23, aircraft proceeding on a course between 120° CW to 260°, fly heading 260° to 1800.

DESTIN-FORT WALTON BEACH, FL
IFR DEPARTURE PROCEDURE: Rwy 14, climb runway heading to 400 before making turn.

EUGENIO MARIA DE HOSTOS —
SEE MAYAGUEZ, PR

FLAGLER COUNTY — SEE BUNNELL, FL

FORT LAUDERDALE, FL
FORT LAUDERDALE-EXECUTIVE
IFR DEPARTURE PROCEDURE: Rwy 8, fly runway heading to 300 feet before turning north. Rwy 26, fly runway heading to 500 feet before turning south.

FORT LAUDERDALE-HOLLYWOOD INTL
Rwys 13, 27L, 300-1
IFR DEPARTURE PROCEDURE: Rwys 9L/R, climb runway heading to 500 before turning.

FORT MYERS, FL
PAGE FIELD . . . . Rwy 31, 300-1
or standard with minimum climb of 280 feet per NM to 300'.

SE-3

▼

▼

Fig. 19-7. *Not for use in navigation.*

▼   ▼

89348

NAME	TAKE-OFF MINIMUMS	NAME	TAKE-OFF MINIMUMS

FORT PIERCE, FL
ST. LUCIE COUNTY INTL . . . . Rwy 18, 600-1*
Rwy 36, 600-1†
*or standard with minimum climb of 300' per
NM to 600 or turn right heading 270° to 600
before proceeding on course.
†or standard with minimum climb of 250' per
NM to 600.
IFR DEPARTURE PROCEDURE: Rwy 14 climb
runway heading to 600 before turning right.

JACKSONVILLE, FL
CRAIG MUNI
Rwy 23, 1100-3 or standard with minimum climb
of 320' per NM to 1100'
IFR DEPARTURE PROCEDURE: Rwy 14, climb
runway heading to 1000' before turning right.

KEY WEST INTL, FL
IFR DEPARTURE PROCEDURE: Rwy 9, climb
runway heading to 200 before turning north.
Rwy 27, climb runway heading to 200 before
turning south.

KISSIMMEE MUNI, FL
Rwy 6, 300-1

LUIS MUNOZ MARIN INTL — SEE
SAN JUAN, PR

MARCO ISLAND, FL
IFR DEPARTURE PROCEDURE: Rwy 35,
aircraft proceeding on a course between
270° clockwise to 030° fly heading 030° to
1000' before proceeding on course.

MAYAGUEZ, PR
EUGENIO MARIA DE HOSTOS
Rwy 9, 700-1 or standard with minimum climb of
350' per NM to 2300.
IFR DEPARTURE PROCEDURE: Rwy 9, climbing
left turn to 2500 direct MAZ VOR/DME, climb in
holding pattern to 2700 for Route-1, 3300 for
G-3. Rwy 27, climbing ritht turn to 2500 direct
MAZ VOR/DME, climb in holding pattern to 2700
for Route-1, 3300 for G-3.

MERCEDITA — SEE PONCE, PR

MIAMI, FL
MIAMI INTL . . . . Rwy 9L, 800-1*
Rwy 9R, 800-1†
Rwy 12, 800-1↓
*or standard with minimum climb of 220'
per NM to 1100 or climb runway heading to
1100 before turning right.
†or standard with minimum climb of 220'
per NM to 1100 or comply with RADAR
vectors.
↓or standard with minimum climb of 220'
per NM to 1100 or climb runway heading
to 1100 before turning left.

OPA LOCKA
Rwys 9R, 9C, 12, 18R, 27L, 27C, 30, 36L, NA.
IFR DEPARTURE PROCEDURE: Rwys 9L and
36R, climb runway heading to 1100 before
proceeding on course.

TAMIAMI
IFR DEPARTURE PROCEDURE: All Rwys climb
runway heading to 1400' before turning south.

NEW PORT RICHEY, FL
TAMPA BAY EXECUTIVE
IFR DEPARTURE PROCEDURE: Rwy 8 climb rwy
heading to 1000 feet before turning. Rwy 26,
departures between 270° clockwise to 360°
climbing right turn heading 320° to 2100 feet
before proceeding on course. Departures between
269° counter clockwise to 180° climbing left turn
heading 200° to 2100 feet before proceeding on
course.

NEW SMYRNA BEACH MUNI, FL
Rwys 2, 6, 300-1

OPA LOCKA — SEE MIAMI, FL

ORLANDO EXECUTIVE, FL
Rwy 25: 500-2 or standard with minimum climb
of 300'/NM to 700'. Rwy 31: 300-2 or standard
with minimum climb of 250'/NM to 700'.

PAGE FIELD — SEE FORT MYERS, FL

PAHOKEE, FL
PALM BEACH COUNTY GLADES
Rwys 7, 35, 400-1
Rwy 17, 300-1

PALM BEACH COUNTY GLADES — SEE
PAHOKEE, FL

PERRY-FOLEY, FL
IFR DEPARTURE PROCEDURE: Rwys 6, 12, 30, 36,
climb on runway heading to 800 before turning.

SE-3

▼   ▼

Fig. 19-8. *Not for use in navigation.*

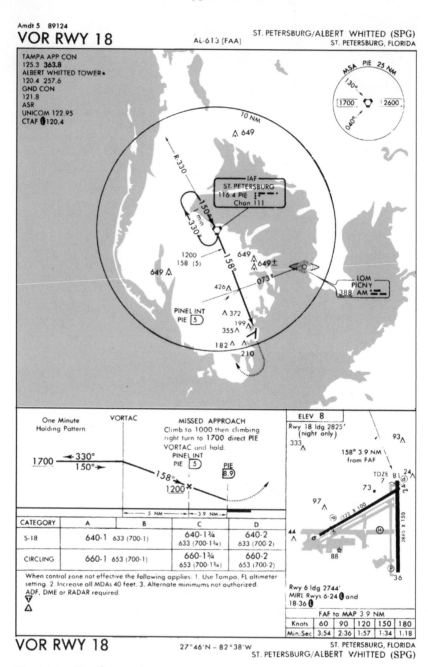

Amdt 5   89124

VOR RWY 18

AL-613 (FAA)

ST. PETERSBURG/ALBERT WHITTED (SPG)
ST. PETERSBURG, FLORIDA

TAMPA APP CON
125.3 363.8
ALBERT WHITTED TOWER *
120.4 257.6
GND CON
121.8
ASR
UNICOM 122.95
CTAF Ⓘ120.4

MSA PIE 25 NM

130° / 1700 ◇ 2600 / 040°

10 NM

Λ 649

R-330

— IAF —
ST. PETERSBURG
116.4 PIE ⋮---⋅
Chan 111

150° min
330°

1200
158 (5)

649 Λ

158°

649
Λ 649±

426 Λ

073°

LOM
PICNY
388 AM ⋅--

PINEL INT
PIE ⑤

Λ 372

199 Λ
355 Λ

182 Λ Λ
210

One Minute Holding Pattern	VORTAC	MISSED APPROACH Climb to 1000 then climbing right turn to 1700 direct PIE VORTAC and hold.

1700 ← 330°
        150° →

PINEL INT
PIE ⑤

PIE
8.9

158°

1200 ✕

← 5 NM →	← 3.9 NM →

CATEGORY	A	B	C	D
S-18	640-1	633 (700-1)	640-1¾ 633 (700-1¾)	640-2 633 (700-2)
CIRCLING	660-1	653 (700-1)	660-1¾ 653 (700-1¾)	660-2 653 (700-2)

When control zone not effective the following applies: 1. Use Tampa, FL altimeter setting. 2. Increase all MDAs 40 feet. 3. Alternate minimums not authorized. ADF, DME or RADAR required.
▽
Δ

ELEV 8

Rwy 18 ldg 2825'
(night only)

333 Λ

93 Λ

158° 3.9 NM
from FAF

TDZE 8
7

24

73 ▪

97 Λ

44 Λ

σ

Ⓗ

88

P

36

Rwy 6 ldg 2744'
MIRL Rwys 6-24 Ⓘ and
18-36 Ⓘ

FAF to MAP 3.9 NM

Knots	60	90	120	150	180
Min:Sec	3:54	2:36	1:57	1:34	1:18

VOR RWY 18

27°46'N – 82°38'W

ST. PETERSBURG, FLORIDA
ST. PETERSBURG/ALBERT WHITTED (SPG)

Fig. 19-9. *Not for use in navigation.*

you that nonstandard takeoff and alternate minimums are specified for this airport, so look at the tabulations in the front of the book.

Figure 19-5 lists the standard alternate minimums at the top of the page. Then look down at ALBERT WHITTED only to find an answer like you'd find in the yellow pages: SEE ST. PETERSBURG, FL. So, turn to St. Petersburg on the next page (Fig. 19-6), where you see that Whitted is not authorized for use as an alternate airport via the VOR Rwy 18 approach when the control zone is not effective.

Then look up the nonstandard takeoff minimums (Fig. 19-7). Once again, the standard minimums are spelled out at the top of the page, and you have to turn the page (to Fig. 19-8) to find the note for AL-BERT WHITTED airport: Runway 6, 18, 24, and 36 (in other words, all of the runways) takeoffs require both a ceiling of 200 feet and visibility of 1 mile.

Review some of the other nonstandard takeoff and landing minimums on these pages to anticipate some of the differences you will experience in instrument flying. Next time you look through a complete set of approach charts, whether they be Jeppesen or NOS, see if you can find the possible reasons for the variations.

Picking a chart apart

Further examination of the Albert Whitted approach chart (Fig. 19-9) reveals that the procedure turn for this approach is a holding pattern over the St. Petersburg VORTAC, which is also the IAF for the airport. The holding pattern, when depicted in lieu of a procedure turn, is mandatory except when radar vectors are being provided.

Note that the holding pattern procedure turn is shown in both the plan and profile views. The inbound leg of the holding pattern should be 1 minute in duration.

The distance from the IAF to the FAF is 5 nautical miles. The FAF is PINEL intersection, which can be identified three ways:

- 5 DME from St. Petersburg VOR (PIE)
- The 073-degree bearing to PICNY (an outer compass locator for another approach)
- A radar fix (Now you can see why ADF, DME, or radar is required for this approach.)

From PINEL to the MAP is another 3.9 nautical miles or 8.9 DME from the VOR. Under the airport diagram you can see a table to estimate time from the FAF to the MAP. Remember, the speed you want to use is ground speed in knots. Another thing to note here is the manner in which the NOS charts depict DME distances. They put the numeral inside a stylized letter D that is attached to the vertical line leading upward from the ground plane that identifies the fix.

An arrow runs from the upper part of the airport diagram down toward Runway 18. The note on it reads "158 degrees 3.9 NM from FAF." This helps you visualize the path the aircraft will fly on the approach.

Airport lighting

The airport diagram in Fig. 19-10 includes a V within a circle alongside the numerals for Runways 18R, 9/27, and 18L, and a VL in a circle next to Runway 36R. The right-hand side of Fig. 19-11 has an explanation that these letters indicate that the runways are equipped with visual approach slope indicators (VASI). The V is for the two-bar VASI that most of us are familiar with, which provides approach slope guidance for most aircraft. The VL indicates a three-bar VASI that uses an upper bar to provide a higher threshold crossing height to accommodate long-bodied or jumbo aircraft.

In the two-bar system, if both upper and lower bars are white you are too high; if both are red, too low; red on top and white on the bottom, on the glide slope. A good way to remember this is by the saying, "Red over white, you're all right. Red over red, you're dead."

Three-bar systems give large aircraft adequate gear clearance when crossing the threshold. Pilots flying light aircraft and even carriers flying other than wide- or long-bodies can use the lower pair of light bars. Pilots of larger aircraft use the upper pair. If you've ever seen a wide-bodied jet on approach and have noticed how high the pilot sits above the landing gear, you can see why the third bar is there. His main gear is 50 feet lower than his eyes, so when following the standard VASI, there would be too great a chance of landing short.

Figure 19-10 shows A1, A3, and A5 inside circles on the airport diagram. These are symbols representing various approach light systems, which are illustrated on Fig. 19-11. The dot at the top of those circles means the runway also has sequence flashing lights. Additionally, at the top of Fig. 19-11 notice that if these circles and numerals are in reverse print, they indicate pilot-controlled lighting.

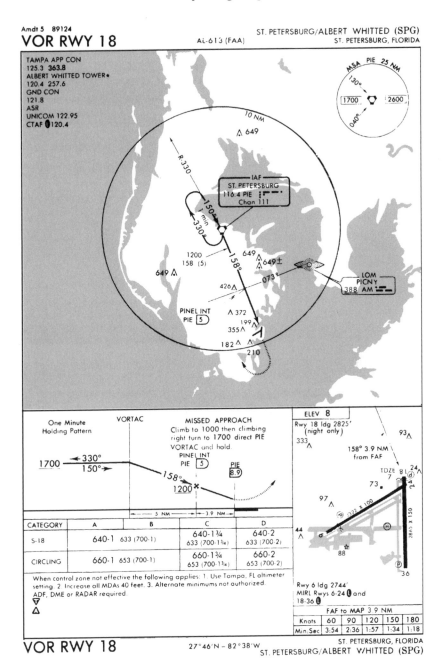

Amdt 5 89124

ST. PETERSBURG/ALBERT WHITTED (SPG)

VOR RWY 18   AL-613 (FAA)   ST. PETERSBURG, FLORIDA

TAMPA APP CON
125.3 363.8
ALBERT WHITTED TOWER*
120.4 257.6
GND CON
121.8
ASR
UNICOM 122.95
CTAF ❶120.4

MSA PIE 25 NM

1700 ◇ 2600

10 NM

⩘ 649

R-330

IAF
ST. PETERSBURG
116.4 PIE
Chan 111

150°

330°

1200
158 (5)

649 ⩘

158°

649
⩘649±

426⩘

073°

LOM
PICNY
388 AM

PINEL INT
PIE ⑤

⩘ 372

199⩘

355⩘

182 ⩘ ⩘

210

One Minute Holding Pattern	VORTAC	MISSED APPROACH	ELEV 8

One Minute
Holding Pattern

VORTAC

MISSED APPROACH
Climb to 1000 then climbing
right turn to 1700 direct PIE
VORTAC and hold.
PINEL INT
PIE ⑤

1700 ←330°
150°→

158°

1200

PIE
8.9

PIE
8.9

ELEV 8
Rwy 18 ldg 2825'
(night only)
333⩘

93⩘

158° 3.9 NM
from FAF

TDZE 81
7

73

97
⩘

1222 X 100

24

21

24

286 X 150

← 5 NM → ← 3.9 NM →

CATEGORY	A	B	C	D
S-18	640-1 633 (700-1)		640-1¾ 633 (700-1¾)	640-2 633 (700-2)
CIRCLING	660-1 653 (700-1)		660-1¾ 653 (700-1¾)	660-2 653 (700-2)

When control zone not effective the following applies: 1. Use Tampa, FL altimeter
setting. 2. Increase all MDAs 40 feet. 3. Alternate minimums not authorized.
ADF, DME or RADAR required.
▽
△

44
⩘

σ

88

36

Rwy 6 ldg 2744'
MIRL Rwys 6-24 ❶ and
18-36 ❶

	FAF to MAP 3.9 NM				
Knots	60	90	120	150	180
Min:Sec	3:54	2:36	1:57	1:34	1:18

VOR RWY 18   27°46'N – 82°38'W

ST. PETERSBURG, FLORIDA
ST. PETERSBURG/ALBERT WHITTED (SPG)

Fig. 19-10. *Not for use in navigation.*

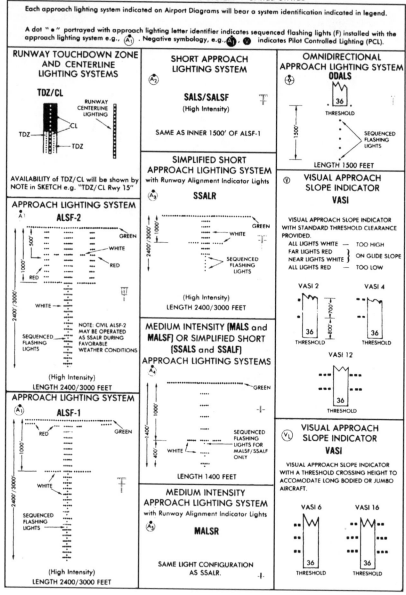

Fig. 19-11. *Not for use in navigation.*

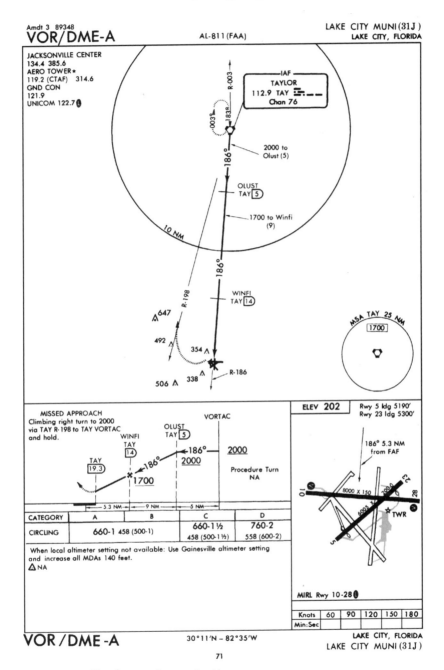

Fig. 19-12. *Not for use in navigation.*

You can find the instructions for activating the pilot radio-controlled airport lighting systems at the front of the approach plate book. If you are flying to a field with such a system, all of the pilot-activated lights are indicated by the symbols in reverse print as noted above or by the specific lights noted in the airport diagram. The VOR/DME-A chart for Lake City, Florida (Fig. 19-12), has the note "MIRL Rwy 10-28" followed by an L in reverse print. You will also see that the VASI symbols for those runways are printed in reverse type. The frequency to be used to activate the lights can be found in the communications grouping in the upper left-hand corner, identified by an L in reverse print. In this case, it's the unicom frequency 122.7. To operate the system, tune the transmitter to 122.7, and key the mike five times within 5 seconds. This will turn on the medium intensity lights.

At some airports, if you were to key the mike seven times within 5 seconds, that would activate the highest intensity lights available, while three keys in 5 seconds would operate the lowest intensity lights available. Check the approach plates to determine the highest intensity lights available and key the mike to bring them up initially to that intensity. This will allow you to see the lights at the greatest distance, which is especially useful in low visibility conditions or around strange airports. Then, as you get closer, when the lights might be too bright, use the keying technique to reduce the intensity.

Another symbol to be aware of is the holding pattern symbol in which the race track is made up of a series of hash marks such as is found over the TAYLOR VOR. This is not a procedure-turn holding pattern. The note in the profile view explains that a procedure turn is not available. The holding pattern depicted by hash marks is a missed approach holding pattern only.

NOS procedure turns

An actual approach procedure begins at the IAF. From this point on, the procedure track will be indicated by a heavy line with arrowheads indicating the proper direction to be flown. These, and other symbols used on the plan view, are in Fig. 19-13. Follow this track on an ILS approach to Orlando Executive Airport (Fig. 19-14).

The procedure turn is indicated in the upper left-hand column of Fig. 19-13 as half an arrowhead with the inbound and outbound magnetic courses (also illustrated on Fig. 19-14).

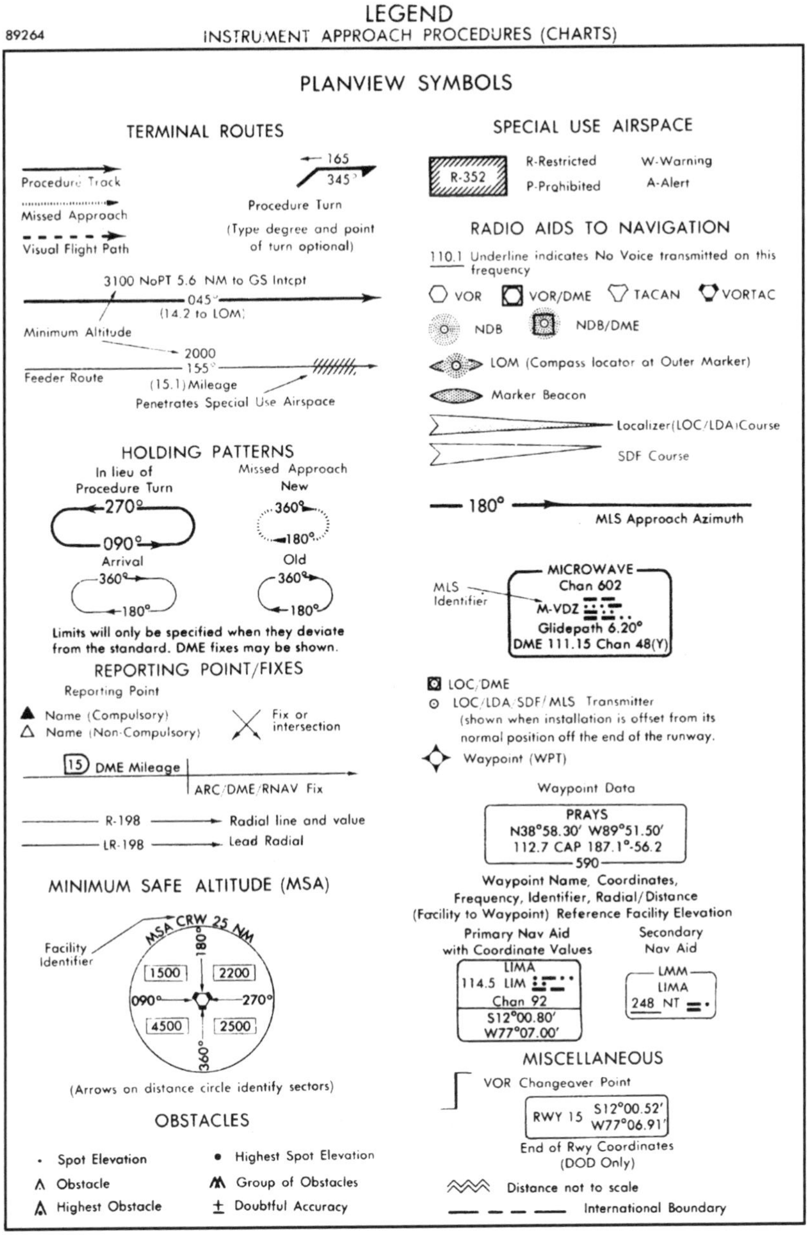

LEGEND

89264

INSTRUMENT APPROACH PROCEDURES (CHARTS)

PLANVIEW SYMBOLS

TERMINAL ROUTES

Procedure Track

Missed Approach

Visual Flight Path

```
          ←— 165
         /  345°
```
Procedure Turn
(Type degree and point
of turn optional)

3100 NoPT 5.6 NM to GS Intcpt
—045°—
(14.2 to LOM)
Minimum Altitude

—2000
—155°
(15.1) Mileage
Feeder Route

Penetrates Special Use Airspace

SPECIAL USE AIRSPACE

R-Restricted	W-Warning
P-Prohibited	A-Alert

R-352

RADIO AIDS TO NAVIGATION

110.1 Underline indicates No Voice transmitted on this frequency

○ VOR   ◧ VOR/DME   ▽ TACAN   ◈ VORTAC

○ NDB   ▣ NDB/DME

◁○▷ LOM (Compass locator at Outer Marker)

Marker Beacon

Localizer(LOC/LDA)Course

SDF Course

— 180° →

MLS Approach Azimuth

HOLDING PATTERNS

In lieu of Procedure Turn

Missed Approach New

←—270°—
—090°→
Arrival

..360°..
..←180°..
Old

—360°—
←—180°—

—360°—
←—180°—

Limits will only be specified when they deviate from the standard. DME fixes may be shown.

REPORTING POINT/FIXES

Reporting Point

▲ Name (Compulsory)
△ Name (Non-Compulsory)

✕ Fix or intersection

⑮ DME Mileage
ARC/DME/RNAV Fix

— R-198 —→ Radial line and value

— LR-198 —→ Lead Radial

MINIMUM SAFE ALTITUDE (MSA)

Facility Identifier

MSA CRW 25 NM

| 1500 | 2200 |
090°—   —270°
| 4500 | 2500 |

(Arrows on distance circle identify sectors)

OBSTACLES

· Spot Elevation   ● Highest Spot Elevation

Λ Obstacle   Ѫ Group of Obstacles

Ѧ Highest Obstacle   ± Doubtful Accuracy

--- MICROWAVE ---
Chan 602
M-VDZ ⋅⋅ ⋅⋅⋅
Glidepath 6.20°
DME 111.15 Chan 48(Y)

MLS Identifier

◙ LOC/DME

○ LOC/LDA/SDF/MLS Transmitter
(shown when installation is offset from its normal position off the end of the runway.

◆ Waypoint (WPT)

Waypoint Data

PRAYS
N38°58.30' W89°51.50'
112.7 CAP 187.1°-56.2
— 590 —

Waypoint Name, Coordinates,
Frequency, Identifier, Radial/Distance
(Facility to Waypoint) Reference Facility Elevation

Primary Nav Aid with Coordinate Values	Secondary Nav Aid
LIMA 114.5 LIM ⋅⋅ ⋅⋅⋅ Chan 92 S12°00.80' W77°07.00'	— LMM — LIMA 248 NT ⋅⋅⋅

MISCELLANEOUS

⌐ VOR Changeover Point

RWY 15 S12°00.52'
W77°06.91'

End of Rwy Coordinates
(DOD Only)

〜〜〜 Distance not to scale

— — — — International Boundary

Fig. 19-13. *Not for use in navigation.*

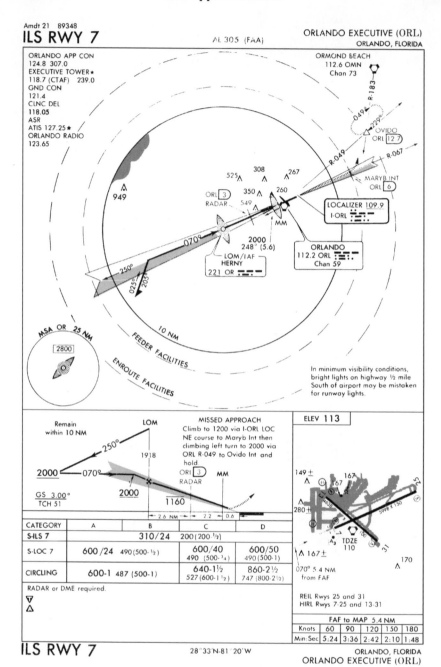

Amdt 21 89348

ILS RWY 7   AL 305 (FAA)   ORLANDO EXECUTIVE (ORL)
                                                 ORLANDO, FLORIDA

ORLANDO APP CON
124.8 307.0
EXECUTIVE TOWER★
118.7 (CTAF) 239.0
GND CON
121.4
CLNC DEL
118.05
ASR
ATIS 127.25★
ORLANDO RADIO
123.65

ORMOND BEACH
112.6 OMN
Chan 73

LOCALIZER 109.9
I-ORL

ORLANDO
112.2 ORL
Chan 59

In minimum visibility conditions,
bright lights on highway ½ mile
South of airport may be mistaken
for runway lights.

MSA OR 25 NM   2800

Remain
within 10 NM

MISSED APPROACH
Climb to 1200 via I-ORL LOC
NE course to Maryb Int then
climbing left turn to 2000 via
ORL R-049 to Ovido Int and
hold

ELEV 113

GS 3.00°
TCH 51

CATEGORY	A	B	C	D
S-ILS 7		310/24	200 (200-½)	
S-LOC 7	600/24 490 (500-½)		600/40 490 (500-¾)	600/50 490 (500-1)
CIRCLING	600-1 487 (500-1)		640-1½ 527 (600-1½)	860-2½ 747 (800-2½)

RADAR or DME required.

REIL Rwys 25 and 31
HIRL Rwys 7-25 and 13-31

FAF to MAP 5.4 NM				
Knots	60	90	120	150	180
Min:Sec	5:24	3:36	2:42	2:10	1:48

ILS RWY 7   28°33'N-81°20'W   ORLANDO, FLORIDA
                                           ORLANDO EXECUTIVE (ORL)

Fig. 19-14. *Not for use in navigation.*

About midway down the left-hand column in Fig. 19-13, note four different holding patterns. The dark, heavy line indicates a holding pattern that is being used in lieu of a procedure turn. The second type, with a finer line, indicates a normal arrival holding pattern. The third type, shorter and finer still, is the old type that depicts a holding pattern used in conjunction with a missed approach, and above that is the new missed approach holding pattern, mentioned earlier, that uses the hash marks to differentiate it. The new missed approach is in Fig. 19-14 at OVIDO.

For some reason, the NOS charts don't have any plan view symbol for the teardrop procedure turn, but they do indicate it on the profile view. The upper left-hand symbol on Fig. 19-15 signifies a teardrop procedure turn, and you can see how it differs from the profile of a normal procedure turn. The teardrop has a continuous line, while the normal procedure turn shows a break at the outbound end.

Review the normal procedure turn illustration at the top center of Fig. 19-15; the procedure turn altitude is 2,400 feet as depicted by the numeral above the line, near the break in the procedure turn track. Note the lightning bolt figure pointing to the inbound leg just prior to the LOM; it has a number (in this case 2,400) above another line. Look down at the PROFILE SYMBOLS in the lower right-hand column of the page and see that this symbol indicates the glide slope intercept point and altitude. Recall from Chapter 16 that the glide slope intercept point is always the FAF on precision approaches.

Note the small 2,156 above the Maltese cross symbol. This indicates the glide slope altitude at the outer marker. This is a good checkpoint when flying an ILS approach to let you know that you are not on a false glide slope.

So you can see that, in this case, you should intercept the glide slope at 2,400 feet, prior to reaching the LOM, then descend on the glide slope, arriving at the LOM when the altimeter reads about 2,156 feet.

Look at the altitudes depicted in the bottom center of Fig. 19-15. These are self-explanatory, but make sure that you understand them. Basically, the line drawn over, under, or both over and under a numeral, indicates a barrier that should not be penetrated; therefore, if the altitude has lines at both top and bottom, it is a mandatory alti-

LEGEND
INSTRUMENT APPROACH PROCEDURES (CHARTS)

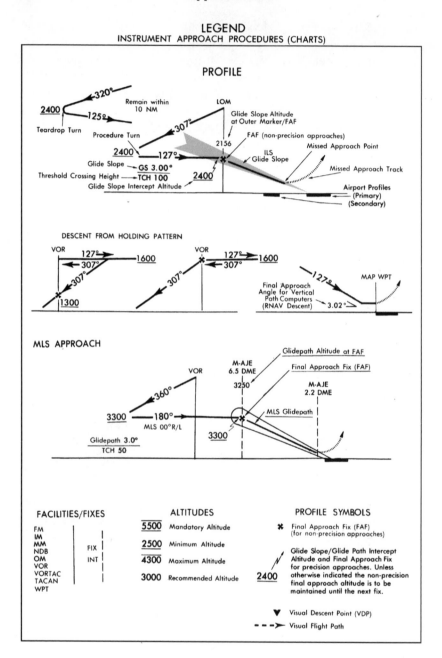

Fig. 19-15. *Not for use in navigation.*

tude. A line above the numeral means that you must stay at or below that altitude; it is a maximum altitude. A line beneath the numeral indicates a minimum altitude, and no lines mean the altitude is merely recommended.

Climb and descent gradients

Remember that the glide slope intercept altitude will give you a clue about whether or not the aircraft is on the true glide slope. Another check would be to determine what the rate of descent should be. You know the indicated approach speed, and when you get the ATIS or the numbers from approach control, you can apply the correction to the airspeed to determine approximate ground speed on approach. Then you look at the symbol at the bottom of the profile view to determine the glide slope angle for the airport. In both the example shown in Fig. 19-15 and the actual approach in Fig. 19-14, the glide slope angle is 3.0 degrees. Taking this angle and the ground speed in knots, you can look at the rate-of-descent table in Fig. 19-16 and determine that the rate of descent for a ground speed of, say, 120 knots, should be 635 feet per minute. Use this recommended rate to also check for wind shear during the approach. If you need to descend at a very high sink rate, you are probably experiencing a tailwind; a very low sink rate would indicate a headwind.

The introductory pages also contain tables to help determine the rate of climb necessary to meet certain airport or runway climb-gradient restrictions (Fig. 19-17).

This chapter was included to give you some insight into the layout of the NOS charts. Comparing NOS to the Jeppesen charts should help you decide which type you prefer. Whichever type you decide upon, remember that the legends and explanatory pages should be reviewed periodically. Each time you review them, you will run across forgotten symbols or explanations or new symbols and explanations.

Remember, pilots never know enough. One mark of a professional pilot is that he or she is willing to open the manuals to review procedures, even when there isn't a check flight coming up or a cloud in the sky.

INSTRUMENT APPROACH PROCEDURE CHARTS
RATE OF DESCENT TABLE
(ft. per min.)

A rate of descent table is provided for use in planning and executing precision descents under known or approximate ground speed conditions. It will be especially useful for approaches when the localizer only is used for course guidance. A best speed, power, attitude combination can be programmed which will result in a stable glide rate and attitude favorable for executing a landing if minimums exist upon breakout. Care should always be exercised so that the minimum descent altitude and missed approach point are not exceeded.

ANGLE OF DESCENT (degrees and tenths)	GROUND SPEED (knots)										
	30	45	60	75	90	105	120	135	150	165	180
2.0	105	160	210	265	320	370	425	475	530	585	635
2.5	130	200	265	330	395	465	530	595	665	730	795
3.0	160	240	320	395	480	555	635	715	795	875	955
3.5	185	280	370	465	555	650	740	835	925	1020	1110
4.0	210	315	425	530	635	740	845	955	1060	1165	1270
4.5	240	355	475	595	715	835	955	1075	1190	1310	1430
5.0	265	395	530	660	795	925	1060	1190	1325	1455	1590
5.5	290	435	580	730	875	1020	1165	1310	1455	1600	1745
6.0	315	475	635	795	955	1110	1270	1430	1590	1745	1905
6.5	345	515	690	860	1030	1205	1375	1550	1720	1890	2065
7.0	370	555	740	925	1110	1295	1480	1665	1850	2035	2220
7.5	395	595	795	990	1190	1390	1585	1785	1985	2180	2380
8.0	425	635	845	1055	1270	1480	1690	1905	2115	2325	2540
8.5	450	675	900	1120	1345	1570	1795	2020	2245	2470	2695
9.0	475	715	950	1190	1425	1665	1900	2140	2375	2615	2855
9.5	500	750	1005	1255	1505	1755	2005	2255	2510	2760	3010
10.0	530	790	1055	1320	1585	1845	2110	2375	2640	2900	3165
10.5	555	830	1105	1385	1660	1940	2215	2490	2770	3045	3320
11.0	580	870	1160	1450	1740	2030	2320	2610	2900	3190	3480
11.5	605	910	1210	1515	1820	2120	2425	2725	3030	3335	3635
12.0	630	945	1260	1575	1890	2205	2520	2835	3150	3465	3780

Fig. 19-16. *Not for use in navigation.*

INSTRUMENT TAKEOFF PROCEDURE CHARTS
RATE OF CLIMB TABLE
(ft. per min.)

A rate of climb table is provided for use in planning and executing takeoff procedures under known or approximate ground speed conditions.

REQUIRED CLIMB RATE (ft. per NM)	GROUND SPEED (KNOTS)						
	30	60	80	90	100	120	140
200	100	200	267	300	333	400	467
250	125	250	333	375	417	500	583
300	150	300	400	450	500	600	700
350	175	350	467	525	583	700	816
400	200	400	533	600	667	800	933
450	225	450	600	675	750	900	1050
500	250	500	667	750	833	1000	1167
550	275	550	733	825	917	1100	1283
600	300	600	800	900	1000	1200	1400
650	325	650	867	975	1083	1300	1516
700	350	700	933	1050	1167	1400	1633

REQUIRED CLIMB RATE (ft. per NM)	GROUND SPEED (KNOTS)					
	150	180	210	240	270	300
200	500	600	700	800	900	1000
250	625	750	875	1000	1125	1250
300	750	900	1050	1200	1350	1500
350	875	1050	1225	1400	1575	1750
400	1000	1200	1400	1600	1700	2000
450	1125	1350	1575	1800	2025	2250
500	1250	1500	1750	2000	2250	2500
550	1375	1650	1925	2200	2475	2750
600	1500	1800	2100	2400	2700	3000
650	1625	1950	2275	2600	2925	3250
700	1750	2100	2450	2800	3150	3500

Fig. 19-17. *Not for use in navigation.*

Section 4

Arrival procedures

20

Radio technique and holding patterns

You can control your airplane by reference to instruments. You're able to plan a flight and decide how to conduct it safely. You can handle changes and emergencies that may come your way. You know how to read and interpret instrument charts and approach plates. Now you actually need to get from cruise flight safely to the ground in poor weather. You need to fly an instrument approach.

Safely flying an approach takes every bit of concentration and skill you can muster. There's very little room for error, especially at the end of an approach, when the tolerances are often the tightest and you're the most fatigued. You'll be better prepared for arrival procedures if you've developed some techniques for flying various kinds of approaches. Let's investigate some methods for flight between cruise altitude and your destination runway.

Before radar coverage was available, radio communications played a much greater role in the progress of an aircraft from point A to point B. Two major types of radio contacts were initiated by the pilot back then. The first was merely a radio contact, such as when a plane was handed off from one controller to the next. The second was the position report that was given when the aircraft crossed a fix that required such a report.

The use of radar has caused the position report to almost become extinct. Because of this, pilots tend to become sloppy in the procedure. (They have the same problem with flying ADF approaches and holding patterns because these maneuvers are becoming more and more scarce in most parts of the country. Actually, this is the time when you should practice these things more.)

The day will come when the center's radar will be down, and you will be forced to make these position reports. In order to cut down on radio frequency congestion, you should be well versed in their use.

Let's take a look at some of these basic communications procedures. Recall the discussion about a radio contact when changing from departure control to center. The same holds true whenever you change from any controller to the next, whether departure, another center sector, or approach. If you are in a radar environment, you will merely state whom you are calling, who you are, your altitude or flight level, and whether you're climbing, descending, or flying level. If climbing or descending, you must also add the assigned altitude or flight level you're flying toward:

Cleveland Center, Cessna 1720Z, out of 5,000 for 8,000.

If you are not in a radar environment, and you do not have to give a *full* position report, your initial contact would include whom you're calling; who you are; your estimated time to the next compulsory reporting point; current altitude or flight level; whether climbing, descending, or in level flight; and again, if climbing or descending, the assigned altitude or flight level that the aircraft is climbing or descending toward:

Cleveland Center, Cessna 1720Z estimating Big Lake intersection at one five [past the hour], level at 8,000.

Proper position reports

In a nonradar environment, a full position report would have to be made whenever you cross a *compulsory reporting point.* A compulsory reporting point is indicated on the en route chart by a solid triangle.

If you were not in a radar environment and were told to make the frequency change when crossing a compulsory reporting point, the initial call, and all subsequent calls when crossing compulsory reporting points, would be simply whom you're calling, who you are, and where you are:

Cleveland Center, Cessna 1720Z, Big Lake intersection.

This will alert the center that a full position report will be forthcoming, and the controller will merely answer:

Cessna 20Z, Cleveland Center, go ahead.

Reply with a complete position report, which will consist of:

1. Who you are.
2. Where you are, which will be the fix you just crossed.
3. The time you were there, in universal coordinated time (UTC) based on a 24-hour clock.
4. The altitude or flight level (and, if necessary, whether you're climbing or descending: if you are, the assigned altitude or flight level you're climbing or descending to).
5. The estimated time you'll cross the next compulsory reporting point.
6. The name of the next compulsory point following that estimated in #5.

Naturally, if numbers 5 or 6 were your clearance limit, you would state that also.

Example: You were to contact Cleveland Center when crossing Big Lake intersection. You crossed it at 10 minutes after the hour and you were climbing to a cruising altitude of 8,000 feet. The next compulsory reporting point on your route is Frosty VOR, which is 40 miles ahead. Your ground speed has been averaging 120 knots. The next compulsory reporting point past Frosty VOR is Smokey Intersection. Smokey Intersection happens to be the clearance limit.

The initial contact would be unchanged, but after Cleveland Center responds, you would give the remainder of your report:

Cessna 20Z, Big Lake at two zero one zero, leaving five thousand to maintain eight thousand, estimating Frosty at three zero (if the hour is going to be the same hour, just like in the first part of the position report you give only the minutes), Smokey next, clearance limit.

As you can see, this is a simple procedure that condenses a lot of information into just a few short words, but it requires a lot of practice to perform it properly and professionally, especially when it's used so seldom within the radar-coverage environment.

Once back in radar contact, you will discontinue making the position reports unless you are informed that radar contact has been lost again or that for some reason radar service has been discontinued.

Required reports to ATC

Additional reports must be made to ATC or an FSS without request:

- When leaving any previously assigned altitude or flight level for a newly assigned altitude or flight level.
- When an altitude change will be made while operating "VFR on top."
- When unable to climb/descend at least 500 fpm.
- When you have missed an approach, in which case you must request a specific action, such as clearance for another approach or to your alternate or perhaps to hold for a while to try another approach when and if the weather improves. (This would be the case if visibility had been restricted by one of a series of rain or snow showers, and you are trying to get in between them.)
- When the average true airspeed at cruise altitude varies by 5 percent or 10 knots (whichever is greater) from what you filed on the flight plan.
- The time and altitude or flight level when reaching a holding fix or the point that you have been cleared to.
- When leaving any assigned holding fix or point.
- When any VOR, TACAN, ADF, LF, ILS, or air/ground communications capability is lost.
- When leaving the final approach fix (or outer marker) inbound on final approach. (Only required when *not* in radar contact.)
- A corrected estimate whenever you realize that a previously submitted estimate is in error in excess of three minutes. (Only required when *not* in radar contact.)

You must also report any weather conditions that have not been forecast and any other hazards to flight, including icing, snow, clear air turbulence, and even flocks of migrating birds.

Communications confusion

The airspace is becoming more congested every day. The ATC system is being overworked in some sectors, and there are many new controllers who are not full performance level. Because of this, we cannot be too careful in what we say when we communicate with them. Although we know enough to repeat all clearances we receive, in the hope that it will provide a second chance to correct any errors in reception or understanding, records now indicate that many clearances that have been read back wrong will be accepted as having been read back correctly by an overworked controller. This means that anytime you have any question regarding a clearance, ask for a specific confirmation. If there is any chance that you made a mistake or the transmission was a little garbled, double-check. Don't just assume that you received it correctly.

Many questions exist regarding semantics, as well as different phraseologies between different countries. For example, in the United States, when cleared to taxi to a runway, you might be told "Taxi to Runway 8L, hold short." In Jamaica, "Taxi to position Runway 8."

They both mean the same thing, but if you didn't know any better, you might taxi into position on Runway 8 in Jamaica. Why is that? Because in the United States, when cleared onto the active runway, you would be told to "Taxi into position and hold."

Hear how that sounds like the Jamaican instruction to position the aircraft at the taxiway hold line? In Jamaica, the instruction to taxi onto the active runway would be "Line up and wait."

An airline accident in New York emphasized problems with semantics and various countries' phraseologies. An Avianca 707 ran out of fuel and crashed. On a number of occasions, the pilot told ATC that he was low on fuel and asked for *priority* handling. Because he didn't use the word *emergency*, ATC merely took him out of a holding pattern and put him into a line for approach. If he had said that he had a low-fuel emergency, they would have put him ahead of everyone on approach. I can't help but wonder if the word *priority* in Colombia means the same as *emergency* in the United States.

No matter how cool, confident, and in control we like to sound when the chips are down, I'm sure that the flight would have had more expeditious handling if the pilot had said "I'm just about out of fuel and if you don't put me ahead of everyone else I'm going to crash. Pan Pan, Mayday, Mayday, Priority, Emergency."

In other words, use as many different phrases as you can think of to alert everyone that you do have a very serious situation aboard the aircraft. Don't ever let the survivors say "Well, I thought the pilot meant. . . ."

More communication failure procedures

The situation will become rather sticky though, if you suddenly lose two-way communications capabilities. The regulations are set up to guide you as best as they can in such a situation, even though emergencies rarely follow a set pattern, and it's hard to write a set of rules telling you how to handle them. The following has been set up as a guide, remembering that FAR 91.3, "Responsibility and Authority of the Pilot in Command," says:

> *(a) The pilot in command of an aircraft is directly responsible for, and is the final authority as to, the operation of that aircraft.*
> *(b) In an emergency requiring immediate action, the pilot in command may deviate from any rule of this subpart or of Subpart B to the extent required to meet that emergency. . . .*

Unless you see fit to exercise your emergency authority as outlined above, FAR 91.185 sets up specific actions to follow if two-way communications capability is lost.

As with many FAA rules, this section begins with the words, "Unless otherwise authorized by ATC, each pilot . . . shall comply with the rules of this section."

The rule goes on to say that if flying in VFR conditions, or if VFR conditions are encountered afterwards, continue VFR and land as soon as practicable. Notify ATC as soon as possible that you are on the ground safely. If you're at a controlled airport, this is a simple trip to the tower, but if it's a noncontrolled field, you'll have to find a phone.

If you are in IFR conditions, however, continue by specific routes and altitudes as follows:

- If already on an airway, you should fly the route as assigned in the last clearance.

- If you were being radar vectored to a route or fix, fly as directly as possible to the route or fix you were being vectored toward. Of course, beware of obstacles, such as high mountains, that you were being vectored around.

- If recleared during flight to a specific point, and told to expect further clearance via a specified route, fly the route you have been told to expect.

- If you haven't been told to expect a specific routing, fly what you filed in the flight plan.

- Your altitude must be the highest of: the altitude last assigned by ATC, the altitude ATC has advised you to expect in a later clearance, or the minimum en route altitude for each segment of the route.

A good example is the situation in Fig. 20-1. A pilot has been cleared via A, B, C, and D to E. While flying between A and B her assigned altitude is 6,000 feet, and she is told to expect a clearance to 8,000 feet at B. Prior to receiving the higher altitude assignment, she experiences two-way radio failure. The pilot would maintain 6,000 feet to B, then climb to 8,000 feet (the altitude she was advised to expect). She would maintain 8,000 feet, then climb to 11,000 at C, or prior to C, if necessary, to comply with a minimum crossing altitude at C. Upon reaching D, the pilot would descend to 8,000 feet (even though the MEA is 7,000) because 8,000 is the highest of the altitude situations stated in the rule.

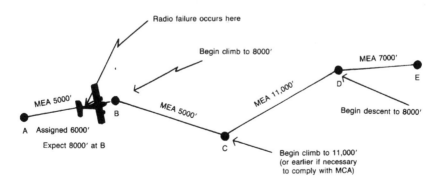

Fig. 20-1.

So far, everything's pretty straightforward, and although the actual wording of the FARs sometimes sounds complicated, if we sit and reason the regs out with an example or two and try to see what the FAA's author(s) had in mind, the FARs tend to simplify themselves.

The only other question would be when to leave a clearance limit, which is answered in FAR 91.185 (c) (3):

> *(i) When the clearance limit is a fix from which an approach begins, commence descent or descent and approach as close as possible to the expected further clearance time if one has been received, or if one has not been received, as close as possible to the estimated time of arrival as calculated from the filed or amended (with ATC) estimated time en route.*
>
> *(ii) If the clearance limit is not a fix from which an approach begins, leave the clearance limit at the expected further clearance time if one has been received, or if none has been received, upon arrival over the clearance limit, and proceed to a fix from which an approach begins and commence descent or descent and approach as close as possible to the estimated time of arrival as calculated from the filed or amended (with ATC) estimated time en route.*

If you have not received holding instructions and are ahead of the flight plan ETA, you should hold at the fix so that you land as close to the ETA as possible. If there is more than one approach at the destination, make the approach of your choice, based on weather forecasts. ATC will have cleared the airspace on all of the approaches.

In the event that two-way communication is lost, ATC will provide service on the basis that you are operating in accordance with the above rules. Naturally, during all of this, you will be monitoring the navaid voice facilities and will follow any instructions you might receive. Try to establish radio contact on the last frequency over which you had two-way communications, which is why you should always write down each assigned frequency, perhaps on the telephone-message notepad, rather than just switch the radio to the new frequency.

Failing to make contact on the last frequency, you should try to make contact on FSS frequencies, and on 121.5 MHz. Squawk code 7600 for the remainder of the flight.

The holding pattern

Back to normal flight operations: As you proceed, ATC might find it necessary to slow you down or stop you to properly sequence all of the flights in the area. Because you cannot actually stop in midair, they will have you fly a racetrack flight path called a *holding pattern*, which serves the purpose. You will enter a holding pattern automatically if, when you come to a clearance limit, you have received no further clearance.

Most holding patterns are depicted on the appropriate charts for the area. If you are required to hold at a fix where no pattern is depicted, you will receive specific information:

- The direction to hold in relation to the holding fix: north, northeast, east, and the like.
- The fix itself.
- The radial, course, magnetic bearing, or airway.
- The length of the outbound leg, if the hold is based on DME distances.
- Instructions to make the turns to the left, if the holding pattern is nonstandard.
- The time to expect a further clearance.

Figure 20-2 shows the terminology used in a standard (right turns) holding pattern. The maximum holding airspeeds are 175 knots IAS for prop-driven aircraft and 200 knots IAS for civil jets under 6,000

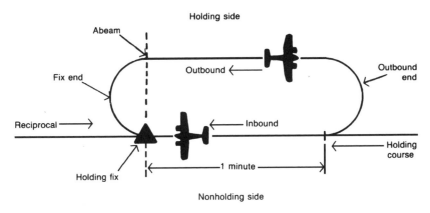

Fig. 20-2. *Standard (right-hand) holding pattern in no-wind condition.*

feet. Civil jets are also restricted to 210 knots IAS between 6,000 feet and 14,000 feet and 265 knots IAS above 14,000 feet.

If an airspeed reduction is necessary to comply with these maximum speeds, you must begin the reduction within 3 minutes of the ETA over the holding fix.

Time the pattern so that the time of the inbound leg will be 1 minute if at or below 14,000 MSL; above 14,000 feet, the inbound leg should be 1½ minutes long. In order to establish this, you should fly the initial outbound leg for these durations and then adjust the outbound leg time to get the proper inbound time. Timing is commenced abeam the fix. There might be times when ATC will designate a different inbound time, and if the holding fix is predicated on a DME distance, you don't worry about the time other than to note what the inbound and outbound times are so that you can cross the fix inbound at any time specified by ATC.

Holding pattern turns should be made at the least bank angle of the following:
- 3 degrees per second (compass/DG).
- 30-degree bank angle (attitude indicator).
- 25-degree bank angle (flight director).

Holding patterns entries

The hardest thing for most pilots to figure out is how to enter a holding pattern. There are only three correct ways to enter a pattern, and you determine which one to use depending on the arrival heading with relation to the inbound leg.

As a guide, the FAA has established one line at 70 degrees to the inbound leg on the holding side and another line at 110 degrees from the inbound leg on the nonholding side.

If you enter the holding pattern from within this sector (Area 3 in Fig. 20-3), merely fly to the fix and turn outbound in the same direction of the hold—turn right for right patterns and left for left patterns: a *direct entry.*

If you enter from Area 1 in Fig. 20-3, cross the fix, make a turn opposite the direction of the holding turns, and then fly the first outbound leg on the nonholding side. At the end of the outbound leg,

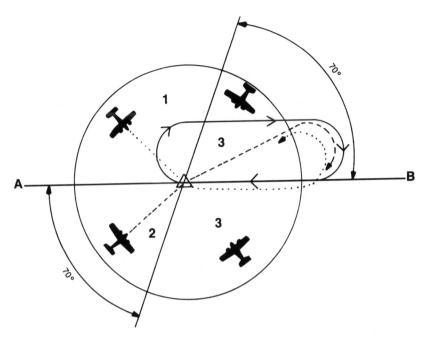

Fig. 20-3. *Entry procedures for standard (right-hand) holding pattern. Aircraft approaching from Zone 1 use a parallel entry; aircraft in Zone 2 use a teardrop entry; and aircraft in Zone 3 enter directly.*

make another turn opposite the holding turns, intercept the inbound leg, cross the fix, and enter the normal pattern: a *parallel entry*, or a parallel entry on the nonholding side.

Finally, if you enter from Area 2 in Fig. 20-3, cross the fix and turn to fly a heading 30 degrees to the outbound leg on the holding side. Fly this heading for 1 minute and then turn in the same direction as the holding turns to intercept the inbound leg: a *teardrop entry*.

In all of these entries, you should make the necessary corrections to compensate for the known wind. Although reams have been written to explain how to know which entry to use, I have found that 90 percent of the time the easiest entry to make from your position is the correct one.

If you're flying an aircraft with a full-faced DG, you might try this idea: Take a piece of thin, stiff, clear plastic, and a felt-tip pen. Trace Fig. 20-3, which depicts the entries for a standard right-hand holding pattern. Lay the line A-B over the face of the DG, with A on the inbound heading, B on its reciprocal, and the fix in the center, and read your entry directly from the pattern. For left-hand patterns, turn the plastic over. Simple? Try it.

STARs

Close to the destination, the ATC clearance might include a STAR (*standard instrument arrival route*): a preplanned instrument flight rule (IFR) air traffic control arrival procedure published for pilot use in graphic and/or textual form. STARs provide transition from the en route structure to an outer fix or an instrument approach fix/approval waypoint in the terminal area.

STARs are usually made for airports in high-density areas, and similarly to the SIDs that are used in departures, the STARs' main purpose is to simplify clearance procedures and cut down on radio frequency congestion. As with a SID, the pilot must possess at least the textual description. The final decision to accept or reject a STAR rests with the pilot, and it simplifies matters a great deal if she makes the remark "NO STAR" in the remarks section of the flight plan if she doesn't want to avail herself of this aid.

Figure 20-4 shows the Downe Four Arrival to Los Angeles International Airport. The chart resembles a SID, except for the letters in the black box in the upper-right corner and the "2" in the numerical listing (10-2), which is used for STARs. Remember, 10-1 is for area charts, 10-2 is for STARs, and 10-3 is for SIDs. Multiple SIDs and STARs at a single airport will have a letter after the numeral such as 10-2B, 10-2C, and the like.

Again, this is a schematic, not to scale, but it does show, in pictorial form, the altitude restrictions en route for each transition to the approach and, in textual form, the exact routes to be flown. It also shows all runway restrictions that are pertinent.

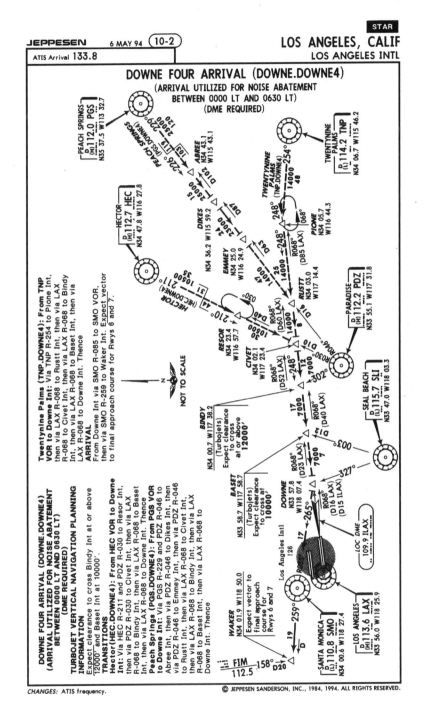

Fig. 20-4. *Reproduced with permission of Jeppesen Sanderson, Inc. Not for use in navigation.*

21

More on holding patterns

VOR holding patterns are introduced as early as possible for two reasons. First, holding patterns around a VOR station provide excellent practice in VOR interception and tracking close to the station. The needle indications are very sensitive; holding pattern practice will quickly sharpen your tracking skills.

Second, something seems to make holding patterns awe-inspiring and difficult. They are really quite simple when you go about them the right way. Nevertheless, many pilots are wary of holding patterns, so I tackle them early in the course so that students will feel comfortable with them later on.

My technique for teaching holding patterns works very well. Initially, you will learn how to fly the pattern while correcting for wind drift. Only after you are comfortable with the racetrack pattern and have mastered wind correction techniques do I then teach pattern entries.

Figure 21-1 illustrates the basic elements of a standard holding pattern around a VOR. The standard pattern has right turns, the inbound leg is 1 minute long. Nonstandard patterns have left turns. Above 14,000 feet the inbound leg is $1\frac{1}{2}$ minutes.

Pick a nearby VOR and fly inbound on any convenient course with the OBS needle centered. At station passage you will be very busy for a few seconds running an important checklist.

Five Ts: time, turn, twist, throttle, talk

Do these at station passage:

1. Note the *time* of arrival at the fix and write the time down on the log or directly on the chart.

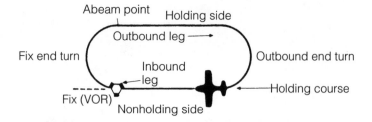

Fig. 21-1. *Elements of a standard holding pattern.*

2. Start a 180° *turn* to the outbound course with a wind correction as necessary.

3. *Twist* the OBS knob to set new inbound VOR course.

4. *Throttle.* Reduce power to conserve fuel while in the holding pattern; you are not going anywhere.

5. *Talk.* Report reaching the holding fix and altitude ("...Approach, 56 Xray entering hold, level at five."). Always ask the question, "Do I need to report to ATC now?" During an instrument approach the answer to that question will most often be yes. (When in doubt, report.)

These are the five Ts of instrument flying: *time, turn, twist, throttle, talk.* In other words, aviate, navigate, and communicate. You will encounter the five Ts over and over, especially during instrument approaches. Use this mental checklist with every holding pattern and keep using it until it begins popping into your head automatically.

After running through the checklist, the next task is to determine when to start timing the outbound leg. This is easy. After station passage, and while you are making the turn to the outbound leg, the TO-FROM indicator will show FROM. When the airplane is abeam the holding fix heading outbound, the indicator will change to the TO position. Start timing when the indicator changes to TO.

Wind corrections

Recall from VOR tracking practice that wind is almost always a factor. How do you correct for the wind in a holding pattern? As you flew inbound to the VOR you kept the needle centered, which should provide a pretty good idea of the wind correction for the inbound leg upon reaching the fix. The wind correction angle on the outbound leg will be *double* and *opposite* the correction on the inbound leg.

If you held a wind correction angle of 4° into the wind on the inbound leg, hold a wind correction of 8° into the wind while flying outbound. The reason for doubling the wind correction angle is to compensate for the effect of the wind on those two 1-minute turns at the fix end and at the outbound end.

Even if you fly for precisely 1 minute on the outbound leg, you will probably find that the inbound leg is not exactly 1 minute. Again, this is because of wind. The correction is simple; adjust the time of the next *outbound* leg to compensate for the difference.

If the inbound leg is only 45 seconds, for example, add 15 seconds to the outbound leg and make it 1 minute and 15 seconds. If the inbound leg is 1 minute 30 seconds, subtract 30 seconds from the outbound leg. The first case compensates for a tailwind on the inbound leg; the second case corrects for a headwind on the inbound leg.

En route holding

The wind correction tips are essentially a description of the procedure used for establishing an en route holding pattern. If there is a delay while en route, ATC might simply issue a hold on the present course, at the same altitude, at a convenient fix.

If no holding pattern is shown on the en route chart, ATC will state:

1. What fix to use and where the holding pattern will be located in relation to the fix (north, south, southwest, etc.).

2. What radial to use.

3. Nonstandard instructions, such as left turns and length of legs in miles if DME is utilized.

4. An *expect further clearance* (EFC) time, or the time to *expect approach clearance* (EAC) if held on a segment of an instrument approach; these "expect" times will be given either as a specific clock time, such as 2045, or in minutes. It is not uncommon to hear ATC say "expect further clearance to XYZ in 10 minutes."

EFCs and EACs are important for two reasons:

- They tell when ATC expects to issue a clearance to resume the flight. You need to know this to adjust the holding pattern to arrive over the fix at the "expect further clearance time."

- EFCs and EACs also tell you when to depart a holding pattern in the event of lost communications. If ATC does not issue an EFC or an EAC, be sure to request it.

A typical holding clearance might be:

"Cessna three four five six Xray, hold southeast of the Huguenot VOR on the one four five degree radial, maintain five thousand, expect further clearance at one two one five."

Many holding patterns are already depicted on en route and approach procedure charts. If the holding pattern is shown, the clearance will be simpler. For example:

"Cessna three four five six Xray, hold as published northeast of SHAFF intersection, maintain five thousand. Expect further clearance at one two one five."

In this case, ATC would expect you to find SHAFF intersection on the L-25 en route chart and fly the depicted pattern northeast of the VOR.

ATC recognizes that holding patterns can be somewhat imprecise because of the wind, the skill of the pilot, and the different airspeeds of different types of aircraft. So they establish a buffer zone on the holding side that is at least double the amount of protected airspace around the pattern.

Holding pattern entry

When comfortable with the patterns, you will have also automatically mastered the *direct* method of entering holding patterns. The direct entry, simply stated, means that you fly to the station or fix and make a turn directly to the outbound leg, which is exactly what you have been practicing. On IFR flights, you will use direct entries for most holds issued by ATC.

The other two ways to establish holding patterns are the *teardrop* and the *parallel* methods (Fig. 21-2). To enter a standard holding pattern with the teardrop method, cross the holding fix and proceed

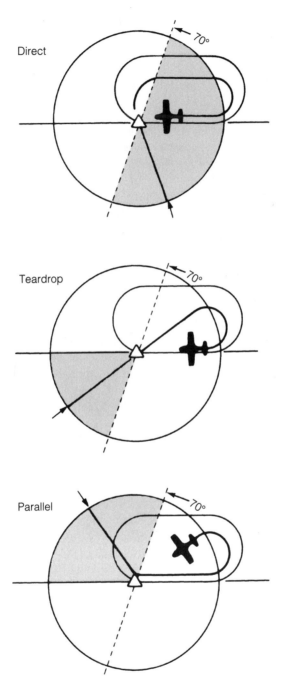

Fig. 21-2. *The three recommended methods of entering a holding pattern.*

outbound at an angle of 30° to the holding course for 1 minute, then turn right to intercept the holding course.

For the parallel method, cross the fix and fly outbound parallel to the holding course for 1 minute, turn left, fly direct to the fix, then turn right to the outbound course.

Before going into further detail, some basic points will make holding entry much easier. Direct, teardrop, and parallel entry methods are *recommendations only*. They are *not* required. The FAA's *Practical Test Standards* require only that you use "an entry procedure that ensures the aircraft remains within the holding pattern airspace for a standard, nonstandard, or nonpublished holding pattern."

One very easy—and completely acceptable—method of entering holding is to fly to the holding fix, then turn on the holding side to the outbound heading. Fly the required time (or distance), then turn inbound on the holding side and reintercept the inbound leg.

Think about this for a minute. No matter how you arrive at the holding fix, simply turn to the outbound heading, fly the time specified, turn back toward the fix, and intercept the inbound leg, making all turns on the holding side. There is no longer the need to make all those confusing, distracting calculations about whether your incoming course is greater or less than 70° to the inbound leg of a standard right-hand pattern (opposite in the case of a left-hand pattern). You won't get into trouble as long as you establish an accurate pattern quickly and efficiently and do not violate the protected airspace.

The entry and first time around the holding pattern are free. ATC understands that it is difficult to fly a perfect one-minute racetrack on the first circuit. Flying slightly off course on the first pattern is tolerated by ATC, provided you maintain the assigned altitude. Better tracking is expected by the second circuit.

You should still become proficient in all three holding entries—direct, teardrop, and parallel. They are the most efficient ways of entering holding that have been devised so far, and your ability to use the three methods with precision will help create an atmosphere of competence and confidence on your flight test. So please read on and learn the three methods as you work toward your goal of becoming a "proud, perfect pilot."

Choosing the correct entry

Note the line that has been drawn in Fig. 21-2 at an angle of 70° to the inbound leg of a standard right-hand pattern. (The pattern would be on the left side of the inbound leg for a nonstandard left-hand pattern, and the 70° line would be drawn in the opposite direction.)

The 70° line helps divide the area surrounding the holding fix into three "entry sectors," as indicated by the shading. The type of entry chosen depends upon the entry sector in which you approach the holding fix. I have flown with a few amazing people who can calculate entry sectors quickly in their heads down to the last degree. But this is the exception, not the rule. Most of us can't "do it by the numbers."

Fortunately, three other ways to plan holding entries have evolved over the years:

- Pencil in the holding pattern on the chart (if it isn't already printed). Draw in an approximate 70° line—it's 20° less than the perpendicular line to the inbound course. Then draw the inbound course to the station. This will show which entry sector you occupy and which type of entry to choose. Remember, you don't have to draw the lines perfectly down to the last degree to set up an acceptable entry.

- Use one of the many plastic holding overlays available. Place the overlay over the holding fix on the chart. Align the holding pattern on the overlay with the inbound leg of the holding pattern. You will see clearly which entry sector you occupy. These clear plastic overlays usually have standard (right-hand) patterns on one side and nonstandard (left-hand) patterns on the other side.

- Use the heading indicator. Visualize the inbound leg of the holding pattern, then take a pencil or other straight object and rotate it 70° counterclockwise to this inbound leg. This will reveal the entry sectors very quickly. For a nonstandard (left-hand) pattern, turn the pencil clockwise 70° to indicate the entry sectors.

Finally, *visualization* is the key! Visualize the holding pattern in relation to the fix and where you are. Upon arrival at the fix simply make the shortest possible turn to join the racetrack of the holding pattern.

Importance of altitude control

ATC is more concerned about altitude control than making a perfect pattern entry because other airplanes might be flying the same pattern at different altitudes above and below. You might be in a pattern at 5,000 feet with planes at 3,000, 4,000, 6,000, 7,000, 8,000, and so on. Stacks such as this are a common occurrence around busy airports during landing delays due to weather. (Fig. 21-3.)

In a stack, planes are cleared to proceed from the bottom, like dealing from the bottom of a deck of cards. Each time a plane is cleared to leave the stack, all the others are cleared sequentially to descend one level at a time. A vertical separation of 1,000 feet is maintained at all times. Obviously, there must be no confusion about altitude assignments, nor any sloppiness in maintaining assigned altitudes, with several planes all holding in the same racetrack pattern around the same fix, separated only by altitude. The goal of 2, 2, and 20 becomes a realistic requirement in this situation.

Holding pattern variations

It is not unusual for an instrument flight to be kept in a holding pattern 20 minutes or more. When this happens it is perfectly all right

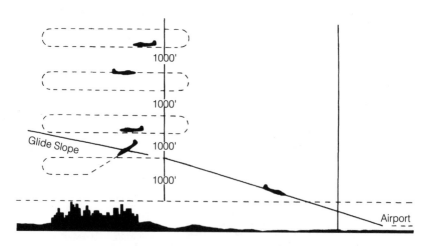

Fig. 21-3. *Stack or shuttle descent in a holding pattern for an instrument approach.*

to ask ATC for a pattern with 2-, 3-, or even 5-minute legs. ATC will usually try to grant the request.

Longer legs make it easier to fly the pattern. More attention can be paid to altitude control and to establishing a precise inbound course with the needle centered. Longer legs are also much easier on passengers. Repetition of a straight leg for 1 minute, followed by a 1-minute 180° turn could cause airsickness.

For planning purposes, a pattern with 2-minute legs will take six minutes to complete, a pattern with 3-minute legs will take eight minutes, and one with 5-minute legs will take 12 minutes. If you expect to be in a holding pattern for any length of time, adjust power and relean the mixture to keep fuel consumption to a minimum.

Intersection holds

ATC will frequently issue en route holds at VOR intersections. Some are depicted on en route charts; be mentally prepared to use them. Intersection holds are easily managed. To reduce cockpit confusion, always set up the holding course on the top (No. 1) nav receiver. Set up the intersecting bearing on the bottom (No. 2) nav receiver.

If, on the No.1 nav, you always set up the course *to the station*, not the radial, the needle will always be located in the same direction as the VOR. If the needle is to the right, the station will also be to the right. This is more a matter of reducing cockpit confusion than anything else. Set up the two VORs as described above and you will always have a clear picture of exactly where you are in the pattern. (These are also good procedures even when no holding is involved and you want to keep track of the intersections along an en route leg.)

You have arrived at the intersection when the needle on the No. 2 nav receiver centers. Do the Five Ts checklist.

Start timing the outbound leg when the No. 2 nav needle moves back on the same side as the VOR indicating that you have reached the abeam position. In short, with the No. 2 nav receiver set to the radial (FROM), if the needle and the VOR are on the same side, you have not arrived. If they are on opposite sides, you have passed the intersection.

One-VOR intersection

It is possible to identify VOR intersections and fly holding patterns around them using only one navigation receiver. Sounds difficult at first, I know, but many pilots are quite accomplished with this procedure. In the early days of VOR navigation having even one VOR in the cockpit was considered a luxury and dual VORs were almost unheard of. So one VOR did the work of two.

It's not a good idea to spend any time on one-VOR intersection holds at this point. If one VOR fails in flight, notify ATC immediately. Consider landing as soon as possible because if the other VOR also fails, you would have a job on your hands getting down safely. ATC would have to do a lot of fast shuffling to reroute the traffic in your vicinity to ensure safe separation.

If you would like to try single-receiver intersection holds, have an instructor coach you through a few in the simulator.

Track the inbound leg with the appropriate wind correction established and the needle centered. One minute before ETA at the intersection, reset the navigation receiver to the frequency of the station providing the cross-bearing and reset the bearing to that of the second station.

On reaching the intersection, proceed normally. When established on the outbound leg, reset the nav frequency and bearing back to the first station. At 1 minute outbound, turn again and get reestablished on the inbound leg with the needle centered. Fly inbound for 30 seconds, then reset the frequency and bearing to anticipate the cross bearing.

Obviously, a lot depends on the student's ability to get established quickly on the inbound bearing with the needle centered.

DME holding patterns

Another holding pattern variation is based upon DME distances, rather than time. If you have indicated on the flight plan that you have DME aboard (code letter A), you can expect to be issued a DME hold. DME holding uses the same entry and racetrack procedures except that distances (in nautical miles) are used in lieu of

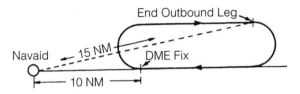

Fig. 21-4. *Holding pattern toward a DME facility.*

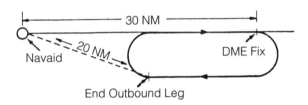

Fig. 21-5. *Holding pattern away from a DME facility.*

time. ATC specifies the distance of the fix from the navaid and the length of the outbound leg.

For example, if heading *toward* the VOR/DME (Fig. 21-4) with the fix distance 10 nm and the outbound leg 15 nm, you will enter the racetrack when the DME reads 10. You will end the outbound leg when it reads 15, and commence the turn back to the inbound leg.

Clearance for a DME hold would be something like this:

"Cessna three four five six Xray, hold 10 north of Carmel VOR on the three six zero degree radial, five-mile legs, expect further clearance one five four five, maintain five thousand."

DME holding patterns can also be established with the inbound leg heading *away* from the VOR/DME, as shown in Fig. 21-5. In this example, the DME fix is 30 nm from the station and the end of the outbound leg is 20 nm from the station.

DME holding patterns are certainly a lot easier to manage, and this should be a factor when considering whether or not to invest in DME equipment. But don't throw away that stopwatch yet! You'll need it to time nonprecision approaches.

22

Approach angles

Many pilots think that instrument flight is the ultimate test of an airman's skill. If this is so, then flying a good instrument approach has to be the epitome of the craft of flying.

Before we get into the actual approaches, let's take a look at some background information, basic concepts, and definitions of terms. Then I shall elaborate on certain pertinent regulations concerning instrument approaches.

The Air Traffic Control Handbook explains the purpose of instrument approach procedures:

> *Instrument approach procedures are designed so as to ensure a safe descent from the en route environment to a point where a safe landing can be made.*
>
> *A pilot adhering to the altitudes, flight paths, and weather minimums depicted on the Instrument Approach Procedure (IAP) chart or vectors and altitudes issued by the radar controller, is assured of terrain and obstruction clearance and runway or airport alignment during approach for landing.*

Sounds simple, yes?

You say no?

Well, it really is; however, just like learning to take off and land, it is first necessary to establish a firm background to build upon. If at this time you are not proficient in straight-and-level flight, level turns, climbs and descents (both straight ahead and while turning), slow flight, speed control, and emergency procedures—all under the hood—you're not ready for approach work. In addition to the maneuvers mentioned previously, you should also be fairly proficient in flying airways, procedure turns, and holding patterns.

Getting it down safely

I suspect that you are asking, "Why? What's so different? Instruments are instruments." It's simply that when you get on an approach, your mind will be so filled with procedures and figures that aircraft control will have to be almost automatic. Not only that, but when you're on the approach, your margin of error becomes progressively smaller and smaller.

You can equate instrument flight with two cones lying on their sides with the large ends connected. The airports are at each small end; one end is the departure field and the other end is the destination. After takeoff, you point the nose skyward and begin to climb. The higher you get, the larger the sphere of allowable error becomes. If your attention wavers, you can go to a full-needle-width deflection of the OBI before you get outside of the airway, which extends 4 nautical miles each side of the centerline.

Sure, the regs say that you should be on the centerline, but if you drift off, you are still in protected airspace. Also, if the altitude starts drifting off and you go up or down 300 feet, it's probably not dangerous, but it's poor airmanship and subject to a citation or a violation. The chances of hitting something up there will be rare, unless someone else is also doing a sloppy job of holding an assigned altitude.

The sphere of allowable error gets smaller each second while on the approach. Close to minimums, the allowable error becomes very small. At this point, a full deflection of the needle isn't 4 miles off, but the width of the airport environment (if flying a VOR or NDB approach) or the width of the runway (if flying an ILS/MLS or localizer approach).

Tall structures can be found outside of this protected area, and running into one of them is not the right way to end a flight. Altitude becomes more critical down here also, and a 300-foot error could mean sudden and serious disaster.

It is assumed that you're fairly good at the basic hood work and that you practice from time to time with a safety pilot. You're ready to begin working on instrument approaches. But first, maybe you should know how the instrument approach procedure charts came about.

In the beginning

Way back in 1930, a 23-year-old barnstormer by the name of Elroy B. ("Jepp") Jeppesen was offered a job as an airmail captain for Boeing Air Transportation Company, which is now known as United Air Lines. This promised a steadier income than that provided by barnstorming his World War I surplus Jenny, so he took the job.

He was flying the rugged mountain terrain between Cheyenne and Salt Lake City. It was a dangerous route with or without the navigational aids, which ranged from few to none. The route claimed many planes and pilots.

One day, Jepp bought a small 10-cent notebook and began jotting down all of the pertinent information regarding his routes. In addition to field lengths, obstructions, and altitudes, he even compiled a list of phone numbers of farmers along his route whom he could call for weather information. Some of this information would be very informative, while some might amount to no more than, "Well, I can see the barn through the snow." It was not unusual on his days off to find Jepp with a pocket altimeter climbing mountains, trees, water tanks, and other obstructions to ascertain their correct heights.

Before long, pilots began talking both of his longevity and his "little black book," and he gave a few copies to his friends. The demand grew so great that he started publishing them in the cellar of his rooming house and selling copies to interested airmen. Business was so good that it was taking all of his time, so he tried to sell it to United Airlines for $5,000. The airline management turned the offer down, a move that ended up making Jepp a millionaire. He took an early retirement from United in 1954.

He opened a facility in Frankfurt, Germany, in 1957 due to the international demand, and in 1958 he opened an office in Washington, D.C., to interface with the government. He sold his publishing firm to the Times/Mirror Publishing Company in 1961. Times Mirror bought Sanderson Films, Inc., in 1968, and in 1970, when Times Mirror took Sanderson Films to Denver, the two company names were merged into today's Jeppesen Sanderson, Inc.

Jepp's "little black book" has grown to volumes of charts and procedures that cover airways worldwide and approaches to more than 16,000 airports. Revisions are issued weekly. Jeppesen manuals are used by all United States commercial airlines and many foreign carriers, as well as by a great majority of private and business pilots.

(I am indebted to Jeppesen Sanderson, Inc., and to the Times/Mirror Publishing Company for the permission they have granted me to use excerpts from the firm's manuals and copies of Jepp approach charts for illustrations.)

The government also publishes a set of charts, but because the airlines in the United States all use Jepps and many instrument students have aspirations to fly for the airlines, I have elected to use the Jepps during most of these discussions. I will, however, devote a chapter to explaining the NOS charts so you will be aware of the similarities and differences between the two.

Precision and nonprecison approaches

There are two basic types of approach procedures: precision and nonprecision. Simply stated, the nonprecision approach is a standard instrument procedure that does not use an electronic glide slope. Three precision approaches use a glide slope: instrument landing system (ILS), microwave landing system (MLS), and the precision approach radar (PAR). Implementation of the MLS has stagnated as a disappointment and will be disregarded in this discussion. PAR approaches are few and far between; they will also be disregarded because the ground controller talks the pilot in, and about all the pilot can do is listen and do what the controller says.

Several components are necessary for an approach to be classified as an ILS. The ground components are a *localizer, glide slope, outer marker, middle marker,* and *approach lights.* If a procedure specifies a visibility minimum based on *runway visual range* (RVR), it will also need *high intensity runway lights* (HIRL), *touchdown zone lighting* (TDZL), *centerline lighting* (RCLS) and markings, and, naturally, the RVR transmissometers for that runway. Of course, before the pilot can accept an ILS approach, her aircraft must have the airborne equipment necessary to receive the ground components.

A few variations exist. The regulations do allow a compass locator or a precision radar to be substituted for the outer marker (OM) or middle marker (MM), and an *airport surveillance radar* (ASR) can usually be substituted for the outer marker.

All other approaches are nonprecision. These primarily consist of the *localizer, back course, VOR, NDB* (ADF), ASR, and the newest of all, GPS approaches—although before long you will see both precision and nonprecision GPS approaches. The VOR and NDB (ADF) approaches are, in effect, designed to bring the aircraft down to the overall airport environment, while all other approaches will bring you right down to the runway.

Approach minimums

One of the most important terms used in conjunction with the instrument approach procedure charts is *minimums*. The minimums are comprised of two factors, the *minimum descent altitude* (MDA)—in the case of a precision approach, the *decision height* (DH)—and visibility. Each approach has its basic minimums that are modified for different airports depending on many factors, including obstacle clearances and terrain.

Visibility is the factor that governs whether or not airliners can *initiate* the approach. When we speak of a field being closed due to weather, the visibility is lower than that allowed by the available approaches. Under Part 91, you can start an approach regardless of visibility, which is discussed later. Prudent pilots, however, would not try to begin an approach when visibility is below minimums. The MDA or DH specified is merely the lowest altitude to which you can descend on the approach unless you have the runway environment in sight and are in a position to make a safe landing. If the airport is reporting a lower ceiling than the MDA/DH specified for the approach, you would still be allowed to commence the approach, although you would not normally expect to see the runway environment when the aircraft arrived at minimums.

The minimums are modified further if certain navigational aids on the ground or in flight are inoperative. These modifications are all specifically spelled out on the instrument approach charts.

On ILS approaches, if the localizer is out, the approach is not authorized. If the glide slope is out, the approach becomes a nonprecision approach and the DH turns into an MDA as specified in the procedure. If the outer or middle marker is out, the DH is increased by 50 feet and the visibility increase will depend on the *aircraft approach category*, which is discussed later in this chapter. On ILS and PAR approaches, if the ALS (*approach light system*) is out, the DH goes up by 50 feet and the visibility increases by ¼ mile. If the SSALSR *(simplified short approach light system* with *RAIL)* or MALSR *(medium intensity approach light system* with *RAIL)* is out, the DH for category A, B, and C increases by 50 feet and the visibility requirements go up by ¼ mile. RAIL stands for *runway alignment indicator lights,* which are the sequenced flashing lights seen on approach light systems that many pilots have nicknamed the *rabbit,* because they tend to run quickly toward the runway threshold.

If you are flying on an ILS approach with a visibility minimum of 1,800 or 2,000 feet RVR, the localizer and glide slope requirements are the same as above. But if the outer or middle marker is out, you must increase the DH by 50 feet for all categories, and for category A, B, and C, the visibility goes up to ½ mile, while in category D, it increases to ¾ mile. If the ALS is out, the DH goes up 50 feet and the visibility goes to ¾ mile. With the HIRL, TDZL, RCLS, or RVR out, there is no change in the DH, but the visibility goes to ½ mile. If only the RCLMs *(runway centerline markings)* are missing, there is no visibility increase, but the DH will change according to the procedure.

In VOR, LOC, LDA *(localizer type directional aid),* and ASR, if the ALS, SSMALSR, or the MALSR is out, the visibility increases ½ mile in category A, B, and C. It will increase ¼ mile in category A, B, and C if the SSALS, MALS, HIRL, or REIL *(runway end identification lights)* are out.

In NDB (ADF), with the ALS, SSALSR, or MALSR out, the visibility increases by ¼ mile in category A, B, and C.

The differences are confusing, but no one expects you to memorize the information. I only brought it up to show you that there are many times when the basic minimums might change, and you should be aware of them and know that you can find the new minimums in the appropriate blocks in the category/minimum section of the IAP chart that you'll be using.

Aircraft categories and descent minimums

The FAA established categories based upon aircraft approach speeds to make primary minima as safe as possible. Here are the actual definitions according to the pilot/controller glossary:

AIRCRAFT APPROACH CATEGORY—A grouping of aircraft based on a speed of 1.3 times the stall speed in the landing configuration at maximum gross landing weight. An aircraft shall fit in only one category. If it is necessary to maneuver at speeds in excess of the upper limit of a speed range for a category, the minimum for the next higher category should be used. For example, an aircraft [that fits] in Category A, but is circling to land at a speed in excess of 91 knots, should use the approach Category B minimums when circling to land. The categories are as follows:

1. Category A—speed less than 91 knots.

2. Category B—speed 91 knots or more, but less than 121 knots.

3. Category C—speed 121 knots or more, but less than 141 knots.

4. Category D—speed 141 knots or more but less than 166 knots.

5. Category E—speed 166 knots or more.

You'll notice the visibility requirements increase as the category speeds increase. That is because the turn radius is greater at the higher speeds and the aircraft will need more room and consequently more visibility to maneuver in.

Two descent minimums, the DH and the MDA, should be defined:

DECISION HEIGHT/DH—With respect to the operation of aircraft, means the height at which a decision must be made during an ILS, MLS, or PAR instrument approach to either continue the approach or to execute a missed approach.

MINIMUM DESCENT ALTITUDE/MDA—The lowest altitude, expressed in feet above mean sea level, to which descent is authorized on final approach or during circle-to-land maneuvering in execution of a standard instrument approach procedure where no electronic glide slope is provided.

The DH is a go-no-go altitude, and you have absolutely no choice in the matter. If you can see enough of the runway environment upon reaching the DH to continue and land *safely*, you do so. If you don't see the runway environment you go around—*immediately*. What constitutes the runway environment is explained in FAR 91.175 below.

The MDA affords a little more time to sort things out. With no electronic glide slope, you will reach the MDA at an uncertain distance from the runway. Once at the MDA, you may fly out the time to the MAP (*missed approach point*). You cannot descend below the MDA until the aircraft is in a position to make a normal approach to the runway. This is spelled out quite clearly in FAR 91.175, Takeoff and landing under IFR (in part):

> *(c) Operation below DH or MDA. Where a DH or MDA is applicable, no pilot may operate an aircraft . . . at any airport below the authorized MDA or continue an approach below the authorized DH unless—*
>
> *(1) The aircraft is continuously in a position from which a descent to a landing on the intended runway can be made at a normal rate of descent using normal maneuvers. . . .*
>
> *(2) The flight visibility is not less than the visibility prescribed in the standard instrument approach procedure being used;*
>
> *(3) . . . At least one of the following visual references for the intended runway is distinctly visible and identifiable to the pilot:*
>
> *(i) The approach light system, except that the pilot may not descend below 100 feet above the touchdown zone elevation using the approach lights as a reference unless the red terminating bars or the red side row bars are also distinctly visible and identifiable.*
>
> *(ii) The threshold.*
>
> *(iii) The threshold markings.*
>
> *(iv) The threshold lights.*
>
> *(v) The runway end identifier lights.*
>
> *(vi) The visual approach slope indicator.*
>
> *(vii) The touchdown zone or touchdown zone markings.*
>
> *(viii) The touchdown zone lights.*
>
> *(ix) The runway or runway markings.*
>
> *(x) The runway lights; and*
>
> *(d) Landing. No pilot operating an aircraft . . . may land that aircraft when the flight visibility is less than the visibility prescribed in the standard instrument approach procedure being used.*
>
> *(e) Missed approach procedures. Each pilot operating an aircraft . . . shall immediately execute an appropriate missed approach procedure when either of the following conditions exist:*

(1) Whenever the requirements of paragraph (c) of this section are not met at either of the following times:

(i) When the aircraft is being operated below MDA; or

(ii) Upon arrival at the missed approach point, including a DH where a DH is specified and its use is required, and at any time after that until touchdown.

(2) Whenever an identifiable part of the airport is not distinctly visible to the pilot during a circling maneuver at or above MDA, unless the inability to see an identifiable part of the airport results only from a normal bank of the aircraft during the circling approach.

That certainly spells things out clearly enough. Sometimes it might take a few readings to get it all set in your mind, but these are some of the most important things to remember about an instrument approach. Failure to comply with these paragraphs has cost us a lot of lives and a lot of aircraft.

23

Mastering minimums

Before we continue leafing through the manuals, there are two general definitions to review: visibility and visual approach. Additionally, there is a difference between ground and in-flight visibility.

Visibility

Visibility on the ground is determined by the human eye or by an instrument known as a transmissometer, which is a device that measures the visibility by determining the amount of light that passes through the atmosphere.

The prevailing visibility is recorded by a trained observer at or near ground level. He looks at objects that are at known distances from his vantage point: trees, buildings, smokestacks, mountains, and radio and TV broadcast antennas; at night the observer looks at lights in the same manner. Prevailing visibility is reported in miles and fractions of miles:

> *Prevailing Visibility—The greatest horizontal visibility equaled or exceeded throughout at least half the horizon circle which need not necessarily be continuous.*

A transmissometer measures two types of visibility:

> *Runway Visibility Value/RVV—The visibility determined for a particular runway by a transmissometer. A meter provides a continuous indication of the visibility (reported in miles or fractions of miles) for the runway. RVV is used in lieu of prevailing visibility in determining minimums for a particular runway.*

> *Runway Visual Range/RVR—An instrumentally derived value, based on standard calibrations, that represents the*

horizontal distance a pilot will see down the runway from the approach end. It is based on the sighting of either high-intensity runway lights or on the visual contrast of other targets, whichever yields the greater visual range. RVR, in contrast to prevailing or runway visibility, is based on what a pilot in a moving aircraft should see looking down the runway. RVR is horizontal visual range, not slant visual range. It is based on the measurement of a transmissometer made near the touchdown point of the instrument runway and is reported in hundreds of feet. RVR is used in lieu of RVV and/or prevailing visibility in determining minimums for a particular runway.

A very important thing to note here is that both RVV and RVR values are for specific runways, and it is sometimes possible to make an approach to another runway when the prevailing visibility is higher. You will find this occurring at airports near the coast, where fog might roll in and cover only part of one runway, and that might be the runway with the transmissometer(s).

It's important for you to know that because the RVR and the human visibility measurements are horizontal range, the visibility might be more or less than what you'll see using slant range from the cockpit. Normally the visibility reported from the ground will be slightly better than what you'll find looking from the windshield on approach.

Visual approach

Occasionally, while under radar control, you might be cleared for a visual approach. Be careful. A visual approach is defined in the pilot/controller glossary:

Visual Approach—An approach conducted on an instrument flight rules (IFR) flight plan which authorizes the pilot to proceed visually and clear of clouds to the airport. The pilot must, at all times, have either the airport or the preceding aircraft in sight. This approach must be authorized, and under the control of the appropriate air traffic control facility. Reported weather at the airport must be ceiling at or above 1,000 feet and visibility of 3 miles or greater.

The trap is the phrase "to proceed visually and clear of clouds." If you enter a cloud, or an area where the conditions revert to IMC *(instrument meteorological conditions)*, you are in violation. So, it pays to look well ahead before you accept a visual approach. You might be able to see and identify the aircraft you are instructed to follow, but look all the way along the approach path, and if there are clouds that must be penetrated, you cannot legally accept the visual approach clearance. It is a useful tool, though, in helping avoid long drawn-out approaches, and it eases the burden on ATC. There is one other possible trap. The *ATC Handbook* amplifies the visual approach:

> *a. When it will be operationally beneficial, ATC may authorize an aircraft to conduct a visual approach to an airport or to follow another aircraft when flight to, and landing at, the airport can be accomplished in VFR weather. The aircraft must have the airport or the identified preceding aircraft in sight before the clearance is issued. If the pilot has the airport in sight but cannot see the aircraft he is following, ATC may still clear the aircraft for a visual approach; however, ATC retains both separation and wake vortex separation responsibility. When visually following a preceding aircraft, acceptance of the visual approach clearance, constitutes acceptance of pilot responsibility for maintaining a safe approach interval and adequate wake turbulence separation.*

Consider this example. Approach control is working four other aircraft. You are told that you are following a Hawaiian Airlines L-1011 at 12 o'clock, 6 miles. You see traffic at your 12 o'clock position, or perhaps 1 o'clock, and it seems to be about the right distance. But before you say, "Cessna 20Z has the L-1011," are you actually sure that it is an L-1011? Is it actually inbound? Is it a Hawaiian plane? Are there any clouds in the vicinity that might block it from your view before you land? You should only acknowledge the traffic if you can identify it positively without a doubt in your mind and be able to keep it in sight from then on until you land. Remember, once you are cleared for a visual approach based on your identifying a preceding aircraft, are told to follow it, and you accept the clearance, you will be responsible for your own wake turbulence separation. Be careful.

I think this is one of the most common mistakes of pilots with whom I used to fly. They seem to be in such a hurry to please ATC that they supposedly positively identify preceding aircraft, even at night, when a dozen other aircraft or lights surround the airport area. When a DC-9 collided with a Cessna approaching San Diego airport some years ago, one of the last sentences on the voice recorder of the DC-9 was that they had been looking at the wrong plane. Don't let this happen to you, no matter how big a hurry you are in to get to the field and no matter how much you'd like to please ATC.

DON'T SAY YOU HAVE A PRECEDING AIRCRAFT IN SIGHT UNLESS YOU CAN IDENTIFY IT POSITIVELY.

Recently the *ATC Handbook* has further amplified these procedures. Sections d and g are identical. One of them is probably an oversight by the authors, but then again, ATC might consider it as being so important that it bears repeating.

These sections state: "Authorization to conduct a visual approach is an IFR authorization and does not alter IFR flight plan cancellation responsibility. . . ." So, be sure to remember to cancel your IFR flight plan after you land. ATC will not do it for you automatically.

There are a few more amplifications to the rules, so you should be cognizant of them. Section e. says:

> *A visual approach is not an IAP and therefore has no missed approach segment. If a go-around is necessary for any reason, aircraft operating at controlled airports will be issued an appropriate advisory/clearance/instruction by the tower. At uncontrolled airports, aircraft are expected to remain clear of clouds and complete a landing as soon as possible. If a landing cannot be accomplished, the aircraft is expected to remain clear of clouds and contact ATC as soon as possible for further clearance. Separation from other IFR aircraft will be maintained under these circumstances.*

This really gives us another reason to be very careful about accepting a visual approach. Even though IFR separation will be provided if you cannot land, the responsibility to remain clear of clouds while maneuvering remains with you.

Contact approach

In addition to the visual approach, there is one other time that you can deviate from the published approach procedure. Perhaps you are executing an approach that has a 1½-mile minimum visibility requirement. The tower informs you that the visibility has just dropped to 1 mile. Also suppose that you are below the base of the clouds and can see a highway below. You identify the highway and know that it runs alongside the runway.

Because the field has just dropped below minimums, you may either execute a missed approach when you reach the MAP or you may request a *contact approach*. From the *ATC Handbook*:

a. Pilots operating in accordance with an IFR flight plan, provided they are clear of clouds and have at least 1 mile flight visibility and can reasonably expect to continue to the destination airport in those conditions, may request ATC authorization for a contact approach.

b. Controllers may authorize a contact approach provided:

(1) The contact approach is specifically requested by the pilot. ATC cannot initiate this approach.

(2) The reported ground visibility at the destination airport is at least 1 statute mile.

(3) The contact approach will be made to an airport having a standard or special instrument approach procedure.

(4) Approved separation is applied between aircraft so cleared and between these aircraft and other IFR or special VFR aircraft.

c. A contact approach is an approach procedure that may be used by a pilot (with prior authorization from ATC) in lieu of conducting a standard or special IAP to an airport. It is not intended for use by a pilot on an IFR flight clearance to operate to an airport not having an authorized IAP. Nor is it intended for an aircraft to conduct an instrument approach to one airport .and then, when "in the clear," to discontinue that approach and proceed to another airport. In the execution of a contact approach, the pilot assumes the responsibility for obstruction clearance. If radar service is being received, it will automatically terminate when the pilot is told to contact the tower.

The contact approach can be very helpful to the instrument pilot. Remember, though, that in order to execute a contact approach, you must be on an IFR flight plan, you can only use the approach to fly to the destination airport as filed on the flight plan, you cannot use it without ATC authorization, and you are fully responsible for obstruction clearance.

Many airports also have DF *(direction finding)* instrument procedures, but these are only for use when the pilot has declared a distress or urgency condition, such as would happen when navigational capabilities are lost.

Straight-in and circling minimums

Straight-in and/or circling minimums are published for most runways; circling minimums are always higher than the straight-in minimums. What constitutes a straight-in or circling approach presents a problem at times. I have had a number of students ask me why a certain airport will have circling minimums when the final approach course is on the same heading as the runway (from the *ATC Handbook*):

> *(b) Straight-in minimums—Straight-in minimums are shown on IAP charts when the final approach course of the IAP is within 30 degrees of the runway alignment and a normal descent can be made from the IFR altitude shown on the IAPs to the runway surface. When either the normal rate of descent or the runway alignment factor of 30 degrees is exceeded, a straight-in minimum is not published and a circling minimum applies. The fact that a straight-in minimum is not published does not preclude the pilot from landing straight-in if he has the active runway in sight and has sufficient time to make a normal approach for landing. Under such conditions and when ATC has cleared him for landing on that runway, he is not expected to circle even though only circling minimums are published. If he desires to circle he should advise ATC.*

> *(d) Circling Minimums—The circling minimums . . . provide adequate obstruction clearance and the pilot should not descend below the circling altitude until the aircraft is in a position to make final descent for landing. . . .*

(1) Maneuver the shortest path to the base or downwind leg as appropriate, under minimum weather conditions. There is no restriction from passing over the airport or other runways.

(2) It should be recognized that many circling maneuvers may be made while VFR or other flying is in progress at the airport. Standard left turns or specific instructions from the controller for maneuvering must be considered when circling to land.

(3) At airports without a control tower, it may be desirable to fly over the airport to determine wind and turn indicators, and to observe other traffic. . . .

Circling approach

It is important to recognize that if you sight the runway too late to make a normal descent to landing, you are required to circle. When circling, you cannot descend below the circling altitude until you are in position to make a normal landing, usually when you turn final. A lot of good pilots have crashed while executing a circling approach because they went too low and hit one of the many obstructions found in the vicinity of most airports, or they developed vertigo when making a tight turn close to the ground and stalled or spun in.

Remember the exception to the aircraft category that puts the aircraft in a higher than normal category if the maneuvering speed in a circling approach is higher than the 1.3 V_{SO} stall speed normally used on a straight-in approach because the radius of turn will be greater with the higher speed. The obstruction clearance at the slower speed and tighter radius might not be adequate for the higher speed; selected visibility requirements increase dramatically as the category of the aircraft increases.

I remember watching a Convair 440 break out of the clouds years ago. He was just over the end of a 6,000-foot runway. The pilot, not wanting to take a wave-off and being below circling minimums, tried to land. The descent was too steep and the airspeed was too high. He was using up way too much runway, and he was determined to get it on the ground, so he pushed the nose over. Halfway down the runway the nosewheel hit and collapsed, and the aircraft slid to a stop about 50 feet from a 300-foot dropoff. Fortunately no one was injured, but the aircraft suffered considerable damage, the field was closed for two days, and an 18-year veteran pilot lost a job. Instrument flying is no time to cut corners.

Figures 23-1 through 23-4 are examples of circling maneuvers. In these maneuvers, the turns are executed using outside visual references and instruments. Keep a close eye on altitude, airspeed, and angle of bank.

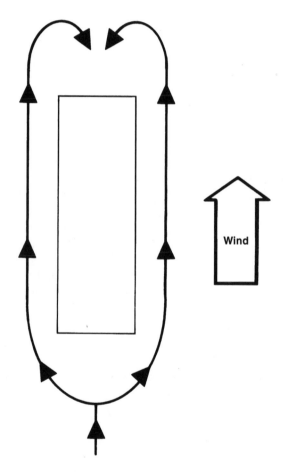

Fig. 23-1. *You are making an approach downwind and must circle to land into the wind. Once you sight the airport, you break right or left to make a standard downwind leg. This way you should never lose sight of the field.*

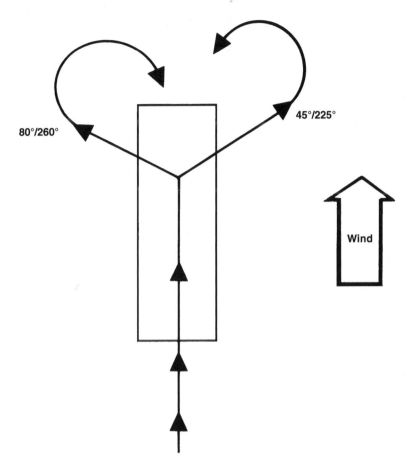

Fig. 23-2. *In this approach, you either sight the field too late to maneuver as shown in Fig. 23-1, or the weather conditions along the normal downwind legs will not allow you the latitude to circle in that manner. In this case, fly directly down the runway (at the circling MDA), and upon reaching the approach end, execute any type of procedure turn that will keep you from losing visual reference with the runway (except when banking).*

Runway not in sight

This brings us down, so to speak, to the *missed approach*. There are three situations in which a missed approach is mandatory. The first is when you reach the DH or MAP depicted on the IAP chart and do not have sufficient visual reference to complete the landing. The second is any time that visual reference to the airport is lost after reach-

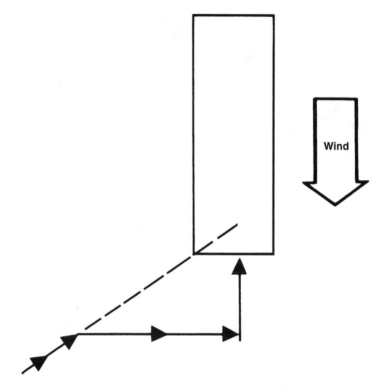

Fig. 23-3. *This shows your most desired break-off from a circling approach. You must sight the runway in time to maneuver normally onto the extended centerline and execute a normal landing.*

ing the DH or MDA. The third is when visual reference to the airport is lost while executing a circling approach.

Whenever you execute a missed approach, you must either comply with the published missed approach procedure as printed on the chart or follow directions from ATC if directed to deviate from the published procedure. If you lose visual reference while executing a circling approach, initiate the missed approach by making a climbing turn toward the landing runway. Continue the turn until established on the proper missed approach procedure.

You should have noticed by now that you'll be pretty busy flying the procedure, and the actual mechanics of flying the aircraft must be practically automatic.

It will be a good idea to practice the maneuvers suggested before—straight and level, climbs, and descents—with and without turns and

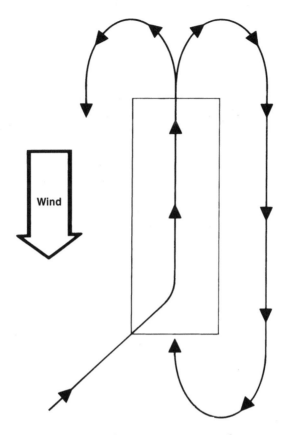

Fig. 23-4. *If you do not see the runway in time, but are approaching into the wind, you should fly down the runway to the departure end, and turn left or right to execute normal downwind and base legs.*

at normal speeds and slow or maneuvering speeds. This time, special emphasis should be placed on the proper use of trim.

Anytime the aircraft is established in a new attitude, all possible control pressures should be relieved by proper trim. Remember that to use trim properly, you must first apply control pressure to establish the new attitude, then use the trim tabs so that the aircraft will maintain the required flight path practically hands off. This will create time to take quick looks at approach plates, retrieve an errant plate off the floor, or jot down a new clearance without the aircraft wandering around. It will also allow you to change attitudes quickly, such as going from descent to a missed approach climb without having to impart excessive pressures to the controls.

24

Approach basics and NDB approaches

One of the most satisfying things you will experience as an instrument pilot is breaking out at minimums on an IFR approach with the runway straight ahead and a comfortable landing assured. The average VFR pilot—and most passengers—considers this nothing short of miraculous!

But it isn't, really. By the time you begin to polish the fine points of IFR approaches, you will have learned to use the ADF and VOR equipment with great precision. And, if your instructor has been following the syllabus, you will have made several unhooded instrument approaches—enough to see the "big picture" of an instrument approach.

This chapter covers the basics of approaches. I will show how to analyze an approach while planning for an IFR flight and move step by step through representative examples of ADF approaches.

Understand, however, that every approach is different. The examples will illustrate the usual sequence of events in nonprecision approaches plus the dialogue with approach control that accompanies this representative sequence. You will be able to apply the approach procedures, techniques, and communications to all nonprecision approaches. But always remember that approaches in the real world of IFR will differ from these representative examples. Adjust your procedures and communications accordingly.

Nonprecision approaches

A nonprecision approach is defined in the Pilot/Controller Glossary as "a standard instrument approach procedure in which no electronic glide slope is provided." ADF, VOR, DME, and several less common types of approaches fall into the nonprecision category because they do not provide electronic glide slopes. The only electronic guidance they provide is for the approach course.

Precision approaches

A precision approach, on the other hand, is defined in the Pilot/Controller Glossary as "a standard instrument approach procedure in which an electronic glide slope/glide path is provided." An ILS provides an electronic glide slope and is thus a precision approach. Precision approach radar (PAR) depicts on the radarscope an electronic glide path along which the airplane is guided by the final approach controller. In the military, this is known as GCA, or "ground controlled approach." Nonprecision and precision approaches have many elements in common.

Common elements

First and foremost, all instrument approaches have an altitude below which you cannot legally descend unless the airplane is in a position to make a safe landing. This altitude is called *minimum descent altitude* (MDA) for nonprecision approaches; *decision height* (DH) for precision approaches. Other common elements include an *initial approach fix* (IAF), a *final approach fix* (FAF), a final approach course, a missed approach procedure, and very often one or more intermediate fixes between the IAF and the FAF.

The term *segment* is used frequently. Here is what the different segments mean:

Initial approach segment. The segment between the IAF and an intermediate fix, or between the IAF and the point where the airplane is established on an intermediate course or the final approach course.

Intermediate approach segment. The segment between the IAF and the FAF.

Final approach segment. The segment between the FAF and MAP (missed approach point).

Missed approach segment. The segment between the MAP, or arrival at the DH, and the missed approach holding fix.

Altitude minimums

Let's explore the question of how altitude minimums are derived. The most important consideration, for obvious reasons, is safe obstacle clearance. This is spelled out in the FAA's *United States Standard for Terminal Instrument Procedures* (TERPS), which is the "bible" on instrument approach tolerances. Along the centerline of the approach course, minimum obstacle clearance is provided for nonprecision approaches as follows:

- NDB located on airport, 350 feet
- NDB off airport with FAF, 300 feet
- VOR located on airport, 300 feet
- VOR off airport, 250 feet
- DME arc as final approach course, 500 feet
- Localizer, 250 feet
- ASR radar (no glide path), 250 feet
- DF steer approach, 500 feet

Obstacle clearance is provided for a "primary" area on either side of the final approach course centerline. The width of the primary area varies with the type of approach and distance from the field. But it is never less than 1 mile on either side of the final approach course centerline of an NDB, VOR, or other nonprecision approach.

These are comfortable obstacle clearances for the precise pilot, but there is not much room for error or sloppy procedures. Coming in from the FAF on an NDB approach, for example, you will clear obstacles a mile on either side of the inbound course by only 350 feet.

One of the most frequent reasons for failing the instrument flight test is going below minimums on an approach. I don't mean just momentarily dipping below a minimum because of turbulence, then correcting right away. What always surprises me are the candidates for an instrument rating who consistently fly 25, 50, or even 100 feet below minimums without taking corrective action, or have not determined the correct minimums to begin with.

Minimums are so basic, yet many pilots seem to have problems with them. Why is this so? I believe it is because pilots do not always use a systematic procedure to analyze the minimums.

Adjustments to MDA

Let's make a step-by-step analysis of a conventional nonprecision approach. If you follow these steps every time you plan a flight, you will develop the good habits that will enable you to quickly size up an unfamiliar approach that might be assigned by ATC at the last minute, perhaps due to a runway change. The example for this exercise is the NDB RWY 26 approach at Pittsfield, Massachusetts, an uncontrolled field in the Berkshire hills of western New England (Fig. 24-1). Let's analyze the Pittsfield approach using a systematic, six-step method. As you will see, this is going to take some detective work.

Fine print

(1) *Read the fine print first.* Don't leave the fine print until last because you might miss something very important. Consider items A and B on the Pittsfield RWY 26 approach chart:

   A. "Inoperative table does not apply. Circling not authorized south of runways 8 and 32."

   B. "Obtain local altimeter setting on CTAF; when not received, use North Adams altimeter setting."

In item A, "inoperative table does not apply" means that the airport has no approach components whose outage would require higher minimums. I'll have more to say about the inoperative components table later in this chapter. But a statement like this should alert you to the fact that the airport has only minimal lighting. Check the lighting information in the airport box at the lower right of the chart and note what lighting there is. Would this be sufficient in terms of your "personal minimums" for an actual IFR approach down to minimums? How about at night?

The second sentence in item A is very important: "Circling not authorized south of runways 8 and 32." This means that obstacle clearance is not provided in this sector at circling minimums. Stray into this area and you might hit something!

Let's turn now to item B, which presents more of a challenge.

You must have an accurate altimeter setting for every instrument approach. More and more uncontrolled airports, including Pittsfield, have an automatic weather reporting system, either "ASOS" or "AWOS." ASOS stands for automated surface observing system. AWOS is the acronym for automated weather observing system.

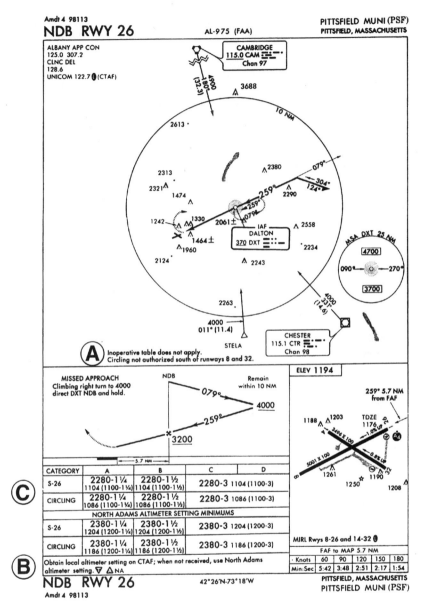

Fig. 24-1. *A typical NDB approach is NDB RWY 26 at Pittsfield, Massachusetts.*

These systems provide highly accurate altimeter settings along with other weather information of concern to incoming pilots. Use ASOS and AWOS information as you would ATIS. Detailed information on these automated systems is found in AIM.

As noted in item B, altimeter settings for Pittsfield are provided through the common traffic advisory frequency (CTAF). But when this is not available, you must get an altimeter setting from North Adams and use higher minimums.

Here is where the detective work comes in. North Adams has no instrument approach, so there is no handy approach chart giving the North Adams frequency for weather information. You must look up North Adams under Massachusetts in the AF/D. There you will find the frequency on which the ASOS information is broadcast. It is 134.775.

Finally, you should look up and learn the meaning of the triangular symbols that follow the fine print on altimeter settings.

The information contained in the fine print for the NDB RWY 26 approach to Pittsfield must be ferreted out on the ground when you plan the flight. Can you imagine what it would be like trying to look up all these things in flight?

And the fine print can be very important. Here is a "gem" from the instrument approach to Indian Mountain Air Force Station, Alaska, as reported by Barry Schiff in *AOPA Pilot*:

"CAUTION: Rwy located on slope of 3,425' mountain...successful go around improbable."

Take a few minutes to browse through the approach charts for your area and highlight the fine print at the airports you are likely to use. There might be some surprises; mark them and they won't surprise you during an approach.

Height of obstacles

(2) *Check the height of obstacles in the vicinity of the airport.* These obstacles determine the MDA. Note how many obstacles rise above 2,000 feet MSL in the vicinity of Pittsfield. Check the airport diagram at lower right to see how many rise above the touchdown zone elevation (TZDE) of 1,176 feet MSL for runway 26. They do not appear to be a problem at Pittsfield.

Aircraft approach category

(3) *Pick the published minimum for your aircraft category and type of approach,* either straight-in or circling (C). Aircraft approach cate-

gories are explained in the front section of every set of NOS approach charts. The explanation is clear and simple:

"Speeds are based on 1.3 times the stall speed in the landing configuration of maximum gross landing weight. An aircraft shall fit in only one category. If it is necessary to maneuver at speeds in excess of the upper limit of a speed range for a category, the minimums for the next higher category should be used. For example, an aircraft which falls in Category A but is circling to land at a speed in excess of 91 knots, should use the approach Category B minimums when circling to land. See following category limits:"

Maneuvering Table

Approach Category	A	B	C	D	E
Speed (Knots)	0–90	91–120	121–140	141–165	Abv 165

Most general aviation propeller-driven airplanes fall in either Categories A or B. Categories A and B are the same at Pittsfied (Fig. 24-1), but this is not always the case.

Straight-in vs. circling. MDAs for each category are further classified by the type of approach, either straight-in ("S-26" at Pittsfield) or circling. Straight-in approaches are allowed when the angle of convergence between the final approach course and the extended runway centerline does not exceed 30° If the angle is greater than 30°, you must use circling minimums. Note that the pilot does not make the decision as to whether an approach is straight-in or not. Yes, you may break off a straight-in approach and circle to land on another runway (using the higher minimums), but the designation of an approach as straight-in or circling is based upon the layout of the airport, the angle between the final approach course and the landing runway, the location of the electronic facilities, and the design of the instrument approach.

With these points in mind, you can establish the basic minimums for each category airplane for Categories A and B for the NDB approach at Pittsfield:

The numbers mean:

- 2280-1 $^1/_4$ are MDAs and the minimum visibilities for both straight-in and circling approaches (using the local altimeter setting).

- 1104 and 1086 are the heights above the airport (HAA) at the MDA.

- (1104-1 $\frac{1}{4}$) and (1104-1 $\frac{1}{2}$) are military minimum ceilings and visibilities and are not applicable to civilian aircraft.

Inoperative components

(4) *Check inoperative component changes in minimums.* If any component of an approach listed on this table (Fig. 24-2) is out of service, the minimums might have to be increased. The table is published on the inside front cover of every set of NOS approach charts. Definitions and descriptions of MM, ALSF, MALSR, etc., are in the front section of the NOS sets on page L1 entitled "Approach Lighting System—United States."

Nonprecision approach visibility minimums increase $\frac{1}{4}$ and $\frac{1}{2}$ mile when certain approach lights and runway lights are inoperative. Check the lighting legend in the front section of the NOS approach chart sets (Fig. 24-3) against the airport diagram to see if the airport has any lighting systems affected by the inoperative components table.

This looks a little intimidating at first, but an instructor can help you sort things out. If you make a habit of checking the destination against the inoperative components table and the lighting legends every time you file IFR, you will soon be able to handle this problem quickly and easily. You will also broaden your understanding of the roles played by the various components, and what the wide variety of approach and runway lights look like.

The inoperative components check for Pittsfield reveals no lights affected by the "Inoperative Components Table." The minimums remain at 2280-1 $\frac{1}{4}$ for both straight-in and circling approaches regardless of lighting.

Approach adjustments

(5) *Make adjustments required by the fine print.* As noted earlier, here is where Pittsfield throws a zinger at the unwary pilot. MDAs must be increased 100 feet at Pittsfield if a local altimeter setting is not available and the North Adams setting is used as a substitute.

Pittsfield should provide the altimeter setting on the CTAF, 122.7. If the Pittsfield altimeter setting is not available for any reason, the fine

INOP COMPONENTS
97198

INOPERATIVE COMPONENTS OR VISUAL AIDS TABLE

Landing minimums published on instrument approach procedure charts are based upon full operation of all components and visual aids associated with the particular instrument approach chart being used. Higher minimums are required with inoperative components or visual aids as indicated below. If more than one component is inoperative, each minimum is raised to the highest minimum required by any single component that is inoperative. ILS glide slope inoperative minimums are published on instrument approach charts as localizer minimums. This table may be amended by notes on the approach chart. Such notes apply only to the particular approach category(ies) as stated. See legend page for description of components indicated below.

(1) ILS, MLS, and PAR

Inoperative Component or Aid	Approach Category	Increase Visibility
ALSF 1 & 2, MALSR, & SSALR	ABCD	¼ mile

(2) ILS with visibility minimum of 1,800 RVR.

ALSF 1 & 2, MALSR, & SSALR	ABCD	To 4000 RVR
TDZI RCLS	ABCD	To 2400 RVR
RVR	ABCD	To ½ mile

(3) VOR, VOR/DME, VORTAC, VOR (TAC), VOR/DME (TAC), LOC, LOC/DME, LDA, LDA/DME, SDF, SDF/DME, GPS, RNAV, and ASR

Inoperative Visual Aid	Approach Category	Increase Visibility
ALSF 1 & 2, MALSR, & SSALR	ABCD	½ mile
SSALS, MALS, & ODALS	ABC	¼ mile

(4) NDB

ALSF 1 & 2, MALSR & SSALR	C	½ mile
	ABD	¼ mile
MALS, SSALS, ODALS	ABC	¼ mile

CORRECTIONS, COMMENTS AND/OR PROCUREMENT

FOR CHARTING ERRORS CONTACT:	FOR CHANGES, ADDITIONS, OR RECOMMENDATIONS ON PROCEDURAL ASPECTS:	TO PURCHASE CHARTS CONTACT:
National Ocean Service/NOAA N/ACC1, SSMC-4, Sta. #2335 1305 East-West Highway Silver Spring, MD 20910-3281 Telephone Toll-Free (800) 626-3677 Internet/E-Mail: Aerochart@NOAA.GOV	Contact Federal Aviation Administration, ATA 110 800 Independence Avenue, S.W. Washington, D.C. 20591 Telephone Toll-Free (800) 457-6656	National Ocean Service NOAA, N/ACC3 Distribution Division Riverdale, MD 20737 Telephone (800) 638-8972

Requests for the creation or revisions to Airport Diagrams should be in accordance with FAA Order 7910.4B.

INOP COMPONENTS
97198

Fig. 24-2. *An inoperative components table is found in the front section of every set of NOS instrument approach procedures.*

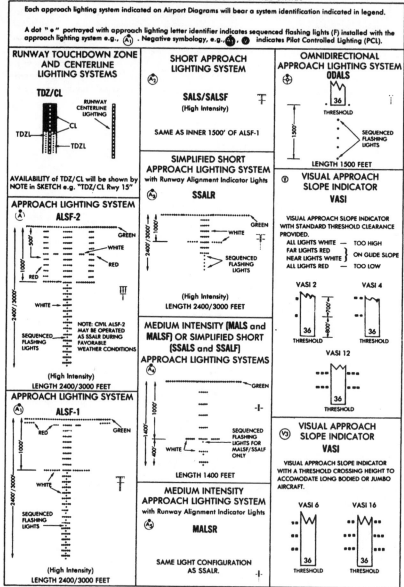

Fig. 24-3. *Approach lighting system codes and descriptions found in the front section of every set of NOS instrument approach procedures charts.*

print will apply; obtain the North Adams setting on 134.775 and use the North Adams limits.

It is a good rule of thumb in your flight planning to automatically add the difference required by alternate altimeter settings (100 feet in this case). If it turns out that you can get a local altimeter setting, it will be a simple matter to glance at the approach chart and drop down to the lower MDA. Better to add the difference in the quiet of the planning room than fumble around for the correct NMA during the approach!

Altimeter error

(6) *Add the altimeter error.* For reasons discussed earlier, always *add* the altimeter error, regardless of whether it is plus or minus. For purposes of illustration, you find an error of 30' when you check the altimeter. Add the altimeter error of 30' to the 100' adjustment if no local altimeter is available. This would yield adjusted MDAs of 2410-1 $^1/_4$ for both straight-in and circling MDAs at Pittsfield.

To summarize the step-by-step method of analyzing minimums:

1. *Read the fine print*
2. *Check the height of obstacles*
3. *Pick the correct minimums for airplane category and type of approach*
4. *Check adjustments for inoperative components table*
5. *Make adjustments required by fine print*
6. *Add the altimeter error*

Operation below MDA

I urge students to fully analyze the MDAs for the approaches they expect to make because a pilot *cannot descend below an MDA at any time during a nonprecision approach unless certain very specific requirements are met* as prescribed in FAR 91.175 (c). The regulation can be summarized:

No pilot may operate an aircraft below the authorized MDA (or continue an approach below the DH) unless:

(1) The aircraft is continuously in a position from which a descent to a landing on the intended runway can be made at a normal rate of descent using normal maneuvers.

(2) The flight visibility is not less than that prescribed for the approach being used.

(3) At least one of the following visual references for the intended runway is distinctly visible and identifiable to the pilot:

   (i) The approach light system, including the red terminating bars or the red side row bars.

   (ii) The landing threshold.

   (iii) The threshold markings.

   (iv) The threshold lights.

   (v) The runway end identifier lights (REIL).

   (vi) The visual approach slope indicator (VASI).

   (vii) The touchdown zone or touchdown zone markings.

   (viii) The touchdown zone lights (TDZL).

   (ix) The runway or runway markings.

   (x) The runway lights.

Visibility minimums required for landing

Even if you can see the runway (or one of the other visual references listed above) as you approach the field at MDA, you may not legally make a landing if the visibility is less than that prescribed for the instrument procedure being used. This is the regulation, and it is stated in FAR 91.175 (d).

Visibility, not ceiling, determines whether or not you can land. MDA establishes the altitude below which you cannot descend unless you have one of the prescribed references in sight. Visibility tells you whether or not you can legally land when you have one of those prescribed references in sight.

Visibility is expressed in miles and fractions of a mile or in feet of runway visual range (RVR). (See Appendix C, Glossary for complete definitions for visibility.) Visibility is the prevailing horizontal visibility near the surface as reported by an accredited observer. ATC tower controllers are qualified to report visibility.

RVR is measured by a transmissometer located alongside a runway. If a runway has a transmissometer, the visibility minimums listed on the approach chart will be expressed as a two-digit figure representing feet of RVR. If there is no transmissometer, the visibility will simply be expressed in miles, as is the case with the Pittsfield NDB approach.

If the transmissometer is out of service, the published RVR minimums must be converted to miles and fractions of a mile according to the table in the front section of the NOS approach chart sets (Fig. 24-4).

Missing RVR also increases the minimum visibility for some precision ILS approaches, as noted on the inoperative components table (Fig. 24-2).

If making an approach at an uncontrolled field, the pilot must decide if the visibility meets the requirements. Check the length of the landing runway, which is a good reference for estimating visibility: a statute mile long (5,280 feet); or half a mile (2,640 feet); or a mile and a half (7,920 feet). If you can see to the end of a mile-long runway when the descent begins, you may legally land when the visibility minimum is one mile. But if you can only see partway down that runway, a landing might be illegal, and you should execute a missed approach. (This is a subject of much controversy and misunderstanding in the aviation community.)

If you are approaching a controlled field, the tower will inform you of the visibility. When it drops below the prescribed visibility minimums, the runway involved—or the entire airport—might be closed to landing traffic.

Missed approach landing

The *missed approach point* (MAP) is no time to fumble for the approach chart and try to figure out what to do next. All attention must be riveted on controlling the airplane during the first few moments of a missed approach—add full power, stop the descent and initiate a climb, raise the flaps and gear, and maintain a steady course.

RVR/Meteorological Visibility Comparable Values

The following table shall be used for converting RVR to meteorological visibility when RVR is not reported for the runway of intended operation. Adjustment of landing minima may be required — see Inoperative Components Table.

RVR (feet)	Visibility (statute miles)	RVR (feet)	Visibility (statute miles)
1600	¼	4000	¾
2000	⅜	4500	⅞
2400	½	5000	1
3200	⅝	6000	1¼

Fig. 24-4. *Table for converting RVR to miles and fractions found in the front section of every set of NOS instrument approach procedures.*

Diverting attention to the fine print of the missed approach procedure at this time could start a chain of events leading to a collision with an obstacle or the ground. While planning the flight, always assume that you will be required to make a missed approach and plan accordingly. Review the procedure again en route, before making the approach.

So plan ahead. I find that students have little trouble coping with a missed approach if they break it into five phases:

1. *Transition* to a stabilized climb. You add full power, stop the descent, raise the flaps in increments, get the gear up, and initiate a normal climb straight ahead. (Or start a turn if directed to in level flight at minimum controllable airspeed. You have been practicing minimum controllable airspeed under the hood; now all that practice becomes very valuable.)

2. *Climb.* Do you climb straight ahead or make a climbing left turn or a climbing right turn? The missed approach for Pittsfield prescribes a climbing right turn (Fig. 24-1). What is the level-off altitude? At Pittsfield it is 4,000 feet.

3. *En route* to the holding fix. What is the holding fix? Is it a facility you already have tuned in, as at Pittsfield? Or is it a VOR fix that might require resetting frequencies and OBS numbers? Do you proceed direct? Do you have to intercept a bearing or radial to get to the fix?

4. *Holding.* What type of pattern entry will you use? What outbound heading do you turn to when you reach the fix? Write it in big numbers on the approach chart.

5. *Departure* from holding. Plan for two alternatives: returning for another instrument approach or diverting to the filed alternate. When approach control asks "What are your intentions?" have your mind made up and respond promptly what you intend to do, including an abbreviated flight plan with route and altitude to the alternate if that's what you decide to do. You cannot depart the missed approach holding pattern until cleared by ATC.

Once again, visualization is the key to success in working out the moves made on a missed approach. Visualization is also the key to success in making the basic approach. A good instrument approach, which always includes the missed approach procedure, begins the night before, along with your planning for the departure and en route phases. Mentally fly the approach step-by-step, or even better, walk through it by placing objects on the floor to simulate the airport and the approach and missed approach fixes.

As the final step in approach planning, run through a MARTHA check:

MA (Missed approach procedures)

R (Radios—nav and com frequencies and OBS settings)

T (Times from FAF to MAP)

H (Heading of final approach course)

A (Altitude of MDAs, adjusted as discussed above)

This abbreviated approach checklist will also come in handy in the air near the destination while preparing for the approach.

With a little practice, you will find that planning an approach takes far less time than reading about it!

NDB approaches

I always introduce students to NDB approaches before VOR or ILS approaches. This might come as a surprise, but it really shouldn't. The two lessons in the syllabus that precede approaches are devoted to ADF procedures; therefore, ADF is still fresh in the student's mind. And because VOR is the backbone of the federal airway system, most students start instrument training with far more VOR experience than ADF. So, I pay extra attention to ADF as we move through the course. By introducing NDB approaches before the others, I can make sure the student is skillful, confident, and comfortable with them. If NDB approaches are introduced later in the course, they might not get the attention they require.

Let's return to the NDB approach at Pittsfield (Fig. 24-1) and talk through the procedure one step at a time. Because this is the first approach discussed in detail in this book, I will also introduce material on approach control, communications, and flight procedures that apply not only to NDB, but also to approaches in general.

Radar vectors

In the real world of IFR, you will be handed off from the ATC center controller to the appropriate approach controller at a comfortable distance from your destination. You will frequently be cleared to a lower altitude just before or just after the handoff to approach control. Leaving an assigned altitude is one of the occasions for a *required* report whenever this occurs during an IFR flight. The readback to ATC will be like this:

"Cessna five six Xray contact approach control, descend to five, report leaving seven."

Remember to use the full call sign on initial contact with approach control. Approach control will give you an expect further clearance or expect approach clearance for use in case of lost communications. The time they give will also help you plan the approach. If you don't get a "further" time, request it.

Approach control will issue vectors to intercept the final approach course (259° at Pittsfield) 1–5 miles outside the final approach fix (DALTON NDB), where you will be "cleared for the approach." This will give you time to establish yourself on the final approach course before reaching the FAF, to slow to approach speed, and to prepare for the final descent and landing. (Sometimes, as at Pittsfield, the FAF and IAF are the same.)

Treat radar vectors as commands. They are issued as required to provide safe separation for incoming traffic; therefore, do not deviate from the headings and altitudes issued by approach control.

Sometimes it becomes necessary for ATC to vector you across the final approach course for spacing or other reasons. This is not unusual at busy airports with a mix of slow traffic and high-speed traffic. It is much easier to move you out of the way of a rapidly closing jet than to have the jet break off the approach. After the jet has passed, you will be vectored back to the final approach course with minimum disruption.

You will normally be informed when it becomes necessary to vector you across the final approach course. If you see that interception of the final approach course is imminent and you have no further instructions, question the controller. Simply give your call number and "final approach course interception imminent, request further clearance." You will be cleared either to complete the approach or to continue on present heading for separation from incoming traffic. Do not turn inbound on the final approach course unless you have received an approach clearance.

The full approach procedure

In the beginning of your intensive work on NDB approaches, skip the radar vectors and request the "full approach procedure" to become completely skilled in all the elements of the approach. Make this request for the full procedure on initial contact with approach control after the handoff from center. You will probably be cleared direct to the FAF and receive an expect further clearance time or expect approach clearance time.

If a direct course to the FAF is within 10° of the final approach course, go ahead and intercept the final approach course and proceed directly to the FAF. Approach control will expect you to do this and will clear you for the approach before reaching the FAF.

Procedure turns

If you can't line up with the final approach course and then proceed directly to the FAF, you will need to execute a *course reversal*. There are two ways of doing this—in a *procedure turn* or in a holding pattern.

At Pittsfield, the course reversal must be made in a procedure turn, as indicated by the arrowhead to the northeast extending out from the 079° bearing from Dalton NDB. Fly outbound on the 079° radial for one minute and make a 45° turn to the right as shown on the chart to a heading of 124°. Reset the OBS to the inbound course to the FAF, 259°. This is the beginning of the procedure turn, an easy, reliable method of course reversal that will return you to the inbound course with a minimum of corrections.

Fly outbound on the 124° heading for one minute, adjusting for the wind, then make a 180° turn to a heading of 304°. Intercept the inbound course using the bracketing procedure described in Chapter 10. Hold the 304° intercept heading until the needle is about three-quarters of the way from full-scale deflection toward the center, then begin a turn to the inbound course, 259°. Correct for the wind and establish a reference heading that will hold the inbound course to the station.

It should be noted that there is no "right" way to make a procedure turn. Nowhere is it written that you must use the 45° procedure published on the approach charts. All that is required is that somehow you must get turned around and headed back on the inbound course within the mileage limit published on the chart, usually 10 nm.

But it makes good sense to use the 45° procedure published on the approach charts. The 45° headings are printed on the chart; so there is no guesswork about headings. And the 45° method will enable you to intercept and get established on the inbound course quickly and easily.

Note that the fine print in the profile section of the Pittsfield NDB 26 approach chart says "Remain within 10 nm." This means that you must complete the procedure turn within 10 nautical miles of the NDB.

How far is 10 nm? Work it out on the circular slide rule while in the planning room. At a ground speed of 90 knots it takes 6 minutes 40 seconds to cover 10 nm; at 100 knots ground speed it takes 6 minutes exactly. You should complete the procedure before the times expire, depending on your ground speed. If you stray beyond the 10-nm radius, obstacle clearance is not guaranteed.

The normal procedure is to fly outbound for 1 minute, depending on the wind, then begin the procedure turn. Descend to the procedure turn altitude (4,000 feet at Pittsfield) while heading outbound and during the procedure turn.

Do not descend below 4,000 feet until you intercept the inbound course of 259°. Again, this is for obstacle clearance reasons. After intercepting the inbound course, you may to descend to 3,200 feet en route to the FAF. On reaching the FAF, do a Five T check and continue descent to the MDA.

Sometimes a holding pattern is mandated for a course reversal instead of a procedure turn. The racetrack pattern shown on the approach chart will be printed with a much darker and heavier line than other holding patterns as shown in Fig. 24-5, the NDB 34 approach at our planning destination, Binghamton, NY. Note that this is listed as an "NDB or GPS RWY 34" approach. This is an example of a GPS "overlay" approach that can be flown with a Global Positioning Satellite (GPS) receiver substituting for an ADF receiver. I'll have more to say about GPS approaches.

If holding patterns are depicted with dotted lines (as is the case at Pittsfield), they are not available for course reversals. Instead, you

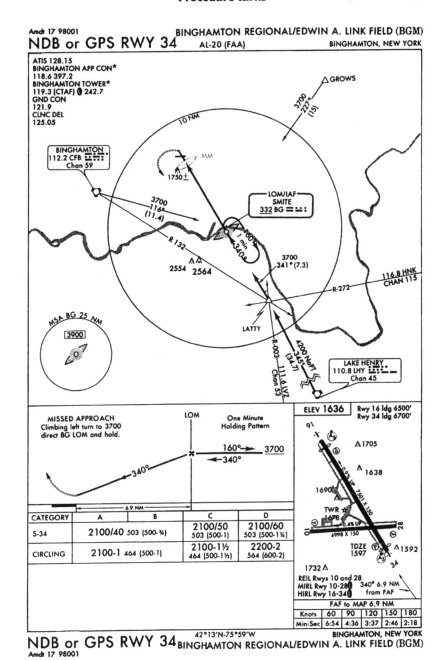

Fig. 24-5. *NDB approach with course reversal in holding pattern.*

must use the procedure turn shown on the chart. Both methods of course reversal—the holding pattern and the procedure turn—are widely used in NDB and VOR approaches. For obstacle clearance reasons, you do not have the option of substituting one type of course reversal for another.

It is very important to the success of an NDB approach to get lined up on the final approach course as soon as possible with an accurate wind correction. If you are not lined up properly at the FAF, the chances of making a successful approach are very slim indeed. A well-executed course reversal is the secret to success in quickly establishing good lineup.

Approach speeds

Slow to approach speed and lower approach flaps, usually one increment, during the course reversal and in steady flight during a full approach or as you head toward the FAF with radar vectors. One hundred knots is a comfortable approach speed for most light airplanes. This will usually result in a 90-knot ground speed in typical winds. The exact speed doesn't make too much difference as long as it is a comfortable speed that you can hold constant throughout the approach, including the descent to MDA after leaving the FAF.

If you are aware of faster traffic behind you, maintain cruise airspeed and keep the flaps up. Plan to make a high-speed final approach at 110 knots, or even 120 knots if safe. Don't worry about coming in too fast and using up too much runway! If the runway is long enough for the jets behind you, that runway will be plenty long enough for you!

On the other hand, it is preferable to use that last one-half mile or the middle marker to transition to your most comfortable airspeed and configuration. This way you will come over the runway threshold in a normal manner. Consistency makes better landings.

Complete the approach checklist as you head toward the FAF, and run through the MARTHA check again. All approach charts for the destination should be on the clipboard with the probable approach chart on top. It is very helpful to clip approach charts to the yoke for quick reference throughout the approach and missed approach. Some airplanes come equipped with a yoke chart clip; you may also purchase a clip at aviation supply companies and many FBOs.

Approach communications

If you have to hold in a depicted holding pattern, you must make another *required* report while entering the hold, as follows:

"Approach control, Cessna five six Xray, (name of fix), entering hold, level at three."

As the expect further clearance time approaches, you can anticipate that approach control will either clear you for the approach, issue a revision of the time, or—at a controlled airport—hand you off to tower. If the latter is the case, you will be given the tower frequency.

At an uncontrolled airport, ATC will ask how you plan to terminate the approach. You have three options:

- Land
- Make a low pass and cancel IFR
- Execute a missed approach

With the first option—a landing—approach control will tell you "report landing or landing assured." You will remain on the approach control frequency until advised "frequency change approved." You must then switch to the CTAF frequency and report your position on the CTAF to alert other traffic about your position and that you are inbound on a specific instrument approach. CTAF is also the frequency to get the weather at the airport and learn the runway in use. When you report landing or landing assured to ATC, the IFR flight plan will be closed by ATC. If the landing is at a remote airport without communications to ATC or a flight service station, a telephone call to an FSS might be required to close the IFR flight plan.

With the second option—low approach and cancel IFR—the IFR flight plan will be canceled when you announce "cancel IFR" to ATC.

With the third option—missed approach—you return to the ATC frequency at the MAP. Executing a missed approach is the occasion for another required report. But you don't have to make this report as soon as you add power for the missed approach. Wait until the climb is stabilized and everything is under control, then report. Always remember: aviate, navigate, communicate!

Expect an abbreviated clearance from approach control for returning to the missed approach holding fix. It is not a good idea to try to copy a clearance while you still have your hands full controlling and cleaning up the airplane in the transition phase of a missed approach. Wait until you are in a stable climb, then contact approach control.

Flying the NDB approach

So far we have discussed elements of the instrument approach that also apply to all nonprecision and precision approaches to one degree or another: radar vectors, approach speeds, course reversals, and terminating the approach. Now let's back up and discuss how you will actually fly the full procedure in the example, the NDB 26 at Pittsfield.

Proceed to the IAF, Dalton NDB, following the clearance from approach control. On the way to Dalton, slow to approach speed and review the MARTHA check. On reaching Dalton, run through the "Five Ts"as you always do at a fix or when making a change in course or altitude:

- Write down the *time* of arrival at Dalton on the approach chart. You will also need to start timing the outbound leg.
- *Turn* to the appropriate heading outbound for the procedure turn.
- *Twist* is not necessary for this NDB approach. Instead, use this item as a reminder to adjust the volume on the identifier to hear it faintly in the background. Monitor the NDB identifier continuously throughout the approach to detect a failure of either the transmitter or receiver.
- Reduce *throttle* for 100 knots if you have not done so already.
- *Talk*: Report as requested by ATC, for example, "Cessna five six Xray, Dalton procedure turn outbound."

When to descend

When approach control has cleared you for the approach, you may begin a descent to the altitude prescribed on the approach chart—4,000 feet at Pittsfield—as soon as you depart Dalton outbound on the procedure turn. If you have not been cleared for the approach, you must remain at your assigned altitude—5,000 feet in this example—until approach control clears you to a lower altitude, or says "cleared for the approach." This reason for remaining at the assigned altitude is obvious because there might be other airplanes in a holding pattern below.

Make a normal, stabilized, constant-airspeed descent. Slow the airplane to the approach speed you have selected, say 100 knots (if you haven't already done so). When stabilized, reduce power 100 RPM (or 1" of manifold pressure) for each 100 feet per minute you want to descend. A reduction of 500 RPM—from 2300 RPM to 1800 RPM, for example—will produce a rate of descent of 500 feet per minute. A reduction of 5" of manifold pressure will also produce a 500-foot-per-minute descent.

It's always a good idea to start a descent as soon as you are cleared to do so. The sooner you get down to the desired altitude, the more time you have to stabilize altitude, airspeed, and heading. This becomes very important as you descend to the MDA on the final approach course. If you have a large amount of altitude to lose, descend at 1,000 feet per minute until 1,000 feet above the desired altitude, then reduce the rate of descent to 500 feet per minute.

Make the procedure turn and descend in the turn to 3,200 feet if you have been cleared for the approach. As you head inbound toward Dalton, intercept the inbound course (259°) and begin bracketing to establish a reference heading that will correct for the wind and maintain an inbound course.

Upon reaching the FAF, follow through with five important steps:
- Start timing the final approach segment
- Adjust heading as necessary
- Make sure the volume is correctly set to faintly hear the identifier throughout the approach
- Reduce power 500 RPM (or 5" of manifold pressure) to begin a 500-foot-per-minute descent to the MDA
- At a controlled airport, contact tower if you haven't already done so. At an uncontrolled airport, report on CTAF passing the FAF. Always report position and intentions on CTAF to alert local traffic. Make frequent additional reports on final as needed.

Timing the approach

Using a stopwatch is recommended to time the final approach segment from FAF to MAP, or a digital timer on the instrument panel that you can start as you pass the FAF. A stopwatch (or digital timer) is also handy for timing the legs of a procedure turn or a holding pattern.

You can also use the sweep-second hand of the conventional clock to time the approach. But many students find it confusing trying to keep track of how many minutes have passed on a long final approach segment. On some long finals the time from FAF to MAP might be more than 5 minutes.

As you concentrate on maintaining the MDA and the final course, it is easy to forget how many times the sweep-second hand of the conventional clock has gone around. (Is it three? No, that was last time. Must be four. But it's taking so long! Maybe I've already gone five minutes!)

Eliminate the confusion altogether. Buy a timer at the beginning of your IFR training and use it on every approach. You will soon find that accurate timing ceases to be a problem. (Occasionally use the panel timepiece—conventional or digital—to maintain proficiency in case a handheld stopwatch fails.)

The times from FAF to MAP at different speeds are located in the lower right corner of the NOS approach charts. These times are based upon no-wind conditions, so you must adjust them for the estimated ground speed. If you are good with numbers you can interpolate and quickly determine the time to match the ground speed. If you make this calculation part of the MARTHA check when approaching the airport, it will save a lot of fumbling at the FAF. None of these calculations will be accurate if you cannot fly a constant airspeed during the descent, then level off at the MDA.

Tip: When the ceiling and visibility are well above minimums, say 600 overcast and 2 miles, use the next faster speed when timing the FAF to MAP segment.

Final approach course

Another problem that I see frequently when giving flight tests for the instrument rating is pilots getting so disoriented on the final approach course after passing the FAF that they cannot find the field. The basic problem here is poor training or lack of practice in tracking and bracketing the NDB. If you understand and practice the NDB procedures discussed earlier, you should have little difficulty in this phase of the approach.

As you head toward the FAF you should have enough time to bracket the inbound course and determine a reference heading that will correct for wind and maintain that course. At station passage, don't chase the needle. Maintain the reference heading outbound from the FAF until the needle settles down and you have completed the "Five Ts" checklist. After that you may make minor adjustments in the reference heading if necessary.

If you have an accurate reference heading when passing the FAF, all you have to do is maintain that reference heading and you will see the landing runway when the time has expired, ceiling and visibility permitting.

There are times, however, when even the best pilots are unable to establish an accurate reference heading as they fly toward the FAF. The wind might be changing rapidly or approach control might turn you inbound so close to the FAF that there is not enough time to get established on course.

Do the best you can heading inbound in this situation. Then at station passage turn immediately to the inbound course, wait 10 seconds for the needle to settle down, note the number of degrees the needle is off to the left or right, and then reintercept and bracket the outbound course. That will get you back on course before you get to the MAP.

Missed approaches

There might be times when even this won't work. If you cannot establish yourself on the final approach course for any reason—or if you have lost track of the timing—you must execute an early missed approach. The procedure for an early missed approach is different from a missed approach at the MAP. In an early missed approach, add full power, clean up the airplane, establish a normal climb, report to ATC, and transition to the published missed approach procedure. Above all, *do not make a turn until you have reached the MAP.* You are not guaranteed the full obstacle clearance associated with that approach if you depart from the final approach course.

A decision to make a normal missed approach at MAP is based upon several variables.

Very often missed approaches are required because you didn't get down to MDA and were unable to establish the visual references spelled out by FAR 91.175 (c) for a further descent. A missed approach is also required if the visibility is below that required for the approach. Or perhaps you weren't lined up properly and just caught a glimpse of a corner of the airport as you flew by.

You might also have to make a late missed approach if you have the required visual references and the visibility needed to land but find, after beginning the descent for landing, that you cannot land for some reason, perhaps another airplane on the runway.

Remember the five phases of a missed approach:

1. *Transition* to a stabilized climb
2. *Climb* straight ahead or in a climbing left or climbing right turn, as prescribed in the published missed approach procedure
3. *En route* to the holding fix
4. *Holding* at the designated holding fix
5. *Departure* from the holding fix for another approach or to an alternate

You should carefully study the missed approach procedure the night before the flight, along with your research on the rest of the approach. Note the missed approach instructions on the approach charts. How will you proceed after pullup? You will have three choices: climb straight ahead, make a climbing left turn, or a climbing right turn, as is the case with Pittsfield. Where will you hold and how will you enter the holding pattern? At Pittsfield there is a holding pattern at Dalton NDB for the missed approach, but not for a procedure turn. You must think these points through in your preflight planning. The cockpit is not the place for original research on missed approaches!

There is no limit, other than the amount of fuel on board, to the number of approach attempts. If you missed the approach for reasons other than weather—poor lineup, for example—go back and try again. But if you reached MDA and the weather was below minimums, or if the weather was obviously deteriorating, the smart move would have been to proceed to the alternate.

Circling approaches

You are on the final approach at MDA and, as you run out of time, hopefully you run out of clouds and the airport is in sight. You do a landing check passing through 500 feet above the airport, lower full flaps, and land.

But somewhere prior to completing the landing check and lowering full flaps, you might find that you have to land on a different runway. If landing at a controlled airport, the tower will make the decision and issue a clearance: "Circle and land Runway (as assigned)." At an uncontrolled field, however, the pilot must make the decision. Several variables affect this decision.

In some cases there is no choice. Many approaches are not sufficiently aligned with the runway to permit a true straight-in approach. When a procedure does not meet the criteria for straight-in approaches, it is designated A, B, C, and so on (NDB-A, NDB-B, NDB-C, etc.), and no straight-in minimums are published. An example of this is the NDB-A approach at Perkasie/Pennridge, Pennsylvania (Fig. 24-6). The final approach course is 181° and the only runway at Perkasie is 8-26. So a circling approach is the only alternative available.

Conditions at the time of reaching the MAP might dictate a circling approach. If the crosswind is too great for the straight-in runway, for example, you should choose a landing runway that is closer into the wind, if one is available. If there isn't a better runway, execute a missed approach.

Sometimes you learn that airplanes in the landing pattern ahead of you are using a different runway. Learn the active runway from unicom, or from other airplanes in the pattern as they report their positions on the CTAF. If a different runway is in use when you arrive, you will have to make a circling approach and fit into the traffic pattern. If there is traffic in the landing pattern, it better be VFR; so you should break out in VFR conditions well above the circling MDA. Nothing says you can't fly a circling approach higher than the circling MDA as long as you are clear of clouds.

You may get right down to the straight-in MDA before deciding to make a circling approach. Perhaps another airplane taxis out and dawdles on the runway just as you are about to land. Add power and go around, just as you would under VFR conditions. But you are still IFR and must circle around again for another attempt at landing. In this case you must climb back up to the published circling minimums, or traffic pattern altitude, in order to continue.

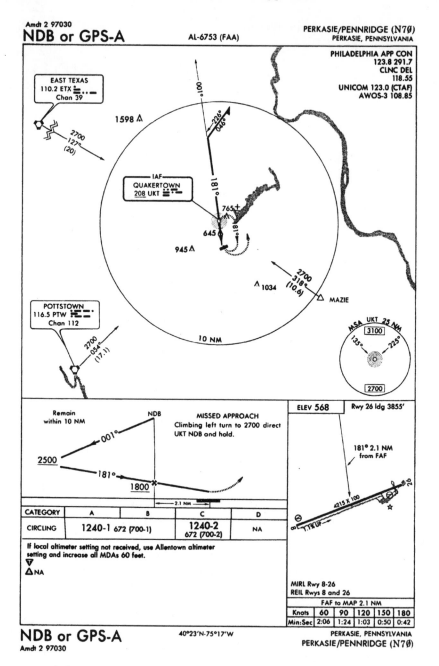

Fig. 24-6. *Typical circling approach procedure when runway is not aligned with final approach course.*

If that puts you back in the clouds, you will have to execute a missed approach. You must keep the runway of intended landing in sight at all times during a circling approach or execute a missed approach.

Circling approach patterns

The recommended circling approach patterns are shown in Fig. 24-7. Pattern A may be used when the final approach course intersects the runway centerline at less than a 90° angle and you see the runway clearly enough to establish a base leg.

If you see the runway too late to fly pattern A, circle as shown on B and make either a left downwind or a right downwind. Fly pattern C if it is desirable to land opposite the direction of the final approach course and the runway is seen in time for a turn to the downwind leg. If the runway is sighted too late for a turn to the downwind as shown in C, fly pattern D.

So far this all sounds very reasonable; however, *the circling minimums might be as much as 500 feet lower than a VFR traffic pattern* for the same runway flown at 1,000 feet above field elevation. Some people call the circling maneuver legal scud running. It takes some very careful maneuvering to make a safe approach and landing from a low altitude.

Furthermore, the circling minimums guarantee an obstacle clearance of *only 300 feet* within the *circling approach area*. This is a very small area and you must remain within it. TERPS describes how the circling approach area is constructed (Fig. 24-8). The circling approach area is based on the same aircraft approach categories A, B, C, D, and E that appear in the minimums sections of instrument approach charts. For category A, which most of us use for instrument training, the circling area has a radius of only 1.3 nautical miles from the end of each runway.

Figure 24-8 shows how arcs drawn from these radii outline the area in which obstacle clearance is provided. *Outside this area, there is no obstacle clearance protection.* Keep the circling approach pattern within the safe area. Use the runway length to help visualize 1.3 nm.

You are required to perform a circling approach on the instrument flight test. And circling approaches are often the only kind allowed at many small airports. Master the skills necessary to carry out this maneuver at circling MDA but realize that circling approaches are imprecise and might be dangerous if not performed properly.

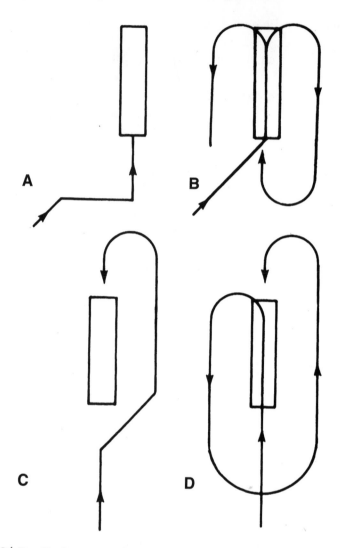

Fig. 24-7. *Circling approach patterns. Use A when final approach course intersects runway centerline at less than 908; use B if you see runway too late to fly pattern A; use C to land in opposite direction from final approach course; use D if you see the runway too late to use pattern C.*

Instructor note. Practice, as always, is the best way to build confidence in circling approaches. In VFR conditions, make it a routine to terminate one instrument approach on every flight with a circling approach at the VFR pattern altitude so that the student will learn to sequence with other traffic.

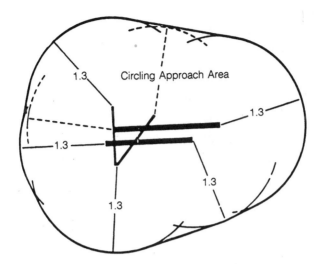

Fig. 24-8. *Circling approach area within which obstacle clearance is provided.*

In actual IFR, take advantage of every opportunity to have the student make circling approaches at circling MDA at an uncontrolled airport. The tower at a busy airport probably won't let you do this for practice, but will insist that you land.

Be sure to brief the student about your intentions. Students won't get much out of the practice if they are totally confused about what is going on.

NDB on airport

Some NDBs are located right on the airport, as seen in Fig. 24-9, the NDB RWY 22 approach at Easton, Maryland, a busy field on Maryland's popular Eastern Shore.

An NDB approach with the NDB on the airport is a very simple approach. Proceed to the NDB, which is the initial approach fix (IAF) and the MAP. (There is often no FAF when the NDB is located on the field.) Turn outbound on the indicated course, in this case 048°, the reciprocal of the inbound course. A procedure turn is indicated rather than a course reversal in a holding pattern.

Do not descend below 1,600 feet until you intercept the inbound course of 228°. Again, this is for obstacle clearance reasons. After

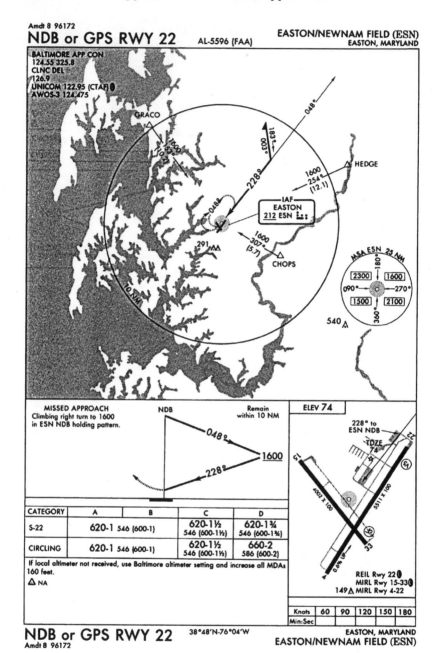

Amdt 8 96172

NDB or GPS RWY 22 AL-5596 (FAA)

EASTON/NEWNAM FIELD (ESN)
EASTON, MARYLAND

BALTIMORE APP CON
124.55 325.8
CLNC DEL
126.9
UNICOM 122.95 (CTAF)
AWOS-3 124.475

GRACO

048°

183°
003°

228°

HEDGE

1600
254°
(12.1)

IAF
EASTON
212 ESN

048°

1600
307°
(5.7)

291

CHOPS

MSA ESN 25 NM
180°
2300 | 1600
090° ————— 270°
1500 | 2100
360°

540

MISSED APPROACH
Climbing right turn to 1600
in ESN NDB holding pattern.

NDB

Remain
within 10 NM

ELEV 74

228° to
ESN NDB

TDZE
74

048°

1600

228°

CATEGORY	A	B	C	D
S-22	620-1 546 (600-1)		620-1½ 546 (600-1½)	620-1¾ 546 (600-1¾)
CIRCLING	620-1 546 (600-1)		620-1½ 546 (600-1½)	660-2 586 (600-2)

If local altimeter not received, use Baltimore altimeter setting and increase all MDAs
160 feet.
△ NA

REIL Rwy 22
MIRL Rwy 15-33
149△ MIRL Rwy 4-22

Knots	60	90	120	150	180
Min:Sec					

NDB or GPS RWY 22

38°48'N-76°04'W

EASTON, MARYLAND
EASTON/NEWNAM FIELD (ESN)

Amdt 8 96172

Fig. 24-9. *NDB approach with station located on the airport.*

intercepting the inbound course, you are free to descend to 620 feet, which is the MDA for straight-in and circling approaches at Easton.

Note that no times are given from FAF to MAP because there is no FAF and the MAP is the NDB itself. Therefore, there is no need to time the inbound leg. Just stay on the inbound course of 228° at the MDA of 620 feet (as adjusted in your planning) until station passage occurs. If you don't have the visual references you need to descend below MDA when the needle reaches the 90° position, you must execute the missed approach.

Planning steps, communications, MARTHA and Five T checks, approach speeds, and missed approach phases will be the same for an on-airport NDB approach and an NDB approach with an FAF some distance from the field.

25

VOR, DME, and GPS approaches

One way to gain a good overview of approaches is to examine them as short cross-country flights. A VOR approach is nothing more than a miniature cross-country. You proceed from the last en route or feeder fix to the IAF, to the FAF, and then to the MAP flying along predetermined courses, making turns and changing altitudes as required.

It might take 2–5 minutes to get from the en route or feeder fix to the IAF and then another 5–10 minutes to arrive at the FAF. Remember that you will be tracking inbound and outbound by VOR during the approach as if tracking VORs inbound and outbound on a cross-country.

The planning, communications, MARTHA and Five T checks, approach speeds, and missed approach phases remain the same for VOR approaches as for NDB approaches.

Flying the VOR approach

Let's work our way step-by-step through the VOR-A approach (Fig. 25-1) at Poughkeepsie/Dutchess County, New York, which is also a GPS "overlay" approach. This approach has one item of special interest: It contains a "dogleg." It also has a procedure turn rather than a course reversal in a holding pattern.

Once again, request the full procedure in order to get the maximum training benefit out of this exercise. Proceed to the IAF, Kingston VOR, do the Five Ts and turn outbound and get established on the 037° radial, the reciprocal of the inbound course, to commence the procedure turn.

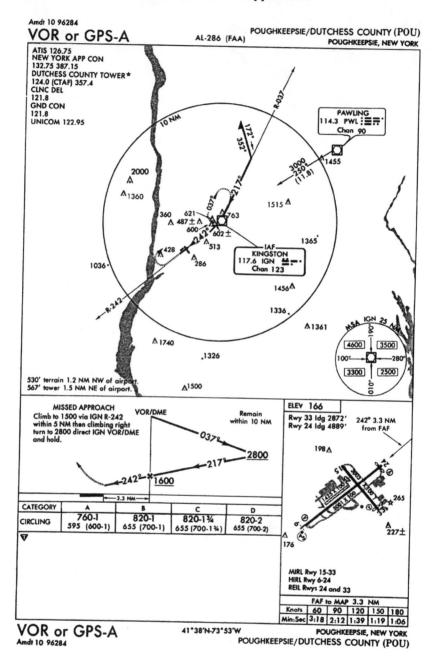

Fig. 25-1. *VOR approach with a dog leg.*

Fly outbound on the 037° radial for one minute and make a 45° turn to the left as shown on the chart. Reset the OBS to the inbound course, 217°.

Fly outbound on the 352° heading for one minute, adjusting for the wind, then make a 180° turn. This will establish the 172° course to intercept the inbound course to FAF, 217°. Hold the 172° intercept heading until the needle is about three-quarters of the way from full-scale deflection toward the center, then begin a turn to the inbound course. Correct for the wind and establish a reference heading that will hold the inbound course.

As noted in the discussion of NDB approaches, a holding pattern might be prescribed in lieu of a procedure turn. If an approach has the note "NoPT," no procedure turn is permitted and *you cannot execute it* without clearance from ATC. A few VOR approaches state flat out, "Procedure Turn NA"—not authorized. Don't even think about it! There is probably a big mountain or a tall radio mast or a power line precisely where you would normally expect to make a procedure turn.

You may commence a descent to the procedure turn minimum altitude as soon as you pass the IAF. The procedure turn minimum altitude is 2,800 feet on the Poughkeepsie VOR-A approach (Fig. 25-1). Do not go below the procedure turn minimum altitude until established on the inbound course. "Established" means a "live" needle, not necessarily centered. When established you may descend to the FAF minimum altitude, 1,600 feet in this case.

On reaching the FAF, begin a descent to the MDA for the approach. Get down to the MDA as quickly as you comfortably can to give yourself the maximum opportunity to see the airport and pick out the landing runway. Make a constant airspeed descent so you do not throw your timing off.

Normally after completing a procedure turn you can expect to fly a straight-line course to the FAF and then on to the MAP. But not on the approach at Poughkeepsie. On reaching the FAF, make a right turn to 242° and proceed toward the MAP on this new course. That's why this is designated an "A" approach; that dog leg does not meet the criteria for a straight-in approach, even though you might end up lined up for a landing on Runway 24, if you fly a perfect approach!

You will be very busy at the FAF as you run through the Five Ts: time, turn, twist, throttle, talk. You must start *timing* the final approach leg; *turn* to intercept 242°; *twist* the OBS to 242°; *throttle* back 500 rpm (or 5" of manifold pressure) to begin a 500 fpm descent at the approach speed; and then *talk* to tower. The report will be "Cessna five six Xray, Kingston inbound." Aviate, navigate, communicate!

The key to coping with complications like this "dogleg" is to spot them while planning the flight, then talk yourself through the approach until you understand clearly the course, turns, descents, and reports. Again, think of it as a miniature cross-country and be sure to include the missed approach procedure as part of the cross-country. Fortunately you won't encounter too many dogleg VOR or NDB approaches, but be prepared to handle them.

A more common variation on the VOR approach is the VOR located on the field. This is the case at Bridgeport, Connecticut (Fig. 25-2). This airport is located on a point of land jutting out into Long Island Sound. The only place approach facilities could be located is on the field or they would be underwater. When a VOR is located on the field, the MAP is reached when the TO-FROM indicator flips to FROM.

The Bridgeport VOR RWY 24 approach also has another frequently seen feature. Many of the fixes are VOR intersections, including the IAF at MILUM and the missed approach holding fix at STANE. Tune Carmel VOR (116.6) on your No. 2 VOR receiver as you approach the area. Carmel VOR will provide the cross-bearings for the IAF, the course reversal in a holding pattern, and the MAP holding pattern.

As you talk through an approach such as this in the planning room, consider writing out a separate sequence of OBS settings for all these fixes on the flight log. Be sure to include other intersections shown on the approach chart (such as BAYYS on the Bridgeport VOR 24 approach chart) in case approach control specifies them in a clearance. (Preparation is 90 percent of the law in the legal world, as is success in aviation.)

DME and DME ARC approaches

Proficiency in the use of distance measuring equipment (DME) is not a requirement of the *Instrument Rating Practical Test Standards*. But ATC expects you to be competent and able to use any equipment in

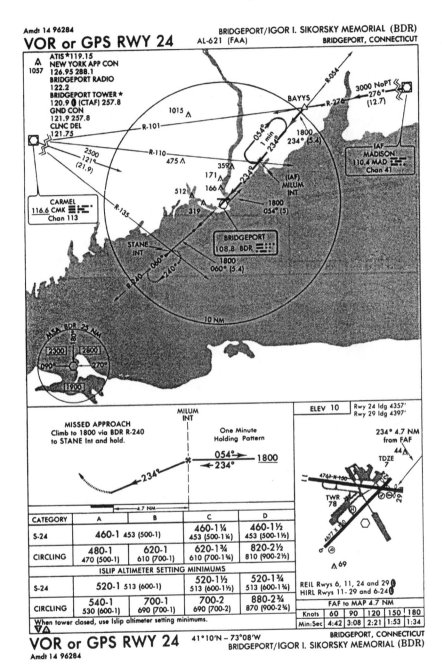

Fig. 25-2. *VOR approach with fixes at VOR intersections.*

the airplane. If you file equipment code A (DME and transponder with altitude encoding capability), ATC will issue clearances with DME points. Be prepared to make DME approaches, some of which might surprise you if you haven't practiced them.

DME indications sometimes appear on VOR approach charts as an aid to making a conventional VOR approach; however, if DME is not included in the name of the approach procedure—if the name of the approach is simply VOR RWY 28, for example—then DME distances are just aids, and DME is not required for the approach. On the other hand, if DME is included in the name of the approach procedure (Fig. 25-3) you must have DME to execute the approach, unless ATC agrees to call out the DME fixes.

Note that the two IAFs for the VOR/DME RWY 15 approach at Johnstown/Cambria County, Pennsylvania, are located where two Johnstown VOR radials intersect the 10 mile DME arc. Then you fly the 10 DME arc around to intercept the 326° radial and turn inbound. The FAF is HINKS intersection, DME 4 on the inbound 146° course.

Flying a DME arc is not as difficult as it looks on the approach chart. You won't have to do the impossible and fly a smooth, continuous, perfect arc. Instead, fly a series of short, straight tangents to the arc, as you would on a time/distance check. These short tangents will keep you close to the 10 miles specified.

Flying the DME ARC

Let's work our way through the Johnstown VOR/DME RWY 15 approach, arriving from the northeast (Fig. 25-3). First, intercept the 074° radial and turn inbound (1) on the reciprocal bearing to the station, 254°. As the DME mileage clicks off, anticipate a turn at 10.5 DME. This will enable you to lead the turn onto the 10-mile arc by half a mile.

At 10.5 DME, turn right 80° (2) to a heading of 334° (Fig. 25-3). When you complete the turn you will be on a tangent to the arc at a distance of 10 miles or very close to it. Rotate the OBS 10° *opposite* to the direction of the first 90° turn (left to a setting of 064°). As you continue on the 344° heading, the DME mileage will begin to increase.

Remember that when you set radials on the OBS, the CDI needle will start out on the *same side as the station* and move to the opposite side

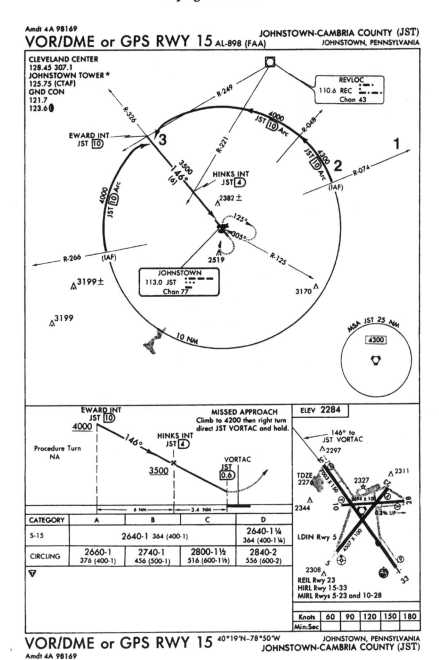

Fig. 25-3. *Procedure for flying VOR/DME RWY 15 at Johnstown, Pennsylvania. Intercept radial 074 at (1), turn onto arc at (2). turn to inbound course at radial 326 (3).*

as you pass the radial. In this case, the CDI needle will move from left to right as you approach and pass the 064° radial.

When the needle centers, turn left 10° to a new heading of 324°. Reset the OBS 10° to read 054°. In a no-wind condition, the DME distance will decrease to 10 miles after the turn, then begin to increase again as you fly the tangent. Continue with these 10° heading and OBS changes as you track around the arc toward the 326° radial.

Naturally, the wind will tend to blow you toward the station or away from it, depending on its direction. If you find the DME distance increasing, you are being blown away from the station. Make the next 10° heading change sooner or make the turn more than 10°. This will bring the airplane back inside the curve.

If the DME distance decreases, you are being blown toward the station. Reset the OBS for 20°. Make the next heading change 10° as usual. You will fly a longer tangent before the needle centers again. This will correct for the wind blowing toward the station.

Lead the turn onto the inbound course by 5°. In the Johnstown example, establish a 146° course inbound to the station. After passing R334, set the OBS to 151° to lead the inbound course (3). Do the Five Ts and start the turn inbound when the needle centers. After completing the turn, reset the OBS to 146° and track this course inbound.

Note how DME distances are used to fix the FAF at HINKS intersection and the MAP at .6 DME. Normally, when the VOR is located on the field, the MAP occurs at station passage. But in this case, you must execute a missed approach before reaching the VOR to avoid obstacles.

With the Johnstown VOR/DME RWY 15 approach, the DME arc is used to position the airplane on a conventional VOR course, and for the FAF and MAP. In some cases, believe it or not, the DME arc *is the final approach course*.

The VOR DME RWY 14 at Baltimore/Martin State, Maryland, (Fig. 25-4) is a fairly simple approach to fly, despite the way it appears on the chart. Intercept the Baltimore 331° radial, fly inbound to the 14.7 DME arc, turn left 90°, then begin making 10° tangents all the way around until the airport lies ahead. The various radials provide the descent points and the MAP.

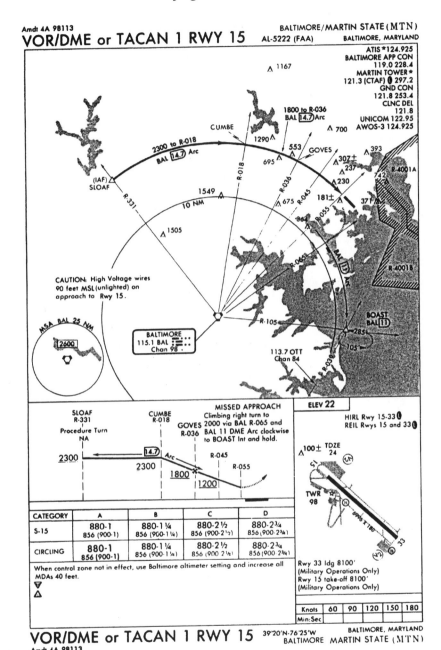

Fig. 25-4. *VOR/DME at Baltimore/Martin State, Maryland, with DME arc as final approach course.*

It should be clear at this point that there is nothing conventional about a DME approach. Every one of them is different, sometimes radically so. But if you have DME aboard and take the time to practice DME approaches and get acquainted with them, you will gain access to a large number of approaches that you might otherwise not be able to use. Believe it or not, DME arcs are easier to fly than to explain in written text, and they are great fun.

Instructor's note. If there are no DME arc approaches available for practice in your area, or if they are too far away, select a nearby VOR en route facility with DME and set up DME arcs around that. Remain VFR at all times and work in a quadrant that will keep you away from all airports and instrument approaches associated with the VOR. And be sure to operate at an altitude that guarantees obstacle and terrain clearance.

In fact, with a little research you should be able to practice a published DME approach (such as those in Fig. 25-3 and 25-4) on a nearby VOR DME. Superimpose the published DME approach on the nearby VOR and see if it will conflict with any other airports and approaches. If there are no conflicts, find the Minimum Obstruction Clearance Altitude (MOCA) for the area as shown in large numerals on the IFR En Route Low Altitude chart. Use the MOCA as the field elevation and add it to the MDA and other altitudes on the approach chart.

GPS approaches

The day will come eventually when the Global Positioning Satellite (GPS) system will replace VOR en route navigation as well as NDB, VOR, DME, ILS, and all other types of approaches except radar. We are in the midst of a revolution in air navigation that will make instrument flying simpler and much safer. Think of it! Instead of learning separate techniques for all of the above, we will only need to learn GPS. And thanks to the wonders of powerful small computers, GPS offers the promise of being much easier to learn and use than anything we have now.

But a few words of caution are in order at this point. We have a long way to go before GPS becomes the standard system for air navigation. Despite the hype surrounding GPS in the last couple of years, the Federal government has not developed a clear policy for the implementation of GPS as the air navigation system of the future. Nor do we have realistic goals on the way toward the achievement of an all-GPS air navigation system. The cockpit equipment is still very costly, and using GPS in a single-pilot, single-engine situation can be extremely work-intensive.

GPS uses timed signals from 24 U.S. military NAVSTAR satellites to provide precise position information through sophisticated, high-tech receiver/processors. (A good discussion of GPS basics may be found in AIM, Chapter 1.) GPS provides two levels of service: "Standard Positioning Service" and "Precise Positioning Service." The standard service is accurate to 100 meters (328.1 feet) or less, which is acceptable for en route navigation and nonprecision approaches. Standard service is available to all users.

The precise service is accurate to 16 meters (52.49 feet), but its use is restricted to military and other national security applications. Even if the precise service was made available for civilian use, the signals would have to be corrected—"augmented"—to meet the course and glide slope requirements for ILS precision approaches.

GPS signal corrections eventually will be provided by "differential GPS" (DGPS). DGPS works through precisely located monitoring stations on the ground that compare the predicted GPS signals for that precise location with the satellite signals actually coming in. The differences between predicted signals and actual signals are processed by the ground stations and converted to differential corrections.

The current plan is to provide corrected GPS signals to airborne receivers through the "Wide Area Augmentation System" (WAAS). "WAAS will consist of 24 monitoring stations that will sample signals from GPS satellites passing overhead," notes the AOPA Air Safety Foundation in its recent "Safety Advisor" booklet, *GPS Technology*. "The data will be sent to three control stations, which will rapidly analyze the information and uplink corrective signals to three geostationary satellites covering the United States. The satellites will broadcast corrected GPS signals data to airborne WAAS-capable receivers."

So we're looking at a new system that—in addition to our present array of 24 GPS satellites—will require 24 additional monitoring stations and 3 control stations on the ground, plus 3 geostationary WAAS satellites, *plus* all new avionics for every aircraft that uses the national airspace system. For the greater precision required of ILS

Category II and III approaches, a supplemental "Local Area Augmentation System" (LAAS) will be installed at selected high-density airports.

AOPA and the Air Transport Association (ATA), which represent the nation's airlines, are supporting the FAA's plans to implement both WAAS and LAAS. And much of the work has already been done. There are now hundreds of GPS nonprecision approaches available throughout the country, with many more on the way.

GPS approach basics

There are two types of GPS approaches in use these days. The most common type is the "overlay" approach that is identical to an existing NDB or VOR approach except that GPS is the means of navigation.

Fixes, courses, frequencies, minimum altitudes, course reversals, and missed approach procedures are the same for GPS as for the underlying NDB or VOR approach.

The second type of GPS nonprecision approach is the "stand-alone" type that may be encountered at airports that have no underlying NDB or VOR approaches, or where there are differences that apply to the GPS approach and not to the others. See Fig. 25-5 and Fig. 26-6, the two approaches to RWY 6 at Potomac Airfield, Friendly, MD. Look closely and you will see that:

- The MDAs are different, with the GPS MDAs being slightly lower.
- The courses are slightly different.
- There are additional waypoints for the GPS approach at IRONS and WOBUB.
- The missed approach instructions are slightly different.

GPS approach planning

Flying the GPS approach is pretty simple—you just intercept and track the courses indicated on your OBS indicator or Horizontal Situation Indicator (HSI). You descend as shown on the approach chart at the various waypoints until you reach MDA and land or execute a missed approach. Sound familiar?

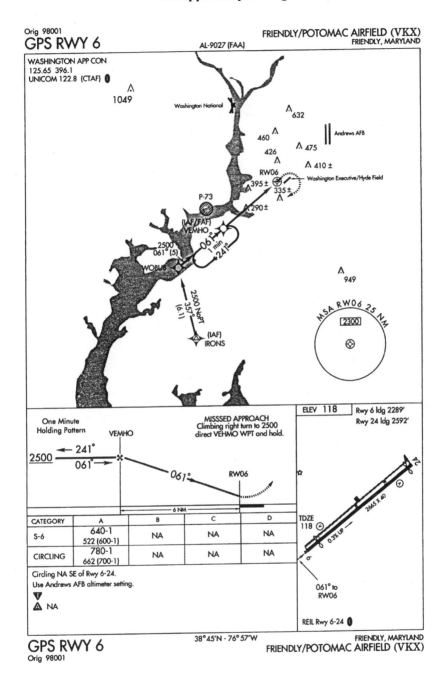

Fig. 25-5. *GPS stand-alone approach to Friendly/Potomac Airfield, Maryland.*

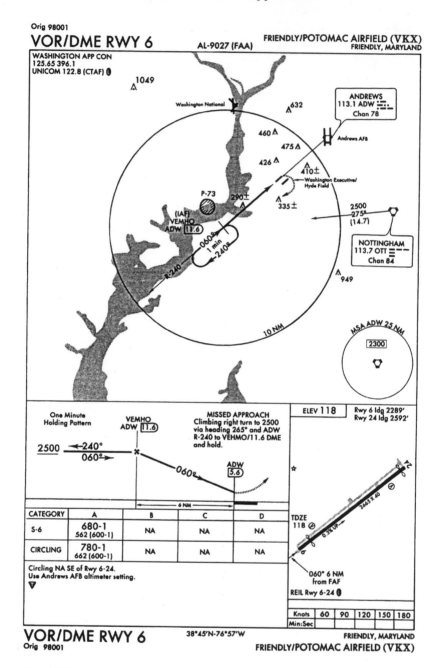

Fig. 25-6. *Conventional VOR approach to Friendly/Potomac Airfield, Maryland.*

The big difference with GPS is that you are dealing with a computer for course and waypoint information, not fixed signals from the ground. The GPS receiver/processor gets its signals from a universal source in the sky then reinterprets these signals according to your instructions. The computer in your GPS system needs to know where you want to go and what you want to do in order to lead you in the right direction. The choices offered by GPS are almost unlimited; so you must enter your instructions very carefully. Or else you might find yourself being taken to some place you don't want to go!

In addition to the approach planning, we must add another layer of planning for GPS. It's easiest, I believe, to think of a set of scenarios, such as the following:

Vectored approaches. How will I set up my GPS computer for vectors to the Final Approach Fix?

Full approaches. What steps do I take to instruct my GPS computer to handle course reversals in a procedure turn? In a holding pattern? (Some systems require that you put GPS tracking on "hold" while executing these maneuvers.)

Changes of clearance. Suppose Approach Control clears me to a different waypoint—or a different runway—than I was planning on. What do I have to do to reset my computer for the new clearance?

Missed approaches. Two things here: How do I instruct my computer to return for another pass? Or what do I need to do when I must proceed to my alternate?

These moves cannot be researched in the cockpit. Each of them must be rehearsed beforehand for every flight, and the key instructions written on your planning log. Talk yourself through each of these scenarios and simulate the "knobology" needed to enter the correction instructions into your GPS computer. Think about setting up a dummy GPS panel to help you make the right moves with the knobs and buttons. Or go to your plane, turn the GPS on, and rehearse the inputs with the real thing while you are on the ground. Some systems have a built-in simulator mode.

Experienced instrument instructors say that it takes 15 to 20 hours of GPS instruction before you are ready to use GPS confidently on an

IFR flight, and sometimes more. Each GPS manufacturer has configured its equipment slightly differently; so you must learn how your particular equipment does the job, in addition to mastering the basic inputs common to all.

In addition to the scenarios above, there are other details that must be considered every time you plan a GPS approach:

Is your database up to date? Revised GPS digital approach databases are issued every 56 days by NOS, the same as your paper NOS or Jeppesen Instrument Approach Procedures. You must have a current database in your GPS receiver/processor. At this point, the major GPS manufacturers have different types of cards for updating chart information. Contact Jeppesen at 1-800-621-5377 for a free catalog listing the different types of data cards currently available and their subscription prices.

How do I tell if my equipment is operating properly? All receiver/processors are required to provide "receiver autonomous integrity monitoring" (RAIM). RAIM checks to see if there is a sufficient number of satellites available for positioning and that their information has not been corrupted. RAIM provides several levels of warning, with the time factor becoming more and more critical in the approach phase. Study your equipment and learn when and how RAIM warnings appear and the actions you should take when RAIM information appears.

This leads us to a final point:

Monitor the underlying NDB or VOR approach while you conduct the GPS approach. Strictly speaking, this is no longer mandatory for GPS overlay or stand-alone approaches. But you must have "alternate means of navigation" aboard your aircraft, such as NDB or VOR. And you must be prepared to use it if you get a RAIM warning, or RAIM capability is lost. Furthermore, if your flight plan requires an alternate airport, this alternate must have an approved approach other than GPS, and you must be prepared to execute this approach in the event of a RAIM problem.

If your GPS has a moving map display—and that is really the way to go these days!—it is easy to become complacent and let GPS do all the work. But the sharp instrument pilot will always cross-check every phase of the flight, especially an approach, with VOR and NDB and be prepared to switch to them instantly if a GPS problem arises. VOR and NDB alternatives should always be a part of your preflight planning, and you can count on your instrument check-ride designated examiner marking you down if you don't do this.

The future for GPS is very bright, and when coupled to such features as moving map displays, HSIs or Flight Directors, and three-axis autopilots, the future promises to eliminate many of the uncertainties, frustrations, and anxieties of instrument flying. And the future might be nearer than you think! These elements are all available now and though expensive, they are seeing increasing acceptance by general aviation. The revolution is here—but the best is yet to come!

Tips on flying approaches

The successful outcome of the approach is usually assured by thorough preflight planning, by carefully studying the approach that will probably be used, and by having all approach charts for that airport readily available.

Know instantly where to look for all significant items on the approach chart.

Be prepared for the next step of the approach. Think ahead about the segment you are about to fly.

Don't try to comprehend or digest the entire approach chart all at once.

Always be prepared for a possible missed approach. "Gotta' landitis" prevents some pilots from growing older!

Keep the approach technique simple.

Slow to approach or holding speed before commencing the approach or during course reversal and lower approach flaps.

Fly "by the numbers" at predetermined airspeeds and power settings to attain a trimmed configuration.

Determine the wind correction before reaching the FAF and fly the reference heading ±2°–5° to maintain the desired track.

Perform a prelanding check prior to reaching the FAF. Lower the landing gear at the FAF or make a power reduction in a fixed-gear airplane. Then note the time over the FAF, check heading and turn as necessary, change OBS if required, and report to the controlling facility.

Know where you are at all times! Continuous situational awareness at all times is the key to confident, safe flying.

26

ILS, localizer, and radar approaches

The ILS approach is the most precise approach available to the general aviation instrument pilot. It is also the easiest to master. It must be easiest because very few instrument students seem to have problems with it! This might seem like a puzzle at first because the ILS approach is fairly complex and requires an extra degree of skill for heading and altitude control. What happens, I think, is that instrument students become enamored with the ILS and practice it more than any other approach. As is the case with everything else in instrument flying, the more you practice something, the better you become at it.

The ILS is a precision approach because it incorporates an electronic glide slope. An ILS approach will bring you in exactly on the runway centerline if you fly the approach properly and it will take you down to within 200 feet of the runway when you break out of an overcast at minimums. To do this, ILS provides very precise indications that you must respond to very precisely.

Needle sensitivity

By the time you have reached the point in instrument training where you concentrate on ILS approaches, you probably will have practiced several without the hood. You know those needles are sensitive. The vertical needle is approximately four times more sensitive when set for the localizer of an ILS than for a VOR. And the horizontal glide slope needle is about four times more sensitive than the localizer indicator needle.

At the outer marker, a displacement of one dot equals approximately 300 feet on the localizer and 50 feet on the glide slope (Fig. 26-1). At

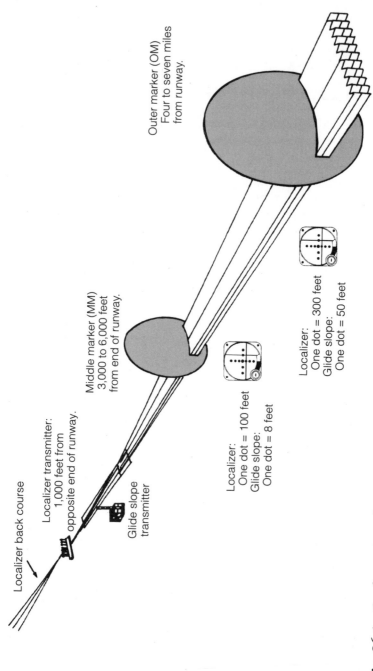

Outer marker (OM)
Four to seven miles
from runway.

Middle marker (MM)
3,000 to 6,000 feet
from end of runway.

Localizer transmitter:
1,000 feet from
opposite end of runway.

Localizer back course

Glide slope
transmitter

Localizer:
One dot = 100 feet
Glide slope:
One dot = 8 feet

Localizer:
One dot = 300 feet
Glide slope:
One dot = 50 feet

Fig. 26-1. *Configuration of a standard ILS approach.*

the middle marker one dot equals 100 feet on the localizer and about eight feet on the glide slope. Only eight feet!

Now more than ever you can begin to understand the importance of the standard of 2, 2, and 20—±2 knots, ±2°, and 20 feet. If you have been working toward these goals throughout your instrument training, you should have no difficulty coping with the sensitivity of the ILS needles.

Flying the ILS

On an ILS final approach segment the basic instrument techniques must be very sharp. Overcontrolling will peg the needles and cause a missed approach. To center the localizer needle, plan the turn onto the final approach course to roll out of the turn just as the needle centers. Quickly establish a reference heading that will correct for the wind, then use rudder pressure alone to make minor heading adjustments to the reference heading. Any bank at all will displace you from the localizer centerline so fast that the needle will probably peg. Keep that localizer needle centered all the time. Avoid the temptation to make heading adjustments with bank.

When you begin the descent, set up a reference descent rate that will maintain the glide slope. The next question is: What is the best descent rate for the approach? How do you determine what rate of descent will keep you on the glide slope? If you can find out, you will know what sort of power adjustment is necessary to set up that rate of descent.

A good method is to take the best estimate of the ground speed, divide by 2, and multiply by 10 for the rate of descent.

(80 knots ÷ 2) × 10 = 400 fpm

(90 knots ÷ 2) × 10 = 450 fpm

(100 knots ÷ 2) × 10 = 500 fpm

(120 knots ÷ 2) × 10 = 600 fpm

When you get the ATIS information for the landing runway, use this rule of thumb to estimate what the ground speed will be for the approach speed. If there is no ATIS and you can't estimate the ground speed accurately, use the indicated airspeed less 10 knots as the next best thing.

Then, as you begin descent on the ILS, reduce power 100 rpm (or 1" manifold pressure) for each 100 feet rate of descent sought. If you estimate that your ground speed will be 80 knots, reduce power 400 rpm (or 4" manifold pressure) to set up a 400 fpm descent.

It's interesting to note that if you are flying a Cessna 172 at an airspeed of 90 knots and you have to maintain a 600-fpm rate of descent to stay on the glide slope, that means the ground speed is 120 knots. You have a strong tailwind and if you have a short runway you might have to circle to land or you could run off the end.

Once you set up a reference descent rate that more or less maintains the glide slope, leave the power alone. (Throttle jockeying is a form of overcontrolling.) Don't worry about the airspeed. Use elevator pressure alone to make minor pitch adjustments. "Pitch to the glide slope—power to the airspeed." Just as easy as flying precise altitude on a cross-country flight.

The glide slope needle becomes the "altimeter" for pitch; if you go above glide slope use forward pressure to decrease pitch slightly and return to the glide slope; if you descend below glide slope use back pressure to establish level flight and reintercept the glide slope. If you go below both the glide slope and the MDA, *execute an automatic missed approach immediately.* Obstacle clearance is not provided below MDA unless you are in a position to make a normal descent to a landing.

Remember how sensitive the glide slope needle is. You don't need to make a large correction to move 8 feet in the vicinity of the middle marker.

If you are flying a retractable, intercepting the glide slope is even simpler: as you intercept, lower the gear. That will automatically produce the proper descent rate to stay on the glide slope, with minor adjustments. It doesn't matter if the plane is a Mooney, Arrow, Aztec, Baron, Seneca, Aerostar, or an Aero Commander, drop the gear and that will set up a good rate of descent to stay on the glide slope.

Analyzing an ILS approach

Let's turn now to the ILS RWY 6 approach at Allentown-Bethlehem-Easton, Pennsylvania, (Fig. 26-2) and analyze it using the step-by-step process applied earlier to ADF and VOR approaches.

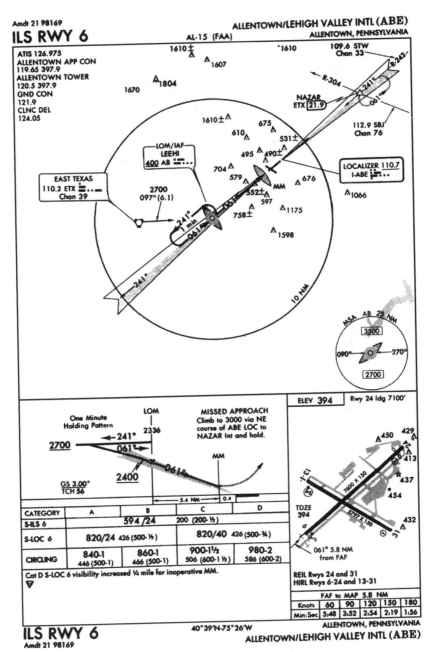

Fig. 26-2. *The ILS RWY 6 at Allentown-Bethlehem-Easton, Pennsylvania, is a typical precision approach.*

1. **Read the fine print.** In the lower left of the profile you will find the glide slope angle (GS 3.00) and the threshold crossing height (TCH 56). This information is provided for all ILS approaches. The threshold crossing height is the altitude in feet above ground level where the glide slope crosses the threshold.

2. **Check the height of obstacles.** Two tall obstacles are within 10 nm of the airport (1,610 feet and 1,598 feet) and several rise above 500 feet in the vicinity of the final approach course.

3. **Pick the correct minimums for airplane category and type of approach.** Now the advantages of the ILS's greater precision become obvious. The altitude minimum for the straight-in approach to Runway 6 is only 594 feet. The visibility minimum is 24. This is a transmissometer-measured visibility of 2,400 feet (less than one-third the length of the landing runway).

Decision height

The 594-foot altitude minimum for the straight-in approach to Runway 6 is a *decision height* (DH), not a minimum decent altitude. Decision height is the height at which a decision must be made during an ILS or other precision approach to continue the approach and land or to execute a missed approach. DH does not allow the maneuvering that is possible with MDA. You cannot level off at DH and continue in the hope of seeing the field and landing. At DH, you *must* decide to land or make a missed approach. These are the only options available at DH.

Even though DH is the minimum altitude on an ILS approach, you must also determine the MDA. If you drop below both DH and MDA on an ILS approach, a missed approach is mandatory. Obstacle clearance is not provided below MDA unless you are in a position to make a normal descent to a landing.

To continue the approach after reaching DH, you must comply with the criteria listed in FAR 91.175 (c) regarding "operation below DH or MDA." Summarized:

- The aircraft must continuously be in a position from which a descent to a landing can be made
- The visibility is not less than that prescribed for the approach in use

- One or more of the nine defined visual references must be distinctly visible and identifiable

Note that there is an additional approach possible with an ILS: a localizer approach shown as S-LOC 6 at Allentown. A localizer approach utilizes the high precision localizer beam for course guidance, but has no glide slope information. If the glide slope transmitter at the airport goes off the air or if the glide slope receiver in the airplane fails, you may continue the approach on the localizer alone. But the approach becomes nonprecision without the glide slope.

The minimum altitude for a localizer approach is a minimum descent altitude (MDA), not a DH. The circling minimums listed below the localizer minimums apply only to the localizer approach. You cannot circle to land out of a full ILS approach with the glide slope, unless the descent is stopped at the circling minimums shown on the approach chart for that specific approach.

Check adjustments for inoperative components table. The inoperable components table is carried in the front section of each set of NOS Instrument Approach Procedures. The visibility minimum increases with the outage of various approach and runway lights. The approach light code for Allentown Runway 6 is shown on the airport diagram (Fig. 26-2) at the approach end of the runway. Use the approach lights table in the front pages of the NOS Instrument Procedures booklet to identify the type of lighting for the ILS landing runway. For Allentown Runway 6, the system is MALSR (A_5). If it goes out, the visibility minimum increases $1/4$ mile for Category A airplanes on the ILS approach. That certainly makes sense. If there is a failure of the lights, you will need more visibility to find the runway, especially at night.

Be sure to examine higher minimums that might be required for inoperative components in the localizer approach. You will find the localizer increases lumped together in Section 3 of the table with many other nonprecision approaches.

Make adjustments required by fine print. None in this case.

Add the altimeter error. If you detect an altimeter error prior to flight, make no adjustments. Just *add* the error to the minimums on the approach.

Now run through the MARTHA check and make sure you understand all these elements as they apply to the intended approach, in this case the ILS RWY 6 approach at Allentown.

MA-Missed approach. "Climb to 3,000 via NE course of ABE LOC to NAZAR Int. and hold." This is different. The procedure calls for you to track outbound on the back course of the localizer. When tracking a localizer back course outbound, the normal tracking procedure is used—turn toward the needle the same as the ILS front course (normal sensing continues when outbound on an ILS back course).

R-Radios. The No. 1 nav will be set on the localizer frequency, 110.7. When you tune the localizer frequency, the glide slope is automatically received. Note that the localizer frequency is underlined. This indicates no voice transmission capability. The identifier is I-ABE. All localizer identifiers have the prefix I to eliminate any confusion between localizers and VORs.

The No. 2 nav will be set for the radial that establishes the holding fix on the missed approach. The station is SBJ (Solberg VOR) on 112.9. You will dial the Solberg 304 radial with the OBS.

ADF will be set to the ILS Runway 6 compass locator, LEEHI, identifier AB.

Marker beacons

Marker beacons send up a very narrow VHF beam to fix an airplane's position on the ILS final approach course. Beacons are tuned automatically whenever the receiver is operating. The outer marker transmits a continuous series of two audible dashes and a light flashes blue when you pass over the marker.

The middle marker transmits a continuous series of audible alternating dots and dashes and an amber light flashes. (Students find it easy to remember the code if they think of it as saying "You're HERE, you're HERE, you're HERE.") Some ILS approaches—mainly at the larger and busier airports—also have an inner marker. The inner marker transmits a continuous series of dots and flashes white.

Back to the MARTHA check.

T-Time. Pick the time from FAF to MAP based upon the best estimate of ground speed. All ILS approaches should be timed. If the glide slope goes out you can continue with a localizer approach without resetting anything. Use the published MDA instead of DH and the MAP will be determined by timing the final approach segment from the FAF to the MAP.

H-Heading. The final approach course heading in this case is 061°.

A-Altitude. DH for the straight-in ILS 6 approach is 594 feet. MDA for the straight-in localizer 6 approach is 820 feet. MDA for circling approach out of the localizer approach is 840 feet.

ILS tips

Request the full procedure where available to get the most out of ILS training. Large, busy airports will probably turn you down because of the heavy flow of traffic. Search out an uncontrolled airport with an ILS where you can practice as many full approach procedures as you wish.

On VFR cross-countries (and at your home airport if it has an ILS) contact approach control and request a "practice" ILS. Remain VFR and fly the approach unhooded when you don't have a safety pilot. (Don't forget collision avoidance—somebody must be looking!) Practice approaches will help you see the "big picture" of how ILS proceeds from step to step at different airports. In spite of common basic elements, all approaches—including ILS—are slightly different.

Practice holding on the localizer course (as shown on the Allentown ILS RWY 6 procedure, for example) without using the compass locator. It takes a little extra practice to set up a holding pattern on that very sensitive localizer needle. Turn outbound when the marker beacon starts to fade.

Always be prepared to switch from the full ILS to the localizer approach at any time, should the glide slope fail.

Be prepared to switch from an ILS or localizer approach to an NDB approach if there is a compass locator at the outer marker. Place the NDB approach chart beneath the ILS chart on the clipboard or yoke chart clip so you can look at it quickly if necessary.

Not all ILS approaches have compass locators or NDB approaches to the same runway. But if there is an NDB approach collocated with the ILS, it is excellent backup in case of transmitter or receiver failure.

When you have tuned and identified the codes of the localizer and any VOR you might need, turn the volume down or the audio off. Failure in these two systems will cause warning flags to appear. On the other hand, adjust the volume on the ADF—after identifying the NDB—to hear the ID faintly in the background. The only way to recognize an ADF or NDB failure is listening to the identifier. As long as you can hear the ID, all is well. (Unless you have inadvertently switched the ADF to REC instead of ADF.)

Remember that a power reduction of 100 RPM (or 1" of manifold pressure) produces a descent of 100 fpm minute; a power reduction of 500 rpm (or 5" of manifold pressure) produces a 500 fpm descent at constant airspeed.

Avoid overcontrolling on the final approach course by using rudder pressure only—no banking—to keep the localizer needle centered. Use gentle elevator pressure to keep the glide slope needle centered. (Heading changes should be limited to 2°, or at most 5°, at any one time. Because the rule of thumb is "never bank more than one-half the degree of heading change," there is no way you can see a 1° bank angle. So, why bother?)

"Pitch to the altitude" and "power to the airspeed" on the glide slope.

ILS/LOC identifier signals are usually not clearly audible until you are at least within 40° of the final approach course. When you are abeam the transmitter site, all you hear is a lot of scratch, which might cause you to miss important communications.

Back course approaches

ILS localizer antennas are located on the runway centerline about 1,000 feet beyond the far end of the approach runway. The localizer signal radiates in two directions:

- The "front course" is used for the ILS approach
- The "back course" (Fig. 26-2) provides a nonprecision approach path to the opposite end of the runway

The back course cannot be used for instrument approaches unless a specific approach procedure has been approved for that back course. Allentown has a back course approach—LOC BC RWY 24 (Fig. 26-3)—based upon the ILS RWY 6 localizer. Note the words **BACK COURSE** printed on the chart in large bold type. The reason for this warning is that back course approaches resemble conventional localizer front course approaches on approach charts. But they cannot be flown like front course approaches because the

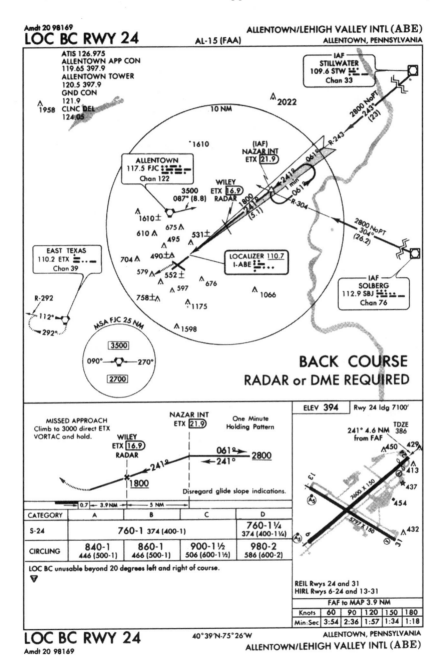

Fig. 26-3. *LOC BC RWY 24 is the back course approach at Allentown.*

needle moves in the opposite direction when heading inbound and no altitude (glide slope) information is available.

The reasons why the localizer needle moves in the opposite direction on a back course approach are fairly complicated questions to fully answer and have to do with the way the localizer signal radiates from the antenna. It is more important to understand that when you make a back course approach, the needle moves *opposite* the way it does on a front course.

Always turn away from the needle to make a heading correction during a back course approach; turn to the left when the needle moves to the right; turn to the right when the needle moves to the left. This is the opposite of VOR or ILS corrections where you always turn toward the needle.

Many students find it simpler to imagine that "they are the needle" and turn toward the bull's-eye at the center of the instrument for correction.

A simple way to remember this is that when you are traveling in the same direction as the course for the normal ILS, you make corrections in the same direction as the needle. When you are traveling in the *opposite direction* from the normal ILS, as you do on a back course approach, you make *opposite corrections*.

One or two practice sessions with back course approaches will make all this clear. Some other points to consider when working with back course approaches are:

- Although a back course does not have glide slope sensing, the glide slope needle might come alive periodically. These are false indications. Ignore them.
- The back course needle will be more sensitive than the front course because the localizer antenna array is usually located at the far end of the ILS front course runway (Fig. 26-2); thus, you will be operating just that much closer to the transmitting antenna during the back course approach.
- Some back course approaches have a marker beacon at the FAF to indicate where the approach descent begins. These back course marker beacons might be coded differently than beacons on the front course ILS. Back course markers transmit a continuous audible series of two dots and the white light flashes.

Localizer, LDA, and SDF approaches

As we have seen, localizer approaches can be made on an ILS system whenever the glide slope is out. Many airports have localizers with no glide slope transmitting equipment. This enables an approach with very precise centerline guidance into an airport with terrain and obstacles that rule out the glide slope required for the full ILS.

Figure 26-4 shows the localizer approach at Pittsfield, Massachusetts. The approach has three features we haven't encountered.

- The fine print below the profile says "Inoperative table does not apply." The minimums are already high because of surrounding mountains; therefore, inoperative component comments are unnecessary.

- The localizer (I-EIF, 108.3) also has DME. See the profile and note how DME is used as a cross-check at the IAF, FAF, marker, and MAP. DME is not required for the approach; however, it would certainly be nice to have DME with high obstacles all around. Note that the lower minimums apply for a straight-in approach to Runway 6 if DME is available.

- The third feature of interest is a *fan marker* rather than a marker beacon between the FAF and the MAP. This one is coded "R," as shown on the approach chart (dit-dah-dit) and activates the white light on the marker beacon panel. Fan markers are similar to other marker beacons but more powerful—100 watts output, whereas ILS beacons have an output of 3 watts or less.

LDA

A *localizer-type directional aid* (LDA) approach is uncommon but you still need to know about it. An LDA is a conventional localizer that is not aligned with the runway, for instance the LDA RWY 2 approach at Hartford-Brainard, Connecticut (Fig. 26-5). The runway heading is 020°, but the localizer approach course is 002°, too great a divergence to qualify for approval as a localizer approach.

Straight-in approaches are allowed with an LDA when the divergence between the localizer course and the runway does not exceed 30°, as is the case at Hartford/Brainard. If the divergence is greater than 30°, only circling approaches may be made.

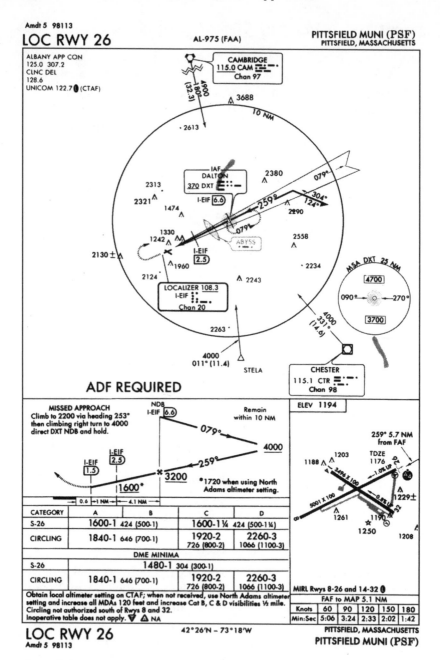

Fig. 26-4. *A localizer approach at Pittsfield, Massachusetts, with two unusual features: DME with the localizer and a fan marker on the final approach course.*

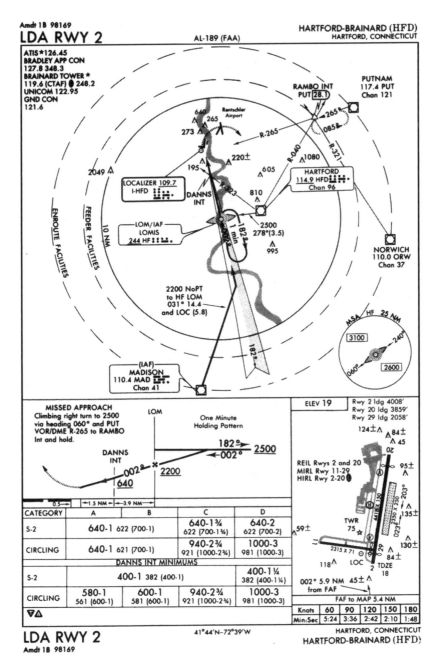

Fig. 26-5. *LDA RWY 2: A localizer directional aid approach at Hartford-Brainard, Connecticut.*

SDF

A *simplified directional facility* (SDF) (Fig. 26-6) transmits a course similar to a localizer but it is not as precise as a localizer. A localizer beam varies between 3° and 6° to produce a width of 700 feet at the landing threshold. The SDF transmitter is fixed at either 6° or 12°. Think of the SDF sensitivity somewhere between a VOR radial and an ILS localizer. The SDF might also be offset from the runway centerline.

You do not need to demonstrate back course, localizer, LDA, or SDF approaches during the instrument flight test. But if any are in your area, especially back courses, you should fly them whenever you have the opportunity. As a rated instrument pilot, you will be expected to execute these approaches whenever they are assigned by ATC. So practice now and avoid surprises and embarrassment later.

Radar assists

Expect radar assists on almost every approach. ATC monitors your en route progress with radar, and then hands you off to radar approach control (RAPCON), which then vectors you to the final approach course for the procedure in use. Approach control will often turn you directly onto the final approach course. (Procedure turns are prohibited on radar approaches.)

This doesn't mean that you shouldn't master the full procedures. On the contrary, you must know the full procedure for every approach you fly in order to visualize what approach control specifies. Be prepared to go to the full procedure if you lose radio communications.

This raises an interesting question. Suppose you are receiving radar vectors to intercept the final approach course of an ILS. But you have not been cleared for the approach itself and you lose radio communications at that point. You don't hear anything on either receiver or any voice frequency you might have tuned in. What do you do?

Carry out the lost communications procedure as specified by FAR 91.185 (Summarized):

- **VFR.** "Continue the flight under VFR and land as soon as practicable." In other words, break off the instrument approach and enter the normal VFR traffic pattern.
- **IFR.** Continue with the "route assigned in the last ATC clearance received." In this case the "route" would be "radar

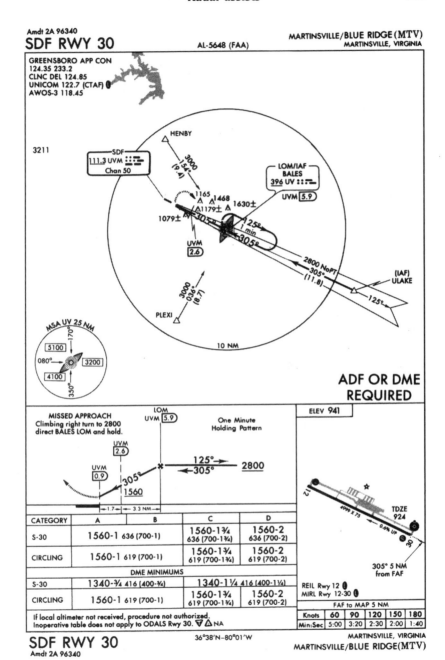

Fig. 26-6. *SDF RWY 30: A simplified directional facility approach at Martinsville/Blue Ridge, Virginia.*

vectors to the final approach course." Turn to intercept the final approach course and complete your approach and land or make a missed approach and depart for the filed alternate.

Radar monitoring and radar vectors are not radar approaches. To get a radar approach you must request it; a radar approach might be offered to airplanes in distress or to expedite traffic.

ASR approaches

The most common type of radar approach is the *airport surveillance radar* (ASR) approach, or *surveillance approach*. Look in the front of any NOS instrument approach procedure booklet to find a section that lists radar approaches available in the area covered by the booklet, along with their minimums (Fig. 26-7). Note the DH/MDA column. ASR approaches don't get very low.

Once approach control assigns an ASR approach, they will tell you *exactly* what to do. If you don't have the ATIS, they will provide it; they will provide lost communications and missed approach procedures; the approach controller will provide radar vectors and runway alignment.

Approach control will issue advance notice of where descent will begin and if the pilot requests it, they also provide recommended altitudes on final approach. You will hear this kind of phraseology:

"Prepare to descend in _____ miles."

"Published minimum descent altitude _____ feet."

"_____ miles from runway. Descend to your minimum descent altitude."

Approach control will keep talking to you on final and inform you if you are deviating from the final approach course:

"Heading _____, on course" (or) "well left" (or) "right of course."

As you get closer the controller will inform you of your distance from the MAP (and the recommended altitude, if requested):

"_____ miles from missed approach point."

"Recommended altitude is _____ feet."

98001

RADAR INSTRUMENT APPROACH MINIMUMS

ERIE, PA Amdt. 7A, OCT 29, 1997 ELEV **733**

ERIE INTL

RADAR- 121.0

	RWY GS/TCH/RPI	CAT	DH/ MDA-VIS	HAT/ HAA	CEIL-VIS	CAT	DH/ MDA-VIS	HAT/ HAA	CEIL-VIS
ASR	24	ABC	**1180**-¾	448	(500-¾)	D	**1180**-1	448	(500-1)
	6	ABCD	**1240**/50	508	(600-1)				
CIRCLING		AB	**1340**-1	608	(700-1)	C	**1340**-1¾	608	(700-1¾)
		D	**1360**-2	628	(700-2)				

Inoperative table does not apply to S-6; Categories C and D visibility increased to 1½ miles for inoperative SSALR.
For inoperative MALSR, increase S-24 categories A,B visibility to 1.
When control tower closed, procedure not authorized.

PHILADELPHIA, PA Amdt. 17, AUG 2, 1984 ELEV **22**

PHILADELPHIA INTL

RADAR- 128.4 343.6

	RWY GS/TCH/RPI	CAT	DH/ MDA-VIS	HAT/ HAA	CEIL-VIS	CAT	DH/ MDA-VIS	HAT/ HAA	CEIL-VIS
ASR	17	AB	**480**-½	469	(500-½)	C	**480**-¾	469	(500-¾)
		D	**480**-1	469	(500-1)				
	9R	AB	**520**/24	499	(500-½)	C	**520**/40	499	(500-¾)
		D	**520**/50	499	(500-1)				
	9L	AB	**480**-1	466	(500-1)	C	**480**-1¼	466	(500-1¼)
		D	**480**-1½	466	(500-1½)				
	35	AB	**540**-1	529	(600-1)	C	**540**-1½	529	(600-1½)
		D	**540**-1¾	529	(600-1¾)				
	27R	AB	**580**/24	569	(600-½)	C	**580**/50	569	(600-1)
		D	**580**/60	569	(600-1¼)				
	27L	AB	**580**/24	569	(600-½)	C	**580**/50	569	(600-1)
		D	**580**/60	569	(600-1¼)				
CIRCLING		A	**580**-1	559	(600-1)	B	**600**-1	579	(600-1)
		C	**600**-1½	579	(600-1½)	D	**640**-2	619	(700-2)

NE-2

RADAR INSTRUMENT APPROACH MINIMUMS

RADAR MINS

98001

Fig. 26-7. *Radar instrument approach minimums are listed in the front section of every NOS instrument approach procedures set.*

Surveillance approach guidance can be discontinued when the pilot reports the runway in sight. Approach will say:

"_______ *miles from runway*" (or) *over missed approach point, take over visually. If unable to proceed visually, execute a missed approach.*"

NO-GYRO approaches

A pilot flying with a partial panel can be given a "no-gyro approach." In this procedure, all turns are started and stopped by approach control. The pilot is expected to make standard rate turns until turning onto final approach, when all turns are half standard rate. Turns should be started immediately upon receiving instructions. The instructions couldn't be easier to follow, consisting of such directions as "turn right," "stop turn," "turning final, make all turns one-half standard rate."

Radar controllers need to practice surveillance and no-gyro approaches, so make a point of requesting this service frequently enough to keep everyone proficient.

Instructor note. ASR approaches might be available for airports other than those listed in the front section of the NOS Instrument Approach Procedures booklet. Telephone the RAPCON serving your area—or go visit the facility—and discuss your training needs with a supervisor. Facility personnel can inform you which airports are best for ASR practice and which ones can't handle practice ASR approaches because of heavy traffic.

PAR approaches

It's too bad that *precision approach radar* (PAR) approaches aren't widely available. They are easy to learn, easy to use, extremely accurate, and no needles have to be centered. Ask any current or former military pilot about PAR—known in the military as *ground controlled approach* (GCA)—and you will hear high praise for this precision approach.

With PAR, an electronic runway centerline and an electronic glide path are transmitted from equipment located alongside the runway in use. The precision radar tracks the incoming plane and shows the plane on a scope in relation to the electronic centerline and glide path. The radar is so sensitive that it can detect and display deviations of a few feet. Experienced controllers monitor the displays and tell incoming pilots what action to take to return to the centerline or to get back on the glide path.

A PAR approach begins like an ASR approach. The airplane is vectored onto the final approach course at a specified altitude. As it nears the electronic glide path, the final controller talks the pilot down. The sequence of instructions from the final controller runs something like this:

"Approaching glide path." (Ten to 30 seconds before final descent.)

"Begin descent." (On reaching the point where final descent is to start.)

"Heading _____. On glide path, on course." (To hold the airplane on course and on glide path.)

"Slightly above glide path, slightly left of course."

"Well above glide path, well left of course."

"Above glide path and coming down."

"Left of course and correcting."

"On course. On glide path."

"Three miles from touchdown."

"At decision height."

"Over approach lights."

"Over landing threshold. Contact tower after landing."

I don't know of any PAR approaches routinely available for civilian pilots to practice, nor did research for this book find one. So there is apparently no opportunity to practice this approach.

However, there are still several military airports that have GCA approaches. Their controllers are so skillful at talking down an airplane that you won't need much practice if you ever have to use a PAR in an emergency. Just call the nearest military airport on 121.5, tell them your emergency, request a "GCA," and do what they tell you. They'll get you down safely in an expeditious manner.

Visual and contact approaches

Be prepared to execute two more types of instrument approaches. The first is the *visual* approach. As you arrive on an IFR flight plan, approach control might clear you for a visual approach to the airport or to follow another airplane. ATC cannot issue a visual clearance unless the approach and landing can be accomplished in VFR conditions. Approach control uses the visual approach to expedite incoming IFR traffic when the airport is VFR.

Some visual approaches are so common around big busy airports that approach charts have been developed. Figure 26-8 is the RIVER VISUAL RWY 18 approach to Washington National Airport, Washington, D.C.

A *contact* approach might also be available in good weather conditions. The visual and contact approaches have a notable difference:

- Approach control assigns visual approaches
- A contact approach must be specifically requested by the pilot

Approach control cannot initiate a contact approach. You may request a contact approach if flying an instrument approach and the airplane breaks out clear of clouds with at least 1 mile flight visibility and you can expect to continue to the destination in these conditions.

"One mile and clear of clouds" rings a bell, doesn't it? Right, these are the minimums for special VFR and they are *extremely marginal minimums*. Legal scud running, some would call it.

Never request a contact approach at a strange airport. Contact approaches should only be used at familiar airports. And they should only be used when there is sufficient ceiling and visibility to depart from the instrument approach and enter a comfortable landing pattern.

Instrument takeoffs

It is best to introduce instrument takeoffs about midway in a training course when the student is able to fly the airplane by instruments in a very competent manner. This is usually the point where the student also begins concentrating on precision approaches, so instrument takeoffs are included in this chapter.

FAR 91.175 (f) covers takeoff minimums under IFR. If you read that regulation carefully, you will see that it does not prescribe any IFR takeoff minimums for aircraft operating under Part 91. That's you

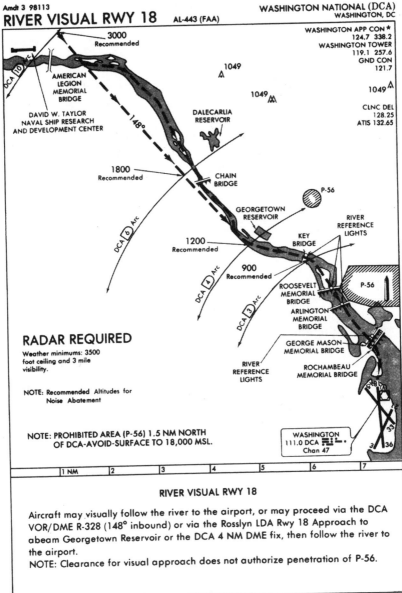

Amdt 3 98113

RIVER VISUAL RWY 18   AL-443 (FAA)

WASHINGTON NATIONAL (DCA)
WASHINGTON, DC

WASHINGTON APP CON *
124.7 338.2
WASHINGTON TOWER
119.1 257.6
GND CON
121.7

CLNC DEL
128.25
ATIS 132.65

3000
Recommended

AMERICAN
LEGION
MEMORIAL
BRIDGE

DAVID W. TAYLOR
NAVAL SHIP RESEARCH
AND DEVELOPMENT CENTER

1049

1049

1049

DALECARLIA
RESERVOIR

148°

1800
Recommended

CHAIN
BRIDGE

P-56

GEORGETOWN
RESERVOIR

RIVER
REFERENCE
LIGHTS

1200
Recommended

KEY
BRIDGE

DCA 6 Arc

900
Recommended

DCA 4 Arc

RADAR REQUIRED

Weather minimums: 3500
foot ceiling and 3 mile
visibility.

NOTE: Recommended Altitudes for
Noise Abatement

ROOSEVELT
MEMORIAL
BRIDGE

ARLINGTON
MEMORIAL
BRIDGE

DCA 3 Arc

P-56

GEORGE MASON
MEMORIAL BRIDGE

RIVER
REFERENCE
LIGHTS

ROCHAMBEAU
MEMORIAL BRIDGE

NOTE: PROHIBITED AREA (P-56) 1.5 NM NORTH
OF DCA-AVOID-SURFACE TO 18,000 MSL.

WASHINGTON
111.0 DCA
Chan 47

33

36

| 1 NM | 2 | 3 | 4 | 5 | 6 | 7 |

RIVER VISUAL RWY 18

Aircraft may visually follow the river to the airport, or may proceed via the DCA
VOR/DME R-328 (148° inbound) or via the Rosslyn LDA Rwy 18 Approach to
abeam Georgetown Reservoir or the DCA 4 NM DME fix, then follow the river to
the airport.
NOTE: Clearance for visual approach does not authorize penetration of P-56.

RIVER VISUAL RWY 18   38°51'N-77°02'W

WASHINGTON, DC
WASHINGTON NATIONAL (DCA)

Amdt 3 98113

Fig. 26-8. *RIVER VISUAL RWY 18: A published visual approach to
Washington National Airport, as depicted on an NOS instrument
approach procedures chart.*

and me. We may legally take off in any kind of weather. But if we do so when the ceiling and visibility are very low, we might be violating FAR 91.13, operating an aircraft in a "careless or reckless manner."

While it might be technically legal to take off when the ceiling and visibility are below IFR minimums, I think it is very poor judgment to do so. You should always be able to return immediately to the departure airport for an instrument approach if a problem develops. Furthermore, when you expect to climb into actual IFR soon after takeoff, you need a ceiling of approximately 200 feet to get established before entering the clouds.

However, practicing an instrument takeoff with a hood on is an exciting and interesting exercise. (It's always a revelation to students that they can do this, and their confidence grows considerably after they have tried a few.) The procedure is simple. When cleared for takeoff, taxi out and line up on the centerline as usual. Hold the brakes and add full power. Release the brakes when you have three-quarters to full power and anticipate the tendency of the airplane to turn left during the roll by applying right rudder pressure.

Use rudder pressure to keep the airplane rolling straight down the centerline by reference to the heading indicator. Don't try to force the plane into the air. Let the airspeed build up 5 knots or so beyond normal lift-off airspeed, then apply back pressure. Pitch up to the first mark above the horizon on the altitude indicator and hold that attitude until a steady, positive rate of climb shows on the VSI and you have reached 500 feet above the airport. Then adjust the attitude for a normal climb.

You have to make corrections promptly to counter the tendency of the airplane to drift left during the takeoff roll. Fortunately, the heading indicator is sensitive enough so that you can detect very slight changes of heading. React quickly and positively to these slight movements of the heading indicator with rudder pressure.

With a little experience you will find that you can glance out the left window and see whether you are drifting left or right. This comes after you have done two or three instrument takeoffs and are confident enough to take your eyes off the heading indicator for a few seconds then look back to it promptly without being distracted.

Flying "away" from the needle

It is time to discuss one last approach, the localizer back course, with an examination of the Kahului, Hawaii, LOC DME (BACK CRS) Rwy 20 approach (Fig. 26-9).

Two main facts to remember concerning back course approaches are (1) you cannot use the glide slope information and (2) the localizer needle movement is reversed. This means that you have to be extraordinarily cautious and think through the approach because you will have to consciously fight your previous training. On a back course, if the OBI moves to the left, don't turn left to bring it back, turn to the right; if it moves to the right, turn to the left. *Turn away from the needle.* As long as you keep that in mind, you should have no trouble.

The approach chart shows that if you are making arc approaches from the east or west, the lead-in radials are indicated. There are two initial approach fixes from the east, OPANA, where the 069-degree radial of the OGG VOR crosses the 13 DME arc, and an unnamed intersection where the 084-degree radial of the OGG VOR crosses the 13 DME arc.

If you are beginning the approach from OPANA, remain at 3,000 feet and fly the arc until passing the 032-degree lead-in radial from the OGG VOR, at which time you can begin the turn inbound to intercept the back course.

If you are beginning the approach from the unnamed intersection where the 084-degree radial from OGG VOR intersects the 13 DME arc, you would have to remain at 5,500 feet until crossing OPANA, then you could descend to 3,000 feet and continue the approach as described previously.

From the other direction, the IAF is PLUMB, which is 8.0 miles northwest of the 13.0 DME fix on the OGG 320-degree radial. The minimum altitude here is also 3,000 feet. The lead-in radial is the 017-degree radial. Remember that you are flying the localizer back course, so the ILS will be tuned and identified on the number 1 VOR receiver, while the OGG VOR should be on the number 2 receiver. In fact, if you read the notes in the profile view, you will see that separate localizer and DME receivers are required, and that you are using the DME information from the OGG VOR. Even

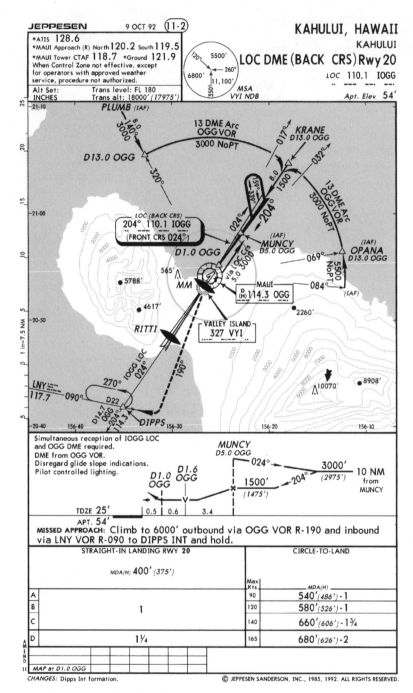

Fig. 26-9 *Reproduced with permission of Jeppesen Sanderson, Inc. Not for use in navigation.*

though the ILS sensing gives you reverse information, I always like to set the OBS to the inbound course (204 degrees in this case) just to keep myself orientated.

You will use the DME to determine letdown points on this specific approach because the approach is made over water. After you turn inbound at KRANE, you can descend to 1,500 feet MSL. You have 1,500 feet to lose, but you have 8 miles (about 4 minutes) to lose them, so if you maintain a reasonable descent rate of between 500 and 600 fpm, you will get down there in plenty of time. Fly at 1,500 feet to MUNCY, which is the 5.0 DME fix from the OGG VOR. It is also the FAF. After MUNCY, descend to minimums, which are 400 feet MSL (377 feet AGL).

The MAP here, 1.0 DME from the OGG VOR, is 0.4 miles from the runway threshold. If you have to execute the missed approach, climb to 3,000 feet MSL on the 190-degree radial of the OGG VOR. You would have set that heading on the number 2 OBS upon turning onto final approach, so now all you have to do is fly it.

You will be holding at DIPPS intersection, which can be identified four ways:

- The intersection of the Lanai (LNY) 090-degree radial and the front course ILS to Kahului
- The LNY 090-degree radial and the OGG VOR 204-degree radial
- The LNY 090-degree radial and the 023-degree bearing to the McGregor Point (MPH) NDB
- The 22.2 DME fix on the LNY 090-degree radial

If, after a missed approach on the back course, you want to try a front course ILS to Kahului, you would probably just turn the number 1 OBS to the inbound (024-degree) ILS course. By the time you're about 10 miles south of the OGG VOR on the missed approach, you should have the heading pretty well tied down, so you can retune the number 2 receiver to the LNY VOR, and set the OBS to 270 degrees TO.

Turn right (toward LNY) as the number 2 needle centers, fly until the number 1 needle (ILS) centers, and make a 30-degree turn to the right for a teardrop entry into the nonstandard (left-hand) holding pattern. While flying the 1-minute outbound leg, you would reset the number 2 (LNY) OBS to 090 degrees FROM, which will be the inbound holding pattern course.

Vectors to final—a warning

On many approaches, not just the approaches discussed in this chapter, you will be provided radar vectors to the final approach course. In such cases, it is crucial that you be aware of the following from the *Aeronautical Information Manual*:

> *After release to approach control, aircraft are vectored to the final approach course (ILS, MLS, VOR, ADF, etc.). Radar vectors and altitude or flight levels will be issued as required for spacing and separating aircraft.* Therefore, pilots must not deviate from the headings issued by approach control. *Aircraft will normally be informed when it is necessary to vector across the final approach course for spacing or other reasons. If approach course crossing is imminent and the pilot has not been informed that he (or she) will be vectored across the final approach course he (or she) should query the controller. . . . The pilot is not expected to turn inbound on the final approach course unless the approach clearance has been issued.*

Here we have a possible disaster. A low-time instrument pilot, suffering from fatigue, is vectored through the final approach course for separation purposes. The controller gets busy with another aircraft. The pilot, not familiar with terrain or perhaps misreading the chart, could conceivably fly into a mountain. This points out a good reason to study the entire approach chart, including the heights of all major obstructions, not just the diagram of the approach itself.

27

Multiengine aircraft approaches

Instrument flying and multiengine flying are two different skills, but in every pilot's career, there comes a time when they merge. It was once assumed that a pilot with an instrument rating who took and passed a noninstrument multiengine practical test could put it all together and fly safely as an instrument/multiengine pilot. The previous instrument training was supposed to "overlap" the new multiengine training that a pilot received.

A series of accidents in the early 1980s, however, changed all that. The accidents involved pilots flying in a multiengine airplane through instrument conditions. In each case the probable cause of the accident was related to a conflict rather than a coordination of instrument and multiengine skills. The FAA began to realize that a person with instrument skills or a person with multiengine skills does not automatically have the combination of instrument/multiengine skills. So the multiengine testing procedures were changed.

Today, you must announce at the beginning of the multiengine checkride whether you are seeking an IFR or a VFR multiengine rating.If you indicate IFR, there is no turning back. If while on the IFR multiengine checkride you fail on an instrument-related task, you cannot change your mind and say, "I really only wanted the VFR multiengine ride." The Practical Test Standards (PTS) are clear on the difference between the IFR and VFR multiengine tests. Following is the note pertaining to IFR and VFR examinations from the PTS:

Note: *If an applicant holds a private or commercial pilot certificate with airplane single-engine land and instrument ratings and seeks to add an airplane multiengine land rating, the applicant is required to demonstrate competency in all TASKS of Area of Operation VI.*

If the applicant elects not to demonstrate competency in instrument flight, the applicant's multiengine privileges will be limited to VFR only. To remove this restriction, the pilot must demonstrate competency in all TASKS of Area of Operation VI. If the applicant elects to demonstrate competency in the TASKS of Area of Operation VI, then fails one or more of those TASKS, the applicant will have failed the practical test. After the test is initiated, the applicant will not be permitted to revert to the "VFR only" option.

So if a person seeks an IFR multiengine rating, he or she must take and pass all the tasks of Area of Operation VI. What are these tasks? Area of Operation VI has three sections:

VI. INSTRUMENT FLIGHT

    A. Engine Failure During Straight-And-Level Flight and Turns

    B. Instrument Approach—All Engines Operating

    C. Instrument Approach—One Engine Inoperative

If a person is only interested in a VFR multiengine rating, he or she takes the entire multiengine test, except Area of Operation VI. Area of Operation VI is the attempt by the FAA to ensure that a pilot can blend instrument and multiengine skills safely. This blend is no longer assumed, but must be tested.

One engine inoperative instrument approach

The biggest challenge of the IFR portion of the multiengine flight test is the "one engine inoperative" instrument approach. The examiner will ask the applicant to perform an instrument approach, and at some point, the examiner will also simulate a failed engine. This simulated engine failure usually occurs at the busiest point in the approach. The examiner will wait until you are in the middle of reading back an approach clearance, selecting a frequency, or making an intercept to pull a throttle back.

Because this is an actual instrument approach that will terminate low to the ground, the examiner should not pull back a mixture control, but rather simulate with the throttle control. The PTS outlines what the applicant is expected to do:

Area of operation VI

Task C: INSTRUMENT APPROACH—ONE ENGINE INOPERATIVE (AMEL)

1. *Objective.* To determine that the applicant:

   a. Exhibits commercial pilot knowledge by explaining the multiengine procedures used during a published instrument approach with one engine inoperative.

   b. Requests and receives an actual or simulated clearance for a published instrument approach.

   c. Recognizes engine failure promptly.

   d. Sets the engine controls, reduces drag, and identifies and verifies the inoperative engine.

   e. Establishes the best engine-inoperative airspeed and trims the airplane.

   f. Verifies the accomplishment of the prescribed checklist procedures for securing the inoperative engine.

   g. Establishes and maintains a bank toward the operating engine, as necessary, for best performance.

   h. Attempts to determine the reason for the engine malfunction.

   i. Requests and receives an actual or simulated clearance for a published instrument approach with one engine inoperative.

   j. Follows instructions and instrument approach procedures.

   k. Recites the missed approach procedure and decides on the point at which the approach will continue or discontinue, considering the performance capability of the airplane.

   l. Descends on course so as to arrive at the DH or MDA, whichever is appropriate, in a position from which a normal landing can be made straight-in or circling.

   m. Maintains the specified airspeed, +/- 10 knots.

   n. Avoids full-scale deflection on the CDI or glide slope indicators, descent below minimums, or exceeding the radius of turn as dictated by the visibility minimums for the aircraft approach category, while circling.

o. Communicates properly with ATC.

p. Completes a safe landing.

As you can see, you would have your hands full. An instrument approach alone is a challenge, and an engine failure is a challenge. When they happen on top of one another, it will be more than an unprepared pilot can handle. Look at letter *h*, above: "Attempts to determine the reason for the engine malfunction." This calls for the pilot in mid-approach to troubleshoot the problem. Imagine holding a localizer against a strong crosswind and a failed engine, reporting the outer marker inbound, memorizing the missed approach procedure, and still finding time to determine what has gone wrong with the bad engine. All the time spent learning the airplane's systems and practicing the emergency procedures will pay off here.

Speaking of the missed approach procedure, look at letter *k*: "Recites the missed approach procedure and decides on the point at which the approach will continue or discontinue, considering the performance capability of the airplane." You cannot assume that with one engine failed, it is even possible to execute a missed approach procedure. The procedure might call for a climb that one engine cannot provide.

In light of this situation, the pilot might need to alter the instrument approach and stay higher than MDA or DH. Of course, by doing this, the pilot runs the risk of not getting under the clouds and ever seeing the airport. It could be a tough call: Do I continue a descent down to an altitude that is in fact below a "point of no single-engine return?" Doing this will increase my chances of seeing the runway, but I better eventually get below the clouds because the low altitude and poor single-engine climb performance makes a missed approach impossible. Or do I stay high at or above a safe go-around altitude and hope the clouds are not down at minimums? The answer to these questions require the pilot to determine where a "point of no single-engine return" is located based on many factors.

How steeply must I climb during the missed approach procedure to miss the terrain? How will the wind and temperature affect the situation? Is the engine failure under control or is there a greater danger from fire? All this and more must be considered within only a few seconds.

For pilots on checkrides, this task is somewhat easier because we know the situation is coming. A single-engine instrument approach will not be an unexpected event to a person on the IFR multiengine test. The applicant could think this through even before takeoff and have a plan of action ready. In reality, we should always be that prepared. Pilots should think about these actions on every multi-engine approach in anticipation of the time when they encounter an actual emergency during, or prior to, an instrument approach.

If the pilot does descend below the clouds and spots the airport while on an instrument approach with one engine failed, he or she will need to face yet another decision. Should I land straight in or circle to land? This should also have been anticipated. If a straight-in landing is possible, the pilot should just continue to descend with landing gear down. If a circle to land is required, this will probably mean that the landing gear must be temporally raised to reduce drag in the level flight portion of the circle maneuver.

The idea of instrument/multiengine involves more than checkride procedures. For most pilots, the instrument/multiengine plateau is the highest elevation reached prior to becoming a professional pilot. The instrument/multiengine level is required to move on to charter flying, corporations, or the airlines, so this level is the master's degree of flying. The master's level requires a higher level of thinking and problem solving.

Real-world training

An important but often unspoken part of the master's level is the fact that you no longer are a student and therefore should not think like a student. This is not to say that you ever stop learning—in that sense, we are all students. However, this idea is different. Instrument/multiengine pilots must function in the real world, not the flight training world, and unfortunately, there is a difference. Are there things that we do in the training environment for the sake of training that we would not do in the real-world environment?

The answer is yes. Hence, real problems can develop. When brought up in the training environment, some students can only function in that environment and do not understand the difference between the training and real-world environments. Here are some examples to illustrate this concept. See if you recognize yourself in any of these circumstances.

After a student and an instructor fly for approximately one hour, conducting instrument approaches and multiengine emergency procedures, they fly back toward the home airport. The home airport has an instrument approach that is incorporated into the flight lesson. The instructor simulates a radar vector to this approach. What is this student thinking about this approach? The student is probably thinking that this approach is simply being used to conclude the lesson. The student might know that the instructor is due back to begin another flight lesson. The student personally might also need to be on the ground so that he or she can make it to work on time.

The student also realizes that the weather is actually VFR and wearing the "hood" is just a game instructors play. The student and instructor are both now solidly in the "last approach syndrome." It would be a real surprise to both of them if they were not able to land at the end of this approach. Has either one of them even looked at the missed approach procedure?

Contrast that training situation with another flight to the same airport using the same instrument approach. This time the pilot has flown several hours to get here. The clouds are low and the outcome of the approach is seriously in doubt. The AWOS or ATIS report of the cloud bases are making the pilot think a missed approach is inevitable. It will be a surprise if the airport is spotted and a landing made.

Do you see how differently the pilots in the two situations are thinking? The pilot in the training environment figures that a landing is assured, and a missed approach would be a surprise. The pilot in the real environment assumes a missed approach is assured, and a landing would be a pleasant surprise. What will the pilot who has only been exposed to the training environment do when he or she is faced with a real-world situation? The "train-brain" pilot will be tempted to fly below the MDA or delay a missed approach past the missed approach point because it has not entered his or her mind yet that a missed approach is inevitable. People get killed when they use a training environment thought process in the real world.

This "train-brain" tendency comes up again and again. One time, I was conducting an IFR check with a student. As a part of the test, we flew an NDB approach about 35 miles away from his home airport. The approach turned out to be a disaster; everything went wrong. When the flight was over, the student said, "You know it really was not fair to make me fly that NDB down there because I have never practiced that one."

How would you have responded? This comment told me more about this student's lack of preparedness for the real world than any blown approach ever could have. In his mind, the only "fair" approaches were the ones that his instructor and he had drilled on over and over again. I guess this guy expected to have an instrument rating someday that was restricted to only the three or four instrument approaches that he had worked on.

Have you ever pulled out a few most often flown instrument approach charts from the chart book? Students do this on checks many times. By the time I get to the airplane, the student has selected about four approaches out of the big book and attached them to a knee board. I can only assume that these are the student's most practiced approaches. I guess these students expect me to have them fly only one of these that they have chosen.

There is a fine line here. Is pulling approach charts out of the book just good planning, or is it a crutch? It might be good planning, but in a real-world environment, how can any pilot know before takeoff which approaches he or she might need to fly on that particular flight? The wind, traffic conflicts, or a missed approach could cause a pilot to end up flying an approach that could not have been planned for. When charts get pulled out the question remains: Could this student find an approach other than what was preselected and fly an unplanned approach with equal accuracy?

Flying the airplane and simultaneously thumbing through the chart book to find a specific approach is not that easy. Students who pull charts beforehand, but who must later find another chart from the book, almost always will lose their heading and altitude while fumbling around in the book. When something does not go according to plan, the student begins to panic and his or her flying skills suffer. This is classic "train-brain."

In the training environment, the situation is controlled and the outcome is known; pulling charts might be good lesson planning. But in the real-world environment, the outcome is unknown and the student's skills should not be limited. Flight instructors start out with students in the training environment, but eventually good instructors must introduce the real environment.

One time on an IFR check, I had a student who pulled off his IFR hood when we reached an approach's MDA. Of course, I had not wanted the student to see the airport yet, hoping to watch him track to the missed approach point. "Why did you take the hood off?" I asked. "Because I'm at MDA. We're done!" the student said, a little agitated that I had even asked such a question. What is wrong with this situation? The student had been exposed only to the training environment where instructors have the power to make clouds go away at convenient times. Was that student ready for the real world of IFR?

Decision training must be a part of real-world training. On many occasions I have kept students in simulated IFR conditions to the missed approach point during an approach that they thought was the last of the day. Early in the approach I will say, "I will let you know if we ever get out of these (simulated) clouds." However, I say that without any intention of saying anything else throughout the remainder of the approach. When the student nears the decision height, the student might fail to make a decision. On several occasions students have descended to DH on the glide slope and arrived at the point where a go-around must begin if the runway is not in sight.

All along, the students expected me to say, "OK, take your hood off" before arrival at DH, but I said nothing. They panic. They actually turn their head and look through the long IFR hood at me with a question on their face. They want to know why I did not make the clouds go away, and at this point, they have not contemplated the possibility of a missed approach. This indecision will cause us to get off the localizer and below DH before a confused student, totally saturated in "train-brain," does anything. Why did the student pause? Why wasn't the student applying full power at the DH? Because the student was not thinking in the real-world environment.

A good instrument/multiengine instructor will foster thinking from the real world. Usually we imagine instrument/multiengine training to be filled with physical skills, such as holding a rudder against a good engine when the other engine has failed or making a localizer and glide slope intercept. But good instruction to the master's degree level should also include thinking skills.

Now you should be ready to tackle instrument flying...if you're mentally and physically up to the challenge.

Section 5

The human factor

28

What causes IFR accidents?

Every aspect of instrument flying—planning, scanning, and interpreting instruments, approach flying, and knowing your airplane and the regulations well enough to correctly respond to emergencies—depends on your physical and mental state at the time your skills are needed. To maximize your safety and minimize the risk of an instrument flight, you need to be familiar with factors that affect your physical and mental capabilities and be able to recognize when you are or may not be at your best. You need to be able to step back from what you're doing and assess whether human factors are preventing you from flyng well or making a good decision. In this section, let's explore the human factor as it applies to flying an instrument airplane.

The IFR accident record

There's little doubt that your instrument rating makes you a safer pilot. Training toward the instrument rating increases your flying precision and familiarity with the airplane, which in turn increases your level of safety even if you never venture into the clouds. The knowledge you gain about aviation weather hazards, and the experience you garner in working with Air Traffic Control and weather sources en route, all can help you be a better pilot even when flying in Visual Meteorological Conditions (VMC). In fact, according to the AOPA Air Safety Foundation,

> *The more pilots fly solely by reference to instruments, the safer they are. Having an instrument rating, and/or having experience flying solely by reference to instruments correlates well with avoidance of weather-related accidents.*

Instrument pilots should remember, though, that

537

> *Having an instrument rating doesn't immunize pilots against weather accidents. 43 percent of weather accidents involve pilots with instrument ratings*

and

> *The weather alone is not to blame for weather-related accidents. According to the NTSB, 85 percent of (weather-related) accidents involved pilot error.*

Safely exercising your instrument flight privileges, then, requires that you obtain current and forecast weather information and that you make good decisions using that information and the weather you actually encounter en route.

Aviation weather hazards fall into four categories: thunderstorms, turbulence, airframe ice, and reduced visibility. Let's take a brief look at each:

Thunderstorms

According to the AOPA Air Safety Foundation (ASF), thunderstorms are involved in only about three percent of all general aviation, weather-related accidents. Thunderstorm penetration, though, is fatal two-thirds of the time. "Many thunderstorm accidents appear," states the ASF, "to have taken place after pilots knowingly flew towards areas of convective activity." Pilots felt pressured, for some external reason, to begin or continue flight even when thunderstorms were reported or forecast. Thunderstorm-related accidents, then, are really the result of pilot decision making.

Turbulence

Turbulence not directly related to thunderstorm activity accounts for a very few general aviation accidents. Those rare accidents are usually not fatal. If you count gusty or strong surface winds near airports, though, turbulent wind contributes to almost half of all lightplane crashes. Don't be overconfident about your ability to compensate for strong or gusty surface winds, or for the airplane's capability to handle them.

Airframe ice

Airframe icing is a factor in only a few more accidents than thunderstorms (approximately five percent of the total), but like thunderstorm mishaps, icing-encounter accidents are very likely to result in the death of pilots and passengers. Airframe ice formation is one of the "great unknowns" of aviation because icing conditions are usually fleeting and therefore difficult to predict. More power and "ice-certified" airplanes tend to be involved in a greater percentage of airframe icing accidents, perhaps because the pilot is lulled into a false confidence that the airplane can fly through an extended ice accumulation.

Flight Service tends to "overpredict" icing because it's hard to forecast cloud and temperature conditions aloft and between weather reporting points. Pilots need to heed warnings of airframe ice and never consider entering an area of suspected ice without a good plan for staying clear of ice-forming clouds.

Reduced visibility

Fog, precipitation, and low clouds contribute to the biggest percentage of weather-related accidents—as much as 70 percent of the total. Strangely, almost half of the "attempt visual flight into instrument conditions" type of crash happens with an instrument-related pilot behind the controls. The rest of the IFR-capable pilots involved in visibility-related accidents crashed as a result of flying improper arrival or departure procedures.

Be confident in your ability to fly on instruments because if you've earned the rating, you've proven your ability. But don't become overconfident and allow yourself into situations you're not ready to handle. Be especially wary if you're tempted to make multiple attempts at laying the same instrument approach. In many of the "improper instrument procedures" accidents, the pilot missed the approach because of poor visibility on the first try, then crashed on the second, third, fourth, or even fifth attempt. Unless you have good reason to believe conditions have improved after your first attempt at the approach, or you can identify a specific error you made the first time that caused you to "miss" an error you know you'll avoid the second time, don't even consider reflying the same approach. If you miss the approach after a valid second attempt, diversion to some other airport, with better weather, should be a mandatory call.

You can see, then, that weather-related instrument accidents are almost always the result of pilot decisions, often caused by stress or some other external factor that affects the pilot's ability to fly safely. Let's look more closely at human factors in aviation.

29

The pilot

Preflighting the pilot

According to reports from the Federal Aviation Administration, the National Transportation Safety Board, and the National Aeronautics and Space Administration, the single most hazardous element in aviation is the pilot. That's right—the pilot.

They have very sterile wording to sum up the probable causes of aircraft accidents/incidents: inadequate preflight planning/preparation; attempted operation beyond experience/ability level; or improper in-flight decisions or planning. These are all very polite ways of saying the same thing—the pilot blew it.

The sad part of all this is that most aircraft accidents could have been and can be avoided. Only a very small portion is caused by mechanical failure. The rest are pilot-induced through overconfidence, ignorance, or a lackadaisical approach to their flight technique. In short, they ignore the ABCs.

Consider this random selection from the files of the NTSB. It is File # 3-3851. On November 12, 1996, near Wise, North Carolina, a Piper PA28 departed from a cloudbank with a portion of one wing missing. The aircraft crashed in an uncontrolled descent, and the pilot and three passengers were killed.

As is the case with many aircraft accidents, the exact cause will never be known. What *is* known is that the aircraft departed Sussex, New Jersey, en route to Myrtle Beach, South Carolina, with the pilot and three passengers on board. The pilot, age 37, had 100 hours total time with only five hours in type. He was not instrument rated. No known weather briefing was obtained prior to takeoff. An in-flight forecast was received that called for marginal VFR (if not IFR) weather conditions. The flight continued until the aircraft impacted the ground at 1313 local time.

The weather at the crash site was 600 feet overcast, four miles visibility with fog and drizzle (definitely IFR). The NTSB lists the probable causes as operation beyond experience/ability level; spatial disorientation (vertigo); exceeded stress limits of the aircraft, resulting in aircraft separation in flight. Type of flight plan—none.

It was a blueprint for a tragedy. A relatively inexperienced, noninstrument-rated private pilot departed into known IFR weather with no briefing or flight plan. Result—four needless deaths.

Where would the prevention of this type of accident start? With the pilot. A noninstrument-rated private pilot has about as much business attempting a four-state cross-country flight in marginal weather as I would racing at Indianapolis. The odds are about the same (believe me, I have absolutely no business racing at Indianapolis). The end results are quite predictable in both instances—tragedy. If you are a person who likes to play the odds and gamble with life, go ahead and gamble with your own, but don't take innocent people with you.

Maybe the people in the aforementioned case knew what they were getting into, but I suspect they didn't. The average passenger places blind trust in the pilot of an aircraft. Often, this trust is not warranted. A pilot's license does not automatically make someone the keeper of all wisdom. It means that he, at the time of the checkride, possessed the *minimum* qualifications for the license or rating. Maybe the skills have not been further developed, or even worse, maybe they have deteriorated. In any case, the pilot should know his/her limitations and those of the aircraft to operate it in a safe, sensible manner.

Let's return to square one, the *preflight*. The total preflight begins and ends with the one who is responsible for the safe conduct of each flight—the pilot.

In any worthwhile human endeavor, there must be preplanning. Depending on the type of flight to be undertaken, the pilot has many plans to make. The flight could be anything from a nice Sunday afternoon hop around the patch to a transcontinental flight. The planning will vary, but not very much. But the first and foremost consideration should be the pilot. Is he or she really ready? For each and every flight, the pilots have to ask some questions of themselves.

Fatigue

The complexities of operating an aircraft are more fatiguing than many people realize. The constant attention to heading, altitude, airspeed, radio tuning, and communications can tire a person rather quickly. This is especially true if you are a student pilot or a relatively new pilot. The physical duties, when combined with the mental stress of learning to fly, can tire even the most robust person in a hurry. This means you must learn to pace yourself. If you notice you are making random errors that you normally would not make, you are probably reaching the saturation point and should end the lesson. If your instructor does not realize that you are tired, tell him. Very little learning can occur if a person is fatigued, and the dangers of flying in a fatigued state cannot be overemphasized. Remember, fatigue is indiscriminate; no one is immune.

If you know that fatigue is dangerous, then what steps should be taken to see that it does not overtake you? This depends a great deal on the type of situation you're in. As mentioned previously, a student should tell the instructor when he becomes tired. Hopefully, the competent instructor will notice the symptoms in the student and end the lesson.

If you are on a long cross-country flight, there are several ways to cope with fatigue. Play with the radios. Take frequent crosschecks of the instruments. Change the seat position. Change seats. But don't punch on the autopilot and go in the back for a rest. You might have a much longer nap than you planned on—like for eternity. If you are in a small two or four-place aircraft, get out of the left seat and slide over to your right until you are sitting on half of each front seat. Now you can play the oldest coordination game known to aviation. You place your left foot on the right rudder pedal in the left-hand side and your right foot on the left rudder in the right-hand side. Put your right hand on the right yoke and your left hand on the left yoke. Now try to fly straight and level. If this doesn't keep you awake, nothing will.

Physical or mental fatigue can be very deceptive. It can sneak up on anyone at anytime, and it is frequently deadly. A number of years ago, a student graduated from our University Flight Technology Program. He graduated with his Commercial, Instrument, Multi, and CFI certificates. This young man decided he would enter the aviation field as an air taxi/charter pilot. He was an average, or maybe slightly better pilot and he found a job flying charter for a firm in the north. I saw him a few months later and inquired as to how everything was going. He told me he was "flying his butt off" building a lot of twin time but was a little tired of going places at all hours of the day and night. I told him I was glad he was happy, but he could keep that job. I'm very content to be at home at night (no knock to you air taxi/charter pilots; it's just my personal preference).

The next time I heard his name mentioned was when I was told that he had died. It seems he had been flying parts to some factory in Canada and had made several trips in succession starting in the evening and continuing through until daybreak. On the last trip, just as a new day was dawning, he fell asleep. He slumped over the yoke and went nearly straight in. It was a needless tragedy induced by fatigue. He needn't have taken that last load. He could have told someone to forget it. But he didn't. He probably thought he could make it. Or maybe he chose not to recognize the signs of fatigue. Maybe he was afraid he would be fired if he refused that one more flight. Which is worse? Being fired or being dead? As long as I am breathing, no one will ever push me into a life-or-death situation such as taking just one more flight when I am too tired to do the job safely.

I realize the above-mentioned case is extreme, but it happened. And it doesn't always happen to the other guy. Be aware of the fatigue factor. Plan for it. Do something about it. Stay alive.

Every case of pilot fatigue does not necessarily end in disaster. It happens to all of us to some degree every time we fly. Not too many years ago, my boss and I set out on a beautiful Saturday morning to pick up a new Decathlon we had purchased for our aerobatic program. We left our home base at Lawrenceville, Illinois, in our Cessna 310 and cruised up the length of Illinois across northern Indiana and well up into Michigan. We stopped there in northern lower Michigan only long enough to sign the purchase papers and have a Coke. We then proceeded across Lake Michigan to a place in northern Illinois where the plane was actually located. It was nearly noon by this time, and as usual, the plane we were to pick up wasn't ready.

Finally, at about 2 P.M., I departed for home in our new aerobatic mount. I was fully fresh in the knowledge that I had at least two hours to play with the new plane during the 225-mile flight home. And play I did.

I leveled at 3,000 feet AGL, tuned the radio to the Pontiac VOR, checked the chart to see that I was clear of any airways, cleared the area for traffic, and upside down we went. Inverted VOR navigation—what fun! Tiring of this, it was time to check all the other little maneuvers that bring such joy. A loop, a slow roll, a snap roll, a hammerhead, an Immelmann (had to reverse my direction from the hammerhead, you know), more inverted flight, and so on. Finally, at about Champaign, I began to come down from the exhilaration of all that fun and games. I looked at my watch and found I had been airborne for over an hour. Champaign was off to my left at least six miles, and according to my route of flight, it should have been directly below me. However, I knew the country well and mentally noted the towns over which I had passed. I tuned in the Lawrenceville VOR, centered the needle, and proceeded home.

As I flew toward home, I noticed a very peculiar feeling beginning to overtake me. My arms felt as though they were made of lead. I couldn't think quite right. The VOR needle kept wandering around (I knew it was just an unreliable radio). Nothing seemed to go right and the last 100 miles seemed to last forever.

As I neared home, a sudden burst of adrenaline hit me and I was overtaken with a case of the "look at me's." Surely I had enough energy left to make at least an inverted pass before landing, I thought. I did—at a safe altitude—and landed.

I taxied the beautiful red, white, and blue machine toward the ramp, eagerly awaiting the accolades from the crowd of people who wanted to get a look at the new plane. As I reached the hangar, I swung the Decathlon around, pulling the mixture as I neared a stop, opened the door, jumped out, and found I could not stand up. My head was swirling and my legs would not support me. My whole body was writhing in agony. Not wanting anyone to know of my plight, I knelt down and pretended to examine the belly of the plane.

I knew that sooner or later I would have to get up or call for help. After a few minutes, I slowly made it to my feet and leaned against the plane. I did my best to try to answer the questions as to how it flew, how it rolled, and so on. My answers were very short.

In the back of my mind, I recalled that I still had to fly our Stearman over to an adjacent field in preparation for an airshow the next day. After an hour or so, I did. I then went home, took two aspirins, and sat down to think.

The things I thought were not very pleasant, and the only reason I have embarked on this self-incriminating expose of stupidity is that

I pray you will learn from it. Profit from my mistake by being smarter than I was.

To put it mildly, I was fatigued. I was so tired that I unknowingly put myself in a dangerous situation. All students are taught early in flight training never to exceed their, or the aircraft's, capabilities. I certainly didn't exceed the Decathlon's, but I exceeded my own physical capabilities about as far as I ever care to. Remember that no one is immune from fatigue. No matter how physically fit you are, or how mentally acute, you can still fall prey to fatigue. It plays a role in all types of aviation: instruction, charter, cross-country, acrobatics, and all the rest. So don't begin your flight tired.

Illness

Illness is another thing the pilot must check himself for before each flight. It can range all the way from the common cold to some disease that could incapacitate the pilot. Most diseases of a serious nature, that is, ones that could cause a total disruption of the pilot's functions, should be caught at the first flight physical and would not allow the prospective pilot to hold a current medical. But if you have flown for a number of years, you might have since developed such an illness. (See Fig. 29-1). The type of illnesses I'm talking about are the everyday cold, the flu, and headaches. Everyone experiences these maladies once in a while. They are not generally incapacitating and people tend to suffer silently and ignore them until they go away. If they are smart, they stay home and take care of themselves. But some pilots think it is okay to go ahead and fly with a cold, flu, or whatever. They are, however, putting themselves in a potentially hazardous situation.

Suppose you have a mild case of stomach flu, and maybe you have a flight you feel you just have to make. It should be no big deal, right? Wrong. Did you ever try to do anything constructive while throwing up? You could be on short final and have to try to land your plane. Stay home or have someone else fly you there if you feel you really must go. Suffering from *get-there-itis* has caused more grief than any other form of aviation-related illness.

Medication

If you are sick and taking some sort of medication, should you fly? The FAA has a very short answer for this. *No.* They say you should

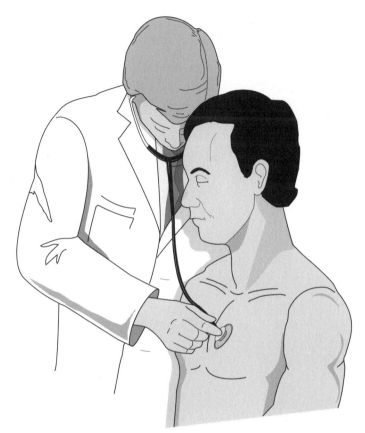

Fig. 29-1. *Your physical condition can affect flight safety.*

obtain a clearance from your doctor, preferably an FAA-certified Aviation medical examiner, before you take anything and fly.

Drugs and flying can be a very dangerous combination that should not be taken lightly. However, I believe the FAA is taking it to the extreme in recommending that you not even take an aspirin without medical authorization. My doctor would probably bill me for an office visit if I called and told him I had a slight headache and wondered if he thought it would be all right if I took an aspirin.

Also, it has been my personal experience that I receive a much better explanation of the do's and don'ts concerning drugs and their possible side effects from my pharmacist. This professional is trained in the compounds and formulations and is, in my opinion, much better qualified to inform me of the correct path to take. I'm not saying that doctors don't know their medicines, only that pharmacists probably know them better.

Amphetamines, barbiturates, tranquilizers, laxatives, and some antibiotics are but a few of the drugs the FAA says should be taken only during a nonflying period. They can cause sleepiness, lack of coordination, and reduced reflex action and should be taken only as directed. Remember, the effects of these drugs might last for hours or days after you have finished taking them, so follow the advice of your doctor or you could wind up in trouble.

You should also use good judgment when taking many over-the-counter types of drugs (Fig. 29-2). I once had a doctor (also a jet fighter pilot) tell me he had a very bad experience from taking a well-

Fig. 29-2. *Even most over-the-counter drugs are forbidden for consumption while flying. If in doubt, check with your Aviation Medical Examiner.*

known over-the-counter antihistamine while flying a jet at 30,000 feet. He related that he was very glad he was not solo at the time. Now he's a doctor and should know when he is apt to get in trouble. Use extreme caution when mixing any sort of drugs and flying.

Here is an example from NTSB File # 3-3883. On December 4, 1976, a 57-year-old private pilot with 2,204 total hours departed the De-pere, Wisconsin, airport in a Cessna 150E. The flight was listed as a local solo, noncommercial, pleasure flight. The pilot was observed performing maneuvers at low altitude before spinning into the ground.

An autopsy revealed .92 Placidyl/100 and salicylate in the pilot's urine. (Placidyl is a barbiturate and can cause extreme drowsiness. Salicylate is more like an aspirin and is often used as a painkiller.) The probable causes listed by the NTSB are physical impairment, failure to obtain/maintain flying speed, and unwarranted low flying. The guy was probably having a good time. Now he's dead. Drugs and flying should be mixed with extreme caution—if at all.

Alcohol

How about alcohol and the pilot? According to one FAA official I talked to, off the record, alcohol is probably a contributing factor in more aircraft accidents than anyone cares to believe. The reason it is so hard to prove is that most alcohol-related accidents are proven through an autopsy. How many nonfatal accidents actually involve alcohol is anybody's guess. If the accident is nonfatal, the FAA might not arrive for hours or days, if they arrive at all. Sometimes they mail out an accident report to the pilot to fill out at his convenience. This allows time for the pilot to make up a good story and swear he hasn't had a drink. Therefore, many of the alcohol-related accidents are pushed to the low end of the spectrum.

Alcohol is a depressant. It slows the reflex action, dulls the senses, and perhaps worst of all, can turn Mike Meakpilot into Ivan Ican. How many pilots, fortified by the remnants of a six-pack, have flown on into deteriorating weather? If you must drink, don't drive. And if you drink and drive, don't drive to the airport.

Here is another example, from NTSB File # 3-3644. On November 6, 1976, a Piper PA-24 departed Evanston, Wyoming, for Salt Lake City, Utah. It carried a pilot and a passenger. The pilot had 488 total hours, 241 in type, and he was not instrument rated. No information is given on the copilot. Near Grantsville, Utah, the plane was observed flying low to the ground. The terrain is listed as containing high obstructions.

The aircraft collided with wires and poles and crashed. Both occupants died. The NTSB lists the probable causes as exercising poor judgment, failing to see and avoid objects or obstructions, and physical impairment due to alcoholic impairment of efficiency and judgment. A subsequent autopsy revealed the pilot had a blood alcohol level of .212 percent and the copilot had .200 percent.

The FAA allows no more than .04 percent alcohol by weight. Anything over this and the FAA considers you as being legally drunk. These pilots were five times that level. They were not able to fly an aircraft with any level of skill and efficiency, but they thought they could and now they are statistics.

Mental preparedness

Before each flight, the pilot should take an objective look at the state of his mental preparedness. Apathy, anxiety, or any other state of mind that could hinder judgment should be laid to rest before a flight is undertaken. A pilot who has his mind on something other than flying is apt to be heading for trouble.

We all have our good days and our bad, our peaks and valleys of mental readiness. Sometimes the cause is just not known. Pilots are people, and people have problems. It could be trouble with your wife, husband, girlfriend, boyfriend, money, or whatever. My point is that these problems have to be dealt with before or after the flight—not during it.

Apathy is a lack of emotion, feeling, or passion. It is a mental state of indifference, lethargy, and sluggishness. What it really is, in plain language, is a state of just not giving a damn. You could care less if the flight is made, the sun comes up, or anything else. If you attempt a flight in this condition, you could be setting yourself up for a bad experience. You must have your mind clear and ready for the sometimes-difficult tasks of flight. If possible, clear up the problem prior to flight time. Or try to put it *completely* from your mind. If you don't, you might be brought back to reality by doing something like landing with the gear up. That is guaranteed to take your mind off your old troubles and put it to work on a new set.

Apathy fostered by poor instruction or from an inadequate training course can usually be corrected. If you feel your instructor is not preparing himself well enough for your dual flight, get another one. Go to a different school, or sit down and have a heart-to-heart talk with the head of flight training at your present school. These actions can usually clear the air and bring desirable results. Then you can get back to the learning you are paying good money for in the first place.

Student/instructor relationship

Let's spend a few moments on the subject of the student/instructor relationship. The learning environment during flight training is a difficult one, at best. Many hours are spent in very close proximity to one other person during which time some very tense moments will occur. The stress of teaching or learning in the closeness of an aircraft cockpit while all sorts of events are unfolding can wear emotions raw. On both sides.

The person has never lived who possessed the ability to get along with everyone in the crowded, stress-filled training cockpit. It is, however, imperative that the student and instructor hold a professional regard for one another, or, at the very least, be able to stand each other. If not, then the natural anxieties produced in any learning situation can be magnified by personality conflicts. When this happens, little learning will occur, and dislikes can become very personal. When this happens, it's time for a change.

Most of the time, the instructor will sense this chill and arrange a trade with another instructor. Then, progress will resume and all will be well. However, some people are so good at hiding their true feelings that one person might not have a clue that the other has a problem. When this is the case, tell him. That's right. Just politely tell the instructor that you are not able to relax with his type of instruction, that you would feel better if you could try it with another instructor. It might be just what the instructor has been praying for!

Anxiety is a much more deeply rooted state of mind. It is a state of mental uneasiness arising from fear or apprehension. The causes of anxiety are much harder to detect, and the cure is usually more difficult than it is with apathy.

A person suffering from anxiety feels that no matter what he or she does, it probably won't be right. It is a self-consuming defeatist attitude and must be overcome before safe flight can take place. Sometimes you must look deeply inward to find the cause of the anxiety. Maybe an open, honest talk with a good friend, flight instructor, or professional psychologist will bring the problem to the fore. Then it can be met head on and conquered. Whatever it takes, it has to be solved before you can perform proficiently as a pilot.

Vertigo

Vertigo, or *spatial disorientation* as it is sometimes called, involves a disorientation in space during which the person involved is unable to sense his attitude accurately with respect to the natural horizon (Fig. 29-3). A pilot suffering from vertigo is unable to perceive by his own senses whether the aircraft is climbing, diving, or turning.

When on the ground, you perceive attitude with respect to the earth by seeing fixed objects around you, by feeling the weight of your body on your feet, and by the vestibular organs in the inner ear. You can orient yourself by any one of these means for short periods of time.

Fig. 29-3. *Unchecked vertigo—look out below!*

While in flight, however, all three of these normal means of orientation can get obscured or confused (Fig. 29-4). The pilot might only be able to see objects that are in or attached to his aircraft when ground references are obscured by clouds or darkness. Your ability to sense the direction of the earth's gravity, by the weight on your body and through your vestibular organs can be confused by accelerations in different directions caused by centrifugal force and turbulence. For example, the senses are unable to distinguish between the force of gravity and the horizontal force resulting from a steep turn.

Because of this fact, gyroscopic instruments are necessary if you are to fly for more than a few minutes without visual access to outside reference points. The use of such instruments does not ensure freedom from vertigo, for no one is immune to vertigo. They do enable a pilot to overcome it, however, if he trains himself to accept the psychological discomfort that results from acting in accordance with instrument indications and to disregard the false impressions received from the senses. This means *disregard your senses and believe your instruments!* It could save your life.

Fig. 29-4. *Vertigo can occur anytime, but you are especially susceptible when the horizon is obscured, such as night or in instrument conditions.*

All student pilots should have the opportunity to experience the sensation of vertigo during their early flight training during maneuvers performed by the instructor. The maneuvers and procedures that can cause vertigo are quite simple. Attempting to read a map or manual while performing coordination exercises or watching the upper wing tip in a prolonged steep turn usually can bring vertigo about. Once experienced and understood, later unanticipated incidents of vertigo can more easily be controlled or overcome. As you gain aeronautical experience, such periods of vertigo will become less frequent, but they can and will persist under instrument flight conditions or at night, so be prepared.

Sometimes closing your eyes for a few moments can help you overcome vertigo, but the best thing to do is to fly according to what you see depicted on the flight instruments and ignoring what you feel.

Remember, pilots are most susceptible to vertigo at night and in any other condition in which the natural horizon is obscured. The best defense against vertigo is to experience it, know you have it, and then fly your instruments. As you will see, even the possession of an instrument rating does not automatically free you from the grasp of vertigo. Although FAA studies have shown that a pilot with an instrument rating is generally better able to overcome vertigo, you are not immune from the effects produced in instrument conditions.

Note this example from NTSB File # 3-3850. On November 28, 1976, a commercial pilot who also held flight instructor and instrument ratings took off with three passengers from the Wilmington, Delaware, airport. They were en route to Savannah, Georgia. The pilot was briefed by weather bureau personnel that the weather was very low IFR. The pilot filed a flight plan and the four departed for Savannah.

Near Oxford, North Carolina, the pilot began to encounter communication difficulties. Radar showed an erratic flight path, and the controller lost the Cessna 210's transponder reply. Although the aircraft was still airborne at this time, the loss of the transponder and the difficulty in communication suggested a possible electrical failure. The pilot apparently was overcome with vertigo while flying in solid IFR conditions and lost part or all of his electrical system. The aircraft went into an uncontrolled descent, crashed, and killed all on board.

The weather at the crash site was listed as one-mile visibility with a ceiling of 100 feet, fog, and light rain. Probable cause: vertigo induced by low visibility and ceiling with the possible loss of the electrical system. The point I'm trying to make with this particular report is that even an instrument-rated pilot can run into compound problems.

The loss of the electrical system while on an actual IFR flight is a very frightening thought. But the situation does not have to be hopeless. A well-trained instrument pilot should be able to keep his aircraft under control after experiencing a total electrical failure. (No one knows exactly what happened in the aforementioned accident, so let us go with some generalities. Not having been there, I do not attempt to say what that pilot should or should not have done because the circumstances might have been far more severe than we know.)

In the event of an in-flight emergency of any kind, the pilot's primary responsibility is to maintain control of his aircraft. This is first and foremost. If all electrical systems are lost while on an IFR flight, the pilot has but two choices. One is to panic, experience vertigo, and die. Or the pilot can use his or her training and the remaining instruments, keep as calm as possible, and control the aircraft. There might not be communication or navigation abilities, but these are secondary. You have to keep control of yourself and the aircraft.

Size up the situation. Where are you? Which way to better weather and more suitable terrain? Use the instruments remaining. If the electrical system is gone, it should still leave the compass, airspeed indicator, and maybe the turn coordinator. I realize many of the new turn coordinators are electric, but the gyro instruments are usually run off of the engine-driven vacuum pump, so they should be operative. Use what you have, and maintain control of the aircraft.

Once control is confirmed, maintain it and start the decision-making process that will give you the best chance for survival. You know the forecast weather for the area in which you are flying. Is it better back where you came from, ahead, or in a completely different direction? Remember the terrain. Select the approximate heading that will take you toward the best of both. *Gently* start to turn in the desired direction and don't give up. Check fuses and circuit breakers. Sometimes a blown fuse can be replaced to give communication and navigation for at least a period of time. If so, see how many fuses you have and use them for quick checks. Then remove them. You might even be able to navigate and shoot an approach, call for some aid, or at least warn ATC of your trouble so they can clear the area of known traffic.

Also, battery-powered NAV/COM radios are now available on the market, and such a radio should be in the possession of any pilot while flying in instrument conditions. These little beauties have a battery life of several hours, can provide you with VOR navigation, and let you talk to ATC at the same time. They are a very affordable backup system capable of providing lifesaving navigation as well as communications with ATC.

If you are not able to remedy the problem, you still have a good chance if you don't panic. Keep control of yourself and the aircraft. Turn toward the best weather and terrain. When the time comes to start down, if it's still solid IFR, slow the aircraft to a safe, slow, descent speed and descend with wings level until you either break out or land. If you have set the aircraft up in landing configuration and control the airspeed with pitch and your altitude with approach power, you just might land in a nice big pasture. If not, you will impact the ground under control with wings level, and as slowly as possible. You can do no more.

More likely, when you break out of the clouds you can circle around and find a suitable place to land, maybe even navigate to an airport and put it on a real runway. It can be done. It has been done. You can do it if you don't panic and let vertigo disorient you.

30

Physiological factors

Combining what you've learned so far, we'll incorporate all of the flight instruments into your scan to enable you to maintain altitude, speed, and heading, and to establish climbs and descents.

Before getting into that, however, let's discuss a few more physiological factors that will affect your flight performance in general and affect your instrument capabilities in particular.

The effects of pressure changes on pilots are well known, and we have all been advised not to fly with a cold or when we have ear or sinus blockage. We must also take care after scuba diving because we can set ourselves up for the bends if we fly at too high a cabin altitude too soon after diving.

Legal drugs

A second problem when flying with a slight cold deals with the effect of medication in your system. Drugs, even the over-the-counter variety, which would have little or no effect on you on the ground, can end up having disastrous results when you fly. This is especially true when you're on instruments.

Antihistamines might cause dizziness, drowsiness, headaches, and/or nausea, and they usually also affect vision.

Barbiturates, as they wear off, bring about sleepiness that often goes unrecognized. They also affect motor and thinking functions.

Diet pills are responsible for nervousness and impaired judgment.

Muscle relaxants, by their nature, cause weakness as well as sleepiness and vertigo.

Stimulants, like diet pills, cause nervousness and impaired judgment as well as blurred vision.

Tranquilizers also cause blurred vision and sleepiness.

Some of these pills can be bought without a prescription, and many times your family doctor will prescribe medication containing these or other ingredients detrimental to the safe operation of aircraft. Ask your doctor about the effects that the medication might have on your ability to fly. If you have any doubt, or if you're thinking of taking an over-the-counter pill in some part of the country where you don't have a doctor, contact the local FAA medical examiner for advice. If, for some reason, you're unable to get medical advice, stay away from the pilot's seat of an aircraft for at least 24 hours after taking any medication.

Alcohol, cigarettes, and hypoxia

Another serious physiological problem stems from the consumption of alcoholic beverages. It takes only 25 percent as much alcohol to impair flying skills as it does to impair driving skills. One or two drinks will decrease your flying ability below safe standards, and in instrument conditions this danger is multiplied. Alcohol, which is readily absorbed into the bloodstream, interferes with the normal use of oxygen by the tissues. As you climb, it causes you to suffer effects similar to *hypoxia* (oxygen starvation). Hypoxia is caused by foreign or toxic substances in the blood and is also caused by reduced atmospheric pressure.

Oxygen is carried through the bloodstream by *hemoglobin*. A certain amount of pressure in the lungs is required to combine the hemoglobin with the oxygen. The higher you go, the lower the pressure, and the less oxygen is absorbed. Not only that, the higher you go, the less oxygen is available in any given volume of air that you inhale.

For all practical purposes, you can consider 10,000 feet MSL as the upper limit for safe, long-range flight for a normal, healthy body without supplemental oxygen. The FAA allows you to fly at 12,500 feet MSL without any supplemental oxygen at all. If you have any physiological problems at all, you might not be able to even handle 10,000 feet safely, and I have already mentioned that alcohol will interfere with your body's normal use of oxygen and this will affect your susceptibility to hypoxia.

What is it about smoking that affects your body? *Carbon monoxide,* that colorless, odorless, tasteless killer chemical. It is absorbed by the hemoglobin in the bloodstream 200 times easier than oxygen. Every bit of carbon monoxide that is absorbed by the hemoglobin takes the place of that much needed oxygen; in the bloodstream, it cannot be cleared out by just a little suck on the oxygen mask. It is more difficult to *remove* carbon monoxide from the bloodstream than it is to absorb it.

Carbon monoxide can be absorbed into the bloodstream from sources other than cigarettes. One prevalent source is exhaust gas from a cracked exhaust pipe. Another is the smoke from a passenger's cigarette. In fact, recent findings tend to support the contention that the nonsmoker suffers more from this passive smoke in a confined area than does the smoker.

The greatest danger of hypoxia, from whatever source, is the way it numbs the brain so the person suffering from it can black out, recover, and have no recollection of the experience, except for a possible headache. In fact, unless you have some definite proof, a victim of hypoxia will argue about ever being unconscious. United Airlines made an interesting training film showing some of their personnel in a decompression chamber. It is a sobering and frightening experience to watch the film, and you subsequently develop a healthy respect for the dangers of hypoxia.

Hypoxia brings about the following symptoms, although not necessarily in the following order:

- Dizziness: not always apparent because it is more often just a little lightheadedness.
- Tingling: again not always noticed because the person sometimes feels as though that part of the body is "falling asleep."
- Blurry vision: often confused with fatigue or tired eyes.
- Feeling of warmth: "who turned on the heater?"
- Euphoria: a feeling of well-being (here's where the brain remains for some reason, feeling nothing else).
- Mental confusion.
- Inability to concentrate.
- Loss of judgment.

- Slow reflexes.
- Clumsiness.
- Loss of consciousness and eventually death, unless oxygen is administered.

Even after recovery, some irreparable damage might have been suffered by the brain. Naturally, through all of the adverse symptoms mentioned above, the person is very susceptible to vertigo.

Another physiological problem, in which many symptoms are the same as hypoxia, is *hyperventilation*. This occurs when you get excited and begin breathing quickly and heavily. In so doing, your body gets rid of too much *carbon dioxide*, a waste product of muscular exertion. A body requires a certain balance between the oxygen and carbon dioxide in the blood. Too much oxygen in relation to the carbon dioxide will cause dizziness and faintness. Take a few fast deep breaths and see for yourself.

People tend to hyperventilate when they get into tight situations involving stress and fear. The best protection against this problem is to learn to relax, calmly appraise the situation, don't allow yourself to panic, don't fall for "sucker holes," don't try flying when conditions are minimal, practice your instrument skills, know your aircraft, know your own capabilities, and calmly force yourself to breathe slowly and steadily.

The red-light-at-night idea is also something that we should think about.

If you're in combat situations, where night vision could mean the difference between life and death, then it is important to maintain your night vision; however, in today's civil aircraft environment, you are constantly encountering white lights from cities, from airports, and from aircraft strobe lights.

Each time your eyes see white lights at night, your night vision will be affected for up to 15 minutes. In the military you're taught to close the same eye immediately so as to preserve at least part of your night vision. This is impractical at best and certainly unnecessary in today's flight environment.

Your eyes are made up of rods and cones. The cones react to white lights and give you the ability to see colors. The rods see only different shades of gray. This is why the night vision is so important in combat. You want to be able to discern movement.

Still and all, if a laser could burn out all of the cones in your eyes, leaving only the rods, you would be left with 20/200 vision. In other words, you would be legally blind.

Finally, with all of the colors and hues on modern-day charts, it is imperative to keep your color vision. For this reason, it is recommended that you keep at least some white light in your cockpit. It shouldn't be so bright as to prevent you from seeing outside, but it should be good enough to allow you to see your important charts and flight documents. Remember, as soon as you see those approach lights at the airport, your night vision is canceled.

Another thing regarding night vision. At night we become relatively nearsighted. This tends to change the runway configuration somewhat, causing the far end to tilt forward. This is another good reason to stay on the glide slope, monitor your rate of descent, and watch the VASI.

Illegal drugs

Before closing this chapter, it is important that you look at the effects of illegal drugs on the body. There isn't room to cover them all, but the most prevalent today seems to be cocaine and its derivative crack.

Recall the tragic death of college basketball star Len Bias. The news media reported that he died of a heart attack *the first time he used cocaine*. Some people said that was impossible, but recent studies are pointing to the very distinct possibility.

Cocaine is a constrictor. It clamps down on your blood vessels. If nothing else, it restricts the blood flow to your brain, causing the same loss of oxygen as smoke. But it also does far worse.

Cocaine has been used in surgery, especially nose surgery, for years. Besides reducing pain, it also constricts the small veins, stopping bleeding. But, a recent study, conducted by doctors at the University of Texas Southwestern Medical Center in Dallas, and published in the *New England Journal of Medicine*, has determined that cocaine is worse than anyone imagined.

The doctors studied the effects on 45 volunteers. They received between 150 and 180 milligrams of cocaine. This is only about two-thirds as much as they would get from normal anesthesia. The results: "Even that small dose caused coronary blood flow to fall by 15 to 20 percent," according to Dr. L. David Hillis, who conducted the test.

When you realize that cocaine users use many times that amount, it is easy to see how some die of heart attacks. By shutting off the flow of blood to the heart, the heart dies.

Hillis said that cocaine stimulates the body's production of adrenaline and other hormones. These trigger the blood vessels to narrow so that less oxygen-carrying blood can get through. At the same time, the drug increases the heart's need for oxygen by boosting the heart rate.

In short, if you've never used cocaine, don't. If you know of someone who does, don't get into any vehicle with him or her, especially an aircraft.

Vertigo awareness

There are only two types of instrument pilots. Those who have suffered from vertigo and those who will. Your eyes are a most important factor in maintaining equilibrium in flight. You will have a tendency to lose your orientation if your supporting senses conflict with what you see.

One source of this conflict (known as *flicker vertigo*) during night and instrument conditions is the reflections of anticollision or strobe lights on clouds or on the aircraft's wings and canopy. These flickers make you feel as if you are turning. You can even be bothered by the airplane's shadow projected onto clouds in an otherwise dark night sky. The easiest way to correct this problem is to turn off the rotating beacon and/or strobe lights until you get over the feeling.

Climbing or descending through an area of broken clouds, either IFR or VFR, can give you the impression that your airspeed and altitude are changing rapidly. This is especially prevalent when flying over water on a bright hazy day or through areas of heavy haze when you lose the horizon. Your best bet in such situations is to go on instruments immediately.

Many times when banking soon after takeoff on a dark night, or when leaving a bright area and proceeding over dark water, desert, mountains, or forests, you will get the impression that the stars are actually lights on the ground. Other times, ground lights that you know are lower than your altitude will appear to be higher.

Both of these impressions will cause you to develop symptoms of vertigo that can be overcome only by going on the gauges at once. (I have long felt that for this very reason, an instrument ticket should be a prerequisite for night flying. There is a long list of night accidents in which the probable cause was listed as the pilot's inability to maintain orientation due to a loss of visual landmarks.)

In addition to sight, we possess the postural senses of touch, pressure, and tension, and the motion-sensing organs of the inner ear. The postural senses cannot detect an unchanging velocity; there must be acceleration or deceleration. Without the sense of sight to help us, the +2G centrifugal force of a steep turn feels identical to a +2G pull-up from a dive, and a wing load of ±1–2Gs often feels the same whether caused by in-flight turbulence or by rough control pressures related to leveling off, climbing, or beginning descents. Most complications with spatial disorientation begin in the inner ear.

The inner ear

Basically, the inner ear consists of a reservoir, called the *common sac*, to which are fitted three hollow rings called the *semicircular canals*. These semicircular canals are orientated in each of three planes, and the entire unit is filled with fluid. Very fine hairs, called *sensory hairs*, project into the fluid within the semicircular canals.

Picture yourself holding a glass of water. If you twist the glass, the water will tend to stay where it is at first. If you were to tape a few threads to the inside of the glass (similar to the sensory hairs of the semicircular canals) and then twist it, the water would stand still at first, and the threads would drag backwards in the water.

The same thing happens in the inner ear. If you begin turning, the semicircular canals begin turning with you, but because of the laws of inertia, the fluid in the canal (in the plane of rotation associated with your motion) resists moving at first, and the sensory hairs are deflected as they drag in the fluid. The hairs send a signal to the brain, telling the brain that you are moving.

If you continue turning (or accelerating, or decelerating) at the same rate, the fluid will begin moving, and after a while it will move at the same rate as the canal. At this time, the sensory hairs will return to their static position and your brain will sense that your turn, and the like, has stopped, even though your aircraft and your body are still turning.

Then, when you begin to roll out of the turn, the opposite will happen. The semicircular canal will move in the opposite direction, and the hairs will deflect to tell the brain that you are moving in the opposite direction. In the case of rolling in and out of a turn, this will work out fine as long as you roll smoothly. But, if you roll out abruptly, when you stop the roll the fluid will continue moving, carrying the hairs in the opposite direction again. Even though you are flying straight and level, your brain will think that you are in another turn. (See Fig. 30-1.)

As long as you can see the outside horizon, or the attitude indicator and the directional gyro, you will not be too confused. Under visual conditions, the outside horizon alone will suffice to tell you that you are turning; you will see objects on the horizon or on the ground as they pass across your view.

When you're on instruments, you need two gauges to give you the same information. The attitude indicator or the turn-and-bank indicator will tell you if you are in a bank, and the directional gyro will confirm that you are actually turning; the numbers rotating past the lubber line will take the place of the objects going by on the outside horizon.

If you close your eyes while making a smooth entry into the turn, roll out suddenly, and then open your eyes and look only at the instruments, the instruments will tell you just the opposite of what your inner ear has been saying, and you will be primed for vertigo. This can happen if you allow yourself to bank too steeply during an instrument turn, then notice your error and overcontrol, snapping quickly back to the desired angle of bank. In such a situation, if you doubt the instruments, you will find that you become dizzy and unable to control your muscle functions. By the time you realize that the attitude indicator is telling you the true attitude, it might be too late. One part of your brain is trying to tell your muscles to roll out of the turn, but another part of your brain has become so confused that it refuses to allow the muscles to respond. In this situation, you are experiencing extreme vertigo, and you are set up for a catastrophe.

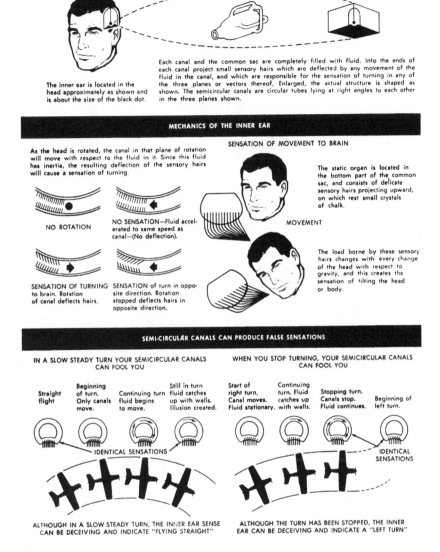

The inner ear is located in the head approximately as shown and is about the size of the black dot.

Each canal and the common sac are completely filled with fluid. Into the ends of each canal project small sensory hairs which are deflected by any movement of the fluid in the canal, and which are responsible for the sensation of turning in any of the three planes or vectors thereof. Enlarged, the actual structure is shaped as shown. The semicircular canals are circular tubes lying at right angles to each other in the three planes shown.

MECHANICS OF THE INNER EAR

As the head is rotated, the canal in that plane of rotation will move with respect to the fluid in it. Since this fluid has inertia, the resulting deflection of the sensory hairs will cause a sensation of turning.

SENSATION OF MOVEMENT TO BRAIN

The static organ is located in the bottom part of the common sac, and consists of delicate sensory hairs projecting upward, on which rest small crystals of chalk.

NO ROTATION

NO SENSATION—Fluid accelerated to same speed as canal—(No deflection).

MOVEMENT

SENSATION OF TURNING to brain. Rotation of canal deflects hairs.

SENSATION of turn in opposite direction. Rotation stopped deflects hairs in opposite direction.

The load borne by these sensory hairs changes with every change of the head with respect to gravity, and this creates the sensation of tilting the head or body.

SEMI-CIRCULAR CANALS CAN PRODUCE FALSE SENSATIONS

IN A SLOW STEADY TURN YOUR SEMICIRCULAR CANALS CAN FOOL YOU

WHEN YOU STOP TURNING, YOUR SEMICIRCULAR CANALS CAN FOOL YOU

| Straight flight | Beginning of turn. Only canals move. | Continuing turn fluid begins to move. | Still in turn fluid catches up with walls. Illusion created. | Start of right turn. Canal moves. Fluid stationary. | Continuing turn. Fluid catches up with walls. | Stopping turn. Canals stop. Fluid continues. | Beginning of left turn. |

IDENTICAL SENSATIONS

IDENTICAL SENSATIONS

ALTHOUGH IN A SLOW STEADY TURN, THE INNER EAR SENSE CAN BE DECEIVING AND INDICATE "FLYING STRAIGHT"

ALTHOUGH THE TURN HAS BEEN STOPPED, THE INNER EAR CAN BE DECEIVING AND INDICATE A "LEFT TURN"

Fig. 30-1

Believing the instruments

The only way to overcome vertigo is to understand it, catch it at its outset, and believe the instruments. You must train yourself to disregard all of the false senses and try to rely solely on the instruments. This takes patience and practice. You must also remember that instruments can fail from time to time, so you must be able to reinforce what some of the instruments are telling you by interpreting other instruments in the same group. The more experience you have, the less chance you'll have of succumbing to vertigo.

There are some specific maneuvers designed to induce vertigo, and though each is designed to produce a specific sensation, any sensation at all will help you learn that you can't trust your senses. Be certain that these maneuvers are only undertaken with an instructor or qualified safety pilot on board.

These maneuvers will show you the absolute need to rely on your instruments and give you an understanding of how head movements, in various aircraft attitudes, cause disorientation:

- While flying straight and level, close your eyes and have the instructor skid right or left while holding the wings level. You will feel as though your body is actually tilting in the direction opposite the skid. As soon as you feel that you are tilting, tell the instructor what direction you're tilting in and open your eyes to see what the aircraft is doing.

- While flying straight and level with the aircraft slowed down to approach airspeed, close your eyes. Have the instructor increase the speed to climb speed while maintaining a straight-and-level attitude. You'll feel as though you're climbing. This is important to remember when executing a missed approach from low airspeeds, especially if you have to divert your head from the instruments to reset a radio or to check a chart for a procedural point.

- From a straight-and-level attitude, close your eyes and have the instructor execute a slow roll into a well-coordinated 45-degree banked turn. Hold this 1.5G turn for 90 degrees. Because of the gravity forces in the turn, you will feel as though the aircraft is climbing. Keep your eyes closed and have the instructor roll out of the turn slowly and with coordinated pressures. Because of the decrease in gravity forces, by the time you have rolled halfway out, you will feel as though you are descending.

The reason for these sensations is that with a lack of visual reference, if the rate of angular acceleration is less than 2 degrees per sec./sec., the body is unable to detect a change of direction in any one of the three planes of motion; therefore, the change of gravity forces, usually associated with climbing or diving, is all that will be felt.

You can experience a false sense of reversal in motion by using the following maneuver:

- With your eyes closed while straight and level, have your instructor smoothly and positively roll into a 45-degree bank while using opposite rudder pressure to keep the nose in a point. If the roll rate is suddenly stopped, you will feel as though you are actually rotating in the opposite direction, and a recovery based solely on your senses could be fatal. This demonstration, when properly executed, will drive home the need for smooth, well-coordinated control usage, and a firm need for a belief only in your instruments.

The next two demonstrations can also be shown by strapping the student in a rotating chair. Whether done in a rotating chair or in flight, the results will be extreme disorientation, causing a quick and forcible, almost violent, movement of the head and body, upward and backward, and to the side opposite the rotation. Naturally if this reaction were applied to the controls of an aircraft in flight, you could get into a lot of trouble:

- From straight-and-level flight, either close your eyes or look down at the floor. The instructor should roll positively into a coordinated bank of 45 degrees (or begin spinning the chair). As the aircraft is rolling into the bank, bend your head and body down and look to either side, and then immediately straighten up and sit normally. The instructor should stop the bank (or stop the chair rotation) just as you reach the normally seated position. (Fig. 30-2.)
- Watch as the plane is rolled into a steep, descending spiral. After 15 seconds or so, bend your head and body down and look to either side as you did in the first maneuver, and again immediately resume the normal seated position.

The disorientation might be severe. You might even become nauseated. This could easily happen in a real-life situation if the aircraft begins turning as you are looking for a chart or book in your flight case. The reaction is usually more severe when you are forced to look down and to the rear. This is one of the reasons for placing all

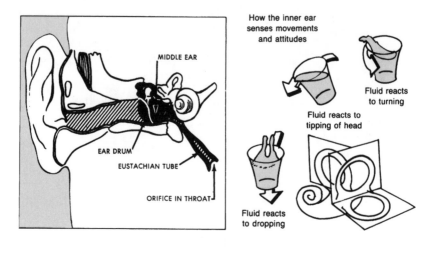

How the inner ear
senses movements
and attitudes

Fluid reacts
to turning

Fluid reacts to
tipping of head

Fluid reacts
to dropping

MIDDLE EAR

EAR DRUM

EUSTACHIAN TUBE

ORIFICE IN THROAT

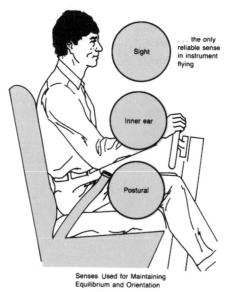

Sight

. . . the only
reliable sense
in instrument
flying

Inner ear

Postural

Senses Used for Maintaining
Equilibrium and Orientation

Fig. 30-2

of the charts and paperwork in some handy location such as on top
of the glareshield where you can reach them without turning your
head and body.

Normal vertigo can only be overcome by familiarity with your instruments, belief in them, and continuous practice. The more you fly
on instruments, the less susceptible you will be to false sensations.

31

Physiology at night

Christmas eve comes quietly, as if everyone is holding their breath for the big day. Southern California sparkled with decorative lights, closed businesses, and the radio playing carols for an occasional long-running party. It was four in the morning, December 24th. Most people rested quietly in their homes. Whether under a blanket of snow or Southwestern warmth, the holidays have an effect on everybody; and for Brett it meant flying almost nonstop for the last 3 days.

The little air freight company won, for the first time, a lucrative UPS feeder contract, for the mountain of packages sent during the holiday season. The pilots at first felt like Santa Claus, shuttling Yuletide cargo hither and yon, enjoying the opportunity to fly and earn money. The opportunity more than matched the size of the company, however, with far more loot to carry than they had planes and pilots for. To win future contracts, the company *had* to satisfy the demand, so the boss made a few demands of the chief pilot. The chief pilot made a few demands of the freight pilots, and Brett had to fly. No fussing about FAA mandated rest periods; no slacking on the job. "The plane has to be in Bakersfield in the morning. *There is no one else available to fly it, so we're counting on you to get it done... Got it?"*

So Brett had to fly. He'd flown all night, every night, and all day for the past 3 days. Sleeping at outstations, between flights, for only a few hours at a time, eating on the plane and loading freight in between. He didn't feel like Santa Claus at all; he was so tired that he couldn't feel anything. Christmas eve, Silent Night—holiday spirit everywhere, and all he cared was that it would soon be over. He didn't want to fly, shouldn't fly, felt way too tired, had been awake and working almost 40 hours, with pitifully small amounts of sleep in between shifts—no more than a couple of hours at a time. He trudged off to the Piper Navajo, waiting quietly in the darkness of the Ontario airport ramp. His limbs felt heavy, making the walking seem difficult. It was not particularly hard for him to stay awake while loading the plane—the physical exertion and cool air helped—but he felt dangerously fatigued. He worked under the floodlights that brightened the ramp around the plane, but night encroached upon the shadows, like dark dreams waiting to jump a tired passer-by. *Dangerous to be flying like this—rather sleep, could pass out right here, on a crate.* He settled heavily into the familiar cockpit, glad to be through with loading the plane—glad it would soon be over. A flight to Bakersfield, and it's Christmas eve, the last day of the contract—heck, it's Christmas eve now, already four-thirty in the morning—get there and the work is done—rest all you want.

It's dark outside, pitch black, early morning dark, and the weather reports indicate low clouds over the route with dense ground fog at Bakersfield—the usual for this time of year. Weather like this makes a pilot earn his keep—approaches to minimums—if the controllers will cooperate and call it at minimums—they'd have to stretch it a bit. A challenge at any time, but with a sleepy pilot, utterly dangerous. Brett launched into the low overcast, instantly comfortable with the throbbing of the two engines outside. The normal mechanical vibrations relaxed him, soon breaking out of the weather, flying above the clouds. The flying soothed him; it felt peaceful, restful…nice. "CALAIR_____, CALAIR_____, SO-CAL APPROACH, LISTEN UP, WAKE UP! TURN RIGHT, HEADING 285, HEADING 285—ACKNOWLEDGE!" The controller shouted pretty loud, practically screaming in the radio—and Brett woke up. *Shouldn't have been sleeping.* He felt momentarily confused at his surroundings—a little surprised to be in a cockpit, and IFR, at that. He tried to concentrate on the task—*right turn, heading 285. Not so hard. Tired.* Brett's voice was slurred in his communications. "Ah, roger—right heading 285." The controller continued to yell on the radio. *Does he have to be so loud? How did the heading get to 240?* He turned back and overshot, rolling out at 300, or so. *Heading 285, ah—285. Wow, tired—feel almost drunk.*

Brett squinted and rubbed his eyes. Sitting comfortably in the cockpit, he didn't have to move much. His body felt like a sandbag, weighing heavily, painfully, in the seat. It ached. The fatigue he felt was pain; a dull, almost overwhelming ache that spread outward from his bones—his body resisted every movement. His eyes felt the heaviest of all. Intellectually, he knew that to fall asleep meant certain death; physically, his eyelids weighed a ton each—impossible to keep open. He rubbed them hard, the pressure of hands against his eyeballs making him see lights. He jerked awake suddenly, kicking one of the rudder pedals and feeling out of balance. *Have to stay awake! Keep my eyes open—don't sleep!* He slapped his knees and adjusted himself in the cockpit—and jerked awake again, almost in the middle of it all. He was falling asleep several times a minute in the process of trying to stay awake, fighting a losing battle.

When he next looked out the windows, he was shocked to find the plane between layers, flying in a 90-degree bank to the right! He flinched on the controls. *Roll out, roll out—reference the attitude indicator and roll out—but the attitude gyro shows wings level. Back outside—definite 90-degree bank; attitude indicator still showing level.* Brett experienced a momentary jolt of adrenaline as he recognized the trouble. *Never had vertigo backward before. Always looks level outside, when the indicator is in a bank. This feels weird. Focus on the attitude indicator. Concentrate on it until it fills your vision. Crawl inside that thing until what it shows, feels like level.* "CALAIR_____, CALAIR_____, SO-CAL APPROACH, WHERE ARE YOU GOING?! MAINTAIN HEADING 290, NOW!" The controller spoke loudly, urgently, knowing that the Calair pilot was in serious trouble. He tried to help, as Brett struggled to fly the plane. *Why is it so hard to keep a heading?*

As the pressure of the flight continued to build, his thoughts fragmented; he had difficulty thinking at all. *Never been so tired. Have no business flying an airplane. So fatigued—my body hurts! Rather sleep than fly. Rather sleep than do anything—such a bother to stay awake. The pesky controller is annoying—won't let me sleep. Don't want to fly the plane…just want to close my eyes.…*

Brett was barely conscious, not quite rational, and losing the battle. The plane arrived over Bakersfield at sunrise, the bright morning light glaring in the fog, which caused the visibility to go down. The tower generously called the visibility at minimums so Brett could attempt the approach. There was a good chance of a missed approach, with the alternate about an hour's flight away—and Brett was about gone. Flying erratic this way, IFR—heavy IFR—*hard to think, it's crazy.* He could barely hold a heading, couldn't help drifting off, almost constantly nodding off—he was in agony; the fatigue of long days focused in his eyelids. He could not keep them open much longer—and stood a good chance of crashing if he flew another hour. He had to land on the first attempt, or…

Somewhere in his sleep-crazed mind, Brett came to grips with the situation. *I make this approach, or crash—so, in a few minutes I'll*

get to close my eyes, one way or another...I'll get to close my eyes.
He didn't really care how the flight turned out, from then on.

Turning onto the localizer, he caught the course and managed, erratically, to fly the approach. The controllers were indeed generous with their estimate of the weather—*nice of them to stretch the minimums like that.* Brett found the runway with his tires and rolled to a stop on the ramp. He opened the cockpit door and felt a cool breeze on his face.

He could finally sleep....

Background information

Brett survived a close brush with fatigue that could easily have been fatal. The most remarkable thing for him, about that experience, was that he arrived at the point where he did not care if he crashed and died, or not—either way, it meant that he could get some rest. It was fatigue-clouded judgment that made an almost suicidal attitude like that possible. Your body has a basic need for sleep—like it needs to eat and breathe. Going long enough without sleep can result in hallucinations, unwilling lapses into sleep, poor motor control, impaired judgment, severe mood swings, and mental psychosis—all of which are bad for flying. But sleep and fatigue are only part of the issue; darkness also affects your eyes (obviously), hearing, balance, speech, emotions, metabolism and decision-making ability, to name a few.

Circadian rhythms and metabolism

From the beating of your heart, to the gentle breathing of your lungs, to the timing of your meals, to the time you awake in the morning; your body demonstrates a remarkable sense of rhythm. These rhythms are important and difficult to ignore. Take your breathing, for example. You breathe in and out constantly throughout the day. When you think about it, you can upset the rhythm by holding your breath, but your body will complain strongly and eventually lapse into unconsciousness, where it will resume normal breathing again. You can breathe faster than normal if you wish, but unless there is a need, such as to compensate for exercise, you will soon get light-headed, and eventually lapse into unconsciousness again, where your body will resume breathing normally, keeping the rhythm. Your heart beats at an even rhythm and we all know what happens if *that* stops.

There is a much slower rhythm to your sleeping. Your body would like to sleep a few hours every day—maintain a circadian rhythm. *Circadian* means "about a day," referring to the daily cycles of your body, and it refers to more than just sleeping. Your bowels, for example, maintain a rhythm of their own—you have a little influence in deciding where to empty them, but *you can't hold out forever.* Sleep is similar to this. You may avoid sleeping altogether, if you wish, but the consequences are similar to restraining your bladder. Eventually, your control of the situation will fade, and your body will sleep anyway. In the process of becoming incapacitated by lack of sleep, you may experience all sorts of symptoms, many of which border on the psychotic.

Your metabolism functions in concert with the rest of your body. *Metabolism* refers to your body's rate of energy consumption—a throttle setting, if you will. A generally high metabolic rate could be demonstrated by a hyperactive child, and a low rate is easily evidenced by a couch potato. Your metabolic rate will change with response to your body's needs. It slows down when you sleep (Fig. 31-1). If you are awake at night, out of rhythm with what your body expects you to do, your metabolic rate may still slow down. This could make you feel more tired than normal, lacking the energy

Fig. 31-1. *As the day draws to a close, your metabolism slows down.*

you'd expect in the daytime. People who skip a night's sleep feel tired at night but then seem to feel some improvement in the morning, because their metabolism is picking up for the daylight hours, as it is accustomed to doing. Because of this, you may assume that your state of alertness will generally be lessened during the dark hours, unless you take pains to change your body's rhythms.

Your body's rhythms are not set in concrete. They can be altered by modifying your habits. You automatically breathe faster when you exercise. Eat bad food, and your bowels may function at warp speed. If you get sick, your body may need to sleep much more than normal. Sleep a lot during the day, and you may stay awake for much of the night. With discipline, you might reverse the normal day-night cycles altogether. Farmers have, for years, awakened at very early hours and retired to bed long before nightfall. Night cargo pilots train themselves to sleep in the day, awaking a bit before sundown, enabling them to keep the same hours as a bat. These things do not come without consequences, however. Brett, for example, quit Calair a few months after his hair-raising flight. He was *tired,* for lack of a better word, of working nights. After adopting a daylight schedule, he needed almost 2 months to fully recover and feel "normal" again. He noticed the reversal of several changes that occurred as he flew the night freight, unawares. The dark circles under his eyes disappeared. His personality changed—he became more vibrant, energetic. He felt healthier. All these things were different when he worked at night; he did not notice them until he had stopped.

Many drugs, to the delight of rock stars and Hollywood, are available to alter sleeping rhythms artificially. Narcotics, alcohol, allergy medicines, weight-control pills, cold remedies, and literally thousands of others can all affect your ability to sleep or stay awake, changing your level of alertness and often clouding your judgment. Some of the popular illegal drugs may also affect a pilot's attitude, causing inappropriate responses to dire situations. For example, some drugged pilots being tested as they fly a simulator actually *laughed* as they impacted the ground. If a pilot expects to maintain an adequate state of alertness for a safe flight, by all means, seek a professional opinion on the viability of questionable drugs in the cockpit.

Fatigue

A pilot who looks or feels generally tired, slow of thought, withdrawn, and unresponsive is probably suffering some of the effects of fatigue. Fatigue, in itself, has proved difficult to describe and even harder to quantify. NTSB investigations often list pilot fatigue as a contributing factor in accidents, and never as a cause, because of the difficulty of measuring or even proving the existence of the condition. An investigator can only assume that fatigue played a part in a crash because of circumstantial evidence: The pilot did not sleep well for days before, was subjected to long hours at work, and appeared to show symptoms of a cold—all factors that could cumulatively lead to performance degradation in the cockpit as a result of fatigue. Some of the causes of fatigue include, but are not limited to:

- Disruption of circadian patterns
- External stresses unrelated to flying
- Inadequate level of preparedness
- Instrument flight
- Irregular/stressful work hours
- Jet lag
- Lack of sleep
- Mechanical vibration and noise
- Multiple flights or legs per day
- Night flight
- Physical discomfort
- Poor or inadequate diet and/or exercise
- Stress-induced anxiety
- Temperature extremes or changes

As pilot fatigue increases, the pilot's ability to fly decreases. In extreme conditions of fatigue, the pilot may become completely incapacitated and incapable of rational decision making. Brett was an example of how severe fatigue can affect a pilot. His condition was brought on by extended stressful work periods with inadequate rest. The effects of fatigue are amplified at night. A tired pilot will experience a greater desire to sleep at night than in the day. As metabolism slows down, the fatigued pilot will experience increased symptoms, possibly demonstrating lousy, irrational judgment and reacting slowly to potentially hazardous in-flight situations.

To prevent the onset of fatigue, a pilot needs only pay attention to personal health—get adequate rest, eat well, manage stress, and avoid harmful conditions/substances. If the pilot notices symptoms of fatigue in spite of taking adequate precautions, then something is wrong, and the pilot should not continue flying and should get assistance.

Vision

Your eyes could easily be compared to a camera. A camera gathers filtered light through the lens, controls its intensity with an aperture, and focuses the image onto a film of chemically treated plastic. The lens may be adjusted fore and aft, changing focal lengths to compensate for the varying distances of the objects photographed. The aperture works like a window of adjustable area, which can open wide, to admit maximum light, or constrict to barely more than a pinprick in very bright conditions, necessary because chemicals on the film function only in controlled lighting conditions. (See Fig. 31-2.)

Light coming into your eyeball passes through the cornea (filter), is modulated by the iris (aperture), and is focused by the lens onto the retina (film). The basic principles are the same, but the eye is remarkably more capable than a camera in a couple of fascinating ways. First, the lens in your eye does not move forward and back, like the camera. Your eyes' lenses are like clear marbles that are soft and flexible. They are surrounded by a ring of fibrous muscle tissue, a little like the rings around planet Saturn. When the muscles pull on the lens, it is stretched thin, when the muscles relax, the lens thickens, thus changing the focal characteristics of the image on the retina. When you get old, that soft, pliable lens may become a bit stiff, making it difficult for you to see objects close-up. The muscles

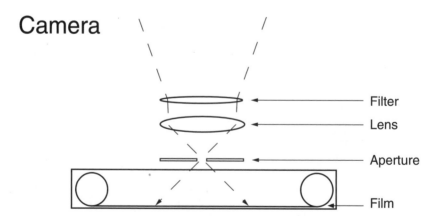

Fig. 31-2. *A camera is basically a mechanical eye.*

that adjust the focus of your lens may suffer from fatigue—anything that exercises your eyes for an extended period, such as reading a lot of fine print, will cause you to feel the effects of this strain, making your eyes feel tired. With this in mind, your eyes will relax, whenever they can, such as when you are sleeping. They will rest while you are awake, as well—particularly if there is nothing to look at, such as a night-blackened sky. Your resting eye will tend to focus at a range of about 4 to 6 ft away, and you won't necessarily be aware of it—a phenomenon called *empty-field myopia*. If you drone along in the plane at night, with no obvious details visible outside the plane, your eyes might relax, causing you to see a range no further distant than the windshield for hours at a time. You can easily solve this problem by simply looking at a distant object, such as the wingtip or a clearly visible light on the ground, and keeping the focus while you scan the view outside.

The second amazing feature of your eye is its reusable, motion picture film, the retina. This is where the eye is wonderfully complicated. The retina is surfaced with millions of tiny nerve endings, which are sensitive to light. Most can distinguish color, while other, somewhat more sensitive nerves see only the raw light intensity—basically black and white. Each nerve ending in the retina is individually wired to your brain and, when hit with light, can fire off a small electrical impulse into the heart of your brain's image processing center. These small electrical impulses are assembled in your brain to form an image, much the way a computer image is created from numerous pixels or a television image is formed from numerous small dots of colored light. (See Fig. 31-3.)

A closer look finds that the small nerve endings in your retinas contain a light-sensitive chemical that breaks down when exposed to light. The chemical reaction produces the electrical impulse in the nerve. Now, once the reaction occurs, the nerve is not sensitive to light any more—it doesn't "see." As with a gun, once the trigger is pulled, it won't fire again until it's reloaded. You can experience exactly this sort of phenomenon by having your picture taken in a dark room with a flash-bulb equipped camera. The flash goes off, and you see nothing but white for a few seconds—the light-reactive chemicals in your eyes have done their job, and for a moment you are blind. But wait, in a few moments, your vision returns! What is going on? There is a massive reloading project occurring within the nerve endings in your eyes. The light-reactive chemicals are being rebuilt with some enzyme action and exposure to large quantities of

Eye
(Top view, right)

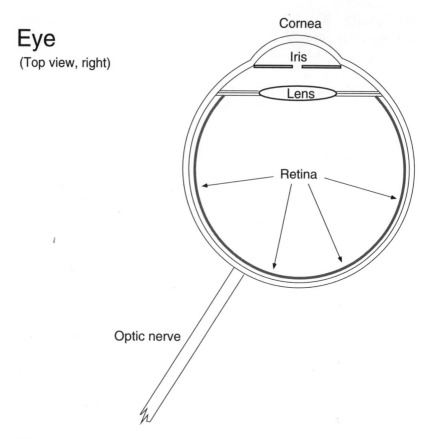

Fig. 31-3. *Although it appears simple—like the camera—your eyes are wonderfully complex.*

oxygen. The blood-borne oxygen comes to your retinas through a large artery to the rear of each eyeball. The blood flow is impressive. If that artery were to break, you could bleed to death in a few minutes. As you sit, reading this book, your eyes are consuming up to 70 percent of the oxygen you breathe, tending to the chemical needs of the nerve endings in your retinas.

Here is where the iris comes into play. Its job is to regulate the light that enters your eye so that your retinas are able to keep up with the exposure. You might think of each nerve as a soldier in a firing line. The iris acts as a battlefield commander, desiring to cover the enemy with continuous fire and thus ordering some "soldiers" to shoot while others are busy reloading. As each gun is fired, it is reloaded while the others shoot. This is in contrast to everyone shooting at the same time, leaving the army momentarily defenseless while reloading. You have so many nerve endings in your retinas that the processes of exposing and recharging the light-sensitive chemicals in your eye happens at the same time, on a continuous basis, so that you may see a constant, moving image. The flashbulb simply surprised the iris, catching it wide open in dark conditions, and overloaded the nerves on your retinas, causing all the nerves to "fire" at once.

The nerve endings in your eye comprise at least two distinct varieties. Close inspection reveals that some resemble small cylinders, others are more conical. They have thus inherited the names, *rods* and *cones.* The cones contain chemicals that are reactive to light of different wavelengths, enabling them to respond to color. The rods contain a similar chemical that overall is much more light-sensitive but does not distinguish wavelength, or color. The rods, because of their sensitivity, have their greatest use in nighttime, low-light conditions. These different types of nerve endings are distributed about the surface of the retina, with cones favoring the areas in direct focus of the lens, rods more to the periphery. This is natural. Since your eye functions primarily in daylight conditions, it is optimized for seeing in rather bright sunlight. So, in daylight, the high density of cones at the central focusing area of your retinas, called the *fovea,* enables you to distinguish color and clarity unrivaled by the very best of camera equipment. Night is another story.

Since the cones are adapted for daylight conditions, they distinguish relatively little at night. The rods, at the periphery of your vision, are much more capable in darkness. The rods are also more sensitive to motion, enabling you to react to a flicker of movement just visible out of the "corner of your eye." At night, these rods provide the majority of the image you see. Unfortunately, the rods are most prevalent in the areas of your retina that correspond to peripheral vision, so your ability to see an object in the dark is actually better if you do not look directly at it. You can experience this phenomenon by looking at the stars in a night sky. Your peripheral vision will sense vast numbers of stars, including many faint details that seem to disappear when you turn to look at them directly. Your color vision does not become dormant at night, however. It still functions but requires relatively bright light to work. If a light source, such as a neon sign, is bright enough, your eye will distinguish it in detail and color, as in daylight. Under normal nighttime conditions, however, lighting is inadequate for the cones to function well, and you have difficulty distinguishing much beyond the muted grays and blues of darkness.

Your eyeball is colored black on the inside, like the black insides of a theater or camera body. The black pigment on the surface of your retina prevents light from glaring reflectively around inside your eyeball. Having no pigment would be a bit like watching a movie in a theater with the lights on. Albinos are born without such pigment in their eyes (which causes the red-colored eyes—you see the blood vessels) and suffer from poor visual acuity because of this. Believe it or not, vitamin A is contained in this pigmented surface, as well as within the cellular fluid of the rods and cones. With a little enzyme action, vitamin A can be converted into more of the light-sensitive chemical that allows your eye to work—it is already very similar in structure. This becomes a real factor in the long term, as your eyes have the capacity to adapt to a dark environment and increase their light sensitivity by producing more of the appropriate chemicals from vitamin A. You could compare this to putting a higher-speed film in your camera. Conversely, the chemicals in your eye can be converted back to vitamin A by simply allowing them to break down as they are exposed to light. Your eye can in this way adjust its sensitivity to the light conditions outside by varying the amount of light-sensitive chemicals in the rods and cones—literally changing the sensitivity of the "film."

If you insist on operating in a dark environment, your eyes will adjust their light sensitivity as best they can to enable you to see—a process called *dark adaptation*. It takes about 2 hours in near-total darkness for your eyes to adjust completely for nighttime. It's a common practice for military pilots to sit in the dark, wearing dark glasses, for a few hours before launching on a night mission—they want to see as well as possible. Any night pilot could benefit from this adaptive capability. Your eyes can adapt the other way, however, and they adjust to daylight conditions very quickly—in only a few minutes. It's a lot easier to break the sight chemicals down than to make them. A pilot who wishes to retain the best of night vision should keep cockpit lights very low and avoid looking at any bright lights (Fig. 31-4).

There is a condition called *night blindness,* which may occur when you have a deficiency of vitamin A. Like the name implies, you would be unable to see in dim-lighting conditions because your eyes do not have sufficient raw materials from which to create the necessary chemicals you need to see. Since vitamin A is usually stored in large quantities in your liver, it would take many months of a vitamin A–deficient diet before you would notice any symptoms in your eyes. Interestingly, night blindness can be reversed within an hour

Fig. 31-4A. *Use low light to read checklists and charts. Bright light will temporarily ruin your night vision.*

by an intravenous injection of vitamin A. For the pilot, however, the lesson is quite clear—eat your carrots.

All the nerve endings in your retina are bundled together into a single, rather thick optic nerve, which trails like an extension cord off the back of your eyeball and into your brain. Coupled to the optic nerve is the main artery and vein that supply blood to the eye. You could picture them as a garden hose and electrical cord attached to a beachball. There is enough slack in the bundle to allow for the articulation of your eyeball. The point on your retina where the optic nerve and artery attach is devoid of nerve endings, forming a small blind spot, called the *optic disk*. The specific location is a little off-center, to the inside of each eye. The blind spot in one eyeball is compensated by the image in the other, and vice versa. If you close one eye, the blind spot is visible, but your

Fig. 31-4B. *Use low light to read checklists and charts. Bright light will temporarily ruin your night vision.*

brain paints over the image, by inferring details from what is visible. You may "see" this blind spot with a simple exercise: Make a couple of small *X*s, about 2 inches apart on a piece of paper. (See Fig. 31-5.) Covering your left eye, look only at the mark on the left, adjusting the paper forward and back, until the other *X* disappears—its image is focused on your blind spot. Covering your right eye, look at the mark on the right. You'll witness the same phenomenon. An interesting variation of this is to draw a line connecting both the marks, extending it a little beyond each. (See Fig. 31-6.) Repeating the exercise, you will find that the mark disappears, but the line is unbroken. Your brain has simply inferred the existence of the line across your blind spot, and "drawn it in" for you. (See Fig. 31-7.)

All of this wonderful anatomy is housed like a ball and socket, manipulated, aimed, and controlled by three pairs of muscles attached to the outside of your eyeballs. These muscles function in a coordinated effort to aim your eyes at things you wish to look at. Typical for musculature in your body, the muscles that guide your vision tend to do so in leaps and jumps, called *saccades*. When you read this page, your eye is literally darting from one point to another on a given line of print, at a rate of two to three times per

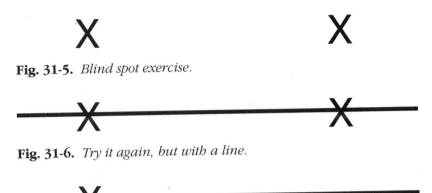

Fig. 31-5. *Blind spot exercise.*

Fig. 31-6. *Try it again, but with a line.*

Fig. 31-7. *Your brain is constantly "touching up" the images you see.*

second, in an action known as *saccadic movement*. Your eyes fixate momentarily at a given point of interest, then leap to another, and so on. It happens so rapidly that less than 10 percent of the time is actually spent moving your eyes, the other 90 percent has them fixated on something. This has the potential of making the image you see appear as though your eye were resting atop an active jack-hammer. This image is stabilized at the visual processing center of your brain, so that you see a rock-steady image of the world around you. Your brain automatically stabilizes the image by reference to objects of known stability. The ground doesn't move, so your brain makes the image of the ground appear steady. If you remove all background images, perhaps looking at a single light against a completely dark background, there is little for your brain to reference to stabilize the image, and you may suddenly become aware of the jerky motions of your eyeballs. The light may appear to jiggle and dance, perhaps darting about like a firefly or even another airplane. This is called *autokinesis,* meaning that the image appears to have a motion of its own. If the horizon were visible behind it, it would just as quickly become stationary again, as your brain picks up on the reference. The system is utterly fantastic.

Depth perception is provided by the two, slightly different images that you see from each eye, as they come into the brain. It is very effective at distances ranging from close-up to about 30 ft or so. Beyond that, your brain compares perspective views and relative motions to perceive distance in the images you see. Subtle details, like the way mountains in the distance appear to move in relation to the stars behind them, would indicate to your brain that the mountains are closer than the stars. If many of these details are removed, as on a dark night, your brain may lose its ability to sense range and distance, making city lights, other airplanes, and stars look like a single, flat image, devoid of perspective. The same is true in landing on a dark runway; you might be surprised by how close—or how far away—the runway really is (Fig. 31-8).

Hearing, balance, and motion sickness

The human ear is among the most sensitive of the animal kingdom. Dogs and bats can hear much higher frequencies, of course, and elephants and whales communicate at much lower frequencies, but as far as hearing acuity is concerned, there are few ears better than your own. The interesting aspect of your hearing is that your brain is forced, for its own sanity, to turn down the volume. At a given time, you are hearing a veritable barrage of background noise that, if you paid much attention to it, would cause difficulty in concentrating on whatever it is that you consider important. Take the pesky leak in the bathroom faucet, for example. You might not notice it at all in the daytime, when ambient noise and activity reduce its constant dripping to unimportance. If you lay awake at night, however,

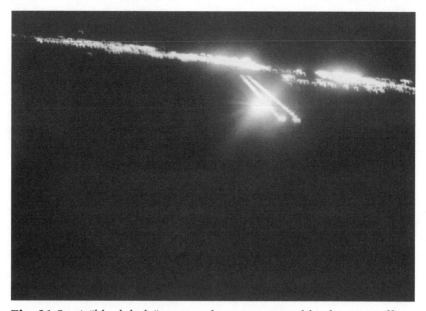

Fig. 31-8. *A "black hole" approach, to a runway like this one, offers little depth information.*

background noise quiets, and your still active mind begins to turn up the volume. You become aware of the chirping of crickets, the rustling of your blankets, and that incessant drip...drip...drip....All these sounds were present before, but you had the volume turned down.

Many years ago, as part of a school fieldtrip, my class was ushered into an echo-proof room at a university physics laboratory. Once inside, with the thick, insulated doors closed behind us, we were instructed to sit quietly and remain very still, for about 5 minutes. There was absolutely no background noise, as the room is completely isolated from all external sound. A little time passed and the student next to me retied his shoes—it was surprisingly loud. A little while later, I heard myself *blink*. We all could hear our own pulse, and even more interesting, the sound our own ears made, due to internal vibrations and blood flow. To some, it sounded like listening to a seashell, others heard a high-frequency tone. We were instructed to listen carefully to the street as we left the room. The sound of traffic was uncomfortable, and, from a good distance, we could hear the switches click as the traffic lights changed. It wasn't long before our hearing returned to normal.

When you fly, your hearing sensitivity is definitely turned down—airplanes are noisy. You may find that it becomes easier to sleep under such circumstances. Of your body's basic senses, hearing is a good indication of your level of alertness. Turn it down, and you may naturally relax. Couple this with the pressure changes due to altitude, and you may feel as though your ears are not working well at all. With the sensitivity of your ears turned down, you are still sensitive to low-frequency vibrations through your skeletal structure. Your bones can actually *hear*. You can demonstrate this to yourself with a little tap on your head—the sound it makes is carried to your auditory nerves through your skull. Sometimes these low-frequency vibrations that come from an airplane and literally rattle your bones can be extraordinarily fatiguing. Children appear to be especially susceptible to this. My 2-year-old son, for example, could hardly stay awake in a light plane after the engine started. This tendency toward relaxation and sleep could be aggravated at night, when your body wants to sleep, anyway.

Since your ears are rather delicate and may be damaged by continuous exposure to loud noise, you should use some form of hearing protection. Exercise care, when choosing a headset, to find a comfortable type; the discomfort caused by the pressure of an ill-fitting headset may contribute to fatigue. However, a good technique for staying alert is to periodically take the headset off. The sudden change in noise seems to help freshen the experience of sitting in a cockpit.

Aside from hearing, there are other sensations that your ears provide. It is your balance that is especially pertinent to night flight. Your sense of balance is determined by a mechanism deep inside your ears. The organ looks like a small, loose tangle of tubing— called the *membranous labyrinth*. It is composed primarily of three loops of tubing, arranged in half-circle arcs that join at their bases at right angles to align with the three spacial axes. Known as *semicircular canals,* these tubes contain fluid that runs about within them in response to movement. The rushing fluid triggers small, hairlike nerve endings whose stimulations produce a sensation of acceleration. Since the tubes are aligned with all three spacial axes, you can sense accelerations in any direction.

Your sense of equilibrium is based upon acceleratory forces, and, because of this, you can be fooled. You can easily become accustomed to steady motion, adopt a new frame of reference relative to that motion, and thus decide you're not moving at all. You are traveling at tremendous speed right now, for example, and you've never felt the sensation of it. Simply living on the earth means that you are traveling something near 1000 mph, because of its rotation. The earth itself is making about 66,600 mph around the sun, and the sun is really clipping along through the galaxy, and so on. You will forever be insensitive to these motions unless they were to change— then you would sense the acceleration or deceleration of the change. Someone once explained this rather succinctly by claiming that a long fall never hurt anybody—it's the sudden stop at the bottom. So you're sensitive to accelerations.

The limitations of your sense of balance are compensated by what you see. It is even possible for a person to maintain balance solely by visual information, but not particularly well:

> *Many persons with complete destruction of the vestibular apparatus have almost normal equilibrium as long as their eyes are open and as long as they perform all motions slowly. But when they move rapidly or close their eyes, equilibrium is immediately lost. (C. Guyton,* Human Physiology and Mechanisms of Disease, *W. B. Saunders, Philadelphia, 1987.)*

However, with normal equilibrium sense, it is simple for a person to maintain balance—and even move about—with their eyes closed. For pilots, a problem arises with the sense of balance. With eyes closed, it is impossible to fly an airplane for very long and maintain an accurate sense of upright. The airplane can create too many accelerations that mimic gravity but are not necessarily oriented toward the earth. A blind pilot, going by internal sense alone, can be easily fooled by normal aerial sensations into flying the airplane in almost any attitude while thinking the plane is straight and level. For this reason, the pilot's eyes referencing the horizon become the primary means of balance and orientation as far as the airplane is concerned. Since night causes some degradation in the pilot's ability to see outside, night may also affect the pilot's ability to keep the airplane upright. In other words, without the aid of attitude instruments, the pilot flying at night is a strong candidate for disorientation and vertigo. It is for this reason that night equipment should include at least some basic attitude instrumentation.

When the internal sensations of balance disagree strongly with what is visually indicated outside, the resulting vertigo can be nauseating. This may be the heart of motion sickness. The night VFR pilot might encounter some difficulty in seeing the horizon outside and thus subconsciously rely on, or pay more attention to, internal indications of balance. When the horizon is perceived, and the pilot feels a discrepancy with her or his internal sense of balance, a moment of disorientation and confusion may cause an upset stomach or other discomfort. Thus, the possibilities for motion sickness and disorientation are greater in the dark.

The FAA publishes an advisory circular on spacial disorientation (60-4A). Much of it is reprinted here, since it is short, direct, and to the point.

> *a. The attitude of an aircraft is generally determined by reference to the natural horizon or other visual references with the surface. If neither horizon nor surface references exist, the attitude of an aircraft must be determined by artificial means from the flight instruments. Sight, supported by other senses, allows the pilot to maintain orientation. However, during periods of low visibility, the supporting senses sometimes conflict with what is seen. When this happens, a pilot is particularly vulnerable to disorientation. The degree of disorientation may vary considerably with individual pilots.*

b. *During a recent 5-year period, there were almost 500 spatial disorientation accidents in the United States. Tragically, such accidents resulted in fatalities over 90 percent of the time.*

c. *Tests conducted with qualified instrument pilots indicate that it can take as much as 35 seconds to establish full control by instruments after the loss of visual reference with the surface. When another large group of pilots were asked to identify what types of spatial disorientation incidents they had personally experienced, the five most common illusions reported were: 60 percent had a sensation that one wing was low although wings were level; 45 percent had, on leveling after banking, tended to bank in opposite direction; 39 percent had felt as if straight and level when in a turn; 34 percent had become confused in attempting to mix "contact" and instrument cues; and 29 percent had, on recovery from steep climbing turn, felt to be turning in opposite direction.*

d. *Surface references and the natural horizon may at times become obscured, although visibility may be above visual flight rule minimums. Lack of natural horizon or surface reference is common on overwater flights, at night, and especially at night in extremely sparsely populated areas, or in low visibility conditions. A sloping cloud formation, an obscured horizon, a dark scene spread with ground lights and stars, and certain geometric patterns of ground lights can provide inaccurate visual information for aligning the aircraft correctly with the actual horizon. The disoriented pilot may place the aircraft in a dangerous attitude. Other factors which contribute to disorientation are reflections from outside lights, sunlight shining through clouds, and reflected light from the anticollision rotating beacon.*

Respiration and oxygen

Your body consumes oxygen constantly. Oxygen passes from the air in your lungs across the membranes in your lungs to your blood and is carried by the blood to your organs and tissues. The oxygen is literally combusted with carbon-based "fuels" derived from the food you eat. The burned carbon, CO_2 is absorbed back into your blood, passed back across the membranes in your lungs, and exhaled into the air. We've already discussed the critical requirement of oxygen in your eyes, but it is equally critical in other parts of your body. If a pilot were to deprive oxygen to the body's systems by pulling excessive "g's," for example, sending blood by the factored force of gravity to his lower extremities, his vision would fade to white—very similar to the picture flash, but in slow motion. Without oxygen to rebuild the light-responsive chemicals in his eyes, vision fades away. The pilot would be blind but conscious for a few moments; once the pilot's brain was sufficiently starved of blood, he or she would be unconscious. If the acceleration were to continue unchecked, the pilot would die from it in a few minutes. This, of course, rarely happens because the airplane, with an unconscious pilot, usually crashes before the "g" loading gets her or him.

Another way to deprive a pilot of oxygen is to fly at high altitude. With altitude increases, the pressure of oxygen in the lungs may be insufficient to cause it to pass across the membrane into the blood. Sudden depressurizations at very high altitudes can cause the oxygen pressure in the body to be greater than that of the surrounding air, causing the pilot to actually exhale oxygen—literally deflate and bring unconsciousness and death in moments.

Under more normal circumstances, the pilot is trained to observe symptoms of hypoxia, as they often occur slowly, gradually depleting the pilot's capabilities over an extended period at moderate altitudes. One of the effects of this hypoxic state is a reduction in visual acuity in the dark. The eyes need oxygen to function. Lack of oxygen will reduce the eye's sensitivity to low light. For this reason, the FAA recommends that a pilot consider using supplemental oxygen at altitudes above 5000 ft, at night. This is completely relative as far as the pilot is concerned. I have lived at an elevation of 4600 ft, or so, for the last 30 years. My body has acclimatized to this altitude. To use oxygen at 5000 ft seems silly for night operations in the traffic pattern. Nevertheless, that is my circumstance, and relative to my condition. Pilots from lower elevations might experience night vision degradation at lower altitudes, and the use of supplemental oxygen would certainly enhance night vision to some degree as altitudes increase.

Psychology

What's going on in a pilot's mind as the plane flies through the night? There are surely as many answers to that question as there are pilots; however, in general terms, there may be a few consistencies. One is a quiet state of denial. Many pilots feel that the night emergency won't happen to them—the engine will keep running, the weather won't close in, and so on. This is a normal response to potentially dangerous conditions that allows pilots to feel more comfortable: Pilots wouldn't fly if they knew their engines were going to fail. Not knowing, and hoping everything will be all right, the pilot goes flying. Denial is a form of optimism, which, when coupled with enough pessimism to inspire the pilot to double-check everything, is okay. The opposite of denial is paranoia, where the pilot feels that everything bad will happen, causing feelings that border on the irrational sense of panic. Neither mode of thinking, in its extremes, is safe. A pilot in severe denial will insist that no emergency will happen, even after one has already occurred. A pilot subject to paranoia might make an emergency response to a normal condition, creating a real emergency in the process. We should strive for the middle ground.

We have a need to be entertained. If you lose interest in what you are doing, doing it well becomes difficult. For pilots, flying the airplane can be boring—especially when the view disappears outside. Long stretches of night or IFR flight can cause a pilot fatigue and much mental stress simply because it isn't fun anymore. If the pilot were strongly fatigued at the same time, sleep may become an irresistible option. Sometimes daydreams will take on an uncanny reality. Lindbergh heard voices behind him, as though a whole group of observers sat in the tail of his plane. Later, he saw visions of whole islands and land-masses pass by below in excruciating detail, when nothing was really there. He was very fatigued, and his mind became distracted with its own entertainment. So many pilots have flown for long stretches, quietly wishing for something different to happen. I can remember flying around the traffic pattern in daylight for "umpteen" thousand times, secretly wishing we'd have an emergency of some kind, just to break up the monotony. Wishing for an emergency? How foolish! And yet these irrational thoughts can arise when the pilot is bored. Night flight, especially for long distances over water, can be exceedingly boring. You might bring along some entertainment, the best of which is another person in the cockpit. If you are alone, find things to occupy your mind with the task of flying—try to enhance the airplane's performance, improve the operation of its systems, eat, listen to music, or strike up a conversation with a controller.

Although many pilots report that night flying is relaxing, most report higher levels of anxiety about the possibility of an engine failure. This anxiety stems from the fact that an in-flight emergency in the dark could require an especially difficult landing; indeed, the odds of surviving an engine failure in darkness are about half as good as in daylight, and a powerless descent into blackness is a frightful prospect for anyone. With this in mind, pilots tend to fuss over the engine operation a bit more than in daylight, watching and listening closely for any sign of impending failure. A natural result of all this attention given to the engine is that the pilot may notice unfamiliar sounds or vibrations. These anomalies have, in most cases, been there all along, but flying at night with a heightened sense of anxiety elevates the pilot's level of awareness and the sounds are suddenly manifest. This is called *automatic rough* and is a perception of roughness caused by the pilot's anxiety over an engine failure. Anxiety levels, conscious or not, can be expected to increase in darkness. They can affect how a pilot responds in an emergency, possibly lowering the panic threshold and causing the pilot to make irrational decisions.

However, the invisibility of much detail may provide a simplified view of the night environment. Occasionally, the only things obviously visible have direct bearing to flying the plane, such as the airport runway and other air traffic. The fact that extraneous details are not visible or distracting to a pilot's attentions gives rise to the impression that night flying is relaxing (Fig. 31-9). This may seem like a contradiction of the above paragraph, but when coupled with denial, the night pilot may easily be lulled into a strong feeling of satisfaction and well-being. The simplicity of the view at night is why most flight simulators favor night environments—it requires far less computer memory to produce the night image than a VFR daylight image. Perhaps for similar reasons, the night pilot feels that less effort is required to keep track of the flight environment in darkness. When an emergency actually occurs, however, the shock of it can be amplified, as the pilot goes from a relaxed state to one of rather severe alarm.

Judgment

The pilot's ability to make decisions can be affected by darkness. Consider this example. The pilot flying a single-engine airplane at cruise altitude is faced with an engine failure. His heart jumps in his throat. His reverie in the cross-country trip is suddenly shattered by

Fig. 31-9. *The view in darkness seems beautifully simple, but an emergency landing in invisible terrain might be a fearful proposition—this city is surrounded by mountains.*

the whispered stillness of a prop windmilling outside. The pilot mentally goes through the three steps of an engine failure: Fly the plane. Pick a place to land. Attempt a restart, if there is time. The pilot begins to slow to best glide speed. Looking outside, there is little to see. The pilot imagines all sorts of horrible features on the ground below, any of which could cause a fiery crash. Flying the plane straight ahead, the pilot focuses attention on the engine. It has to work. It *has* to. He works the throttle, the mixture, switches mags, tanks, checks carburetor heat. The engine doesn't respond. As the plane descends, the pilot still hasn't selected a place to land. His anxiety is increasing rapidly now—he's becoming irrational. Since the pilot cannot visualize the emergency landing, he becomes focused on getting the sick engine to run again. As his attentions narrow on the engine, flying the plane becomes less important, and its glide steepens; the pilot has not given much thought to the direction of flight. Busily working with the engine controls, now approaching panic, the pilot has practically left the plane to fly itself. And that's the way he dies, hunched over the engine controls, pulling, switching, and tugging, while the plane flies over numerous survivable landing sites and crashes, uncontrolled, into the ground. Sadly, this scenario has been duplicated many times.

The pilot cannot afford to become irrational. Almost all the physiological aspects of night flight can affect the pilot's rationality. Of these, perhaps the greatest is fatigue, but the others should not be severely discounted. Stress levels can accumulate rapidly during critical night maneuvers. At many rural airports, the simple act of flying the pattern can approach the complexity of an IFR circle-to-land procedure, requiring the pilot to divide attentions rapidly from instruments to outside, to configuration changes and the radio all at nearly the same time. In extreme situations, such as the simultaneous occurrence of an emergency or visual illusion, the pilot's capacity to perform might be surpassed. In Brett's example, his capacity was saturated by the simple act of trying to stay awake—it occupied his full attention and grew in severity until he almost gave up the fight.

Consider a juggler at the circus. She is comfortable juggling, say, three balls. While doing this, she can entertain the crowd with jokes and alter her manner of juggling to bounce the balls and adjust their order in her hands. Suddenly she adds another ball. With four, she moves faster, touching each ball for an instant as she sends it flying to the other hand. At five, her face is a picture of concentration, her hands a blur. She has stopped talking and begun to sweat. This is like a pilot during a night landing approach into a difficult airport. The juggler is still there, but an enterprising clown adds a chainsaw to the fray. Now the juggler is really busy—more than that, she is saturated. No matter how skillful a juggler she is, there is a limit to the number of items she can juggle, especially the pressing ones, like the chainsaw. It is the same with pilots. No amount of skill can put a pilot above the possibility of becoming saturated with tasks. With the addition of a chainsaw, the juggler drops one of the balls. Her capacity is five—period—and that will change from day to day, depending upon her condition. The chainsaw required that she lighten her workload, so she dropped something relatively menial. Pilots at night need to prioritize their tasks, as well. For the hapless pilot who crashed while trying to start the engine, he focused on the menial task and was killed by an important one. A pilot demonstrates good judgment when personal limits are well understood and the situation managed to prevent exceeding them, for a pilot who "drops the balls" dies.

The effects of night can easily reduce your capacity to fly the plane, effectively lowering your saturation levels. For example, daylight VFR flight provides an easy reference horizon with which to reference the plane's attitude. In that condition, you might easily cope with a number of factors including some nasty emergencies. Night-time, on the other hand, might force you to handle the problem while referencing instruments, limit your options of landing sites, and simultaneously raise your anxiety levels. You could reach your saturation level much sooner.

Technique

Your physiological needs are always a factor when you fly a plane. They cannot be ignored. The point of this section is to highlight a few preparations you should make to help you cope with your body's needs in darkness.

Eyes

If you plan to fly in dark conditions, you can maximize your vision by sitting in darkness for at least 2 hours before you go fly. Artificial darkness works fine. You may shut drapes, close the door, and wear sunglasses with the lights off. Listen to music or sleep. Once your eyes have fully adapted to darkness, it becomes important to avoid light conditions. If you see bright lights, you will destroy the past few hours' effort in a couple of minutes. On the flight line, avoid looking directly into landing lights. Since this is sometimes unavoidable, or perhaps closing your eyes would be foolish, close one eye to protect at least half your night vision. Use red lights in the cockpit. Studies indicate that red light does not alter your night vision yet still provides enough illumination for you to see charts and instruments.

When looking for obscure details outside, try not to look directly at the object you wish to see, but look to one side of it. This will put the image on a part of your retina that is optimized for night vision, enabling you to perceive the object with greater clarity.

Sleep

It may seem obvious to suggest that you get adequate sleep in preparation for a night flight, but the practicalities here can easily prevent it. Your body is probably not accustomed to sleeping in the day, preventing you from sleeping at all. The demands of work and other daylight time constraints might also interfere with your opportunities to take a nap. You must consider all these factors and manipulate your schedule to allow for adequate rest before you fly off into darkness. Pilots who make regular night flights find that it takes many days, even weeks to train their bodies to sleep adequately in the daytime. If you consider a long night trip, you had best take advantage of a sleep opportunity beforehand. Ignoring your need for rest might severely hamper your ability to fly—it could even kill you.

When you fly, ask yourself honestly, how do you feel? Be prepared to skip the flight if you don't feel up to it. Brett knew that he shouldn't be flying, but to help the company, he flew anyway. It was a good time to simply say no. Look your mental state over carefully, discounting what might be normal excitement to be going flying. Ask yourself these questions:

- Have you had enough sleep?
- Have you slept recently?
- Are you mentally tired?
- Stressed about anything?
- Are you angry?
- Are you in a hurry?
- Feel overloaded?
- Hungry?
- Healthy?
- Does the flight make you nervous?
- Are you scared?

You might make up several questions of your own. The point here is that you honestly assess your own condition and make a decision about your ability to fly. If you have doubts, don't go. If you're not sure about some aspect of your condition but can't place your finger on exactly what, you might seek a friend to go along, one who is competent in the airplane.

You might use coffee or other caffeinated beverages to help you stay awake at night. That's fine, as far as staying awake is concerned, but you will need to make some preparations for your bladder. The pit-stop options are numerous: You could plan a flight with shorter legs, allowing you the opportunity to visit the bathroom periodically. You could bring along a urinal, although the opinions vary widely on the effectiveness of these items, especially among women. *Aviation Consumer* published an article describing how their staff sampled the use of various brands of urinals and "in-flight range extenders" in effort to find out which varieties worked best. Their findings were interesting, to say the least, ranging from delight to rabid hatred. Surprisingly, the item that seemed to interfere the least with the task of flying the plane was a simple diaper. Most airline flight crews have the benefit of an on-board bathroom. I work as a pilot for a regional airline. During one night flight into Rapid City, South Dakota, the captain, who drank several cups of coffee, got up to use the restroom and left me alone on the flight deck. He was gone for a long time. I had already begun the descent into the airport area, and wondered what had become of the him. About then, I got a call on the interphone. It was the captain. "Open the %*@! door! Didn't you hear me knocking?!" He'd locked himself out of the cockpit. He struggled with the door a minute, in full view of the passengers, then had to ask the flight attendant how to work her phone. I let him in, but I think the passengers heard me laughing. You'll need to plan for the bladder contingency.

Staying Awake

In flight, the techniques pilots use to stay awake are numerous. Coffee is one, but eating carrots or chewing other noisy, crunchy food works well, too. I mentioned sucking on ice-cubes in a previous chapter. Breathing oxygen can help you stay awake. Listen to the radio, talk to yourself, sing, chew gum, slap yourself, clap your hands, adjust your seating position, maneuver the plane, fly lower, read a chart, write down the indications of each of the plane's instruments, make a performance log, find out how far it is from your position to the nearest big city or how far you are from Nome, Alaska, and, if you must, turn up the interior lighting.

Exercising in the cockpit helps keep you alert, as well, by elevating your heart rate and giving you something to do. On the down side, it can also help you relax—so when you finish exercising, find something else to occupy your attention or simply plan to exercise for the rest of the flight.

There are several exercises that you can do while sitting and flying an airplane. Here are some examples:

- Start with your neck. Tip your head back as far as it can go, stretch your neck muscles, then tip your head forward—this is not to be confused with nodding off—to stretch the back of your neck. Do the same thing from side to side. Move slowly and hold the extreme positions for a few moments. You should feel some stretching in your neck muscles. Don't forget to keep flying the plane, and move slowly enough to prevent getting vertigo.

- Move down to your shoulders. Rotate your shoulders around while keeping your hands where they are. Move slowly at the limits of your shoulder's travel. You should feel some stretching here, as well.

- With each arm in turn, reach as far behind you as you can, then over your head, then to the floor. With the opposite hand, pull your elbow across your chest as far as it will go.

- Clasp your hands together and stretch them both over your head and behind you as far as you can reach. Clasp your hands together in front of your chest and pull as if you were pulling them apart, hold the pressure for a moment, then relax.

- Press your hands together as though praying and push hard, holding the strain for a few moments before relaxing again.

- As you sit, twist your torso from side to side, holding your elbows together in front of your chest while moving them from side to side, with your hips stationary.

- Keeping your back against the seat, reach one elbow at a time toward the bottom seat cushion, stretching your torso from side to side.

- Arch your back, pressing your shoulder blades against the backrest and pushing your stomach out as far as it can go. Reverse the motion by shrugging your shoulders forward, bowing your back against the seat and sucking your stomach inward.

- Take a deep breath and hold your breath as long as you can, then exhale. Repeat.

- Press your hands on your seat and lift both feet off the floor, such that your thighs are not supported by the seat cushion. Hold that position for a while, then relax. Do it again.

- Loosen your safety belt, press your hands on the armrests or seat cushion, and lift your whole body off the seat. Hold that position for a time then relax.

- Lift each leg individually, bend and straighten the knees, roll your ankles around, and make a circling motion with your feet.

- Stomp your feet on the floor several times.
- You might also soak a towel in ice water and put it on your face, over your head, up your shirt or in your lap. Pour cold water on your hair. Open the cockpit vents. Keep cool. It is more difficult to sleep when you are just cool enough to be uncomfortable while resting.

With all the commotion you might be making in the cockpit, don't forget to fly the plane. If you have a companion in the cockpit who is competent in flying the plane, caution him or her to stay awake, then make yourself comfortable and take a nap. If you are alone and are having difficulty staying conscious or even interested in flying, you'd best pick a nice airport and land the plane.

Vertigo, disorientation, and motion sickness

One of the finest students I ever had was a victim of motion sickness. Try as we could, we couldn't shake it. Early on, during his training, it became evident that only short flights would do. We would fly until his stomach became upset, then discontinue the lesson and try again. He learned to fly with a smoothness the likes of which I haven't seen since. We needed two or three lessons to get through stalls. On his solo cross-country flight, there was enough light turbulence to make the poor guy fly with one hand and hold a bag over his mouth with the other. Over the course of the training, his condition improved a bit but did not go away. Thankfully, we were careful enough that he never actually vomited during any lesson, until the very end—a night flight.

He made nine beautiful landings at an outlying airport. On the way back, we played with the airplane, making some steep climbs and turns. He had fun until his stomach began to throw in the towel. We immediately settled down and flew toward home. The approach controller advised us of oncoming traffic, 500 ft lower, against the city lights. We'd been able to locate all the earlier traffic but had difficulty finding this one, disguised among the lights below. We discussed that for a while, until I noticed the airplane passing below, out my side and down. It was still quite difficult to see, even 500 ft away with its lights blinking. I banked the plane to the right, to point it out to my student. He saw the plane, but when we rolled back to level, he broke sweat....

"Quick, get the bag!"

"Where is it?" I said.

"In the back....Hurry!"

I couldn't reach his flight bag, it was positioned too far aft in the baggage area of the C-152. I quickly unbuckled my safety belt and began climbing over the seat...

"Quick, take the controls!"

I flopped back into my seat and took hold of the yoke. There was a blast of air as he opened the window and leaned outside. A vague smell of vomit permeated the cabin. I felt bad about banking the plane, and began to apologize.

"Hey, no problem, it wasn't bad—but we're gonna' have to clean mashed potatoes off the airplane."

This student went on to pursue the instrument rating but was unable to continue because he would get sick on almost every flight. He needed to see the horizon in order to keep his stomach settled. This is common. Lack of outside reference seems to contribute to motion sickness. If you plan on flying at night, it would be wise to position a sick bag somewhere handy in the cockpit.

The causes of vertigo were discussed in the previous section. The cures are simple. If you become disoriented, stare at the horizon or attitude indicator and fly by that reference until the vertigo goes away. If you don't focus on a realistic reference like that, the effects of vertigo tend to be prolonged. Vertigo generally strikes while the plane is turning and the pilot's head moves simultaneously. You can avoid getting vertigo in the cockpit by making your movements slow, particularly your head—where your inner ear is located. While reaching for items or controls on the floor, such as the fuel selector, try to keep your head upright and an eye on the attitude indicator.

There are other ways of getting disoriented. Getting lost, for example. The ease with which you can get lost at night is highly dependent on your geographic location. Flying over the ocean presents few lighted references, and the few ships that are there tend to move. Certain areas over land can also be difficult, but they are few, and, conversely, in VFR conditions it is often possible to see your destination from tremendous distances at night. So, generally speaking, you will probably get lost more often in daylight than at night.

The following is from AC 60-4A again:

a. *You, the pilot, should understand the elements contributing to spatial disorientation so as to prevent loss of aircraft control if these conditions are inadvertently encountered.*

b. *The following are certain basic steps which should assist materially in preventing spatial disorientation.*
   (1) Before you fly with less than 3 miles visibility, obtain training and maintain proficiency in aircraft control by reference to instruments.
   (2) When flying at night or in reduced visibility, use your flight instruments, in conjunction with visual references.
   (3) Maintain night currency if you intend to fly at night. Include cross-country and local operations at different airports.
   (4) Study and become familiar with unique geographical conditions in areas in which you intend to operate.
   (5) Check weather forecasts before departure, en route, and at destination. Be alert for weather deterioration.
   (6) Do not attempt [VFR] flight when there is a possibility of getting trapped in deteriorating weather.
   (7) Rely on instrument indications unless the natural horizon or surface reference is clearly visible. You and only you have full knowledge of your limitations. Know these limitations and be guided by them.

Skills to practice

Everyone manifests different symptoms of fatigue. This is an exercise to help you become aware of the symptoms that are yours. For this exercise to work, you need to get tired. Pick a long day at work, or if you want, plan to stay awake for a night and day. Early on, when you are refreshed and alert, roll dice 25 separate times and total the figure in your head. Have someone time you and verify your math. At the end of the long day, repeat the task and compare your performance. You may find that you are unwilling to perform the task the second time, that it took much longer, or that the whole process was irritating. Take note of the way you feel. In the future, if you feel that way in the cockpit, your performance is probably suffering and you should land.

Try this experiment with your vision. Set your alarm to go off early in the morning, while it is still very dark. Without turning on a light, wake up enough to read the headlines of a newspaper, or even its finer print, if you can. While you were sleeping, your eye became adapted to darkness—large quantities of photoreactive chemicals exist in your eyes. Then turn on the light and look at it until you don't have to squint anymore—you just caused your eyes to adapt themselves to light—the chemicals have been changed back to vitamin A. Turn the light off and try to read the paper again. I'd be surprised if you could see it at all.

Further reading

FAA Advisory Circular AC 60-4A, U.S. Department of Transportation, February 9, 1983.

32

High-altitude physiology

Respiration

Every cell in the human body needs oxygen to survive. Without oxygen, we die; it is that simple. The body, in order to live, must feed oxygen to the cells. It must deliver oxygen from the environment outside the body to the environment of tissues inside the body. The delivery system sustains life and is made up of three phases (Fig. 32-1).

The outside air is brought into the body when we inhale. All gases of the atmosphere get inside, but it is only the oxygen that is important for this process. The air passes through the mouth or nose, down the trachea (windpipe), and into the bronchial tubes, where the air is divided up between the two lungs.

The lungs are great mazes with seemingly endless passages. Each passage is smaller than the last until reaching the end at the *alveoli*. The alveoli are tiny air sacks and are the link between the outside world and the bloodstream. Adults have between 250 and 350 million of these air sacks.

The blood arrives through microscopic capillaries, and when capillary meets alveoli, you have a capillary junction. It is at this junction where the air and the blood get together to make an exchange of gases.

Figure 32-2 shows the air giving the blood its oxygen while the blood deposits its carbon dioxide in the air. The oxygen travels on in the blood, and the carbon dioxide is swept out in the next exhale. The transfer of these gases is called *diffusion*. This diffusion must occur efficiently or the gas transfer does not take place properly. In that event, oxygen cannot get in and carbon dioxide cannot get out. This situation will rapidly cause problems for the person involved.

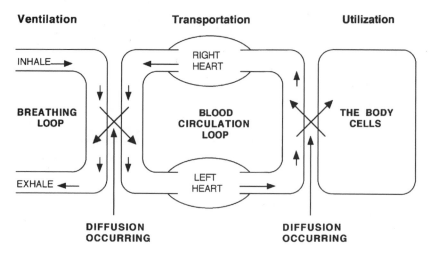

Fig. 32-1. *The three phases of respiration.*

The blood drops off its carbon dioxide in the lungs and loads up with oxygen bound for the cells. In order for the blood to move to the cells for the delivery, it gets pushed along by the left side of the heart muscle. When the oxygen-filled blood arrives at the cells, another diffusion takes place. The cells get the oxygen they need and discard into the blood the carbon dioxide waste they have used up. The blood delivers the good stuff and carries away the bad stuff. This is called *metabolism*. The blood gets a second push from the right side of the heart into the lungs, where the process starts all over again.

This works great in theory, but other factors are involved that can reduce this system's ability to do its job. Some of these factors are common to pilots who routinely leave the safe confines of the atmosphere near sea level and travel to places that the body is not used to. Going up makes the efficiency of the respiratory system go down.

Hypoxia

Hypoxia is a lack of oxygen that is being metabolized by the body's cells. The problem comes from some breakdown in the links of the respiration system. To really understand hypoxia we must get microscopic. We must look at the capillary junction where the actual transfer of oxygen occurs.

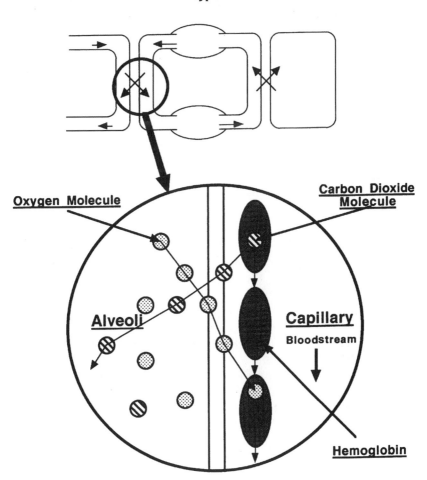

Oxygen Molecule

Carbon Dioxide Molecule

Alveoli

Capillary

Bloodstream

Hemoglobin

Fig. 32-2. *The capillary junction.*

The blood itself has many components, but not all of these carry the oxygen. The oxygen carrier is called the *hemoglobin*. It is the hemoglobin that actually catches the oxygen and makes the delivery to the cells. The oxygen arrives in the alveoli, and the blood with its hemoglobin arrives in the capillaries on opposite sides of a very thin wall. The actual transfer requires a force to push the oxygen through the wall and into the blood where the hemoglobin can soak it up. The force required is the pressure of the air itself.

Air pressure is sometimes hard to understand and accept. We always set our altimeters for barometric pressure before each flight; however, since we truly cannot feel the air pressure, we take it for granted. Though we have heard that on a standard day at sea level there is a pressure of 14.7 psi from the air's weight above pushing down, we do not actually notice it. The reason we do not feel the pressure from above is because it is also from all around and even inside our bodies pushing out. There is equal pressure inside and outside of our bodies, so we feel no net effect.

But we would feel a difference if the air were water. You can feel the weight of water on the body when swimming. Refer to Fig. 32-3. If you dive to the bottom of the deep end of the pool, you notice the force of the water against the body. The shallow end does not provide as much pressure because there is simply less water above you to bear down on you. Picture the Earth's atmosphere as a swimming pool. The surface of the water is the edge of space where there is no longer a usable atmosphere. The bottom of the pool is the surface of the Earth. The bottom of the deep end of the pool is like sea level. The shallow end is Mount McKinley. When at the "deep end" of the atmosphere (sea level), there is greater pressure from the air above because there just is a greater amount of air above. When higher in the atmosphere, there is less pressure because there is less air above.

This air pressure from above is what pushes the oxygen through the alveoli wall and into the bloodstream. When we fly to high altitudes where the pressure is less, there is less force to push the oxygen across, and consequently, less oxygen gets to the cells. This is hypoxia.

The air is only approximately 21 percent oxygen. The other 79 percent is made up of all the other gases found in the atmosphere of the Earth. Since oxygen is only part of the air, it is also just one part of the pressure that makes up the air. If the total air pressure were 14.7

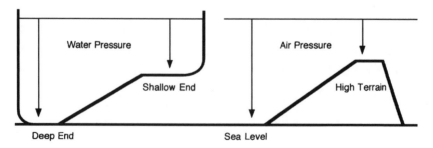

Fig. 32-3. *Air pressure at sea level is like being at the bottom of the deep end of a swimming pool.*

psi, then the portion of that pressure that is from oxygen would be only about 3.0 psi or 21 percent of 14.7 psi.

This means that on a standard day at sea level, there is approximately a 3-psi force that pushes the oxygen through the wall and into the blood. At 10,000 feet above sea level the standard pressure is about 10 psi, which would mean that the oxygen push would only be 2.1 psi. At 18,000 feet the total pressure is 7.34, and therefore, the part of the pressure that is oxygen would be down to 1.5 psi.

You can see that as we go up, the partial pressure of the life-giving oxygen goes way down. There must be a pressure gradient that is forceful enough to do the job of pushing the oxygen into the bloodstream. If you ever fly so high that the force pushing the oxygen is not as strong as the alveoli wall, then the oxygen simply will not get through.

We humans have made several attempts to artificially replace the pressure so that the cells will again receive the oxygen. We can breath 100 percent oxygen from a canister rather than 21 percent from the atmosphere. The pressure from the canister does not need to be as great as the atmosphere because, being 100 percent, there is no partial pressure; it's all oxygen. So even 3 psi would be like the normal atmosphere at sea level.

We have tried "pressure breathing," using a device that blows the air into the lungs and increasing the pressure to acceptable levels. But the most common device (and most comfortable) to artificially provide breathing pressure is the pressurized airplane cabin. The only reason we need to pump up the inside of an airplane is to allow enough partial pressure to push the oxygen over the edge and into the hemoglobin.

This reduction of oxygen partial pressure is often called *altitude hypoxia*, but there are other forms. To understand the different forms, let's use an analogy. Let's say that the air sacks (alveoli) where air collects is like a subway station. The capillaries are the tubes that the subway train runs through, and the hemoglobin cells are the subway train cars. The people who want to ride the train will represent the air molecules.

Under normal circumstances, when the train pulls up to the station, the doors open and people rush off the platform and into the subway cars. This represents the situation at sea level; there is plenty of oxygen (people) and plenty of pressure for the people to get on board.

What would happen if when the train pulled up to the station, the doors opened, but the people were too slow and did not get on the train in time. The doors would close and the train would move on, but with fewer people (oxygen molecules) on board. This would be the effect of breathing at a high altitude, and the result would be altitude hypoxia.

What would happen if, when the train stopped at the station and the doors opened, some "bad" people pushed onto the train first and kept the "good" people (oxygen) from even getting on the train. In this case the train would pull away from the station without the needed oxygen, but with pollutants instead. The "bad" molecules in this example could be carbon monoxide, alcohol, cigarette smoke, or over-the-counter drugs such as cold remedies and antihistamines. These molecules can take the place of the oxygen in the blood-stream. The hemoglobin can become filled will unwanted, unhealthy, and even dangerous molecules.

The hemoglobin will actually soak up the toxic molecules faster than it will oxygen molecules, so when oxygen attempts to get on board the train car, the train car could be already full of the bad stuff. The toxic molecules are poison to the cells. When the cells get a delivery of poison rather than oxygen, they begin to malfunction. This form of hypoxia is called *histotoxic hypoxia*, and it is much more common than you might think.

Histotoxic hypoxia is what causes intoxication. Being drunk is simply the restriction of oxygen to the brain. When a pilot drinks alcohol, it brings on histotoxic hypoxia, and brain functions slow down. When this is combined with a flight in an airplane, altitude hypoxia is added to compound the problem. This is why two drinks will have the effect of six drinks if you are at high altitude. Histotoxic and altitude hypoxia together are killers.

Then there is the case when the subway train enters the station, but the train does not have the usual amount of cars. There is plenty of oxygen and plenty of pressure, but not sufficient room to carry all the oxygen that is needed. This is called *hypemic hypoxia*, or *anemia*. When a person is anemic, they do not have enough healthy hemoglobin cells to deliver the oxygen to the tissues. People who suffer from anemia will often feel tired, run-down, and without energy. They simply do not get the proper amount of oxygen, and the body's cells operate in slow motion.

What if the train pulled into the station, the doors opened, the people rushed on board, but the train did not move on. If the blood flow stops, the hemoglobin also stops, and the oxygen is no longer transported. This is *stagnate hypoxia.* This form is a problem when pilots pull heavy G loads. The force of the G factor can be stronger than the heart's ability to pump the blood upward to the brain. When the blood cannot be pumped to the brain against a G force, a blackout will result. And, of course, the most serious source of stagnate hypoxia is when the heart stops pumping altogether. This will cause the oxygen to be stranded at the station, while the cells throughout the body begin to die.

Finally, what if the train pulled up to the station platform, the doors opened, but there were absolutely no people to get on board. Eventually, the doors close and the train moves on, but with no oxygen at all. This lack of oxygen is called *anoxia,* which leads to suffocation.

Effects of hypoxia and TUC

No matter which type of hypoxia a pilot might be exposed to, reversing the trend immediately is vitally important. The problem is that hypoxia can sneak up on a pilot. The symptoms might be so subtle that the pilot is unaware of the danger until it is too late. It is possible for a pilot to lose consciousness and therefore be unable to take life-saving action. The time span from first exposure to hypoxic conditions until unconsciousness is called the *time of useful conscience* (TUC). The TUC varies from person to person and from altitude to altitude, but the following table shows the averages:

Altitude above MSL	TUC w/o activity	TUC w/ activity
22,000	10 minutes	5 minutes
25,000	5 minutes	3 minutes
30,000	1 minute	45 seconds
35,000	45 seconds	30 seconds
40,000	25 seconds	18 seconds

These times are shockingly fast. If a pilot is exposed to an altitude of 30,000 feet while sitting quietly in the airplane, on a good day he or she has only 45 seconds to figure out what to do. Solving this problem will require some quick action and thought, but response time and clear thinking are among the first faculties to be robbed by hypoxia. A pilot really only has about 15 seconds to effect some sort of corrective action before he or she will no longer be able to think straight.

Hypoxia is hard to self-diagnose. The symptoms that tell the pilot of the onset of their own hypoxia vary from person to person. Some common symptoms are the following:

- Tingling in fingers and toes
- Reduced peripheral vision
- Exhaustion and fatigue
- Warm or cold sensations
- Dizziness
- Perspiration
- Weak muscles and sluggish movements
- Loss of muscle coordination
- Slurred speech
- Change of attitudes: overaggressive, overconfident, or timid
- Euphoria

Each person will experience different symptoms and in a different order. I placed "tingling in fingers and toes" first on the list because that happens to be my first hypoxia symptom, but it might not be your first symptom. It is very important to know what symptom your body will display first, second, and third so that when these symptoms do occur, you can start getting on oxygen.

The only way to learn what your hypoxic symptoms are and the order the symptoms present themselves is to experience hypoxia. The safest place to do this is during a joint FAA-Military-NASA sponsored physiological training program. The programs are available to any person who is at least 18 years old and holds a current airman medical certificate. The course consists of classroom lectures on physiological topics such as hypoxia, vision, disorientation, and survival. The focus of the training is a "flight" to high altitude in an altitude chamber. The following is a list of facilities that offer the course:

- FAA Aeronautical Center, Oklahoma City
- Andrews AFB, Maryland

- Barbers Point NAS, Hawaii

- Beale AFB, Texas
- Brooks AFB, Texas

- Brunswick NAS, Maine

- Cherry Point MCAS, North Carolina
- Columbus AFB, Mississippi

- Edwards AFB, California
- Ellsworth AFB, California

- El Toro AFB, California
- Fairchild AFB, Washington

- Jacksonville NAS, Florida

- Laughlin AFB, Texas
- Lemoore NAS, California
- Little Rock AFB, Arkansas

- MacDill AFB, California
- Mather AFB, California
- NASA Johnson Space Center, Texas
- Norfolk NAS, Virginia
- Patuxent River NAS, Maryland
- Pease AFB, New Hampshire
- Peterson AFB, Colorado
- Point Mugu NMC, California
- Reese AFB, Texas
- San Diego NAS, California
- Sheppard AFB, Texas
- Vance AFB, Oklahoma
- Whidbey Island NAS, Washington
- Williams AFB, Arizona
- Wright AFB, Arizona
- Wright-Patterson AFB, Ohio

Figure 32-4 is a copy of my own physiological training card from Cherry Point NAS, near the coast of North Carolina. To schedule a course you must first contact your nearest Flight Standards District Office of the FAA and receive an application for the training. The pilot then sends the application to the Mike Monroney Aeronautical Center in Oklahoma City for processing.

PHYSIOLOGICAL TRAINING

This is to certify that the following
person has met the requirements
for the Physiological Training Program as prescribed by the Federal
Aviation Administration.

NAME

AIRMAN CERTIFICATE NUMBER

DATE OF TRAINING

PHYSIOLOGICAL TRAINING UNIT

SIGNATURE OF PHYSIOLOGICAL
TRAINING OFFICER

FAA Form 3150–1 (3–67)

Fig. 32-4. *Certificate for completion of the physiological training course.*

The course is inexpensive. The courses are taught to about 20 people at a time, so the individual sites where the course is offered collect the applications until they have enough for a course to be scheduled. Applicants are notified of a course date within 30 to 60 days after applying.

You should plan to go with a group. Consider taking a CAP chapter, flight school members, or the entire charter department. By scheduling an entire group, you will not have to wait long for a class date. If you put just one or two names in, you will have to wait until 18 others have also signed up. That might take some time and planning might be difficult.

Second, do not even think about going into the altitude chamber unless you are feeling good, have eaten well for a few days, had plenty of sleep the night before, and do not have a head cold! Last, plan to stay overnight after the class in the city where the class is given. Do not attempt to drive several hours home after a bout with full-blown hypoxia. The chamber will make a believer out of any skeptic that hypoxia is serious. Hypoxia leaves you spent and worn out. Afterwards, you will want to do nothing but eat, sleep, and feel normal again.

The chamber is a long, narrow, mobile home-sized box with thick windows. The students and instructors climb in and sit down at various stations inside. Each station has an oxygen mask. Great care is taken by the instructors inside and to observers outside to ensure safety. When everything is ready, the air is slowly sucked out of the chamber to simulate high altitude. While everyone is on oxygen, the instructors give several demonstrations of altitude effects.

The chamber eventually arrives at the target high altitude. This is where the fun begins. Half the students in the room are instructed to take off their masks and perform simple duties such as writing on a tablet, stacking blocks, playing patty-cake, or inserting pegs through a hole.

The hypoxia effects are almost immediate. I have seen students continue writing without the pencil ever touching the paper. I have also seen students, who only moments before had normal coordination, slap each other instead of each other's hand in a game of patty-cake. Smokers drop like flies. I even saw one pilot attempt to punch an instructor. Later, the pilot said that he thought the instructor was coming to take his mask away, when in reality, he was coming to replace the mask. Hypoxia had completely stolen his ability to see what was actually taking place.

Each station is numbered, and an outside observer is assigned to watch the person through the windows at that station. When the observer thinks the student has had enough, he or she radios, for instance, "Get number 7!" to the inside instructor.

Later, the second half of the room gets their shot at hypoxia. When it's all over, air is slowly added to the room to simulate a descent back down to real-world pressure.

Once outside the chamber, you can review the symptoms and the order of the symptoms that overcame you at altitude. It seems that everyone has a different first symptom, but we are all in agreement that we would hate to be in that condition and still be required to fly an airplane. The instructors and doctors keep you around for about an hour after leaving the chamber just to make sure there are no additional effects of the altitude change.

Other high-altitude sicknesses

The effects of low atmospheric pressure can be harmful and even painful on other parts of the body. Whenever you climb to higher altitudes, the atmosphere that is inside the body will want to come out. If this expanding gas ever gets trapped and cannot get out, the pilot will experience discomfort and pain. This air is all through the body and can cause several problems.

Air exists inside the middle ear behind the eardrum. A small tube called the *eustachian* connects the middle ear with the nasal cavity. When we climb, the air inside the ear starts experiencing more pressure than the air outside. Eventually, the air comes out and equalizes. We say that our ears have "popped." When we descend, the air pressure outside will start increasing, and the air will want to go back into the ear through the tube to equalize.

A problem arises when the eustachian tube becomes blocked. This can happen when you get a head cold and the air pressure cannot equalize naturally. Even with a head cold, the ear usually pops on the way up because the small pocket of air works its way out from inside. But the air has a harder time going back through the swollen tube. Therefore, coming down can be painful and hearing can be affected. The medical term for this is *barotitis*.

The same type of problems exist with *barosinusitis*, which is a blockage of the sinuses due to nasal congestion, and *barodontalgia*, which is tooth pain. Gum or root abscess and air trapped behind a tooth filling can cause great pain. If you experience any unusual tooth pain when flying, see a dentist and tell him or her you are a pilot. Last, pain and cramps can be caused by expanding gases anywhere in the digestive system.

Causes and effects of gas expansion and bubble formations

One of the most dangerous potential effects of high altitudes is decompression sickness. As indicated earlier in this chapter, the atmosphere is made up of more than just oxygen. The largest portion of the air is nitrogen. Usually the nitrogen that we breath into our bodies gets pushed right back out when we exhale. But some of the nitrogen is absorbed into our tissues. Normally the nitrogen is dissolved into a liquid and held by the cells. This does not cause any problems unless the nitrogen rapidly changes back to a gas in the form of bubbles.

The bubbles form according to Henry's Law, which says that "the amount of gas dissolved in a solution is directly proportional to the partial pressure of the gas over the solution." This law can be observed when you open a soft-drink bottle. When the cap is on the bottle, there are no bubbles in the drink. When you take the cap off, however, the bubbles appear from the drink, and the drink might overflow. When capped, the pressure was contained over the soft drink, and this held the gas in the liquid state. But when the pressure was released, there was no force to hold the gas in the liquid state, so it quickly bubbled up.

When this bubbling-up occurs in the tissues throughout the body, it can cause decompression sickness, or "the bends." This extremely painful, potentially fatal disorder is associated with scuba diving as well as flying. When a diver comes up to the surface, the pressure changes on the body from heavy pressure under water to reduced pressure above water. Pilots likewise travel from high pressure at sea level to reduced pressure at altitude.

These gas bubbles will first become evident in the joints. Bending the joints seems to give temporary pain relief, hence the name. In severe cases the victim will experience pain in the chest and begin to cough. There will be the feeling of "air hunger." The gas bubbles in the skin might cause tingling, itching, and rashes. Finally, the bubbles can affect the central nervous system, causing sight problems, loss of muscle control, seizures, and paralysis.

The treatment for decompression sickness is recompression. For the pilot, this means quickly getting to a lower altitude. Usually symptoms will subside when the pressure is increased. If they continue, however, an airman medical examiner should be contacted. The examiner might recommend that the victim be treated in a repressurization chamber.

The altitude chamber, discussed earlier as the means to induce hypoxia, draws air out of a chamber. A repressurization chamber is just the opposite; it blows air in and holds it like a balloon that cannot expand. This additional pressure around the body will force the gas bubbles back into the liquid state.

As if you did not need another reason to lose weight, fatty tissue contains more nitrogen than all other tissue. So being overweight makes decompression sickness a greater threat. In addition, mixing flight after scuba diving is a very bad idea. You or anyone you fly with should wait an adequate length of time between diving and flying.

Vision

Finally, vision is affected at high altitude. The higher you climb, the farther from moisture and the associated haze you get. Without haze acting as a filter, the sky seems to be darker and the sun's rays are more intense. Because the sky is darkened, shadow areas in the cockpit are even harder to see, while sunlit areas are more glaring. When you are flying in bright sunlight, shaded instrument panels with overhanging dashboards might make instruments unreadable.

All this makes wearing the proper sunglasses important. Do not wear the type of glasses that change with illumination. The glasses won't know what to change to when you are looking at both dark and brightly lit areas. Dark, single-color, graded-density sunglasses will cut down on the glare yet not black out the shaded areas.

The anatomy of the eye is also affected by high altitude. The eye requires large doses of oxygen to work properly, especially at night. The eye is able to see because it uses a film called the *retina* to transfer light to nerve impulses that the brain can interpret. The retina actually has two types of film: daytime and nighttime. The daytime films are called *cones*, and humans, because they have adapted to daylight, have more of these than other creatures. The nighttime films are the *rods*. Rods produce a light-sensitive substance called *rhodopsin*, or visual purple, that aids in night vision.

The production of rhodopsin requires oxygen. Since we do not rely on rhodopsin in the day, oxygen is less critical in the daylight. But at night our vision is greatly diminished when we have a reduced oxygen supply, as would be the case at higher altitudes. Pilots flying at night need supplemental oxygen at lower altitudes than during the day, or they can plan to fly lower at night (obstruction clearance provided, of course).

33

Pilot judgment

We have discussed accident avoidance in many different forms so far in the book, but at this point we are going to take a closer look at what can set you up for accident situations, and how to avoid them. A large portion of this discussion will involve what is known as the *poor judgment (PJ) chain,* which is an interesting way of saying how you make decisions. No pilot intentionally sets himself or herself up for an accident or makes poor decisions with that intent in mind. But accidents do happen, and there are ways to minimize your chances of ending up in one. The PJ chain plays a role in these situations and for that reason we will include it in our discussion.

FAA publication FAA-P-8740-53, "Introduction to Pilot Judgment," discusses the poor judgment chain in great detail. We have covered some of the concepts contained in the document throughout the book so far, but essentially the FAA document states that most accidents are the result of a series of events. The example used in this document is a pilot that is noninstrument rated, has a schedule restraint, and is running late. This pilot has limited adverse weather experience but decides to fly through an area of possible thunderstorms at dusk. Due to his lack of instrument experience, combined with the darkness, turbulence, and heavy clouds, he becomes disoriented and loses control of the airplane.

As you read that little example, you probably identified several poor decisions the pilot made that got him into trouble. Time was a big motivator for the pilot and his decision to press on when the weather conditions were questionable. He was behind schedule and had the classic case of "get-home-itis" that afflicts every pilot from time to time. He also had minimal instrument experience and made a conscious decision to fly into a known thunderstorm area at dusk. The pilot had several opportunities to avoid the accident that resulted from the series of events and decisions that were made along the way.

First, the pilot could have just decided to wait until conditions improved and the flight could have been made safely. This would have necessitated that the pilot overcome the pressure of meeting the schedule, but it is always better to arrive late than to never arrive. Second, the pilot could have altered the route of flight in order to fly around the area of thunderstorms. Like waiting, this would have resulted in arriving later than desired, but at least they would have arrived. Once the pilot encountered weather, he or she could have altered course to get out of that area, either turning back or in a direction away from the storm. Too often, though, pilots "lock on" to an idea and are hard pressed to consider other options as an alternative to the one they have decided on. Finally, once the pilot encountered inclement weather, he should have trusted the instruments and maintained control of the plane, as opposed to becoming disoriented and losing control of the aircraft.

Two major principles play a role in the poor judgment chain: Poor judgment increases the probability that another will follow, and judgments are based on information pilots have about themselves, the aircraft, and the environment. Pilots are less likely to make poor judgments if this information is accurate. Essentially, this means that one poor judgment increases the availability of false information, which might then negatively influence judgments that follow.

As a pilot continues further into a chain of bad judgments, alternatives for safe flight decrease. One bad decision might prevent other options that were available at that point from being open in the future. For example, if a pilot makes a poor judgment and flies into hazardous weather, the option to circumnavigate the weather is automatically lost. By interrupting the poor judgment chain early in the decision-making process, the pilot has more options available for a safe flight. Through delaying making a good decision, the pilot may reach a point at which there are no good alternatives available. Get into the habit of making the best decision you can based on the information available, and do not be afraid to change your mind as additional information becomes known.

Three mental processes of safe flight

The same FAA document also covers three mental processes related to safe flight. These processes include automatic reaction, problem resolving, and repeated reviewing. Good pilots are actually performing many activities at the same time while they fly. Altitude, heading, and attitude are all constantly monitored and the airplane is adjusted to maintain the desired values. After a period of time, pilots no longer think consciously about what they need to do with their hands and feet to make the airplane do something; it just happens as they automatically make the controls move in the proper way to fly the plane the way they want. This is known as *automatic reaction*.

Problem solving is a three-step process that includes:

1. Uncover, define, and analyze the problem.
2. Consider the methods and outcomes of possible solutions.
3. Apply the selected solution to the best of your ability.

Through taking these steps, you will improve your ability to understand problems and resolve them. By correctly determining the actual cause of a problem, rather than misunderstanding it, you can aid in making better decisions to resolve it. The poor judgment chain can be avoided or broken through the use of good problem-solving skills.

The last mental process is repeated reviewing. This is the process of "continuously trying to find or anticipate situations that might require problem resolving or automatic reaction." Part of this skill includes using feedback related to poor decisions. As you fly, you need to constantly be aware of the factors that affect your flight, including yourself, the plane, weather, and anything else that could be a factor. Through remaining "situationally aware," you will be better informed and able to analyze the actual conditions you are flying in.

Five hazardous attitudes

The last topics in the "Introduction to Pilot Judgment" we will cover are five attitudes that can be hazardous to a safe flight. These attitudes include:

1. Antiauthority: "Don't tell me!"
2. Impulsivity: "Do something—quickly."
3. Invulnerability: "It won't happen to me."
4. Macho: "I can do it."
5. Resignation: "What's the use?"

Each of these attitudes can get in the way of a pilot making a good decision. Overinflated ego, lack of confidence in their abilities, and the need to prove themselves are just a few personality traits that pilots can exhibit that can cloud their judgment. This poor judgment mindset is not because these pilots intend to make bad decisions, but because they are influenced by behavioral traits that negatively influence their ability to make a good decision. We all have these traits to one degree or another, but how much we let them influence us is a big factor in whether we can see through them to make a good decision. While you are monitoring the plane, weather, and other aspects of your flight, also monitor yourself. If you find that any of these traits are getting in the way of a good decision, it's time to take a step back and rethink the situation.

Accident prevention

The poor judgment chain we just covered is a big influence in how many pilots get into accident situations. Through poor planning and an inability to recognize the need to reevaluate a situation, pilots can put themselves in harm's way. We know we need to use sound decision-making guidelines, but exactly what does this mean? In this section we will cover some of the steps you can take to avoid accidents or emergency situations. While this is not all encompassing, the ideas in this section should give you a foundation to build on as you begin to think about some of the habits you may have developed.

First, think about the risks associated with an action before you take it. Take two pilots who decide to fly under a railroad bridge over a river. The first hops in the plane, heads out to the bridge, and flies underneath it. The second takes a boat out to the bridge, inspects it for wires that may run under it, measures its height and width, and sees how the terrain is around the area. He then goes back and measures the plane's dimensions, checks the winds, and then when he feels it's safe, goes and flies under the bridge. Both pilots executed the same maneuver, but the second pilot planned his actions, and waited until he knew it was safe to fly. The first showed very poor judgment in flying under the bridge because he did not have all the facts.

How many pilots don't have all the facts when they fly? Are they within weight and balance limits? How much fuel will they actually burn during the flight? What is the weather along the route of flight? The list of the factors that a pilot should understand is very large, but manageable. Before you fly, understand what it is you want to do, decide what the limiting criteria for the flight are, and adhere to them. Whether the flight is a hop around the patch or a cross-country from New York to Los Angeles, you need to determine what you want to accomplish. This may seem like common-sense stuff, but not enough pilots take advantage of all the information available to them and weigh the results of their actions.

Use your checklists when you fly. We all forget things, and using a checklist will help us avoid making embarrassing, if not serious, errors in operating the airplane. Every retractable-gear airplane has a checklist with some reference to the gear being down before landing, yet every year pilots manage to land with the gear up. We become comfortable with a plane that we fly on a regular basis, and it is not uncommon for anyone to become somewhat complacent. Checklists can help us avoid missing something we need to do as we take off, cruise at altitude, or land.

You should also know the airplane you are flying. As much as familiarity can generate complacency, lack of experience with a plane can get a pilot into trouble. If you are learning to fly an aircraft that is new to you, read the operations manual. Learn the airspeeds, proper engine operation techniques, and every other bit of information regarding the plane. General Chuck Yeager once stated that knowing the systems of the test aircraft he flew gave him the ability to recover from situations other pilots may not have been able to. The same holds true for all of us. Learn about the fuel system, how the emergency gear extension works, and everything else you can read about. Poke around the airplane to verify how systems work. This knowledge could be very useful in an emergency situation, or in just making you a more competent pilot.

Finally, know your limitations. Many accidents are caused because pilots push the airplane or themselves beyond what they are capable of. If you are not comfortable with your proficiency in a certain area, get instruction from a qualified flight instructor to gain experience and confidence. Do you feel good about your crosswind landing techniques, or does the plane tend to stray across the runway as you land? How about stalls? I fly with quite a few pilots who are not really interested in aerobatics as much as how to recover from a full stall or an unusual attitude. Most of us fly in a routine manner, to a small number of airports that we become accustomed to. Are you comfortable flying into a runway that has a real obstacle at the end of it or is truly a short runway? These are the things that cause accidents. Pilots get into situations that are not really dangerous but are beyond their level of experience.

Realize that when you are tired, sick, or under stress, your mind may not be clearly focused on flying, and you are more prone to making mistakes under these conditions. Don't fall prey to the personality traits we covered in the poor judgment chain sections. If you don't feel good about making a flight, for whatever reason, don't make it. Wait until the conditions become satisfactory before you fly. Real pilots are those who can say they will stay on the ground until conditions become safe.

34

Stress

Flight instructors might not always realize it, but they know more about stress than most other pilots. Some have even experienced it! Consider this scenario: ATC says, "Roger your missed approach. You are cleared to the alternate...expect an ILS back course approach to Runway 35 left. Ceiling is now 800 feet overcast, visibility 2 miles in light rain and fog..." The pilot thinks, "A back course approach to minimums! I haven't done one of those in years. I thought the weather was supposed to be improving!"

The instrument pilot heading for an alternate to execute an unfamiliar procedure down to minimums is certainly under a lot of stress. So is the white-knuckled student pilot on a second or third flight with thoughts such as *what happens if the engine quits?* Or *please don't let me get sick.* Experienced instrument instructors know also that a student's first time in the clouds for any extended period is very stressful. And we know we might be seeing the effects of stress when a competent instrument student begins getting behind the airplane during routine procedures, or unexpectedly blows an easy approach.

What is stress?

We are all familiar with stress in daily life, and we often use terms such as "under stress" and "stressed out" to express our feelings when we run into difficulties or have a bad day. Stress is a state of physical, mental, or emotional tension brought on by factors, often beyond our control, that upset our equilibrium.

Stress can come in many forms. Hunger, for example, is a form of stress. When our physical equilibrium is upset by going too long without food, we feel the stress of hunger, which causes us to get something to eat that will relieve the hunger. Anger is another form of stress and is somewhat more complicated than hunger because anger produces the classic "fight or flight" reaction. If your boss unfairly criticizes you during an important meeting it is only natural to feel anger. You will be impelled to "fight or flight" by either responding to the criticism immediately or keeping your thoughts to yourself at the meeting and discussing them later. Some stress can cause mental and emotional tension so severe that a person might never recover from it completely, as in the death of a loved one. On the other hand, stress can be a positive factor, producing physical reactions that can help you cope with the source of the tension. The quick release of adrenaline in a tight situation speeds up the heart and breathing rates so that the body—and especially the brain—receives more oxygen. This gives you an extra boost to help cope with the problem. Navy carrier pilots and other military pilots unconsciously learn to harness stress so that their sensory perceptions become sharper and their thought processes move faster when they are in a tough spot. Or else they don't last long as military pilots.

Flying stress

You hear talk every now and then about someone being a "natural born pilot," but I have never met one. Yes, some people have a greater aptitude for flying than others do. But *nobody* is free of tension while learning to fly. No one is completely comfortable. Instructors and students need to understand this so that the stress of flying can be minimized. Learning will simply not take place if there is too much stress.

Fortunately, causes of flying stress are easy to understand if you think about it a bit. And once you understand the causes you can then take some simple, common-sense steps to minimize their effect. I'm not a psychologist, but over the years I have found the following to be the greatest sources of stress. And I have developed techniques to ease my students' tensions in these situations.

Fear of the unknown

Anytime we face something we have never experienced before, our bodies prepare us for "fight or flight." Flying in actual IFR for the first time sets up an almost classic scenario for fear of the unknown because there is absolutely no way to prepare a person for this experience other than actually doing it.

And doing it is the best way to dispel a student's apprehension about flying in actual IFR. After a few good sessions in actual IFR with an instructor aboard, most students begin to enjoy it and look forward to it. The secret is simple: Expose them to actual IFR on a routine basis whenever the opportunity presents itself. As a designated examiner, I am amazed by the number of students who show up for the checkride who have *never* experienced actual IFR. What do you think will happen when that pilot gets an instrument rating and flies into the clouds for the first time? *World-class stress.*

Other instrument situations evoke fear of the unknown: the first approach to minimums under a hood or in actual IFR, first partial panel work, first experience of vertigo, and any kind of emergency. Again, as in the case of the first actual IFR experience, the way to reduce stress in these situations is for the instructor to expose the student to them routinely during the course. And to keep doing it until the unknown becomes so commonplace that it no longer causes any significant amount of stress.

Instructor note. A good instructor will be alert for other situations that a particular student might be apprehensive about. We're all different and what one person might take in stride might scare the daylights out of someone else the first time it is encountered. The solution, again, is to present these situations routinely and frequently until the student becomes comfortable with them.

Fear of failure

None of us likes to appear foolish or clumsy, and we have a natural tendency to avoid situations that might be embarrassing. And let's face it—some instrument procedures are very intimidating when you first encounter them. This could set up a stressful situation that might make it even more difficult for the student to master the procedure.

The instructor's attitude is very important in overcoming fear of failure. If the instructor makes the student feel like a klutz, the student will perform poorly. In the normal course of instruction, the student will achieve success more often if the instructor is patient and supportive. This attitude will go a long way toward minimizing fear of failure.

Good preflight and postflight briefings are essential for reducing a student's fear of failure. If a student has a clear idea of what's expected in the way of performance on each flight, those intimidating procedures will be a lot easier to handle. But the instructor can't do it all. Students who are conscientious about doing the assigned reading and preparing the background briefings in this book will be much better prepared for new procedures and much less apprehensive about them.

Practice is perhaps the best prescription for fear of failure; once you master something new, you lose your anxiety about it. But it also is important not to get "hung up" on a problem. Sometimes it is better to move on to something else, then return to the problem later when you are feeling more confident.

Fear of catastrophe

Student pilots have no difficulty imagining disaster at every turn. By the time a pilot starts training for an instrument rating many disaster scenarios have been put to rest by practicing or thinking through such emergencies as engine failure, electrical failure, and off-field landings.

Instrument flying gives rise to a whole new set of scenarios: lost communications, partial panel, diversion to an alternate at minimums with low fuel, sudden, unexpected icing in the clouds, and so forth. If these scenarios are not confronted and dealt with, you can end up carrying a pretty heavy load of stress even on routine flights.

Just talking over "worst-case" scenarios with your instructor can help dispel stressful anxiety. For example, it is possible to make a successful landing with 0/0 ceiling and visibility. It's been done in extreme emergencies and there are ways of doing it that work. (Hint: Practice descents at 125 fpm.) But if you never discuss this with your instructor, you'll never be able to dispel the anxiety. Talk about any flying situation that worries you. You will be surprised to find that other pilots, even instructors, have been apprehensive about the same things that bother you and have come up with many imaginative and successful solutions.

Be prepared to respond when your instructor simulates emergencies in flight. On your instrument rating flight test, the examiner might well give you a steady stream of "verbal" emergencies to see how well you respond, such as:

"I see a rapid buildup of ice on the wing."

"You have just lost radio communications."

"The weather is getting worse at the destination."

"You have just lost navigation instruments. Where is the nearest precision approach radar?" Practice and familiarity will ease the anxiety that thinking about these emergencies can cause. This is true, as well, after you obtain the instrument rating. Practice the elements of instrument flying you are *least* familiar with whenever the opportunity arises, with an instructor aboard for retraining if you have lost proficiency or want to try a new or unfamiliar procedure. Then, when you get diverted to that field with an ILS back course approach at minimums, you should have little trouble heading inbound with skill and confidence.

Physical factors

Altitude, noise, motion, and many other physiological factors produce stress by making it more difficult for your body to function at peak effectiveness. They all grind away as you fly along, and the longer the flight, the greater the stress.

Excellent sections in AIM Chapter 8, "Medical Facts for Pilots," deal with the many physiological factors affecting pilot performance; and I urge you to become thoroughly familiar with this material before the instrument flight test. The discussions of the self-imposed adverse effects of medication and alcohol are particularly important. AIM makes these two points forcefully:

- "The safest rule is not to fly as a crewmember while taking any medication, unless approved to do so by the FAA."
- "An excellent rule is to allow at least 12 to 24 hours between bottle and throttle, depending on the amount of alcoholic beverage consumed."

Effects of stress

Stress can cause numerous problems for a pilot:

- Produce impatience that might deepen almost to the point of paranoia. "They're really out to get me with all these amended clearances."
- Cause enough fatigue over a period of time to produce dangerous lapses of attention. "Um, center, were you talking to me?"
- Slow down thought processes enough to fall "behind the airplane" on a routine approach.

- Interfere with judgment. "No need to file for the alternate. We'll come back around and pick up the lights this time for sure and go on in."
- Ultimately develop into full-blown panic.

Stress is thus a serious problem that affects not only your passengers, but also everyone else sharing your airspace. You should be able to recognize that impatience, lapses of attention, and getting behind the airplane are consequences of stress. Then you can decide on a less stressful course of action, such as returning to your home field if on a local flight, making an intermediate stop on a long cross-country, or voluntarily making a missed approach at the final approach fix or sooner if you are having trouble maintaining courses and altitudes or are unable to keep up with the sequence of events.

The first step is to recognize and accept the fact that you are stressed out, then back off from your current course of action and substitute something less stressful. If you wait too long to do this, panic might take over and rob you of the ability to control the situation. When this happens, the only alternative is to call for help.

Instructor note. Experienced instructors have seen most or all of these problems on training flights, but we have been slow to associate them with stress. Instead we often conclude that the student just hasn't caught on yet, and we repeat the situation in which the problem occurred without considering other alternatives. Either way it's our job to get at the root of the problem and straighten it out regardless of whether the cause is stress, lack of preparation, unfamiliarity, or a combination of factors.

Nonflying stress

Stress has been recognized as a factor in daily life for years and there have been many studies of how stress affects performance, relationships, physical well-being, and mental health. The nonflying stress that is brought to the airport might have more impact on your flying than you think.

A classical tool for measuring stress in daily life is the Holmes/Rahe Life Change Scale (Fig. 34-1) developed several years ago by Dr. Thomas Holmes and Dr. Richard Rahe of the University of Washington.

Rank	Life Event	Mean Value
1	Death of a spouse	100
2	Divorce	73
3	Marital separation	65
4	Jail term	63
5	Death of close family member	63
6	Personal injury or illness	53
7	Marriage	50
8	Fired at work	47
9	Marital reconciliation	45
10	Retirement	45
11	Changes in family member's health	44
12	Pregnancy	40
13	Sex difficulties	39
14	Gain of new family member	39
15	Business readjustment	39
16	Change in financial state	38
17	Death of close friend	37
18	Change to different line of work	36
19	Change in number of arguments with spouse	35
20	Mortgage over $10,000	31
21	Foreclosure of mortgage or loan	30
22	Change in work responsibilities	29
23	Son or daughter leaving home	29
24	Trouble with in-laws	29
25	Outstanding personal achievement	28
26	Wife begins or stops work	26
27	Begin or end school	26
28	Change in living conditions	25
29	Revision of personal habits	24
30	Trouble with boss	23
31	Change in work hours, conditions	20
32	Change in residence	20
33	Change in schools	20
34	Change in recreation	19
35	Change in church activities	19
36	Change in social activities	18
37	Mortgage or loan under $10,000	17
38	Change in sleeping habits	16
39	Change in number of family get-togethers	15
40	Change in eating habits	15
41	Vacation	13
42	Christmas	12
43	Minor violation of the law	11
	TOTAL:	

NOTE: You can also use the life change scale to project future stress based on expected changes in the upcoming year.

Fig. 34-1. *The Holmes/Rahe Life Change Scale. Reprinted from Journal of Psychosomatic Research, Vol. 2, pp. 213–218, T. H. Holmes and R. H. Rahe, "The Social Readjustment Scale," 1967, with permission from Elsevier Science.*

Start at the top of this list and total the "mean values" of the changes that have occurred in your life over the past year. If a change has occurred more than once, increase the value accordingly. (Add an additional 0 to the dollar amounts in items 20 and 37 to reflect current financial realities.)

Here is the way the Air Safety Foundation interprets the scores:

- Below 150: Little or no problem—you probably won't have any adverse reactions to the changes in your life.
- 150–199: Mild problem—a 37 percent chance you'll feel the impact of the stress with physical symptoms.
- 200–299: Moderate problem—a 51 percent chance of experiencing a stress-related illness or accident.
- 300 and above: Danger! Stress is threatening your well-being. An 80 percent chance of a stress-related illness or accident.

The AOPA Air Safety Foundation concludes: "If your score alarms you, *do something about it.* Postpone a move or a job change, or even going on a diet (any change that's under *your* control) even flying, if necessary, until your score settles down. A good person to consult is your aviation medical examiner."

In my opinion it is not necessary to discontinue flying if you recognize that the nonflying stress in your life is on the high side. But be careful about placing yourself in stressful flying situations.

Raise your own personal instrument minimums for a while. One good rule of thumb for reducing the stress of a single-pilot, single-engine IFR flight is do not go if the ceiling is below 1,000 feet anywhere along your route.

Then stick to your new higher minimums even if it means not getting back home on time. "Gethomeitis" is a still a major factor in general-aviation accidents despite years of case histories and warnings about it.

Flight test stress

I have placed this discussion of stress at this point in the book because you are about to face one of the most stressful events in your pursuit of flying—the instrument rating flight test.

The good news is that if you have absorbed the lessons in this book, you will be so well prepared for the flight test that there will be no surprises. You will have no problem coping with the stress of the checkride. You will be thoroughly familiar with *everything* that will come up, no matter how ornery the examiner might be.

P.S. Get a good night's sleep before the flight test and arrive early! Good luck!

35

Cockpit resource management

Cockpit resource management (CRM) has incorrectly been considered a "big airplane" concern. A distinction should be made between crew resource management and cockpit resource management. Traditionally, CRM was a topic involving several people working together on a flight deck. For those of us who fly single-pilot airplanes, CRM did not seem necessary; you have no need for crew coordination if there is no crew. CRM to the single pilot once was nothing more than folding your charts correctly before takeoff and having the next radio frequency stored. That notion of single-pilot CRM is simply not up to date.

The idea of CRM is to effectively combine all resources available to the pilot so that the greatest safety and efficiency can be maintained. What is a resource? A resource is any person, reference manual, chart, or aircraft system that provides information or assistance to pilots in doing their job. Of course, the pilot's job is to operate the airplane with safety, which comes from complete situational awareness.

The pilot in command is an orchestra conductor. The string section, woodwinds, and percussion instruments are all the conductor's resources that must be blended together in the correct proportions to make the music. In the final analysis, it is the conductor that uses all the different parts, together with his or her own experience and judgment, to make the work excellent.

The goal of CRM is to reach a synergy. Synergy is a system where the output of all of its parts is greater than the sum of its parts. In other words, if two people working together can produce more in cooperation than they could working individually, a synergy is achieved.

Our aviation heritage is very antisynergy. The image of the single-pilot hero-warrior became part of the romance of flight. When you are flying an airplane all alone with no outside contact, you have only yourself to rely on. This self-reliance became a way of life and formed the personality of many pilots. If you were Charles Lindbergh flying across the ocean with no copilot, no radio, and no human contact, it was necessary to develop a me-against-the-world attitude. But today that attitude is out of place and dangerous.

Unfortunately, old habits die hard. Many pilots find it difficult to move from a pilot-only system to a pilot-resource system. This is not to say that the responsibility of pilot in command has changed. Whether flying single-pilot or as a captain of a three-person crew, the PIC is the final decision maker. The hope, however, is that those decisions will be informed and will consider all factors involved.

Working together as a flight crew

Anytime people work together, there are bound to be some conflicts. Most airline interviews contain several questions that are aimed at evaluating the prospective pilot-employees ability to work with a crew. "What would you do if your captain flew below the decision height without having the runway in sight?" "What would you do if you smelled alcohol on your captain's breath?" "What would you do if your captain missed an ATC instruction?" These and many other questions like them attempt to place the applicant in a difficult position—one that might be encountered on the job.

So, how *do* you correct other members of your crew without being combative? How do you change the actions of your captain without undermining the authority of the captain? I suggest you use the phrase: "Captain, I'm not comfortable with. . .". This statement will alert the other pilot to a concern you have without coming across as judgmental. It might be that the other pilot, who has more experience and savvy than you, will now give you a lesson on why the procedure is in fact correct. Or maybe the other pilot will use your statement to change the course of action. The statement gave the other pilot an "out" without accusing the pilot of being wrong.

But what about the single-pilot operation? The important fact to remember is that in today's system, the pilot is never alone. There might only be one pilot in the airplane, but he or she can interact with hundreds of others. Yes, it is important for the pilot to arrange the cockpit efficiently with charts, flashlight, pencil, and clipboard within reach, but CRM is more. It is the effective use of all resources, the largest segment of which is outside the airplane.

In-flight cockpit management

Most pilots get weather information by telephone or DUAT, plan their flights, then get in the airplane and never access the weather information system again. They isolate themselves from the wealth of information available on their radio if they only knew how to get it. Pilots even will call in flight plans but be unable to activate the plan once airborne because they do not know how to get into the system.

Many VFR pilots use their radios when departing an airport, but then the radio goes essentially unused for the remainder of the trip. Pilots then fly across the land, listening to downwind reports at distant airports and idle chatter on the unicom frequency when vital information passes them by on the FSS frequencies.

Before takeoff, you should write down the FSS frequencies that you can receive along your route of flight. Then while in flight, monitor the frequency that will do you the most good. In this way, vital decision-making information will come to you. You won't have to scramble for it.

If the weather broadcast in the AIRMETs, SIGMETs, and CWAs (air-route traffic control Center Weather Advisories) are so important, why are they only broadcast at certain intervals, allowing for the possibility that a pilot misses an alert? This question is being addressed by a new system that brings all these alerts under one umbrella: the Hazardous Inflight Weather Advisory Service (HIWAS).

HIWAS is a continuous broadcast that summarizes information from all existing AIRMETs, SIGMETs, CWAs, and PIREPs. When a HIWAS alert is issued, the announcement can be heard on all ARTCC, FSS, and airport terminal frequencies. The announcement instructs airborne pilots to contact the continuous HIWAS frequency in their area. The pilot can then switch frequencies (or use the "both" feature of the audio panel) and hear the recorded message.

HIWAS is not yet a nationwide service. In areas where HIWAS has been installed, the local FSS and ARTCCs will stop broadcasting alert messages at time intervals around the clock and rely on the HIWAS system. In your preflight briefing, you should ask if the area along your route of flight has HIWAS service.

If you are monitoring a frequency and an alert is broadcast, your next step is to talk with someone. With the exception of an FSS phone call, all the services discussed thus far involve a pilot listening to a recording or a computer-generated voice. But to get the best information, you need to ask questions unique to your situation.

If hazardous weather is in your area or along your route of flight, you'll be faced with the classic decision: whether to continue the flight as planned, stop at an interim airport, or turn around and go back to where you came from. FAR Part 91.5 speaks to the required pilot action in the event that something causes you to consider changing plans: "For an IFR flight, or for a flight not in the vicinity of an airport. . . [the pilot must familiarize himself with] alternatives available if the planned flight cannot be completed." If while in flight you hear a weather alert or actually see bad weather ahead, you need information immediately to help you make your decision.

The best thing to do is to talk to someone who has access to real-time information. If the situation concerns precipitation and/or thunderstorms, you want to talk to someone who is watching a radar screen. Your first attempt at communicating should be to a flight service station.

The easiest way to contact a flight service station is on a discrete frequency. A *discrete frequency* means that you dial in the correct numbers and broadcast, then in return, the person at the FSS talks back to you on that same frequency. A control tower is another example of a discrete frequency.

These frequencies can be found on the top of a thick-lined FSS information block on a sectional chart (Fig. 35-1). If an FSS and a VOR are co-located, then the entire VOR information block is outlined with a thick line. The frequencies of the FSS are located outside the block on the top thick line. The discrete frequencies are those without the letter *R* beside them.

Every flight service station has 122.2 and 121.5 as standard frequencies. So even if you could not find the information block, you could try to contact an FSS on 122.2. The emergency frequency is 121.5, and all FSSs listen in on that channel as well.

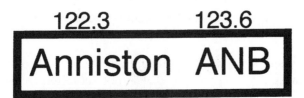

Fig. 35-1. *The FSS "thick line" box.*

When you speak to the FSS with the discrete frequency tuned in, you can address them with the term "radio," followed by the name of the FSS. The briefer will reply back to you on the same frequency. Now you can get the valuable information on which to base your decision.

What if, due to the line-of-sight limitations of the VHF radio, you are unable to contact the briefer on the discrete frequency? The answer is certainly not to give up trying to get a personal briefing. You must now fall back to plan B. The FSS system anticipated situations where you would be out of range and designed ways in which the long arm of the FSS could be extended. The range of the FSS and therefore your ability to receive information in flight is extended at certain VOR stations. When a VOR is capable of providing a communications link from your position to the FSS, the VOR information block on the sectional chart will indicate a parent station. Under the VOR block is a bracket. Inside the bracket is the name of the flight service station that monitors that VOR.

Figure 35-2 shows the top of the box, and a frequency is shown with the letter *R* beside it. The *R* stands for "receive." In other words, the FSS can receive your transmission if you talk on the frequency indicated. For this example, the frequency of 122.1 would be placed in the "communications" side of the radio and 115.5 in the "navigation" side of the radio. When you do this, your transmission does not have

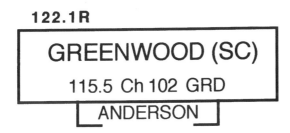

Fig. 35-2. *The VOR link information box.*

to travel a long distance to the location of the FSS but only to the nearby VOR.

Your voice is then relayed by a land line to the FSS. The *land line* is just a dedicated telephone line that is not subjected to line-of-sight limitations, and the message gets through. When the briefer responds to your request, he or she will talk back to you on the navigational frequency of the VOR. Again, the briefer's voice travels over mountains and through valleys on the land line to the VOR, and then through the air to your airplane.

You must remember two things when using this link: (1) Mention to the briefer on which VOR you would like him or her to respond to you and (2) turn up the volume on the navigation side of your radio. Now you can talk on one frequency, and the briefer talks back on the VOR. The valuable communication has now been established over a longer distance.

What if the VOR is out of service or there is no VOR in range that provides the link? You must fall back to plan C. In addition to monitoring VORs, many flight service stations will monitor remote communications outlets (RCOs). The RCO is another radio/land line combination, but it is a discrete frequency and has no navigational function. The RCO is illustrated on sectional charts by a thick blue box with location of the RCO antenna and the letters *RCO*. Figure 35-3 shows an RCO box. The RCO frequency is on the top and the parent station is in the bracket below.

When you call an RCO frequency, your voice is transmitted through the air to the RCO antenna. Then your voice travels across the land line for the rest of the way to the FSS. The briefer returns the favor, using the land line for the first part of the trip back to the RCO site and then back through the air to your radio. The actual through-the-

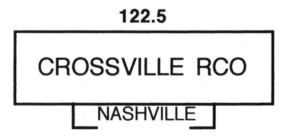

Fig. 35-3. *The RCO information box.*

air part of the transmission can be up to 50 nautical miles, depending on the line of sight. As the flight service stations consolidate, there will be even more reliance on VOR and RCO communication links.

Air route traffic control centers use RCOs as well. The center's controller might be controlling traffic in an area that is hundreds of miles from the actual location of the center. Just like the FSS, the controller cannot communicate using line-of-sight transmissions across those miles to the place where his or her radar screen has coverage, so the controller uses an RCO that is located somewhere under the radar coverage.

With the direct communication possibilities to a flight service station, together with VOR and RCO relays, the network of coverage is almost nationwide. If you are approximately 3,000 feet AGL or higher, you should be able to use some method to reach weather information. Someday all these communications will be delivered via satellite, and all the present communication links will no longer be necessary. Rather than using land lines to carry voices over mountains and around the curvature of the Earth, we will be able to transmit and receive through an orbiting relay station.

Flight watch

If for some reason plans A, B, and C fail to reach a flight service station, you should call the Enroute Flight Advisory Service (EFAS). EFAS goes by another name as well, "Flight Watch." The Flight Watch briefer is an expert in en route weather; therefore, you should not try to open or close a flight plan on this frequency. The Flight Watch information is no longer shown on sectional charts because now the service is standard nationwide.

Anytime you are flying over the United States at 5,000 feet AGL or higher, you can reach Flight Watch (Fig. 35-4). For flights below 18,000 feet MSL, Flight Watch has a universal frequency: 122.0. If you are not sure which Flight Watch station you are closest to, just call on 122.0 and say "Flight Watch" with your aircraft number.

Basically, the area in which a Flight Watch station has jurisdiction is the same as the ARTCC areas. The Flight Watch specialist is located at an FSS but can communicate by way of RCO over a wide area of responsibility. Flight Watch is usually operated from 6 A.M. until 10 P.M. every day. Complete diagrams of Flight Watch stations and their coverage areas are shown in the Airport/Facility Directory.

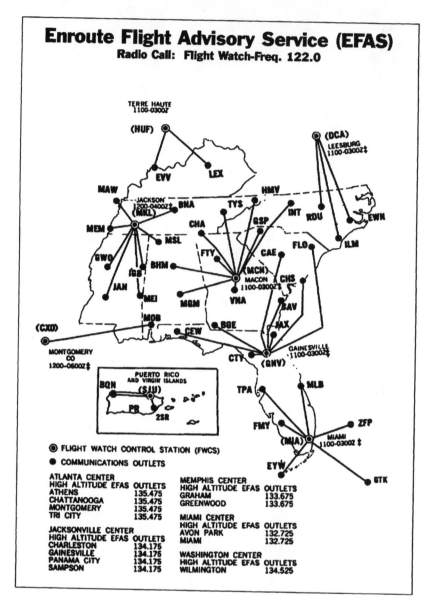

Fig. 35-4. *The "Flight Watch" network.*

The Flight Watch specialist is the "teller" at the bank of pilot reports (PIREPs). The Flight Watch specialist can relay PIREPs, which contain firsthand pilot report information. More often than not, that specialist is the person who recorded the pilot report in the first place. The Flight Watch specialist will often solicit a PIREP from you when you call for information.

When you talk to Flight Watch, you are talking to someone who is looking at all the weather information possible. If you see bad weather ahead on your route of flight, you should call Flight Watch because the person at the other end is already looking at that bad weather on radar. The Flight Watch specialist is in the best position and has the best information to give you decision-making data. How wide is the bad-weather pattern? Which way is it moving? Can I go around the weather safely? If I can go around, which direction is best? When you get the answers to these questions, you'll be able to make important weather-related decisions with confidence.

Do not forget the radar controllers at the nation's air route traffic control centers, as well as the approach and departure controllers. These people can help find the nearest airport in an emergency, call out precipitation echoes, give you a frequency, tell you a minimum vectoring altitude, or give you any number of other valuable bits of information. Remember, you are pilot in command, and controllers are additional people who work in your crew.

With all these ways to access information, there should never be a time when the pilot is isolated. With all these controllers, briefers, and specialists in your crew, your crew is larger than the flightcrew in the nose of a DC-10. Remember, you are not alone even when you fly alone.

Also, do not forget your maintenance crew members. I recently entered an uncontrolled airport's traffic pattern with a CFI applicant. When he placed the landing gear handle in the DOWN position, the landing gear did not come down. We exited the traffic pattern and ran through the emergency gear extension checklist. The gear was successfully pumped down, and we got a green light on the indicator. We made a normal and uneventful landing, but not before I talked to an A&P mechanic on the unicom frequency. I put that mechanic on my crew even though he was in his office.

Get people involved when there are problems and you have time to talk. The more ideas, the better. You must build situational awareness in order to make the best decisions. People working together have the best opportunity to create situational awareness. This is not piloting by committee. The pilot in command is still in command; we simply want the commander to be well informed.

Demonstrating your ability to master instrument flight on your initial instrument flight test and Instrument Proficiency Checks (IPCs) means little if you're not able to fly to those same standards day-to-day. To maintain that level of precision and proficiency, you've also got to master the human element, to be able to access your personal physical and mental well-being for that day's flight, and to recognize in yourself when you need more information with which to make a safe flight.

Section 6

Instrument training syllabus

36

Instrument training syllabus

Flight lesson 1: Introduction to IFR

If you're just beginning your instrument training, you're an instructor helping another pilot toward his or her instrument rating, or if you're an experienced pilot wanting a solid refresher of instrument flying basics, you'll need a plan of action. The Federal Aviation Administration's Practical Test Standards (PTS) for the Instrument Rating is a good place to start for information because it contains the criteria that FAA and designated examiners use in bestowing the instrument rating on worthy pilots. Your instrument flight training, teaching, or review needs to include a reading of the applicable Federal Air Regulations, both the rating requirements and the privileges (Part 61 or 141, as appropriate) and the operating rules (Part 91 or, for commercial operators, Part 135 or 121, again as appropriate). No instrument training or review is complete without information from the Aeronautical Information Manual (AIM), a plain language set of recommended procedures that serves as the "rule of the road" for pilots.

Want some guidance about receiving or setting up an instrument training program? Here's some help.

With instructor assistance the student will plan, file, and fly a short instrument cross-country flight to a destination 51 to 75 nautical miles distant and return. The student will learn how to plan an IFR flight, prepare a flight log, obtain a thorough weather briefing, file the IFR flight plans, obtain clearances, and conduct the flight.

The instructor will demonstrate an unhooded instrument approach to a full stop landing at the destination. A new clearance will be obtained for the return flight. Ideally, the student will be able to fly 60 percent to 90 percent of the flight with some coaching from the instructor.

Preflight briefing

Review:

   1. Communications procedures and frequencies

Introduce:

   1. IFR flight planning

   2. IFR weather briefings

   3. Filing IFR flight plans

   4. Copying clearances

   5. Instrument preflight checks

   6. Departure, en route, and approach procedures

Completion standards:

The lesson is complete when the student has a sufficient overview of the planning, filing, and conduct of an IFR flight to begin planning and filing IFR flight plans with a minimum of assistance from the instructor. The student will meet private pilot standards for holding headings within ±10°, airspeeds within ±10 knots, and altitudes within ±100 feet.

Postflight critique

Background briefing 1-2: Introduction to basic instruments

After Flight Lesson 1, the instructor should plan to give several hours of ground instruction covering basic instruments, their purpose, interpretation, and appropriate pilot actions. The emphasis in this briefing is on the relationships and interactions of the control, primary, and support instruments and how to scan them properly using the heading, altitude, and attitude scanning pattern.

The briefing is complete when the student knows what the instruments show during straight and level flight, turns, climbs, and descents; how control, primary, and support instruments relate to each other during these fundamental maneuvers; and how to scan the instruments properly.

References:

FAA Instrument Flying Handbook. Chapter II, Instrument Flying: Coping with Illusions in Flight and Chapter IV, Basic Flight Instruments.

Questions:

1. What is the difference between a control instrument and a primary instrument?

2. Give your definition of a primary instrument.

3. Under what conditions would the needle of a primary instrument be moving?

4. What is the primary instrument for:

   Bank during straight and level flight?

   Bank during a standard rate turn?

   Bank during a climbing turn?

   Bank during a straight ahead descent?

   Pitch during level flight?

   Pitch during a constant speed climb?

   Pitch during a constant rate descent?

   Pitch during an ILS approach?

   Power during the transition from a climb to straight and level?

   Power during a standard rate turn?

5. Describe at least six errors in the use of the heading indicator.

6. Describe at least six common errors in the use of pitch instruments.

7. Describe four common errors in power management.

8. Describe four common errors in the use of trim.

9. The vacuum pump has failed. Describe what happens and what instruments are affected.

10. What causes vertigo? How can you remedy this problem?

11. What causes incorrect airspeed indications? What are preventative measures?

12. Describe the recovery from nose low and nose high unusual attitudes.

13. What indications confirm that you have indeed recovered from an unusual attitude?

14. Describe VOR accuracy requirements prescribed by the FARs and how the regulations may be satisfied.

15. Make a list of the items added to a VFR checklist for the instrument pilot.

16. What are the fuel requirements for an instrument flight?

17. List the equipment checks required by the FARs and when they must be done.

18. What erroneous indications can the attitude indicator show and why?

19. When do you reset the attitude indicator?

20. Describe the errors of the magnetic compass and how to compensate for them.

21. Describe the errors of the VSI and how to compensate.

22. Describe the errors of the fuel gauges and how to compensate.

Flight lesson 2: Maneuvering Solely by Reference to Instruments—Part I

The student will plan, file, obtain the clearance, and depart IFR for the same destination as in Flight Lesson 1. Approximately 20 nautical miles into the flight, the instructor will direct the student to cancel IFR and continue VFR for training in maintaining heading and altitude solely by reference to instruments during straight and level flight, climbs, turns, speed transitions, and descents. The student will monitor a practice IFR approach conducted by the instructor and try to fly as much of it as possible on return to the airport.

Preflight Briefing

Review:

1. Flight planning, weather briefing, filing flight plans, copying clearances

2. Preflight instrument checks

3. IFR communications

4. Departure and en route procedures

Introduce:

   1. Precision straight and level flight

   2. Speed transitions

   3. Standard rate turns

   4. Minimum controllable airspeed

Supplementary Exercises:

   1. Pattern A

Completion standards:

The lesson is complete when the student can maintain headings within ±2°, airspeed within ±2 knots, altitudes within ±20 feet for periods of 30 seconds to one minute throughout all maneuvers introduced in this lesson. The student should also begin to develop a general understanding of the instrument approach procedure.

Postflight critique

Flight lesson 3: Maneuvering Solely by Reference to Instruments–Part II

The student will plan, file, obtain a clearance, and depart on a short IFR cross-country flight to a destination other than that selected for the first two flight lessons. When well-established on the en route portion of the IFR cross-country, the instructor will direct the student to cancel IFR. The student will practice precise heading, altitude, and airspeed control in straight and level flight, climbs, descents, turns, and speed transitions. The student will learn precise heading, altitude, and airspeed control in constant airspeed and constant rate climbs and descents. The student will monitor a practice IFR approach conducted by the instructor on return to the airport.

Preflight Briefing

Review:

   1. Flight planning, weather briefing, filing flight plans, and copying clearances

   2. Preflight instrument checks

   3. IFR communications

4. Departure and en route procedures
5. Straight and level, standard rate turns, and speed transitions
6. Minimum controllable airspeed

Introduce:

1. Constant airspeed climbs
2. Constant rate climbs
3. Constant airspeed descents
4. Constant rate descents

Supplementary exercises:

1. Step climbs and descents
2. Vertical S
3. Pattern B
4. Pattern C

Completion standards:

The lesson is complete when the student can maintain headings within ±2°, airspeed within ±2 knots, and altitudes within ±20 feet throughout all maneuvers in this lesson for periods of one or two minutes.

Postflight critique

Flight lesson 4: VOR tracking and bracketing

The student will plan, file, obtain a clearance, and depart on an IFR cross-country flight to a destination suggested by the instructor. The student will practice precise control in straight and level flight, climbs, descents, turns and speed transitions as they occur during the IFR cross-country phase of the flight. When directed by the instructor, the student will cancel IFR.

The student will practice flight at minimum controllable airspeed and practice establishing position in relation to a VOR station. The student will learn to intercept, bracket, and track VOR courses and radials with the needle held within the center circle at all times. The student will monitor an unhooded approach conducted by the instructor on return to the home airport.

Preflight Briefing

Review:

1. Constant airspeed climbs and descents

2. Minimum controllable airspeed

Introduce:

1. VOR orientation

2. VOR tracking and bracketing

Supplementary exercises:

1. 16-point orientation exercise

2. VOR time-distance checks

3. Pattern A around a VOR

4. Pattern B around a VOR

Completion standards:

The lesson is complete when the student can (1) maintain the course deviation indicator needle within one-half of its full deflection throughout all the maneuvers in this lesson, except for station passage, and (2) maintain desired headings within $\pm 2°$, airspeed within ± 2 knots, and altitudes within ± 100 feet during the VOR work and then ± 20 feet for extended periods throughout the rest of the flight.

Postflight critique

Flight lesson 5: VOR holding patterns

The student will file and depart on an IFR cross-country flight to a destination suggested by the instructor using VOR stations as outbound fixes so the student may practice VOR tracking and bracketing; when directed by the instructor, the student will cancel IFR. The student will learn to enter a holding pattern at a VOR station or VOR fix using direct, tear-drop, and parallel methods. The student will learn how to correct for the wind to maintain standard and nonstandard holding patterns. On return to the home field, the student will monitor a practice instrument approach conducted by the instructor.

Preflight Briefing

Review:

    1. VOR tracking and bracketing, inbound and outbound

Introduce:

    1. Holding pattern entry

    2. Holding patterns

Completion standards:

The lesson is complete when the student can (1) maintain the course deviation indicator needle within the center circle bull's-eye on all inbound courses, (2) maintain inbound legs of holding patterns within ±15 seconds of the desired one-minute length, and (3) maintain headings within ±2°, airspeed within ±2 knots, and altitudes within ±100 feet during the VOR work and ±20 feet for extended periods during the rest of the flight.

Postflight critique

Flight lesson 6: Unusual attitudes, partial panel

The student will file and depart on an IFR flight to a destination suggested by the instructor. When directed by the instructor, the student will cancel IFR. The student will practice turns to headings with the magnetic compass. The student will learn power-off stalls and steep turns and will learn to recover from unusual attitudes. The student will learn precise control of the aircraft under partial panel conditions with the attitude indicator and directional indicator covered to simulate a vacuum system failure. The student will fly a no-gyro instrument approach on return to the airport.

Preflight Briefing

Review:

    1. Turning to a heading with magnetic compass

Introduce:

    1. Power-off stalls

    2. Steep turns

    3. Critical attitude recovery

    4. Partial panel

Completion standards:

The lesson is complete when the student can (1) recover from power-off stalls with a loss of altitude of 50 feet or less, (2) maintain altitude within ±100 feet, airspeed within ±10 knots, desired angle of bank within ±5°, and roll out within ±10° of the specified heading during 45° banked turns in either direction, (3) recover to straight and level flight without the use of the attitude indicator after an unusual attitude, (4) consistently maintain headings within ±2°, airspeed within ±2 knots, and altitudes within ±20 feet while maneuvering on partial panel.

The student will demonstrate magnetic compass errors, recognize imminent stalls, and use correct control pressures and movements in proper sequence in unusual attitude recovery in order to complete the lesson.

Postflight critique

Flight lesson 7: ADF orientation, tracking, and bracketing

The student will file and depart on an IFR cross-country flight to a destination selected by the instructor along a route in the vicinity of a strong NDB or standard AM broadcast station. The student will practice control by partial panel in straight and level flight, climbs, turns, descents, and speed transitions as they occur during the IFR cross-country phase of the flight.

The instructor will direct the student to cancel IFR in the vicinity of the NDB. The student will practice steep (45° bank) turns under the hood. The student will learn to orient the airplane around an NDB and to intercept, bracket, and track inbound and outbound bearings. Pattern A may be used to practice interception, tracking, and bracketing.

Assigned reading:

Review:

    1. Partial panel

    2. Steep turns

Introduce:

    1. ADF orientation

  2. ADF tracking and bracketing

  3. Pattern A around an NDB

  4. Pattern B around an NDB

Supplementary exercise:

  1. ADF time-distance checks

Completion standards:

The lesson is complete when the student maintains ±100 feet while performing steep turns and is able to predetermine magnetic bearing to an NDB within 10 seconds, then turn to the station and intercept a bearing and track within ±5° of the course.

Postflight critique

Flight lesson 8: ADF holding

The student will file and depart on an IFR cross-country flight to a destination selected by the instructor along a route in the vicinity of a strong NDB. When directed by the instructor, the student will cancel IFR and practice ADF orientation, tracking, and bracketing. The student will learn to enter a holding pattern at an NDB by direct, teardrop, and parallel methods. The student will learn how to correct for the wind to maintain holding patterns.

Preflight Briefing

Review:

  1. Partial panel

  2. ADF orientation

  3. ADF bracketing and tracking

Introduce:

  1. ADF holding patterns

Completion standards:

The lesson is complete when the student demonstrates proficiency in predetermining bearing to an NDB, entering holding patterns by the correct method, and intercepting, bracketing, and tracking a magnetic bearing to and from the NDB within ±5°.

Postflight critique

Background briefing 8-9: Instrument approach procedures

Immediately after completing Background Briefing 1-2, the student should commence work on Background Briefing 8-9, writing as many answers as possible. The student should allow adequate time to prepare for this briefing because it covers a large body of information of great importance, especially on the instrument flight test.

The briefing covers weather minimums, approach charts, alternate airports, communications and clearances, ADF, VOR, and ILS procedures, holding patterns, straight-in and circling approaches, at uncontrolled airports, and missed approaches.

The briefing is complete when the student can talk through ADF, VOR, and ILS approaches—and the appropriate missed approach procedures—at nearby airports.

Questions:

1. Instrument approach procedures are based on criteria established in what publication of the U.S. government?

2. What publication serves as the instrument pilot's reference for transitioning from the en route phase to the landing phase in instrument conditions?

3. Basic weather minimums are prescribed for what two broad categories of approaches?

4. Name six factors that change the published minimums.

5. What does the phrase "vectors to the final approach course" mean?

6. What is the lowest forecast ceiling permitted for an instrument approach at your home airport?

7. How is the highest obstruction at an airport depicted on an approach chart?

8. What is the determining factor in whether or not a legal approach may be attempted?

9. Takeoff weather minimums are found in what publications?

10. Alternate airport weather minimums are found in what publications?

11. When is an alternate airport required? When is an airport authorized as an alternate?

12. In case of communications failure, when and where can an instrument approach be commenced?

13. What are the obstacle clearance altitudes on:

    VOR approaches?

    Localizer approaches?

    ASR approaches?

    NDB approaches?

    DF approaches?

    VOR approaches with FAF?

14. What is the maximum permissible distance from the airport during a circling approach in a Cessna 172?

15. Describe the procedure for executing an early missed approach.

16. Describe holding pattern protection and variations in holding patterns that a pilot might encounter; describe the reasons for these variations.

17. Describe five different acceptable procedures when executing a circling approach.

18. Position reports are not required when in radar contact, except in five specific instances. What are they?

19. When can you descend below the glide slope on an ILS approach?

20. When can you descend below the MDA on a nonprecision approach?

21. Can a takeoff be legally and safely executed when the current METAR weather is 1/8SM FG VV006?

22. What are you giving up during a takeoff at 1/8SM FG VV006?

23. What are your personal weather minimums for takeoff and for landing? Why?

24. Describe the clearance delivery procedures at your home airport.

25. Describe at least six variations in procedures in receiving IFR clearances.

26. When issued a cruise clearance, how do you get clearance to commence the approach?

27. When in VFR conditions executing an approach, why would you want to cancel IFR?

28. Describe the method of activating pilot-controlled lighting.

29. Describe the purposes of holding patterns depicted by a heavy solid line, a light solid line, and a light dotted line.

30. When stabilized on an ILS approach, you find you are flying 15 knots too fast. What is one popular method of correcting this excess speed?

31. What is the recommended procedure when intercepting the ILS glide slope in a fixed gear, single-engine airplane? In a high-performance retractable?

32. When landing at an airport without a control tower, when should you attempt contact on the CTAF? Where do you get the local altimeter setting, and what does it buy you?

33. Where and when should you use that old memory jogger checklist, time, turn, twist, throttle, talk? Explain its use and significance.

34. Must you time an ILS approach? Why? Why not?

35. What special actions must you take when flying into a Class B or Class C airspace? Into a restricted area or an MOA?

36. Where and how can you tell quickly if your destination airport has an approved instrument approach procedure?

37. Talk your way through several NDB, VOR, DME, ILS, and LOC/BC approaches, from the feeder fix through the missed approach.

Flight lesson 9: NDB approaches—I

The student will file and fly a short IFR cross-country flight to a nearby destination with a published NDB approach. The student will fly the NDB approach, conduct a missed approach, cancel IFR, then make additional approaches as directed by the instructor. ADF holding patterns in both the approach and the missed approach procedure will be included, if possible.

Preflight Briefing

Review:

    1. Partial panel

Introduce:

    1. NDB approaches

    2. Missed approaches

Completion standards:

The lesson is complete when the student demonstrates an understanding of the NDB approach, tracks the inbound and outbound bearings within ±5°, and maintains altitudes within ±100 feet to MDA. The MDA must be maintained to +100/−0 feet.

Postflight critique

Flight lesson 10: NDB approaches—II

The student will file and fly a short IFR cross-country to a different destination with a published NDB approach. The student will fly the NDB approach, conduct a missed approach, cancel IFR, then make additional NDB approaches, with ADF holding patterns, as directed by the instructor. At least one of the additional NDB approaches will be made with partial panel.

Preflight Briefing

Introduce:

    1. No-gyro approaches

Review:

    1. Partial panel

    2. NDB approaches

    3. Missed approaches

    4. ADF holding patterns

Completion standards:

The lesson is complete when the student demonstrates an understanding of the NDB approach, tracks the inbound and outbound bearings within ±5°, and maintains altitudes within ±100 feet to MDA. The MDA must be maintained to +100/−0 feet.

Postflight critique

Flight lesson 11: VOR approaches—I

The student will file and fly a short IFR cross-country flight to the nearest destination with a published VOR approach. The student will fly the VOR approach, conduct a missed approach, cancel IFR, then make additional VOR approaches as directed by the instructor. VOR holding patterns in either the approach or the missed approach procedure will be included.

Preflight Briefing

Review:

1. Partial panel
2. Missed approaches
3. VOR holding patterns

Introduce:

1. VOR approaches

Completion standards:

The lesson is complete when the student demonstrates an understanding of the VOR approach, tracks VOR radials within ±5°, maintains altitudes within ±100 feet to MDA, then +100/–0 feet, and maintains the desired airspeed within ±10 knots.

Postflight critique

Flight lesson 12: VOR approaches—II

The student will file and fly a short IFR cross-country to a different destination with a published VOR approach. The student will fly the VOR approach, conduct a missed approach, cancel IFR, then make additional VOR approaches, with VOR holding patterns, as directed by the instructor. At least one of the additional VOR approaches will be made on partial panel. The student will practice recovery from unusual attitudes.

Preflight Briefing

Review:

1. Partial panel
2. VOR approaches
3. Missed approaches
4. VOR holding patterns
5. Critical attitude recovery

Completion standards:

The lesson is complete when the student demonstrates competence in performing VOR approaches and generally tracks VOR radials within ±2°, maintains altitudes within ±50 feet to MDA, then +50/–0 feet, promptly executes the missed approach, and properly enters a missed approach holding pattern.

Postflight critique

Flight lesson 13: VOR approaches—III

This repeats Flight Lesson 12. The student will file and fly a short IFR cross-country to a different destination with a published VOR approach. The student will fly the VOR approach, conduct a missed approach, cancel IFR, then make additional VOR approaches, with VOR holding patterns, as directed by the instructor. At least one of the additional VOR approaches will be made on partial panel. The student will practice recovery from unusual attitudes.

Preflight Briefing

Review:

1. Partial panel
2. VOR approaches
3. Missed approaches
4. VOR holding patterns
5. Critical attitude recovery

Supplementary exercise:

1. DME arc approaches

Completion standards:

The lesson is complete when the student demonstrates competence under full and partial panel conditions and generally tracks VOR radials within ±2°, maintains altitudes within ±50 feet to MDA, then +50/–0 feet, maintains airspeed within ±5 knots, promptly executes the missed approach, and properly enters a missed approach holding pattern.

Postflight critique

Flight lesson 14: ILS approaches—I

The student will file and depart on an IFR cross-country flight to a nearby destination with an ILS approach. The student will fly the ILS approach, conduct a missed approach, cancel IFR, then make additional ILS approaches as directed by the instructor. Holding patterns in either the approach or the missed approach procedure will be included. Preflight Briefing

Review:

    1. Missed approaches

Introduce:

    1. ILS approaches

Completion standards:

The lesson is complete when the student demonstrates an understanding of the ILS approach and maintains altitudes within ±100 feet, tracks the localizer and glide slope without exceeding full scale deflections, maintains the desired airspeed within ±10 knots, and takes prompt action at DH.

Postflight critique

Flight lesson 15: ILS approaches—II

The student will file and fly a short IFR cross-country to a different destination with published ILS and ADF approaches. The student will fly the ILS approach, conduct a missed approach, cancel IFR, then make additional ILS approaches, with holding patterns, as directed by the instructor. At least one of the additional approaches will be made on partial panel.

One of the ILS approaches will be conducted if possible, without the glide slope or using only localizer minimums to simulate the loss of the glide slope receiver. The student will also make one ADF approach to simulate total failure of the aircraft's ILS receiver.

Preflight Briefing

Review:

1. Partial panel

2. ILS approaches

3. ADF approaches

4. Missed approaches

Introduce:

1. Loss of radio navigation equipment

2. Localizer approaches

Supplementary exercise:

1. Instrument takeoff

Completion standards:

The lesson is complete when the student maintains altitudes within ±50 feet, tracks localizer and glide slope within $1/_2$ scale deflections, maintains desired airspeed within ±5 knots, recognizes and copes with instrument failures such as loss of glide slope, ILS/localizer, and attitude indicator.

Postflight critique

Flight lesson 16: ILS back course, localizer, LDF, SDF, and radar approaches

The student will file and depart on an IFR cross-country to a destination with a published ILS back course approach. The student will fly the ILS back course approach, conduct a missed approach, cancel IFR, then make additional back course approaches, with holding patterns, as directed by the instructor. At least one approach will be made on partial panel.

If an airport with a radar approach is available within the local flying area, a second IFR cross-country should be filed and flown, and several practice radar approaches should be made. If radar approaches are not available, the instructor should simulate the radar approaches by providing vectors and other standard radar approach instructions to a nearby airport.

If airports with SDF and LDA approaches are available within the local flying area, additional IFR cross-country flights should be filed and flown to these airports, time permitting, for practice with these distinctive approaches. Otherwise, they must be covered in ground instruction.

Preflight Briefing

Review:

1. Partial panel
2. Loss of radio navigation equipment
3. ILS approaches
4. Missed approaches

Introduce:

1. ILS back course approaches
2. Localizer, LDA, and SDF approaches
3. Radar approaches

Completion standards:

The lesson is complete when the student demonstrates competence in ILS back course approaches and tracks the localizer within $\frac{1}{2}$ scale deflections, and competence in following directions on radar and no-gyro approaches, maintaining headings within $\pm 2°$ and altitudes within ± 50 feet.

Postflight critique

Background briefing 16-17:
IFR cross-country procedures

Upon completion of Background Briefing 8-9 the student should commence working on Background Briefing 16-17, again writing as many of the answers as possible. Considerable emphasis is placed on planning an IFR cross-country flight and its ramifications.

As a designated FAA flight test examiner, I see cross-country planning surfacing again and again as one of the weak areas. Some candidates even expect to conduct their instrument flight test with only a low altitude en route chart, an approach chart or two, and little or no orderliness. A well-organized flight log, thoroughly worked out, is the key to mastering instrument flying.

This briefing is based on the planning for a 200 nautical mile (nm) IFR cross-country. To obtain maximum benefit from the training, it should be a different flight from the 250 nm IFR cross-country flight required by FAR 61.65 (d)(iii), which is conducted in Flight Lesson 17.

The briefing is complete when the student is ready in all respects to plan and conduct the 250 nm IFR cross-country required by FAR 61.65 (d)(iii).

Questions:

1. What are four sources of information for determining the IFR route for your flight plan?

2. Plan a 200 nm IFR cross-country flight and work out all the details on your flight log. (Each section of this flight log should be thoroughly explored with your instructor to ensure that you have a complete understanding of the use and value of the flight log.)

3. You have just picked up your clearance and you find that your routing for a segment of the flight plan has been substantially changed. You have at least five courses of action. What are they? When and why would each option be appropriate?

4. Explain the purpose of the TEC and when you would use it.

5. Explain the "preferred route" system and how it operates.

6. Give the two main reasons to note the time when you start a takeoff roll on an IFR flight.

7. Give at least two methods for maintaining a record of changes in assigned altitude. Why is this necessary?

8. How are you guaranteed terrain clearance when flying direct (VOR to VOR or off airways)? Who is ultimately responsible?

9. What do the following give you? Explain. MEA, MOCA, MCA, MRA, MAA, MSA.

10. Review all the symbols on the en route low altitude chart legend. Ask your instructor to explain any that you don't understand.

11. What is the advantage in recording the time when you reach each fix during an IFR cross-country?

12. What advantage is there in "visualizing" where you are at all times?

13. What is the purpose in writing your clearance limit?

14. Explain when, how, and why you might use the frequency 122.0 MHz.

15. How might you use the frequency 121.5 MHz? Why? When?

Flight lesson 17: Long IFR cross-country flight

This flight will satisfy the requirements of FAR 61.65 (d)(iii) for one 250-mile cross-country flight in simulated or actual IFR conditions, on federal airways, or as routed by ATC, including three different kinds of approaches.

The student will file and fly IFR flight plans for each leg of the 250-mile cross-country. Each approach will be made to a full-stop landing, and the student will refile an IFR flight plan after each full-stop landing en route. One approach and landing will be made at an uncontrolled airport.

Preflight Briefing

Review:

1. IFR departure and en route procedures

2. ADF approaches

3. VOR approaches

4. ILS approaches

Introduce:

1. Lost communications procedures

2. IFR departures from uncontrolled airports

Completion standards:

The lesson is complete when the student demonstrates competence in "putting it all together" and generally maintains airspeed within ±2 knots, headings and VOR radials within ±2°, and altitudes within ±20 feet. The approaches shall meet the FAA practical test standards as stated in the completion standards for Flight Lessons 10, 13, and 15.

Deficiencies on this flight will not normally require repetition of this lesson, but will be corrected with extra work in Flight Lessons 18 and 19.

Postflight critique

Flight lesson 18: Progress check

This flight lesson will be conducted by another instrument flight instructor. The student will file and depart on an IFR cross-country flight to a destination selected by the instructor. When directed by the instructor, the student will cancel IFR and practice maneuvers chosen by the instructor to determine the student's proficiency in carrying out the tasks required by *Instrument Rating Practical Test Standards*. At least one of the instrument approaches will be on partial panel.

Preflight Briefing

Review:

1. Flight planning
2. Obtaining and analyzing weather information
3. Filing an IFR flight plan
4. IFR departure and en route procedures
5. Partial panel
6. Lost communications procedures
7. Loss of radio navigation equipment
8. ADF, VOR, and ILS approaches
9. Holding patterns
10. Missed approaches

Completion standards:

The lesson is complete when the instructor determines what deficiencies, if any, require additional practice.

Postflight critique

Flight lesson 19: Flight test preparation

The student will file and depart on an IFR cross-country flight to a destination selected by the instructor. When directed by the instructor, the student will cancel IFR and practice any maneuvers that may require further attention to attain the proficiency required by *Instrument Pilot Practical Test Standards*. At least one of the approaches will be on partial panel.

Preflight Briefing

Review:

1. Flight planning
2. Obtaining and analyzing weather information
3. Filing IFR flight plans
4. IFR departure and en route procedures
5. Partial panel
6. Lost communications procedures
7. Loss of radio navigation equipment
8. ADF, VOR, and ILS approaches
9. Holding patterns
10. Missed approaches

Completion standards:

The lesson is complete when the student has corrected any deficiencies noted in Flight Lessons 17 and 18.

Postflight critique

Background briefing 19-20: Preparation for the Instrument Flight Test Oral Exam

This very important final background briefing covers material the student can expect on the oral examination that the designated examiner will give prior to the flight test.

The briefing is complete when the student can promptly and accurately answer the questions and work out the problems that can be expected on the oral examination prior to the flight test.

Questions:

(Appropriate reference at end of each question.)

1. What are the IFR currency requirements? (FAR 61.57)
2. If instrument currency expires, how do you become current? (FAR 61.57)
3. What is an "appropriate" safety pilot? (FAR 91.109)
4. What is legally considered "instrument flight time?" (FAR 61.51)
5. Who has direct responsibility and final authority for the operation of an aircraft? (FAR 91.3)
6. What certificates must pilots carry in their personal possession for flight? (FAR 61.3)
7. What documents must be on board the aircraft for an IFR flight? (FAR 91.9, FAR 91.203).
8. What is the fuel requirement for flight in IFR conditions? For VFR flight? (FAR 91.167, FAR 91.151)
9. Under what conditions must you list an alternate when filing IFR? (FAR 91.169)
10. What are alternate airport weather minimums? (FAR 91.169)
11. What restrictions apply regarding the operation of portable electronic devices on board an aircraft on an IFR flight? (FAR 91.21)
12. What are the four methods of checking VOR accuracy and the required records? (FAR 91.171)
13. How often must VOR accuracy be checked for IFR operations? (FAR 91.171)
14. At what point can you cancel an IFR flight plan? (AIM 5-1-13)
15. How frequently should you check your altimeter setting? (91.121)
16. What are the minimum weather conditions for IFR takeoff? (FAR 91.175)

17. What additional instruments and equipment are required for IFR over VFR? (FAR 91.205)

18. Explain DH versus MDA. (FAA *Instrument Flying Handbook*, FAR 1)

19. Explain MEA, MOCA, MRA, MAA. (FAA *Instrument Flying Handbook*, FAR 1)

20. How should you navigate your course on an IFR flight? (FAR 91.181)

21. Name the components of the ILS system. (AIM 1-1-10)

22. Can anything be substituted for an outer marker on an ILS approach? (FAR 91.175)

23. When is a procedure turn prohibited on an instrument approach? (FAR 91.175, Instrument Approach Procedures legend)

24. When may you descend below DH or MDA? (FAR 91.175)

25. Explain the terms "straight in" versus "circling" minimums. (AIM "Pilot/Controller Glossary")

26. How do you determine the minimum safe altitude on a "direct" off-airway flight? (FAR 91.177)

27. When must the pitot-static system and altimeter be inspected for IFR operations? (FAR 91.411)

28. How often must the transponder be inspected? (FAR 91.413)

29. Give the appropriate cruising altitudes when operating IFR below 18,000 feet. (FAR 91.179)

30. Describe clearance to "VFR on top." (AIM 4-4-7)

31. Do the FARs require an alternate static source? (FAR 91.205)

32. How will the alternate static source affect the instruments? (FAA *Instrument Flying Handbook*)

33. When should pitot heat be used? When is it recommended?

34. Explain HAA and HAT. (AIM "Pilot/Controller Glossary")

35. Explain "maintain" versus "cruise" in an IFR assignment. (AIM "Pilot/Controller Glossary")

36. List the inspections required on an aircraft to be operated IFR. (FAR 91.409, FAR 91.411, FAR 91.413)

37. Describe the operations and limitations of the gyroscopic instruments. (FAA *Instrument Flying Handbook*)

38. Discuss the purpose and use of SIDs and STARs. (AIM 5-2-6, AIM 5-4-1, "Pilot/Controller Glossary")

39. Describe a contact approach. (AIM 5-4-22, AIM 5-5-3, "Pilot/Controller Glossary")

40. Describe a visual approach. (AIM 5-5-11, "Pilot/Controller Glossary")

41. Describe minimums as determined by aircraft approach category. How do you know in which category your aircraft is classified? (Instrument approach procedure chart)

42. In a radar environment, what radio reports are expected from you without being requested by ATC? (AIM 5-3-2)

43. Outline your actions if you lose radio communications with ATC. (FAR 91.185, AIM 6-4-1)

44. What would you do if *all* radio equipment failed?

45. How can you determine where restricted areas are located along your route and what are your actions? (En route low altitude chart legend)

46. How can you identify the boundaries between ATC centers? (En route low altitude chart legend)

47. What does the symbol "x" on a flag on an en route chart indicate to the IFR pilot? (En route low altitude chart legend)

48. Describe mandatory changeover points on an airway. What is their purpose? (AIM 5-3-6, En route low altitude chart legend)

49. Describe the different altitudes shown on an airway. (En route low altitude chart legend)

50. What does the symbol "r" on a flag on an en route chart indicate to the IFR pilot? (En route low altitude chart legend)

51. Describe the mileage markings on an airway. (En route low altitude chart legend)

52. Describe the correct ways to identify intersections. (En route low altitude chart legend)

53. Explain the purpose of aural signals carried by VORs and NDBs and when they are used. (FAA *Instrument Flying Handbook*)

54. What is the significance of a "T" bar on an airway at an intersection versus the absence of such a "T" bar on an airway at an intersection. (En route low altitude chart legend)

55. When can you use DME to identify an intersection? (En route low altitude chart legend)

56. Is there any significance to a solid triangle in the center of a VORTAC? (En route low altitude chart legend)

57. What is the purpose of "Flight Watch?" What is the frequency? (AIM 7-1-4)

58. How can you contact the nearest flight watch as indicated on the chart? (En route low altitude chart legend)

59. What is the difference between holding patterns depicted in fine lines versus those depicted in dark lines on approach charts? (Instrument approach procedure chart)

60. Describe the intent of the circle on approach charts. (Instrument approach procedure chart)

61. What are the different ways you might identify an outer marker. (Instrument approach procedure chart)

62. How do you identify the highest obstruction on an approach chart? (Instrument approach procedure chart)

63. Describe the holding pattern entry on the missed approach at several nearby NDBs. Show examples of where direct, parallel, and tear drop entries must be used. (Instrument approach procedure charts, AIM 5-3-7)

64. Describe the same entries prescribed in No. 63 on several nearby VOR approaches. (Instrument approach procedure charts, AIM 5-3-7)

65. Explain feeder routes and show how they are used on several nearby approaches. (AIM "Pilot/Controller Glossary," Instrument approach procedure charts)

66. Describe how an NDB is depicted on an approach chart and on a low altitude en route chart. (Instrument approach procedure chart, en route low altitude chart)

67. What is the significance of altitudes associated with feeder routes? Show one on a nearby approach. (Instrument approach procedure charts)

68. What is the significance of altitudes marked on the profiles of nearby VOR, VOR/DME, NDB, and ILS approaches. (Instrument approach procedure charts)

69. In case of communications failure, when and where can you begin an approach? (FAR 91.185, AIM 6-3-1)

70. Describe the use of the magnetic compass and common errors associated with this instrument. (FAA *Instrument Flying Handbook*)

71. On an instrument approach, describe the criteria for making a missed approach. (FAR 91.175)

Flight lesson 20:
Flight test recommendation flight

This flight is a dress rehearsal for the FAA flight test; your instructor will play the role of the FAA examiner. The instructor will direct the student to plan and fly a short IFR cross-country flight. The student will carry out all the tasks specified in the Instrument Pilot Practical Flight Test Standards.

Preflight Briefing

Review:

1. Flight planning
2. Obtaining and analyzing weather information
3. Filing IFR flight plans
4. IFR departure and en route procedures
5. Partial panel
6. Lost communications procedures
7. Loss of radio navigation equipment
8. ADF, VOR, and ILS approaches
9. Holding patterns
10. Missed approaches

Completion standards:

The lesson is complete when the instructor is confident the student will pass the FAA instrument flight test and so indicates by endorsing the student's logbook and signing the "Airman Certificate and/or Rating Application," FAA Form 8710-1 (7-95).

Postflight critique

Flight instructor endorsements

Here is wording approved by the FAA for the various endorsements flight instructors must make in student logbooks to sign a student off for an instrument flight test:

Aeronautical Knowledge:

I have given Mr./Ms. _____ the required ground training on the aeronautical areas required by FAR 61.65 (b) and certify that he/she is prepared for the instrument rating knowledge test.

(signed by instructor)

(flight instructor certificate no. and exp. date)

(date of endorsement)

Flight Proficiency:

I have given Mr./Ms. _____ the training in an (airplane/training device) required by FAR 61.65 (c) and certify that (he/she) is prepared for the instrument rating practical test.

(signed by instructor)

(flight instructor certificate no. and exp. date)

(date of endorsement)

Instructors will also review written test questions that were missed, as listed on the written test report, and sign and date the report in the space provided to indicate that this review has been completed.

Finally, the instructor must sign and date the "Instructor's Recommendation" portion of the "Airman Certificate and/or Rating Application" (FAA Form 8710-1) (7-95) after checking to make sure that the applicant has supplied all the required information and filled out the applicant portion of the form correctly.

37

Putting it all together: The long IFR cross-country

Flying the long cross-country trip Iis the culmination of all your efforts so far, as well as a pregraduation introduction to the real world of IFR. It meets the requirements of FAR 61.65 (d) (iii), which calls for "at least one cross-country flight...that is performed under IFR and consists of—

(A) A distance of at least 250 nautical miles along airways or ATC-directed routing;

(B) An instrument approach at each airport; and

(C) Three different kinds of approaches with the use of navigation systems."

When you appear for the instrument flight test, one of the first things the examiner will do is check your logbook to make sure you have accomplished this long IFR cross-country as required.

If this long cross-country is attempted just to meet the bare bones requirements of the FAR, you're missing the true benefit and purpose of the flight. You can meet the qualifications by covering the distance and stumbling through the three different approaches, but you'll shortchange yourself.

This is the point at which instrument students demonstrate to their instructors and themselves that they can do all the planning, fly from one place to another using the ATC system effectively, and make the different approaches down to minimums. Then all the instructor has to do after this lesson is review and polish up those elements that were found a bit lacking on the long cross-country. The student should then be ready for the flight test.

The value of actual IFR

It doesn't make too much difference whether you have an actual IFR day or whether you have to simulate IFR with a hood. The ideal long cross-country has at least one leg of fairly heavy actual IFR. If the entire flight has to be completed in actual IFR conditions, great. It certainly builds confidence in the student to experience the real conditions of a cross-country flight.

It is almost criminal to find that a number of instrument-rated pilots have never seen the inside of a cloud. They file IFR and everything goes just fine until they see a bank of clouds ahead of them. Some will cancel IFR and duck under the clouds, which is just the opposite of what they were trained to do.

Students really haven't earned their instrument ratings unless they have experienced some actual IFR conditions, not merely flown through a few clouds. Actual IFR builds confidence and shows that all those hours under the hood were not make-believe or theory.

Once the initial shock wears off, it's much easier to fly IFR in actual conditions than to simulate it with a hood. The hood is restrictive and uncomfortable, and if you can legitimately get rid of it, the flight will seem more normal. Also under actual IFR the skies are certainly less crowded.

The long IFR cross-country required by the FAR is a basic, straightforward flight. If you have followed the syllabus carefully, you will have planned, filed, and departed on many IFR cross-countries already.

Uncontrolled airports

File three separate flights at the outset, one for each of the three different types of approaches required. Ideally, one approach should be to an uncontrolled airport. One of the most common errors I see on instrument flight tests is a candidate making an approach at an uncontrolled airport without making traffic advisories on unicom/CTAF. The candidate invades the territory of other pilots and comes barging into their airspace unannounced, then leaves without so much as a hello, thank you, or good-bye. It certainly gets the locals perturbed if they find themselves sharing the final with some pilot who has a hood on! Or worse yet, who sets up a collision course with the traffic using the opposite runway.

One approach to an uncontrolled field will bring out any deficiencies in uncontrolled airport procedures. Most examiners make it a point to evaluate the way an instrument candidate conducts IFR approaches at an uncontrolled airport.

If you haven't had much practice picking up a new clearance in the air, the long IFR cross-country is a good opportunity to do so. An instrument rating should not be attached to a person's pilot certificate without some acquaintance with air filing and picking up a clearance in the air.

Plan to execute a missed approach, which will keep the IFR flight plan open. When ATC requests intentions before commencing the approach, tell them you have an IFR flight plan on file to your next destination. Be sure to have that flight plan handy so you can read it to ATC in case it has gone astray in the computer.

Controllers are usually very considerate of pilots in the air when reading a clearance. They'll read it piecemeal and try not to issue more than two or three parts of a clearance at one time.

Void time clearances

The long IFR cross-country also provides a good opportunity to practice void time clearances over the telephone when departing from an uncontrolled airport. Your flight plan has already been filed, so you will telephone flight service or some part of the ATC system and ask for the clearance. They might ask you to call back at a certain time for the IFR clearance, release, and void time. Void time means exactly that—the clearance is void after that time. If you see that you can't take off before the void time, you will have to call flight service again and request a more practical release time.

Void time clearances take a bit of forethought. You have to complete the preflight and get the charts and radios all set for departure before calling for the clearance and release time. Position the airplane for a prompt departure.

Partial panel

I also like to do some partial panel work during a leg that is long enough so the student is well caught up on the cockpit workload. One thing I like to see on partial panel is the student tuning the ADF to an NDB or commercial broadcast station up ahead that will serve as a backup heading indicator, giving a constant relative bearing in combination with the magnetic compass.

As you plan the flight, ask what you would do if given a simulated vacuum system failure at various points along the way, then note what NDBs or commercial broadcast stations can function as a backup for the heading indicator at these different points. The examiner will certainly be impressed if you demonstrate this on the instrument flight test.

The fine points of IFR cross-country flying are well covered by Background Briefing 16-17 that precedes this flight lesson; however, I would like to discuss two other aspects of cross-country flying in greater detail: fuel management and lost communications.

Fuel management

Year after year, running out of fuel continues to be a major cause of general aviation accidents. There is no excuse for this. Everyone knows, or should know, how much fuel is in the tanks at the beginning of the flight, the gallons or pounds per hour that will be consumed at various power settings, and how long it will take to complete the flight. Fuel requirements must be computed for every cross-country flight. The FARs require this. Fuel is time in the tanks. When will the fuel be exhausted?

What is the problem? There are really two problems.

First, many pilots rely too heavily on fuel gauges. The fuel gauges in general aviation airplanes are not reliable, especially below one-quarter of a tank when the fuel is sloshing and surging around.

Second, you must break the habit of thinking in terms of "full tank," "half tank," "quarter of a tank," etc. Instead, think of hours and minutes of fuel remaining.

To do this, it is essential that you keep track of the time when tanks are switched on and off. Keep an accurate log of takeoff time and arrival time at all fixes and checkpoints. This isn't very hard because you have been logging these items on all cross-country flights, VFR as well as IFR, since student pilot days.

Logging the flight

Figure 37-1 shows the filled-out log of the flight to Binghamton, New York, that was previously planned. It contains all the information that was entered during the flight.

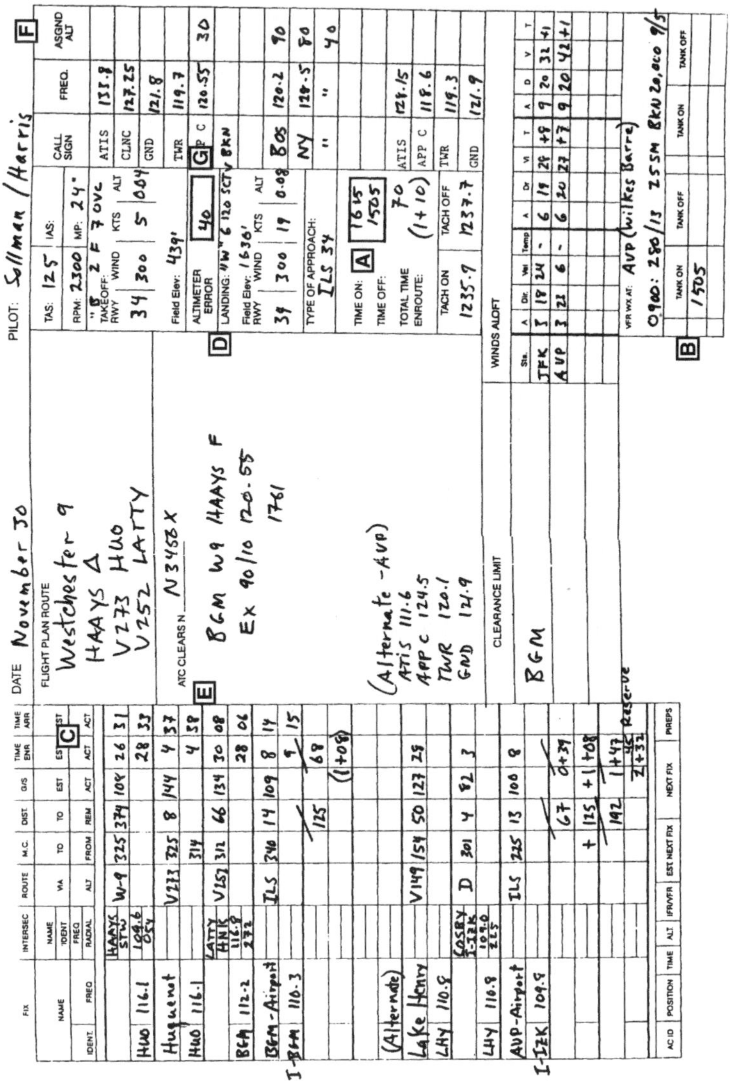

Fig. 37-1. *The completed flight log to Binghamton showing how items were entered during the flight.*

Note the block for entering takeoff time (A), and the blocks for logging the times when tanks are normally switched (B). There was no need to switch tanks because we flew an airplane that feeds from both tanks.

The two columns under (C) provide blocks in which to log estimated and actual times en route and arrival for each leg and fix. No actual times were logged for the flight to the alternate because the weather at Binghamton was VFR by the arrival time. Notice the landing ATIS "Whiskey" (D); above that is ATIS information "Bravo" for the departure airport, Westchester County.

While looking at the log, note how the clearance was copied (E), and how frequency and altitude changes were recorded (F) and (G).

Also, in block (A) the time of arrival was logged and the total time en route was computed and entered. To compute fuel remaining, subtract total time en route (1:10) from the hours and minutes in the tank on departure: 5:30 − 1:10 = 4:20. Think hours and minutes—in this case 4 hours and 20 minutes fuel remaining—not full and half-full tanks.

Obtaining weather information in flight

It would be nice if the weather on a cross-country flight always turned out to be the same as forecast when you worked out your flight plan. But this is rarely the case. In many parts of the country and in many seasons of the year, weather can change rapidly, and sometimes violently.

As your flight progresses, you should systematically check out the weather up ahead to make sure you can get through to your destination and make a safe approach and landing, and that your alternate is still available if you can't land where you want to.

The best way to do this is to contact "flight watch" on 122.0 MHz. Use the name of the ATC facility serving the area—Boston, New York, Cleveland, etc.—and be sure to get permission from ATC that you are leaving that ATC frequency for a few minutes. ATC might want you to delay your switch to flight watch in order to amend a clearance or hand you off to another facility.

Flight watch is the call sign for the FAA's *en route flight advisory service* (EFAS). When you talk to flight watch you will speak directly to a

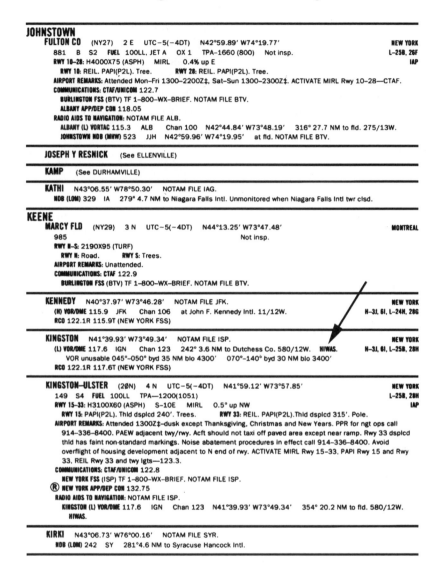

JOHNSTOWN
   FULTON CO   (NY27) 2 E UTC–5(–4DT) N42°59.89' W74°19.77'   **NEW YORK**
       881   B S2 **FUEL** 100LL, JET A OX 1 TPA–1660 (800) Not insp.   L–25B, 26F
      RWY 10–28: H4000X75 (ASPH) MIRL 0.4% up E   **IAP**
       RWY 10: REIL. PAPI(P2L). Tree.   **RWY 28:** REIL. PAPI(P2L). Tree.
      AIRPORT REMARKS: Attended Mon–Fri 1300–2200Z‡, Sat–Sun 1300–2300Z‡. ACTIVATE MIRL Rwy 10–28—CTAF.
      COMMUNICATIONS: CTAF/UNICOM 122.7
        BURLINGTON FSS (BTV) TF 1–800–WX–BRIEF. NOTAM FILE BTV.
        ALBANY APP/DEP CON 118.05
      RADIO AIDS TO NAVIGATION: NOTAM FILE ALB.
        ALBANY (L) VORTAC 115.3   ALB   Chan 100 N42°44.84' W73°48.19'   316° 27.7 NM to fld. 275/13W.
        JOHNSTOWN NDB (MHW) 523   JJH N42°59.96' W74°19.95'   at fld. NOTAM FILE BTV.

   JOSEPH Y RESNICK   (See ELLENVILLE)

   KAMP   (See DURHAMVILLE)

   KATHI   N43°06.55' W78°50.30'   NOTAM FILE IAG.
      NDB (LOM) 329   IA   279° 4.7 NM to Niagara Falls Intl. Unmonitored when Niagara Falls Intl twr clsd.

KEENE
   MARCY FLD (NY29) 3 N UTC–5(–4DT) N44°13.25' W73°47.48'   **MONTREAL**
       985   Not insp.
      RWY N–S: 2190X95 (TURF)
       RWY N: Road.   **RWY S:** Trees.
      AIRPORT REMARKS: Unattended.
      COMMUNICATIONS: CTAF 122.9
        BURLINGTON FSS (BTV) TF 1–800–WX–BRIEF. NOTAM FILE BTV.

   KENNEDY   N40°37.97' W73°46.28'   NOTAM FILE JFK.   **NEW YORK**
      (H) VOR/DME 115.9   JFK   Chan 106   at John F. Kennedy Intl. 11/12W.   **H–3J, 6I, L–24H, 28G**
      RCO 122.1R 115.9T (NEW YORK FSS)

   KINGSTON   N41°39.93' W73°49.34'   NOTAM FILE ISP.   **NEW YORK**
      (L) VOR/DME 117.6   IGN   Chan 123   242° 3.6 NM to Dutchess Co. 580/12W.   **HIWAS.**   **H–3J, 6I, L–25B, 28H**
        VOR unusable 045°–050° byd 35 NM blo 4300'   070°–140° byd 30 NM blo 3400'
      RCO 122.1R 117.6T (NEW YORK FSS)

   KINGSTON–ULSTER (20N) 4 N UTC–5(–4DT) N41°59.12' W73°57.85'   **NEW YORK**
       149   S4 **FUEL** 100LL TPA—1200(1051)   L–25B, 28H
      RWY 15–33: H3100X60 (ASPH) S–10E MIRL 0.5° up NW   **IAP**
       RWY 15: PAPI(P2L). Thld dsplcd 240'. Trees.   **RWY 33:** REIL. PAPI(P2L).Thld dsplcd 315'. Pole.
      AIRPORT REMARKS: Attended 1300Z‡–dusk except Thanksgiving, Christmas and New Years. PPR for ngt ops call
        914–336–8400. PAEW adjacent twy/rwy. Acft should not taxi off paved area except near ramp. Rwy 33 dsplcd
        thld has faint non-standard markings. Noise abatement procedures in effect call 914–336–8400. Avoid
        overflight of housing development adjacent to N end of rwy. ACTIVATE MIRL Rwy 15–33, PAPI Rwy 15 and Rwy
        33, REIL Rwy 33 and twy lgts—123.3.
      COMMUNICATIONS: CTAF/UNICOM 122.8
        NEW YORK FSS (ISP) TF 1–800–WX–BRIEF. NOTAM FILE ISP.
     ® **NEW YORK APP/DEP CON** 132.75
      RADIO AIDS TO NAVIGATION: NOTAM FILE ISP.
        KINGSTON (L) VOR/DME 117.6   IGN   Chan 123 N41°39.93' W73°49.34'   354° 20.2 NM to fld. 580/12W.
        HIWAS.

   KIRKI   N43°06.73' W76°00.16'   NOTAM FILE SYR.
      NDB (LOM) 242   SY   281° 4.6 NM to Syracuse Hancock Intl.

Fig. 37-2. *HIWAS at Kingston VOR indicated by arrow.*

qualified weather briefer who will answer your specific questions without having to read out all the details. On the flight to Binghamton, for example, you could call New York flight watch half an hour or so before landing and get the latest terminal weather for Binghamton and Wilkes-Barre, the alternate. Be sure to advise ATC that you are back on their frequency after you are finished with flight watch.

Selected VORs also broadcast hazardous in-flight weather advisory service (HIWAS) information. (See Fig. 37-2.) HIWAS is continuous. It disseminates severe weather forecast alerts, SIGMETs, convective SIGMETs, center weather advisories (CWAs), AIRMETs, and urgent PIREPs. Look up the VORs in your area in A/FD and see which provide HIWAS, then listen in a couple of times to find out what is available.

Lost radio contact

Complete loss of communications while airborne on an IFR flight is extremely rare. What is more likely to happen is a loss of radio contact. Many times lost radio contact is self-induced. The squelch might be tuned down too low or a wrong frequency might have been inadvertently tuned. Double-check all the navcom settings and the audio panel settings. Also try transmitting on the other radio. It is possible for one to fail and the other to work just fine.

You might be flying through a quiet zone along the airway, or flying low enough so that the VHF line-of-sight transmissions are blocked by mountains or other obstructions.

Here is a typical scenario: an ARTCC to ARTCC handoff. Let's say Boston ARTCC calls you with a transmission: "Cessna five six Xray, contact New York Center, one two eight point five."

You acknowledge the handoff by reading back the clearance: "New York Center one two eight point five, five six Xray," then you tune the new frequency and report, using the full N number because this is an initial contact, and you report your altitude: "New York Center, Cessna three four five six Xray at niner thousand."

No response. Suddenly all is quiet.

You try two or three more times and still get no response. What do you do? Make sure you haven't made some mistake such as switching from speaker to headphones.

Return to the frequency you just left and report Cessna "five six Xray, unable New York Center one two eight point five." Boston should answer and might tell you to remain with them, or to attempt to contact New York Center on another frequency, or to stay with Boston and try New York again in a few minutes or a few miles.

Now suppose you had returned to the Boston Center frequency and couldn't raise them again. Nothing doing on either New York ahead or Boston behind you. Now what?

1. Check for other center frequencies in the area (Fig. 37-3). In this case, there is another New York Center frequency available: 134.65.

2. Attempt to raise another aircraft to relay the message. If you hear a specific N-number on the frequency, call that aircraft. If nothing is ever heard, transmit in the blind: "Any aircraft, Cessna three four five six Xray, requesting a communications relay." If someone answers, have the pilot try to reach the new center—New York in this case—and relay their instructions to you.

3. Try contacting the nearest approach control or even a nearby tower and ask them to relay your message to the center.

4. Call flight service on 122.1, and give them a nearby VOR identification and frequency with voice capability that you can hear clearly. Be patient. It might take them a couple of minutes to complete another transmission before they can get to you.

5. Come up on the emergency frequency 121.5, say you're having communications difficulties, and if anyone replies, contact them with your next transmission. Many pilots are intimidated by the emergency frequency. They are reluctant to use it because they think they will have to write a lengthy

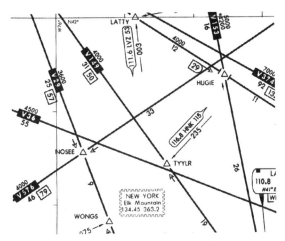

Fig. 37-3. *Box containing sector frequencies for an air route traffic control center (ARTCC).*

letter to the FAA in Washington justifying the action. This is usually not the case. You don't have to report to anyone just because you used the emergency frequency. The emergency frequency is intended for lost radio contact as well as for more serious situations. When all else fails, try to reestablish communications on this frequency.

6. When you switch to 121.5 as the last resort, set the transponder to squawk 7600—the code for radio communications lost—until the conclusion of the flight or until ATC communications have been restored. (The squawk code for a flight emergency is still 7700. *See* a current edition of AIM for any additional amendments.)

7. Try to repair the radio. You might have a hot microphone or a mike button stuck in the keyed position. You can find out very quickly by unplugging the mike from its jack. If you start hearing other transmissions again, the mike is the problem. A muting relay engages when you depress the microphone key. It cuts out incoming signals when you transmit because you can't talk and listen at the same time without getting a howling, screeching feedback that will blast your eardrums out of action. So if the mike button is stuck, the muting relay will be engaged and you won't hear anything.

Because a stuck mike transmits continuously it also blocks everything on that frequency within miles. When you unplug the stuck mike, you will begin hearing normal transmissions again, plus a lot of unkind remarks about the pilot with the stuck mike. So help everyone out when you have a radio problem and check for a stuck mike. If it sounds like you are still transmitting after you take your thumb off the mike button, check for a stuck mike first before trying any of the other alternatives.

Also try giving the transceivers a firm push back into the racks. Sometimes they come loose and connections at the back become disengaged.

Two-way radio communications failure

If you have tried every method of restoring radio communications and nothing works, then you must assume a two-way radio communications failure. FAR 91.185 spells out very clearly the procedure you must follow when this happens.

Importance of logging times

Note how important it is in FAR 91.185 to know the estimated time of arrival (ETA) at your destination if you have not received an "expect further clearance time." ATC will expect you to show up for an approach at an estimated time of arrival (ETA) based upon your filed or amended estimated time en route (ETE).

ATC knows when you took off from the departure airport—do you?—and they know the estimated time en route from your filed or amended flight plan. They will add the time en route to the departure time and come up with an ETA. ATC will reserve all approaches at the destination for this ETA plus 30 minutes. Until that time expires, you "own" that airport.

So it becomes extremely important to know what time you took off because without that information, you'll never know when to commence the approach. ("You can't tell when you're going to get someplace if you don't know when you left someplace.") This seems so basic, yet many pilots forget to log their takeoff time. When I notice this on an instrument flight test, you can bet that at some point I will ask the applicant to tell me what the destination ETA will be in the event of lost communications or to detail the steps to take in case of communications failure.

Emergency altitudes

A second point that always raises questions is the altitude for completing the flight in the event of a two-way radio communications failure. ATC expects the flight to continue at (1) the last assigned altitude, *or* (2) the minimum altitude for IFR operations, *or* (3) the altitude that ATC has advised you to expect in a further clearance. You will fly at the highest altitude of the three choices for any given leg.

You will have to make some choices. These choices hinge on how the "minimum altitude for IFR operations" is determined, and whether or not it is higher than the altitude assigned or advised to expect.

Along an airway, the minimum for IFR operations is the MEA or minimum en route altitude. If the en route low altitude chart shows that you are approaching a route segment with an MEA that is *higher* than that assigned by ATC or what ATC advised to expect, you would begin a climb to reach the new altitude so that you reach it prior to the point or fix where the new altitude begins. If the MEA drops down below the last assigned altitude, you would descend to the last altitude assigned by ATC or one that you were advised to expect, whichever is higher.

But suppose you were cleared direct. What is the minimum altitude off airways where there are no MEAs? Consult the appropriate sectional chart that covers the flight area and pick out the "maximum elevation" figure for the latitude-longitude square you occupy, then add 1,000 feet to comply with FAR minimum altitude regulations (2,000 feet in mountainous terrain—see FAR 91.177). The maximum elevation figures are the big numbers in the center of each latitude-longitude square. You would use the "maximum elevation plus a thousand" figure as a substitute for the MEA.

The likelihood of having to use the two-way radio communications failure procedures is remote, but you have to know them. And "Loss of Communications" is a required task specified in the "Emergency Operations" section of the *Instrument Rating Practical Test Standards*.

In the real world of IFR you can avoid all the complications of lost communications by investing in a hand-held portable transceiver and carrying it on every flight. It will also be invaluable in the next scenario.

Complete electrical failure

Single-engine airplanes are vulnerable to complete electrical failure. There is usually only one alternator—driven by one inexpensive V-belt—and only one voltage regulator. So there is no backup if one of these key components fails.

The first step in the event of a complete electrical failure is to turn off all electrical equipment to minimize drain on the battery. Control the airplane by partial panel. At night you will need a flashlight to do this. Careful pilots carry two flashlights for this purpose—one as a backup in case the other starts to fade—plus extra batteries.

If the failure occurs when you are VFR, maintain VFR and land as soon as practicable, just as you would if you had two-way radio communications failure.

If you are in actual IFR conditions when a complete electrical failure occurs, the problem is equally simple, although your work is cut out for you. Find VFR conditions and land as quickly as possible.

When planning for the flight, you filled in the VFR WX AT: block to quickly seek a safe haven in this circumstance (Fig. 37-4). You know where you want to go. If you have been logging the actual time of arrival at each fix along the way, you have an accurate idea of where you were when the failure occurred. Draw a straight line to the new destination from the point you intend to turn. Calculate a new magnetic heading and turn to that heading. Later, with more time, calculate an ETA at the new destination. It's like those "diversions" you worked on so hard during the cross-country phase of private pilot training.

When everything is under control, set the transponder to 7700 and turn it on. Then turn on the transceiver on which you last talked to ATC. You have only a few minutes of battery power left; make sure ATC understands your predicament and intentions, then shut everything off. You can put the remaining battery power to good use at the VFR destination to contact a tower or flash some lights.

Again, a portable transceiver will make life a lot easier. The range of these units is vastly improved if connected to an aircraft antenna; consider installation of an antenna jack.

IFR cross-country tips

Avoid marking up the chart with a highlighter. This can add to confusion on subsequent flights in the same area. I make it a game to see how much of an IFR flight I can complete without consulting the low altitude en route charts except for an amended clearance. A

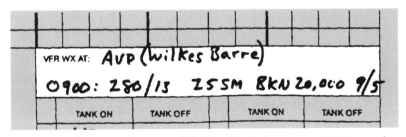

Fig. 37-4. *Flight log showing filled-in block for nearest VFR weather.*

previously marked up chart can be hazardous, especially if you have made a series of flights over the same area since the last chart revision.

After planning the flight, talk yourself through it. Make sure you understand where all those obscure intersections are and note the airports along the way, especially those with instrument approaches.

Carry VFR charts to cover the route. Highlight the IFR route on the VFR charts for quick reference in case VFR landmarks are needed.

Save the flight logs. You might frequently use the same routes, or portions of them, over and over again. ATC often has preferred routings and some of them might be unpublished; old logs will show these preferences and you can plan for them and avoid the frustration of filing for one route and receiving a clearance for another.

Visit the center and the approach control that handles your local area. Call ahead for an appointment; you will find the supervisors and controllers most cooperative in showing you around and answering questions, especially if you can make an appointment for one of their less busy times.

38

The instrument pilot's library

Like any other highly technical activity, flying, and especially instrument flight, requires pilots to keep up with a regimen of "professional reading." Books, videos, and magazines should be a part of every instrument pilot's recurrent training. Most pilots, luckily, have voracious appetites for anything to do with airplanes, so it's not difficult to get them to review techniques and regulations.

The easiest way to stay abreast of changes, and to review the basics, is to subscribe to and read magazines. Aviation periodicals have myriad articles about a number of topics, which over time tend to review about everything you'll need to know to stay on top of the rules. They offer fantastic insight and tips for flying specific types of procedures.

I hesitate to recommend specific magazine titles for fear of leaving one out, or for including one that might not be applicable to your personal needs. I do recommend that you investigate magazines put out by pilot associations, and any that might be produced by "type clubs," or groups of pilots flying the same basic model airplane as yours. I also suggest you look into magazines that may not contain lots of glossy pictures, but that relate tremendously useful details about instrument flying technique. Ask around at the FBO, or do a publications search at your favorite aviation Web site for titles and contact information.

To increase your understanding of the instrument flight environment, read everything you can about aviation meteorology and weather hazard avoidance. Even what might not on the surface look directly applicable to you may give you surprising insights. For instance, your airplane might not have weather radar, but if you're familiar with how radar works, you can better evaluate radar images on computer and television screens, better understand the limitations of Air Traffic Control, Flight Service, and Flight Watch radar reports, and comprehend what radar-equipped pilots are saying "on frequency" while you navigate in the vicinity of storms. In short, everything you can learn about aviation weather will make you safer and better able to evaluate weather hazards before and during flight.

Read accident reports. As one of my mentors puts it, "Experience is learning from what you yourself do. Education is learning from other people's experience." Reading "near miss" and actual accident write-ups will help you better understand how pilot decision making can lead down the wrong path. This in turn may help you make safer, better decisions if you ever face a similar situation. You can find incident and accident reviews on the FAA and NTSB Web sites, in the National Aeronautics and Space Administration (NASA) "Callback" reports, and in most aviation magazines.

And lastly, at least once a year, reread the applicable Federal Air Regulations (FARs) and the Aeronautical Information Manual (AIM). As the attitude indicator is the center of your instrument scan, these publications are the focus of your continuing instrument education. A look at the FARs and AIM help you sharpen your orientation and tell you a lot about where you're going as an instrument pilot.

Whether you read printed materials, access data by computer, or watch information on video, you need a continually updating library of instrument knowledge to safely fly in the instrument environment.

Afterword

Instrument flying gives you tremendous capability and freedom to travel by lightplane. It's rare when an instrument-rated pilot, flying an IFR-equipped and certified airplane, has to cancel a flight outright or even delay a trip for more than a few hours.

To take advantage of this expanded capability, you must commit to extensive training and retraining, and to a high standard of excellence to safely transport yourself and your passengers. There's no allowing for a "bad day" when flying an ILS to 200 feet and half a mile of visibility. If your temperament, recent experience, or training don't allow for this level of precision, you need to establish more conservative "personal minimums" and strictly hold yourself to them.

After learning the basics of instrument flight, your biggest task is in making good decisions about when and where you'll fly. So long as you continue to practice and learn about basic instrument flight, navigation, and system emergencies, and you commit to learning all you can about weather and your own decision-making process, then you should enjoy endless, safe, and enjoyable hours of flight through, around, and above the clouds

Index

1-800-WX-BRIEF, 219–221, 219

abbreviated briefings, 234–236
abeam, holding pattern, 407, **407**
acceleration in directional gyro, 47
acceleration in magnetic compass, 10
accident reporting, 537–538
accidents, 537–540
  accident reporting, 537–538
  causes, 537–540
  hazardous pilot attitudes for flight,
    619–620
  icing, 539
  judgment, 617–622
  mental process of safe flight, 618–619
  poor judgment (PJ) chains, 617
  prevention techniques, 620–622
  stress, pilot, 623–631
  thunderstorms, 538
  turbulence, 538
  visibility and fog, 539–540
  weather-related, 538–540, 541, 542
AccuWeather, 213
advection fog, 203
adverse conditions, in weather briefing,
  221, 224–225
Aeronautical Information Manual (AIM),
  205, 690
ailerons, 49
Air Chart System, 267
air route traffic control center (ARTCC), 69,
  163, 244
Air Traffic Control (*See also* clearances;
  Ground Control; radio
  communications), 158, 257–258
  approach, 467–468
  avoiding confusion, 403–404
  call signs, 259
  canceling IFR, 262–263

Air Traffic Control (*Cont.*):
  compulsory reporting points, 400–402
  emergency vs. priority handling, 403
  enroute holding, 415–416
  expect approach clearance (EAC) time,
    415–416
  expect further clearance (EFC) time,
    415–416
  frequencies used, 171, 260–261
  Ground Control, 259–260
  handoffs, 260
  inoperative radio, 263
  phraseology, 258, 403–404
  position reporting, 400–402
  preflight planning, 145
  priority handling request, 403
  radio failure, 404–406, **405**
  remote communications outlets (RCOs),
    254, **255**
  required reports, 261–262, 402
  uncontrolled airports, 676–677
  who, who, where, what sequence,
    258–259
air-pressure effects on human body, 606,
  606
aircraft, 429
  approach categories, 207–208, 429,
    430–433, 452–454
aircraft altitude, 152
Aircraft Evaluation Group (AEG), 178
Aircraft Owners and Pilots Association
  (AOPA), 212–213
aircraft performance restrictions,
  66–68
aircraft requirements for IFR, 148–149
AIRMETs, 198–200, 226–227
Airplane Flight Manual (AFM) (*See also*
  equipment required for IFR), 185,
  186–188, 193

Note: **Boldface** numbers indicate illustrations; *italic t* indicates a table.

airport charts (*See also* charts and maps),
    351–368, **352, 353, 354**
  additional information, 355–357, **356,
    358, 359**
  airport reference point (ARP), 351, 355
  alternate minimums, 362–366, **363**,
    364–**365**
  displaced thresholds, 359–360
  expanded charts, 366, **367, 368**
  frequencies used, 351
  identifiers, 351
  jet runways, 360
  lighting information, 355–357, **356,
    358, 359**
  runway length, 357–360, 359
  runways and taxiways, 352, **354**
  symbols, 352–355, **353**
  takeoff minimums, 360–362, **361**
  visibility minimums, 362
airport reference point (ARP), 351, 355
airport services, 174–176
airport surveillance radar (ASR), 428, 429,
    430, 516–517, **517**
Airport/Facility Directory (A/FD), 151,
    163, 174
airports:
  on charts and maps, 276, 286, 294
  displaced thresholds, 359–360
  jet runways, 360
  lighting, 384–388, **385, 386, 387**,
    429–430, 454, **456**
  lighting information, 355–357, **356,
    358, 359**
  runway length, 357–360, 359
airspace:
  on charts and maps, 288, 294
  Class B airspace, 295–297, **296**
  equipment requirements vs., 190–191
  restricted airspace, 270–271, **271**, 274,
    276
airspeed, 31–32, 38, 48–50
  altitude vs., 48–50
  approach, 466
  "chasing the needle," 57, 115–116
  climb airspeed (Vy), 81
  climbs, 54–55
  constant-airspeed approach, 60–61
  constant-airspeed climbs, 54
  constant-rate climbs, 56–57
  constant-rate descents, 59–60
  constant-speed descents, 58–59
  control pressure, 50
  estimating time enroute, 236–237
  high-speed final, 61
  indicated airspeed (IAS), 171
  minimum controllable, 74–76

airspeed (*Cont.*):
  never-exceed airspeed (Vne), 84
  pitch settings, 50
  power vs., 48–50, 48
  practice maneuvers, 77–79, **78**
  required reports, 261, 402
  stalls, 80–81
  transitions, 32–33
airspeed indicator, 3, 16–18, 19, 184, 185
  climbs, 52–53
  descents, 54
  errors, 11, 241
  straight and level flight, 53
  takeoffs, 52
airway charts, 146
airways, 151–153, 162
airworthiness, 177, 184
Airworthiness Certificate, 193
airworthiness directives (AD), 189–190
alcohol use, 549–550, 558–561
alternate airports, 208–209, **210**, 211, 226,
    229–232, **232**
  airport requirements, 153–155
  alternate minimums, 362–366, **363**,
    364–**365**
  approach categories, 207–208
  FAR Part 91.169, 154
  minimums, 372, 375–383, **376**, 378–**382**
alternate minimums, 362–366, **363**,
    364–**365**
alternators/generators, 148, 184
altimeter error, 505
altimeters, 3, 6, 17, 19, 20, **20**, 25, 148,
    184, 185, 241
  climbs, 53
  descents, 54
  errors, 10–11, 457
  straight and level flight, 53
altitude, 7
  aircraft altitude, 152
  airspeed vs., 48–50
  approach, 426, 468–469
  area minimum altitudes (AMAs), 291, 294
  cabin altitude, 152
  cabin pressure altitude, 152
  on charts and maps, 278–288, **282**
  circle of equal altitudes (COEA), 136
  circling minimums, 347–349, **348**
  climbs, 38–40
  constant-rate descents, 59–60
  control pressure, 28
  controlling, 28–30, **29**
  decision altitude (DA), 341, 345
  decision height (DH), 340–341, 429,
    431–433, 448, 504–506
  descents, 38–40

altitude (*Cont.*):
  elevation of airfield, 172–174, **173**
  emergency altitudes, 685–686
  equipment requirements vs., 190–191
  FAR 91.177, minimums, 278
  height above touchdown (HAT), 341
  high-speed final, 61
  holding patterns, 419–420, **420**
  landing minimums, 345–347, **346**
  maximum authorized altitude (MAA),
    287
  minimum, 151–153
  minimum controllable airspeed, 76
  minimum crossing altitude (MCA),
    152, 278
  minimum descent altitude (MDA), 429,
    431–433, 448, 450–458, **449**, 450,
    470, 502, 504–506
  minimum enroute altitude (MEA), 151,
    280, 286, 287
  minimum obstruction clearance altitude
    (MOCA), 152, 280, 283, 490
  minimum off-route altitude (MORA),
    280, 283, 286, 288, 291, 294
  minimum reception altitude (MRA), 152,
    278, 286
  minimum safe altitude (MSA), 320
  minimum sector altitude (MSA), 320
  missed approach point (MAP), 340,
    344–345, 349
  obstruction clearance altitude (OCA), 340
  operation below MDA, 457–458
  required reports, 261, 262, 402
  steep turns, 82
  step climbs and descents, 28–30, **30**
  takeoff minimums, 360–362, **361**
  threshold crossing height (TCH), 340
  transition altitudes, 321
  vertical S maneuver, 77, **77**
altitude hypoxia, 607–608
ambiguity indicator (*See* To/From
  indicator, VOR)
amended clearance, 89, 250–252
anemia, 608, 609
angle of attack, stalls, 80–81
angles for approach, 425–435
anoxia, 609
apathy vs. preparedness for flight, 550–551
approach and landing, 58, 60–61, 146,
  163–169, 655–657
  adjustments to approach altitude,
    454–457
  air traffic control, 467–468
  aircraft, 429
  airport surveillance radar (ASR), 428,
    429, 430, 516–517, **517**

approach and landing (*Cont.*):
  airspeed, 466
  alternate minimums, 154, 362–366, **363**,
    364–**365**
  altimeter error, 457, 505
  altitude minimums, 426, 449, 468–469
  angles for approach, 425–435
  approach charts, 309–332, **310, 311**
  Approach Control, 467–468, 516
  approach light system (ALS), 429–430
  approach procedure flight track, 329
  artificial horizon, 61
  attitude indicator, 61
  automatic direction finder (ADF), 117,
    118–119, 399, 429
  back course approaches, 429, 508–510,
    509, 523–526, **524**, 662–663
  CAT II and CATIIIA approaches, 341,
    342, 343
  categories, approach categories,
    207–208, 429, 430–433, 452–454,
    504
  on charts and maps, 276
  charted visual approaches, 304–307,
    305, 306
  circling approach, 441–443, **442, 443,
    444, 445**, 453–454, 472–475, **472**
  circling minimums, 347–349, **348**,
    440–441, 473
  clearance changes, 495
  clearances, 467–468
  compulsory reporting points, 400–402
  constant-airspeed, 60–61
  constant-rate descents, 59–60
  constant-speed descents, 58–59
  contact approach, 438–440, 519–520
  conversion tables, 347, **348**
  course reversals, 329
  decision altitude (DA), 341, 345
  decision height (DH), 340–341, 429,
    431–433, 448, 504–506
  descent timing, 468–469
  direct entry, holding pattern, 408–410,
    416–418, **417**
  dirty descent, 59
  distance measuring equipment (DME),
    422–423, **423**, 484–490, **485, 487**,
    511
  dogleg in approach, 481–484, **482**
  elevation of airfield, 172–174, **173**
  equipment requirements, 454, **455**
  fan markers, 511
  final approach course, 470–471
  final approach fix (FAF), 328–329,
    335–336, 383, 469–470, 448, 511
  final approach segment, 448

approach and landing (*Cont.*):
Five T sequence–time, turn, twist,
   throttle, talk, 413–414
fixed approaches, 495
"full approach procedure," 463
glide slope, 338, 428
global positioning system (GPS) overlay
   approach, 481–484, **482, 485**, 492,
   493, 494
global positioning system (GPS),
   129–130, **131, 132, 133**, 429,
   490–497
ground controlled approach (GCA),
   518–519
height above touchdown (HAT), 341
high-speed final, 61, 61
holding patterns, 276, 278, 383, 391,
   399, 407–410, **407, 409**, 413–424,
   414
initial approach fix (IAF), 167–169, **168**,
   328–329, 383, 448
initial approach segment, 448
inoperative equipment, 454, **455**,
   505–511
instrument approach procedures (IAP),
   425
instrument landing system (ILS), 167,
   168, 428, 499–526, **500, 503**,
   661–663
intermediate approach segment, 448
intersection holds, 421–422
landing gear retraction, 81
landing gear use, 61
landing minimums, 345–347, **346**
level-off after descent, 60
lighting, airport, 384–388, **385, 386,
   387**, 428, 429–430, 454, **456**
localizer, 428, 429, 430, 506, 507, 511,
   662–663
localizer back course approach,
   523–526, **524**
localizer type directional aid (LDA),
   430, 511, **512, 513**, 662–663
low approach and cancel IFR, 467
maneuvering or approach flaps, 347
maneuvering table, approach
   categories, 453
microwave landing system (MLS), 428
middle marker, 428
minimum descent altitude (MDA), 429,
   431–433, 448, 450–458, 470, **449**,
   502, 504–506
minimum obstacle clearance altitude
   (MOCA), 490
minimums, 60, 426, 428, 429–430,
   435–446, 449, 511

approach and landing (*Cont.*):
missed approach, 75, 331, 344–345,
   432, 443–445, 459–461, 467,
   471–472, 495, 502, 504, 506
missed approach point (MAP), 58, 340,
   344–345, 349, 432, 458, 459–461,
   469–470, 471–472, 511
missed approach segment, 448
multiengine aircraft (*See* multiengine
   aircraft approaches)
no procedure turn (NoPT) sectors, 327
no-gyro approaches, 518
nondirectional beacon (NDB), 117–118,
   429, 461, 468, 477–479, **478**, 507,
   657–659
nonprecision approach, 58, 333–335,
   428–429, 447–448
NOS approach charts, 369–396
obstacle/obstruction clearance, 62–65,
   299, 426, 449, 452, 475–**477**, 504,
   511
obstruction clearance altitude (OCA), 340
operation below MDA, 457–458
outer marker, 428
overlay approach, GPS, 129–130, **131,
   132, 133**, 481–484, **482, 485**, 492,
   493, 494
parallel entry, holding pattern, 409,
   416–418, **417**
pitch settings, 61
position reporting, 400–402
power setting, 61, 508
precision approach radar (PAR), 428,
   518–519
precision approaches, 337–340, **339**,
   428–429, 448
procedure turns, 327, 329, 340–341,
   388–393, **389, 390, 392**,
   463–466
profile descents, 297–302, **298, 300,
   301**, 302–304, **303**
radar, 399, 514–516, 526, 662–663
radar approach control (RAPCON), 514
radar vectors, 461–462, 461
radio communications, 399–412, 467–468
radio failure, 404–406, **405**
runway visibility value (RVV), 435
runway visual range (RVR), 428,
   435–436, 458–459
safety first, 426–427, 497–498
segments of approach, 448
simplified directional facility (SDF), 514,
   515, 662–663
standard terminal arrival route (STAR),
   63, 163, 167, 297–302, **298, 300,
   301**, 410, **411**

approach and landing (*Cont.*):
standard terminal procedures (TERPs), 312, 449
straight-in approach, 453–454
straight-in approach, minimums, 440–441
teardrop entry, holding pattern, 409, 416–418, **417**
threshold crossing height (TCH), 340
timing the approach, 469–470, 506
tips for flying approaches, 497–498
touchdown zone elevation (TDZE), 173–174, 335
transitioning for approach, 327–329, **330**
uncontrolled airports, 676–677
Universal Approach Chart Update*italic* notice, 313, 314–**317**
vectored approaches, 495
vectors to final, 526
visibility, 429, 435–436, 458–459
visual approach, 436–438, 519–520, **521**
visual descent point (VDP), 336–337
VOR approach, 58, 429, 430, 481–484, **482, 485**, 659–661
VOR/DME approach, 488–490, **489**
weather minimums, 205–208
approach charts (*See also* NOS approach charts), 309–332, **310, 311**
airports, 320
approach procedure flight track, 329
CAT II and CATIIIA approaches, 341, **342, 343**
circling minimums, 347–349, **348**
conversion tables, 347, **348**
course reversals, 329
creation of charts, 312
decision altitude (DA), 341, 345
decision height (DH), 340–341
effective dates, 312
filing a chart, 318–321, 318
final approach fix (FAF), 328–329, 335–336
fixes, 324, 327
frequency used, 321
glide slope, 338
Ground Control, 321
height above touchdown (HAT), 341
Howie Keefe updates, 312–318
identifiers, 324
initial approach fix (IAF), 328–329
landing minimums, 345–347, **346**
localizer, 324
LORAN/GPS Navigator Atlas*italic*, 318
minimum safe altitude (MSA), 320
minimum sector altitude (MSA), 320
missed approach, 331, 344–345

approach charts (*Cont.*):
missed approach point (MAP), 340, 344–345, 349
no procedure turn (NoPT) sectors, 327
nonprecision approach, 333–335
obstacle clearance, 331
obstruction clearance altitude (OCA), 340
plan view (bird's eye view), 321–327, **322, 323, 325, 326**
precision approaches, 337–340, **339**
procedure identifications, 320
procedure turns, 329, 340–341
profile view, 333–350, **334**
standard terminal procedures (TERPs), 312
terrain information, 331, **332**
threshold crossing height (TCH), 340
touchdown zone elevation (TDZE), 335
transition altitudes, 321
transitioning for approach, 327–329, **330**
Universal Approach Chart Update notice, 313, 314–**317**
Universal Enroute Chart Update notice, 313, 314–**317**
visual descent point (VDP), 336–337
Approach Control, 467–468, 516
approach flaps, 347
approach light system (ALS), 429–430
Approved Flight Manuals, 177
arc to distance relationship, 134–135
area charts, 291–295, **293**
area minimum altitudes (AMAs), 291, 294
arrival (*See* approach and landing)
artificial horizon (*See also* attitude indicator), 17, 18, **18**, 20, 22
approach and landing, 61
climbs, 52
descents, 53–54
level-off, after climb, 57–58
straight and level flight, 53
ATC delays, in weather briefing, 222
attitude, attitude control, 3, 27–28
control pressure, 37
straight-and-level flight, 36–38
trim, 33–34
unusual attitude, 83–85, 652–653
attitude indicator (*See also* artificial horizon), 3, 6, 36, 38
approach and landing, 61
climbs, 55, **55**, 56
constant-speed descents, 58–59, **59**
errors, 9, 47
failure of instrument, 85, 87
minimum controllable airspeed, 76

attitude indicator (*Cont.*):
  takeoff setting, 51
  taxi checks, 256
attitude, heading, altitude method,
    43–44
attitude, of pilot, 619–620
autokinesis illusion, 583
automated flight service stations (AFSS),
    212
automatic direction finder (ADF), 88, 117,
    118–119, 399, 429, 653–654
  bracketing, 123–126, **125**
  "chasing the needle," 123
  holding patterns, 128–129
  homing, 122–123, **124**
  instrument landing system (ILS), 508
  intercept angle, 120–121, **122**
  intercepting a bearing, 120–121, **122**
  outbound bearings, from NDB, 126–127
  preflight check, 247
  time/distance checks, 119–120
  tracking and bracketing, 123–126, **125**
  wind drift/wind correction, 121
automatic rough illusion, 591
automatic terminal information service
    (ATIS), 171, 240, 244, 259
autopilots, 3
Aviation Weather Services, 199, 215

back course approach, 429, 508–510, **509**,
    523–526, **524**
balance, 584–588
ball (*See* turn and bank coordinator)
ball flags, on charts and maps, 288
barodontalgia (tooth pain related to
    altitude), 614
barosinusitis (sinus pain related to
    altitude), 614
barotitis (earache related to altitude), 614
bearing
  intercepting ADF bearing, 120–121, **122**
  magnetic bearing, 118
  relative bearing, 118
books on aviation, 689–690
bracketing, 115, 123–126, **125**, 650–651
brakes, takeoffs, 52
briefings, 157–158, 174–176, **175**, 198–200,
    212, 221–226, 234–236, 680–682

cabin altitude, 152
cabin pressure altitude, 152
call signs, 259
callback reports, 690
canceling IFR, 262–263
CAR 3.655(d) miscellaneous equipment,
    186–187

CAR 3.655, required basic equipment list,
    184–185
carbon monoxide poisoning, 559, 560
carburetor icing, 150
cardinal heading, 127
CAT II and CAT IIIA approaches, 341,
    342, 343
categories, approach categories, 207–208,
    429, 430–433, 452–454, 504
ceiling (*See also* visibility), 199, 204–205,
    207–208, 211, 217, 218, 219
celestial navigation, 135–137
  circle of equal altitudes (COEA), 136
  circle of equal radius (COER), 136, 138
  circle of position (COP), 136
  geographical position (GP), 135
  line of position (LOP), 136, 138, 139
  nadir, 135
  three-star fix, 136–137, **137, 138**
  timing, 138–139
  zenith, 135
  zenith distance (ZD), 135
center of gravity, 48
Certified Flight Instructors, Instruments
    (CFIIs), 3
changeover points (COPs), 283–284
charted visual approaches, 304–307,
    305, 306
charts and maps (*See also* airport charts;
    approach charts; NOS approach
    charts), 157, 162, 267–290
  airport charts, 351–368, **352, 353, 354**
  airport diagrams, 167
  airports, 276, 286, 294, 320
  airspace, 288, 294
  altitudes, 278–288, **282**
  approach charts, 276, 309–332, **310, 311**
  area charts, 291–295, **293**
  area minimum altitudes (AMAs),
    291, 294
  ball flags, 288
  changeover points (COPs), 283–284
  charted visual approaches, 304–307,
    305, 306
  Class B airspace, 295–297, **296**
  climb gradients, 302
  creation of charts, 312
  cross-country flight, 687–688
  distance measuring equipment (DME),
    294
  early attempts at charting, 427–428
  effective dates, 312
  elevations (MSL), 268, 288, 294
  enroute charts, 267–290
  enroute fixes, 169–171, **170**
  filing a chart, 318–321

charts and maps (*Cont.*):
  fixes, 278–279, 286, 324, 327
  flight service stations (FSS), 274
  frequencies used, 351
  global positioning system (GPS), 272
  Ground Control, 321
  holding patterns, 276, 278
  Howie Keefe updates, 312–318
  identifiers, 324, 351
  index maps, 268, **269**
  inertial navigation system (INS), 272
  initial approach fixes (IAFs), 167–169,
    168
  instrument landing system (ILS), 167,
    168, 294
  intersections, 284
  Jeppesen NavData identifiers, 286
  Jeppesen's beginnings, 427–428
  KRM, 272
  L/MF facilities, 274, 287
  landing aids, 272
  latitude and longitude, 284, 288
  LDA, 272
  localizer, 272, 324
  LORAN/GPS Navigator Atlasitalic, 318
  loran-C, 272
  maximum authorized altitude (MAA),
    287
  microwave landing systems (MLS), 272
  mileage between facilities, 283, 286, 288
  minimum crossing altitude (MCA), 278
  minimum enroute altitude (MEA), 280,
    286, 287
  minimum obstruction clearance altitude
    (MOCA), 152, 280, 283, 490
  minimum off-route altitude (MORA),
    280, 283, 286, 288, 291, 294
  minimum reception altitude (MRA),
    278, 286
  minimum safe altitude (MSA), 320
  minimum sector altitude (MSA), 320
  navaid facilities, 272, **273**, 286–287
  no procedure turn (NoPT) sectors, 327
  nonprecision approach, 333–335
  NOS approach charts, 369–396
  obstacle clearance, 331
  organization of, in cockpit, 243–244,
    247–248, 267–268
  plan view (bird's eye view), 321–327,
    322, 323, 325, 326
  plotters, 271
  precision approaches, 337–340, **339**
  preflight planning, 146–147
  profile descents, 297–302, **298, 300,**
    301, 302–304, **303**
  profile view, 333–350, **334**

charts and maps (*Cont.*):
  radar, 276
  radio communication notes, 270, **270**
  remote communications outlets (RCOs),
    274
  reporting points, 276, 279–280, 284
  restricted airspace, 270–271, **271,**
    274, 276
  runway checks, 257
  scale, 271
  simplified directional facility (SDF), 272
  standard instrument departure (SID),
    297–302, **298, 300, 301**
  standard terminal arrivals (STAR), 163,
    297–302, **298, 300, 301**, 410, **411**
  standard terminal procedures (TERPs),
    312
  symbols, 268, 272, **273, 275, 277, 281,**
    283–284, **285, 289, 290, 292,**
    352–355, **353**
  TACAN stations, 272
  telephone communications facilities, 274
  terrain features, 294, 331, **332**
  time zone boundaries, 288
  touchdown zone elevation (TDZE), 335
  transition altitudes, 321
  Universal Approach Chart Update
    notice, 313, 314–**317**
  Universal Enroute Chart Update
    notice, 313, 314–**317**
  VOR, 274, 276, 284, 286, 287
  waypoints, 280, 290
  Zigdex, 268, **270**, 268
"chasing the needle," 57, 115–116, 123
checklists (*See also* pilot readiness for
    flight; preflight planning), 239–248,
    2401
  ADF check, 247
  airspeed indicator errors, 241
  altimeter error, 241
  automatic terminal information service
    (ATIS) frequency, 240–241
  cockpit organization, 243–245, 247–248
  electrical equipment, 241–242
  fuel quantity, 239–240
  gyro instrument check, 248
  ILS checks, 246–247
  preflight planning, 145
  radio communications check, 244, 248
  rolling engine run-up, 256–256
  runway checks, 257
  taxi checks, 256
  transponder check, 247
  VOR check with VOT, 245–246, **245**
  VOR checks, 242–243, 246
cigarette use, 558–561

circadian rhythms, night flying vs., 572–574
circle of equal altitudes (COEA), 136
circle of equal radius (COER), 136, 138
circle of position (COP), 136
circling approach, 441–443, **442, 443, 444, 445**, 453–454, 472–475, **472**
circling minimums, 347–349, **348**, 440–441, 473
Class B airspace, 295–297, **296**
clear ice, 150
clearances, 68, 158, 244, 249–263
  air route traffic control center (ARTCC), 69
  amended, 89, 250–252
  approach, 467–468, 495
  canceling IFR, 262–263
  "cleared direct" on VOR, 113
  control towers, 69
  discrete clearance delivery, 69
  expect approach clearance (EAC) time, 415–416
  expect further clearance (EFC) time, 415–416
  frequency used, 171
  ground control, 69
  IFR clearance, 69–71, 69
  obtaining clearance, 252–253
  practice, 249–250
  radio failure, 88–90
  remote communications outlets (RCOs), 254, **255**
  requesting a clearance, 253
  shorthand notes, 250, **251**
  takeoff and departure, 56, 70–71
  taxiing, 69
  unacceptable clearances, 253–254
  uncontrolled airports, 69
  void time, 70, 254, 677
cleared as filed..., 68, 238
"cleared direct" on VOR, 113
climb airspeed (Vy), 81
climb profiles, 71–72, 71
climb speed (Vy), 54
climbs, 6, 25, 38–40, 52–53, 522
  airspeed, 54–55
  airspeed indicator, 52–53
  altimeter, 53
  artificial horizon, 52
  attitude indicator, 55, **55**, 56
  climb airspeed (Vy), 81
  climb gradients, 72
  climb speed (Vy), 54
  constant-airspeed, 54
  constant-rate, 56–57
  departure climbouts, 55–56

climbs (*Cont.*):
  directional gyro, 53
  estimated climbout time, weather, 228
  gradients, depicted on charts, 302, 393, **394, 395**
  ground speed vs., 302
  initial climb, 56
  level-off, 57–58
  pitch angle setting, 52
  pitch setting, 54–55
  power settings, 52, 54
  practice maneuver, 79
  profiles, climb profiles, 71–72
  rate of climb, 54, 56, 302
  required reports, 261, 402
  step climbs, 28–30, **30**
  takeoff minimums, 360–362, **361**
  turns and banks, 49
  unusual attitude recovery, 84
  vertical S maneuver, 77, **77**
clock, standard rate turns with clock, 25–26
clocks, 148
cockpit organization, 243–245, 267–268, 633–642
cockpit resource management (CRM), 633–642
code letters for equipment requirements, 188
cold fronts, 218
compass (*See* magnetic compass)
compulsory reporting points, 400–402
computerized weather services, 212–213
constant-airspeed approach, 60–61
constant-airspeed climbs, 54
constant-rate climbs, 56–57
constant-rate descents, 59–60
constant-speed descents, 58–59
contact approach, 438–440, 519–520
control displacement, 7–8, 40, 41
control instruments, 4–7, 5–6(t), 13(t), 87–88
control pressure, 6, 7–8, 28, 35, 37, 40, 41, 50, 87, 522
control towers, takeoffs, 69
conversion tables, approach charts, 347, **348**
correction cards for magnetic compass, 182
course acquisition code (C/A-code), 139
course deviation indicator (CDI), 92, 99, **99**, 101–102
course reversal, 329, 464–466, **465**
course selector, VOR, 98
cross-country flight, 95, 663–666, 675–688
  charts and maps, 687–688
  electrical failure, 686–687

cross-country flight (*Cont.*):
emergency altitudes, 685–686
estimated time enroute (ETE), 685
estimated time of arrival (ETA), 685
flight logs, 678–680, **679**, 685
fuel management, 678
lost radio contact, 682–684
partial panel flight, 677–678
radio failure, 684
time logging, 685
tips for safe flight, 687–688
uncontrolled airports, 676–677
void time clearances, 677
weather briefings, 680–682
cruise, level-off, after climb, 57–58
current conditions, in weather briefing, 221
cylinder head temperature indicator, 184

dark adaptation of eyes, 580
decision altitude (DA), 341, 345
decision height (DH), 340–341, 429,
431–433, 448, 504–506
decompression sickness (bends),
614–616
departure (*See* takeoff and departure)
departure climbouts, 55–56
Departure Control, 70, 171, 244, 257
departure procedures, 69–72
descent
high-speed final, 61
descents (*See also* approach and landing),
6, 25, 38–40, 53–54, 58
airspeed indicator, 54
altimeter, 54
artificial horizon, 53–54
constant-rate descents, 59–60
constant-speed, 58–59
directional gyro, 54
dirty descent, 59
dives, 84
gradients, on charts and maps, 393,
394, 395
level-off, 60
never-exceed airspeed (Vne), 84
obstacle clearance, 299
pitch setting, 54
power settings, 54
power vs. descent, 508
practice maneuver, 79
profile descents, 297–302, **298, 300,
301**, 302–304, **303**
required reports, 261, 402
step descents, 28–30, **30**
unusual attitude recovery, 84
vertical S maneuver, 77, **77**
vertical speed indicator, 54

destination forecasts, in weather briefing,
221, 225
deviation in magnetic compass, 10
dew point, 203–204
differential GPS (DGPS), 139–140
dip error in magnetic compass, 14
direct entry, holding pattern, 408–410,
416–418, **417**
direct user access terminal service
(DUATs), 212, 213–214
direction finder (DF) loop, 117–118
directional gyro (DG), 19, 22–23, **23**, 89,
148
acceleration errors, 47
climbs, 53
descents, 54
errors, 47
precession, 47
straight and level flight, 53
takeoff setting, 51, 52
unusual attitude, 83–85
dirty descent, 59
discrete clearance delivery, takeoffs, 69
discrete frequency, 636–639, **637, 638**
displaced thresholds, 359–360
distance, arc to distance relationship,
134–135
distance measuring equipment (DME),
148
approach, 484–490, **485, 487**, 511
on charts and maps, 294
holds, 422–423, **423**
distractions from instrument scanning, 44
dives, 84
DME fix, 66
dogleg in approach, 481–484, **482**
Doster, John, FAA GADO, 49–50
downbursts, 197
downdrafts, 7, 25, 196
drag, 49
drift, wind drift, 121, 248, 522
drug use, 546–549, 557–558, 561–562

ears (*See* hearing; inner ear)
electrical system
preflight check, 241–242
required, 184
system failure, 686–687
elevation of airfield, 172–174, **173**
elevations (MSL), on charts and maps,
268, 288, 294
embedded thunderstorms, 151, 197
emergency altitudes, 685–686
emergency locator transmitter (ELT), 185
emergency procedures, 237, 555–556
accidents, 537–540

emergency procedures (*Cont.*):
  altitude, emergency altitudes, 685–686
  electrical failure, 686–687
  emergency vs. priority handling, 403
  failure of instruments, 85–87, **86**
  no-gyro approaches, 518
  one engine inoperative instrument
    approach, 528–531
  partial panel flight, 677–678
  partial-panel procedures, 85–87, **86**
  priority handling request, 403
  radio failure, 88–90, 237, 263, 404–406,
    405, 684
emergency vs. priority handling, 403
empty-field myopia, 577
engines
  equipment required, 184
  one engine inoperative instrument
    approach, 528–531
  rolling engine run-up, 256–256
  sound of, as diagnostic tool, 25
enroute fixes, 169–171, **170**
Enroute Flight Advisory Service (EFAS),
  639–642, **640**
enroute forecasts, in weather briefing,
  221, 225
enroute holding, 415–416
enroute low-altitude charts, 146, 162,
  164–**165**
equipment required for IFR, 148
  Airplane Flight Manual (AFM), 186–187,
    193
  airspace requirements, 190–191
  airworthiness, 184
  airworthiness directives (AD), 189–190
  altitude vs., 190–191
  approach, 454, **455**
  CAR 3.655(d) miscellaneous equipment,
    186–187
  CAR 3.655, required basic equipment
    list, 184–185
  code letters for equipment requirements,
    188
  correction cards for magnetic compass,
    182
  FAR 91.205, 185–186, 185
  FAR 91.213 (airworthiness directives),
    189–190
  FAR 91.213 (inoperative equipment),
    187–188
  flight plan equipment codes, 235
  four-step test for required equipment,
    191–192
  fuel reserve, 182
  global positioning system (GPS), 496
  inoperative equipment, 182, 187–188,
    191, 261, 402, 454, **455**, 505

equipment required for IFR (*Cont.*):
  letter of authorization (LOA), for MEL
    use, 179, **180**
  master minimum equipment list
    (MMEL), 178
  minimum equipment lists (MEL), 177,
    178–181
  no-gyro approaches, 518
  operating limitations, aircraft, 185–186
  operations without MEL, 182–183
  Pilot and Aircraft Courtesy Evaluation
    (PACE) program, 193
  pressurized cabins, 190–191
  process of obtaining authorization for
    MEL use, 179–181, **181**
  proposed master minimum equipment
    list (PM-MEL), 178
  ramp checks, 192–193
  receiver autonomous integrity
    monitoring (RAIM), GPS, 496
  supplemental oxygen, 190–191
  supplemental type certificate
    (STC), 179
  transponders, 190–191
  type certificate data sheet (TCDS), 185
  weather conditions, 190–191
  weight and balance data sheet, 193
errors in instrument readouts, 9–12, 40,
  47–48, 241, 457, 505
  global positioning system (GPS),
    139–140
  partial panel flight, 677–678
  vertigo or spatial disorientation vs.,
    552–556, 565–568, **568**
estimated climbout time, weather, 228
estimated time enroute (ETE), 685
estimated time of arrival (ETA), 685
exercises for the cockpit, 596–598
expect approach clearance (EAC) time,
  415–416
expect further clearance (EFC) time,
  415–416

failure of instruments, 85–87, **86**
fan markers, approach, 511
FAR 91.103, preflight planning, 146
FAR 91.177 (minimum altitudes), 278
FAR 91.205 (equipment requirements),
  185–186
FAR 91.213 (airworthiness directives),
  189–190
FAR 91.213 (inoperative equipment),
  187–188
FAR Part 91.167 (fuel requirements),
  153–155
FAR Part 91.169 (alternate airports), 154
FAR Part 91.527 (icing conditions), 149

fatigue, pilot, 200, 543–546, 570–572, 574–576, 594–596, 600–601
Federal Air Regulations (FARs), 690
Federal Register, 312
filing a flight plan, 145, 234
final approach, high-speed final, 61, 61
final approach course, 470–471
final approach fix (FAF), 328–329, 335–336, 383, 448, 469–470, 511
final approach segment, 448
Five T sequence–time, turn, twist, throttle, talk, 413–414
fix end, holding pattern, 407, **407**
fixed approaches, 495
fixes, 278–279, 286, 324, 327
  DME, 66
  enroute fixes, 169–171, **170**, 169
  final approach fix (FAF), 328–329, 335–336, 383, 448, 469–470, 511
  initial approach fix (IAF), 167–169, **168**, 328–329 383, 448
flags, VOR, 99
flaps, 4, 75–76, 347
flare guns, 185
flicker vertigo, 562
flight crews and cockpit resource management (CRM), 634–635
flight instructor endorsements, 673
flight kit for preflight planning, 146–148
flight log, 159, 160–**161**, 169, **169**, 678–680, **679**, 685
Flight Operations Evaluation Board (FOEB), 178
flight plans (*See also* preflight planning), 68, 70
  alternate airports, 229–232, **232**
  cleared as filed, 238
  equipment codes, 235
  estimating time enroute, 236–237
  filing the flight plan, 234
  flight log, 169, **169**
  in-flight notations, 232–236
  one-call weather briefing/flight plan filing technique, 236
  outlook briefings, 236
  total time enroute, 237–238
  weather briefings, 212
flight proficiency, 673
flight service stations (FSS), 140, 199, 212, 274
Flight Standards District Offices (FSDO), 178
flight test preparation, 667
flight test recommendation flight, 672
Flight Watch, 639–642, **640**
flotation gear, 185
fog, 196, 203–205, 211, 227–228, 539–540

freezing level, 202–203
freezing rain, 218
frequencies used in radio communications/navigation, 171, **172**, 240, 244–245, 248, 260–261, 270, **270**, 321, 636–639, **637, 638**
  on charts and maps, 351
  discrete frequency, 636–639, **637, 638**
  Enroute Flight Advisory Service (EFAS), 639–642, **640**
  Flight Watch, 639–642, **640**
  instrument landing system (ILS), 506
  lost radio contact, 682–684
frequency selector, VOR, 98
fronts, 218
fuel gauges, 11–12, 184
fuel pressure indicator, 184
fuel requirements, 153–155, 678
  calibrate-to-empty rule, 182
  FAR Part 91.167, 153–155
  fuel quantity, 239–240
"full approach procedure," 463

generators/alternators, 148, 184
geographical position (GP), 135
glaze ice, 150
glide slope, 338, 428
global positioning system (GPS), 3, 129–130, **131, 132, 133**, 140, 141, 158, 272, 429
  approach, 490–497
  approach planning, 492–497
  clearance changes, 495
  course acquisition code (C/A-code), 139
  databases, revisions to, 496
  differential GPS (DGPS), 139–140, 491
  errors, 139–140
  fixed approaches, 495
  horizontal situation indicator (HSI), 492
  LORAN/GPS Navigator Atlasitalic, 318
  missed approach, 495
  monitoring underlying VOR/NDB approaches, 496
  moving map display, 496–497
  overlay approach, 481–484, **482, 485**, 492, **493, 494**
  precision code (P-code), 139
  receiver autonomous integrity monitoring (RAIM), 496
  satellites, 491–492
  selective availability (SA), 139
  standard vs. precision positioning services, 491
  three-star fix, 136–137, **137, 138**
  vectored approaches, 495
  VOR vs., 102

global positioning system (*Cont.*):
   wide area augmentation system
      (WAAS), 491
go/no-go decisions, 145, 195, 226–228
great circle navigation, 130, 134–135, **134**
Greenwich time (*See* universal
   coordinated time)
Ground Control, 174–176, 244, 259–260
   chart/on charts and maps, 321
   frequency used, 171
   takeoffs, 69
ground controlled approach (GCA),
   518–519
ground speed, 302
   wind vs, 228–229, **230, 231**
ground visibility, 205
gust fronts, 197
gyro instruments
   preflight check, 248
   unusual attitude, 83–85

hail, 196–197
handoffs, 260
hazardous pilot attitudes for flight,
   619–620
heading, 3, 6, 30–31
   cardinal heading, 127
   magnetic compass, 14–15, **15**, 118
   reference (holding) heading, 113–115
heading indicator, 3
   automatic direction finder (ADF), 88
   errors, 9–10, 102–104
   failure of instrument, 85, 88
   magnetic heading, 118
   runway checks, 257
   takeoff and departure, 522
   taxi checks, 256
headwinds, 111
hearing, 584–588
height above touchdown (HAT), 341
high intensity runway lights (HIRL), 428
high/low (H/L) charts, 268
high-altitude (HI) charts, 268
high-altitude flight physiology (*See also*
   physiology of flight), 603–616
high-speed final, 61, 61
histotoxic hypoxia, 608
holding course, holding pattern, 407, **407**
holding fix, holding pattern, 407, **407**
   required reports, 261, 402
holding patterns, 276, 278, 383, 391, 399,
   407–410, **407, 409**, 413–424, **414**
   abeam, 407, **407**
   altitude control, 419–420, **420**
   automatic direction finder (ADF),
      128–129, 654

holding patterns (*Cont.*):
   choosing correct entry, 418–419
   course reversal, 464–466, **465**
   direct entry, 408–410, 416–418, **417**
   distance measuring equipment (DME)
      holds, 422–423, **423**
   enroute holding, 415–416
   entry, 408–410, **409**, 416–418, **417**
   entry sectors, 419
   expect approach clearance (EAC) time,
      415–416
   expect further clearance (EFC) time,
      415–416
   Five T sequence–time, turn, twist,
      throttle, talk, 413–414
   fix end, 407, **407**
   holding course, 407, **407**
   holding fix, 407, **407**
   holding side, 407, **407**
   inbound leg, 407, **407**
   intersection holds, 421–422
   non-holding side, 407, **407**
   outbound end, 407, **407**
   outbound leg, 407, **407**
   parallel entry, 409, 416–418, **417**
   reciprocal point, 407, **407**
   reference (holding) heading, 113–115
   required reports, 261, 402
   teardrop entry, 409, 416–418, **417**
   variations, 420–421
   VOR, 651–652
   wind correction, 414–415
holding side, holding pattern, 407, **407**
Holmes/Rahe Life Change Scale, 628–630,
      629
homing, ADF, 122–123, **124**
horizon, 6
horizon, artificial (*See* artificial horizon)
horizontal situation indicator (HSI), 492
human eye, 578–584, **578**
humidity, 197
hurricanes, 219
hypemic hypoxia, 608, 609
hyperventilation, 560
hypoxia, 558–561, 589, 604–614

ice fog, 203
icing conditions, 149–151, 195, 199,
      201–203, 211, 218, 227, 539
   carburetor icing, 150
   FAR Part 91.527, 149
   freezing level, 202–203
   pitot heating systems, 242
identifiers, on charts and maps, 324, 351
illness, 546
in-flight notations, 232–236

inbound leg, holding pattern, 407, **407**
index maps, 268, **269**
indicated airspeed (IAS), 171
inertia, 49
inertial navigation system (INS), 272
initial approach fix (IAF), 167–169, **168**,
   328–329 383, 448
initial approach segment, 448
inner ear, balance, and vertigo, 563–565,
   565
inoperative equipment, 182, 187–188, 191,
   454, **455**, 505, 511
   required reports, 261, 402
inspections
   Pilot and Aircraft Courtesy Evaluation
      (PACE) program, 193
   ramp checks, 192–193
instructors
   flight instructor endorsements, 673
   student/instructor relationship,
      551–552
instrument approach procedures (IAP),
   425
instrument fixation, 39, 43, 85
instrument landing system (ILS), 167, **168**,
   369–372, **371**, 428, 499–526, **500,
   503**, 661–663
   altimeter error, 505
   analyzing ILS approaches, 502–504
   automatic direction finder (ADF)
      volume, 508
   back course approaches, 508–510, **509**
   categories, approach categories, 504
   on charts and maps, 294
   decision height (DH), 504–506
   descent rate, 501–502
   glide slope, 501–502
   ground speed estimation, 501–502
   inoperative equipment, 505
   localizer approach vs., 507
   localizer frequency, 506
   marker beacons, 506–507
   minimum descent altitude (MDA), 502,
      504–506
   missed approach, 502, 504
   needle sensitivity, 499–501
   nondirectional beacon (NDB) approach
      vs., 507
   obstacle clearance, 504
   power vs. descent, 508
   practice, 507
   preflight check, 246–247
   timing the approach, 506
   tips for ILS approaches, 507–408
instrument meteorological conditions
   (IMC), 436

instrument panel layout, 15–16, **16**
   failure or partial-panel procedures,
      85–87, **86**
*Instrument Pilot Practical Test
   Standards*, 667
Instrument Rating Practical Test, 666
instrument training syllabus, 643–692
   ADF holding patterns, 654
   ADF orientation, tracking, bracketing,
      653–654
   cross-country procedures, 663–665
   flight instructor endorsements, 673
   flight test preparation, 667
   flight test recommendation flight, 672
   ILS approaches, 661–663
   instrument approach procedures,
      655–657
   Instrument Flight Test oral exam,
      preparation, 667–672
   *Instrument Pilot Practical Test
      Standards*, 667
   Instrument Rating Practical Test,
      666
   introduction to basic instruments,
      646–648
   introduction to IFR, 645–646
   long IFR cross-country flight, 665–666,
      675–688
   maneuvering solely by reference to
      instruments, 648–650
   NDB approaches, 657–659
   progress check, 666
   unusual attitudes, partial panel, 652–653
   VOR approaches, 659–661
   VOR holding patterns, 651–652
   VOR tracking and bracketing, 650–651
intercept angle, ADF, 120–121, **122**
intermediate approach segment, 448
interpreting instrument readings, 44–45
intersection holds, 421–422
intersections, on charts and maps, 284

Jeppesen Company, 267, 268, 271, 276,
   312, 321, 374
Jeppesen NavData identifiers, 286
Jeppesen, Elroy B., 427–428
judgment, 591–594, 617–622
   accident prevention techniques, 620–622
   hazardous pilot attitudes for flight,
      619–620
   mental process of safe flight,
      618–619
   poor judgment (PJ) chains, 617

Keefe, Howie, 267, 312–318
KRM, 272

L/MF facilities, 274, 287
land lines, 638
landing (*See* approach and landing)
landing aids, 272
landing gear, 4
   approach and landing, 61
   minimum controllable airspeed,
      75–76
   position indicator, 185
   retracting gear, 81
landing minimums, 345–347, **346**
lapse rate, 197
latitude, 284, 288
legal airworthiness, 177
letter of authorization (LOA), for MEL use,
      179, **180**
level-off, after climb, 57–58, 57
level-off, after descent, 60
lift, 49
lighting, airport, 257, 355–357, **356, 358,
      359**, 384–388, **385, 386, 387**, 428,
      429–430, 454, **456**
line of position (LOP), 136, 138, 139
localizer, 272, 324, 428, 429, 430, 506
localizer approach, 507, 511
localizer back course approach, 523–526,
      524
localizer type directional aid (LDA), 272,
      430, 511, **512, 513**
logbooks
   flight log, 159, 160–**161**, 169, **169**,
      678–680, **679**
   organization of, in cockpit, 243–244,
      247–248
   VOR checks, 149
longitude, 284, 288
LORAN/GPS Navigator Atlas, 318
loran-C, 272, 318
lost radio contact, 682–684
low approach and cancel IFR, 467
low-altitude (LO) charts, 268
low-level wind shear, 197–198

magnetic bearing, 118
magnetic compass, 184, 185
   acceleration, 10
   correction cards, 182
   deviation, 10
   dip error, 14
   errors, 10, 12–15
   procedures and exercises, 14–15, **15**
   turns using compass, 12–15
   variation, 10
magnetic heading, 118
maneuvering table, approach categories,
      453

manifold pressure indicator, 184
   minimum controllable airspeed, 76
   pressure gauge, 185
   stalls, 80–81, 80
maneuvering or approach flaps, 347
marker beacons, instrument landing
      system (ILS), 506–507
master minimum equipment list (MMEL),
      178
master switch, 184
maximum authorized altitude (MAA), 287
Medical Certificate, 193
medical facts (*See* physiology of flight;
      pilot's readiness for flight)
medication use, 546–549, 557–558
medium intensity approach light system
      (MALSR), 430
mental preparedness for flight, 550–551,
      618–619
metabolic effects
   of high-altitude flight effects, 604
   of night flying, 572–574
Meteorology Aviation Routines (METARs),
      214–217
microbursts, 151, 197–198, **198**
microwave landing systems (MLS),
      272, 428
middle marker, 428
mileage between facilities, on charts and
      maps, 283, 286, 288, 383
minimum controllable airspeed, 74–76, 74
minimum crossing altitude (MCA), 152,
      278
minimum descent altitude (MDA), 470,
      502, 504–506
minimum descent altitude (MDA), 429,
      431–433, 448, 450–458, **449**
minimum enroute altitude (MEA), 151,
      280, 286, 287
minimum equipment lists (MEL), 177
minimum obstruction clearance altitude
      (MOCA), 152, 280, 283, 490
minimum off-route altitude (MORA), 280,
      283, 286, 288, 291, 294
minimum reception altitude (MRA), 152,
      278, 286
minimum safe altitude (MSA), 320
minimum sector altitude (MSA), 320
minimums, 429–430, 449
   altitude, 449
   approach, 60, 426, 428, 435–446,
      458–459, 511
   circling approach, 441–443, **442, 443,
      444, 445**, 473
   circling, 440–441, 473
   contact approach, 438–440

minimums (*Cont.*):
personal minimums, 542
runway visibility value (RVV), 435
runway visual range (RVR), 435–436
straight-in approach, 440–441
takeoff and departure, 520, 522
visibility, 435–436, 458–459
visual approach, 436–438
miscellaneous equipment, 186–187
missed approach, 75, 261, 331, 344–345,
402, 432, 443–445, 459–461, 467,
471–472, 495, 502, 504, 506
missed approach point (MAP), 58, 340,
344–345, 349, 432, 448, 459–461,
469–470, 471–472, 511
Mode C transponders, 190–191
motion sickness, 584–588, 598–600
moving map display, GPS, 3, 496–497
multiengine aircraft approaches, 527–534
Area of Operation IV, tasks, 528,
529–531
one engine inoperative instrument
approach, 528–531
Practical Test Standards (PTS), 527–528
training, 531–534

nadir, 135
National Aeronautics and Space
Administration (NASA), 690
National Data Flight Digest, 312
National Oceanic Service (NOA), 267, 268
NOS approach charts, 369–396
National Transportation Safety Board
(NTSB), 189, 690
National Weather Service (NWS), 197, 202
Nautical Almanac, 136
NavData identifiers, Jeppesen, 286
navigation (*See also* VOR), 3, 4, 95–142,
158
arc to distance relationship, 134–135
automatic direction finder (ADF), 117,
118–119, 653–654
bracketing, 115, 123–126, **125**, 650–651
celestial navigation, 135–137
changeover points (COPs), 283–284
"chasing the needle," 115–116
circle of equal altitudes (COEA), 136
circle of equal radius (COER), 136, 138
circle of position (COP), 136
"cleared direct" on VOR, 113
course deviation indicator (CDI), 99,
99, 101–102
course selector, VOR, 98
cross-country flights, 95
direction finder (DF) loop, 117–118
discrete frequency, 636–639, **637, 638**

navigation (*Cont.*):
distance measuring equipment (DME),
294
emergency procedures, 555–556
enroute fixes, 169–171, 170
equipment required for IFR, 148, 184
fixes, 278–279, 286
flags, VOR, 99
flight service stations (FSS), 140
frequency selector, VOR, 98
geographical position (GP), 135
global positioning system (GPS), 102,
129–130, **131, 132, 133**, 140, 141,
158, 272
great circle navigation, 130, 134, **134**
heading indicator errors, 102–104
holding patterns, ADF, 128–129
homing, ADF, 122–123, **124**
inertial navigation system (INS), 272
instrument navigation procedures,
141–142
intercept angle, ADF, 120–121, **122**
intercepting ADF bearing, 120–121, **122**
Jeppesen NavData identifiers, 286
KRM, 272
L/MF facilities, 274
landing aids, 272
LDA, 272
line of position (LOP), 136, 138, 139
LOC, 272
LORAN/GPS Navigator Atlas, 318
loran-C, 272
microwave landing systems (MLS), 272
nadir, 135
navaids on charts and maps, 286–287
nondirectional beacon (NDB), 117–118
pilotage, 96–98
practice patterns, 127–128, **127**
preflight planning, 96–102, 145
radar navigation aids, 140–141, 276
radials, VOR, 100–101, **101**, 109–111,
110
radio navigation aids, 140–141
reference (holding) heading, 113–115
reporting points, 279–280, 284
route planning, 159–162
sectional charts, 97–98
simplified directional facility (SDF), 272
station passage, VOR, 116
TACAN stations, 272
three-star fix, 136–137, **137, 138**
time/distance calculations, 90–93,
108–109, **109**
time/distance checks, ADF, 119–120
timing, in celestial navigation, 138–139
To/From indicator, VOR, 99, **100**

navigation (*Cont.*):
    tracking ADF, 123–126, **125**
    tracking VOR, 100–102, 113
    VOR, 90–93, 98, 242–243, 274, 276, 284,
        286, 287, 636–639, **637, 638**
    VORTAC, 383
    waypoints, 280, 290
    wind correction, 111–112, **114**, 121
    wind drift, 121
    zenith, 135
    zenith distance (ZD), 135
needle/ball (*See* turn and bank coordinator)
never-exceed airspeed (V$_{ne}$), 84
night blindness, 580–581
night flying (*See also* physiology of flight),
    569–602
night vision, 560–561, 576–584, **576**, 594
no-gyro approaches, 518
non-holding side, holding pattern, 407,
    407
nondirectional beacon (NDB), 429
    homing, 122–123, **124**
    intercept angle, ADF, 120–121, **122**
    outbound bearings, 126–127
    time/distance checks, ADF, 119–120
nondirectional beacon (NDB) approach,
    117–118, 461, 468, 477–479, **478**, 507,
    657–659
nonprecision approaches, 58, 333–335,
    428–429, 447–448
NOS approach charts, 369–396, 428
    additional information listings, 372–375,
        373, 374
    alternate minimums, 372, 375–383, **376**,
        378–**382**
    amended information, 374
    climb and descent gradients, 393,
        394, 395
    final approach fix (FAF), 383
    holding pattern, 383, 391
    initial approach fix (IAF), 383
    instrument landing system (ILS),
        369–372, **371**
    lighting, airport, 384–388, **385, 386, 387**
    mileage between facilities, 383
    procedure turns, 388–393, **389, 390, 392**
    takeoff minimums, 372, 375–383,
        375–383, **376**, 378–**382**
    trailers, 370
    VORTAC, 383
NOTAMs, 220, 222, 369

obstacle/obstruction clearance, 62–65,
    299, 331, 475–**477**, 504
    approach, 426, 449, 452, 511
    instrument landing system (ILS), 504

obstacle/obstruction clearance (*Cont.*):
    minimum obstacle clearance altitude
        (MOCA), 152, 280, 283, 490
    obstruction clearance altitude (OCA),
        340
obstruction clearance altitude (OCA), 340
oil pressure gauge, 12, 185
oil pressure indicator, 184
oil quantity indicator, 184
oil temperature indicator, 184, 185
omni bearing indicator (OBI), 92, 101
omni bearing selector (OBS), VOR, 104,
    245
omnidirectional radio range (*See* VOR)
"one, two, three rule," 208
operating limitations, aircraft, 185–186
oral exam, Instrument Flight Test,
    preparation, 667–672
outbound end, holding pattern, 407, **407**
outbound leg, holding pattern, 407, **407**
outer marker, 428
outlook briefings, 236
over-water flight, 185
overcontrolling, 41
overlay approach, GPS, 129–130, **131,
    132, 133**, 481–484, **482, 485**, 492,
    493, 494, 492
oxygen consumption by human body,
    588–589
oxygen, supplemental, 152, 190–191

parallel entry, holding pattern, 409,
    416–418, **417**
partial-panel flight, 85–88, **86**, 677–678
performance, aircraft, restrictions, 66–68
periodicals, aviation, 689–690
personal minimums, 209–211, 542
phraseology of radio communications,
    258, 403–404
physiology of flight, 557–568
    air-pressure effects on human body,
        606, **606**
    anoxia, 609
    autokinesis illusion, 583
    automatic rough illusion, 591
    balance, 584–588
    barodontalgia (tooth pain related to
        altitude), 614
    barosinusitis (sinus pain related to
        altitude), 614
    barotitis (earache related to altitude), 614
    carbon monoxide poisoning, 559, 560
    circadian rhythms, 572–574
    dark adaptation of eyes, 580
    decompression sickness (bends),
        614–616

physiology of flight (*Cont.*):
  drug use, 557–558, 561–562
  empty-field myopia, 577
  exercises for the cockpit, 596–598
  fatigue, 570–572, 574–576, 594–596,
    600–601
  hearing, 584–588
  high-altitude flight, 603–616
  human eye, 578–584, **578**
  hyperventilation, 560
  hypoxia, 589, 604–614
  inner ear, balance, and vertigo,
    563–565, **565**
  judgment, 591–594
  medication use, 557–558
  metabolic effects, 572–574, 604
  motion sickness, 584–588, 598–600
  night blindness, 580–581
  night flying, 569–602
  night vision, 560–561
  oxygen consumption by human body,
    588–589
  "popping" of ears at altitude, 614
  psychological effects of night flying,
    589–591
  respiration, 588–589, 603–604, **604, 605**
  saccadic movement of eyes, 583
  sleep (*See also* fatigue, pilot), 594–596
  spatial disorientation, 562–568, 584–588,
    598–600
  staying awake during night flight,
    596–598
  stress, 623–631
  time of useful consciousness (TUC),
    609–613
  vertigo, 562–568, 584–588, 598–600
  vision, 576–584, **576**, 594, 616
Pilot and Aircraft Courtesy Evaluation
  (PACE) program, 193
Pilot Certificate, 193
pilot readiness for flight (*See also*
  physiology of flight), 541–556
  alcohol use, 549–550
  apathy vs. preparedness for flight,
    550–551
  drug use, 546–549
  fatigue, pilot, 543–546
  illness, 546
  medication use, 546–549
  mental preparedness for flight, 550–551
  personal minimums, 542
  spatial disorientation, 552–556
  student/instructor relationship, 551–552
  vertigo, 552–556
Pilot Reports (PIREPs), 201 202–203,
  226–227, 640

pilotage, 96–98
Pilot's Automatic Telephone Weather
  Answering Service (PATWAS),
  219–221
pink temporary registration slip, 193
pitch, 3, 50, 248
pitch angle, climb setting, 52
pitch setting
  approach and landing, 61
  climbs, 54–55
  constant-speed descents, 58–59
  descents, 54
  failure of instrument, 87
  minimum controllable airspeed, 75–76
pitot heat, 242, 257
pitot-static system, 17–18, **17**
plan view (bird's eye view) of charts and
  maps, 321–327, **322, 323, 325, 326**
plotters, 147, 271
poor judgment (PJ) chains, 617
"popping" of ears at altitude, 614
position reporting, 262, 400–402
power, 4, 33
  airspeed changes, 32–33
  airspeed control, 31–32
  airspeed vs., 48–50
  approach and landing setting, 61
  climb settings, 52
  climbs, 54
  constant-rate descents, 59–60
  constant-speed descents, 58–59
  cruise, 57–58
  descents setting, 54, 508
  level-off, after climb, 57–58, 57
  minimum controllable airspeed, 75–76
  pitch, 50
  stalls, 80–81
  takeoff and departure, 522
Practical Test Standards, 418, 527–528
practice maneuvers, 73–74, **74**, 77–79, **78**,
  79–80, **79**
  navigation, 127–128, **127**
  VOR, 116–117
precession in directional gyro, 47
precipitation-induced fog, 203
precision approach radar (PAR), 428,
  518–519
precision approaches, 337–340, **339**,
  428–429, 448
precision code (P-code), 139
preferred IFR routes, 162–163
preflight planning (*See also* flight plans;
  pilot's readiness for flight), 96–102,
  145–156
  ADF check, 247
  Air Traffic Control, 145

preflight planning (*Cont.*):
  aircraft requirements, 148–149
  airspeed indicator errors, 241
  airways, 151–153
  alternate airport requirements, 153–155
  altimeter error, 241
  altitude minimums, 151–153
  automatic terminal information service
    (ATIS), 240–241
  charts and maps, 146–147
  checklists, 145, 239–248, **240**
  cockpit organization, 243–245, 247–248
  electrical equipment, 241–242
  FAR 91.103, 146
  filing a flight plan, 145
  flight kit, 146–148
  fuel requirements, 153–155, 239–240
  go/no-go decisions, 145
  gyro instrument check, 248
  icing hazards, 149–151
  ILS checks, 246–247
  landmarks and navigation, 145
  pilotage, 96–98
  pilot's readiness for flight, 541–556
  radio communications check, 244, 248
  rolling engine run-up, 256–256
  runway checks, 257
  sectional charts, 97–98
  taxi checks, 256
  thunderstorms, 149–151
  transponder check, 247
  VOR check with VOT, 245–246, **245**
  VOR checks, 98, 242–243, 246
  weather, 145
pressure (*See* control pressure)
pressurized cabin, 152, 190–191
prevailing visibility, 206, 435
primary instruments, 4–7, 5–6(*t*), 13(*t*), 87
priority handling request, 403
procedure turns, 327, 329, 340–341,
    388–393, **389, 390, 392**, 463–466
  no procedure turn (NoPT) sectors, 327
proficiency, 673
proficiency checks, VOR, 104–105
profile descent, 297–302, **298, 300, 301**,
    302–304, **303**
profile view, on charts and maps,
    333–350, **334**
profiles, climb profiles, 71–72
proposed master minimum equipment list
    (PM-MEL), 178
psychological effects of night flying,
    589–591
pump driven vacuum system, **19**

quality-of-turn instrument, 3

radar, 276, 399
  airport surveillance radar (ASR),
    516–517, **517**
  approach, 461–462, 514–516, 526
  ground controlled approach (GCA),
    518–519
  navigation aids, 140–141
  precision approach radar (PAR), 518–519
  required reports, 262, 402
  weather radar, 199
radar approach control (RAPCON), 514
radial intercepts, VOR, 90–93, 109–111, **110**
radials, VOR, 100–101, **101**, 109–111, **110**
radiation fog, 203
radio communications, 257–258, 399–412
  approach, 467–468
  avoiding confusion, 403–404
  call signs, 259
  canceling IFR, 262–263
  charts and maps, notations for, 270, **270**
  compulsory reporting points, 400–402
  departure sequence, 244
  discrete frequency, 636–639, **637, 638**
  emergency vs. priority handling, 403
  Enroute Flight Advisory Service (EFAS),
    639–642, **640**
  equipment required for IFR, 148
  Flight Watch, 639–642, **640**
  frequencies used, 171, **172**, 244–245,
    248, 260–261, 270, **270**, 321, 351,
    636–639, **637, 638**
  Ground Control, 259–260
  inoperative radio, 88–90, 237, 263,
    404–406, **405**, 684
  land lines, 638
  license, 193
  lost radio contact, 682–684
  phraseology, 258, 403–404
  position reporting, 400–402
  preflight check, 244, 248
  priority handling request, 403
  remote communications outlets (RCOs),
    254, **255**, 274
  requesting a clearance, 253
  required reports, 261–262, 402
  runway checks, 257
  uncontrolled airports, 676–677
  who, who, where, what sequence,
    258–259
radio navigation aids, 140–141
rain, 218
ramp checks, equipment operation,
    192–193
rate of climb, 54, 56, 302, 522
receiver autonomous integrity monitoring
    (RAIM), 496

reciprocal point, holding pattern, 407, **407**
recovery from stalls, 80–81
reference (holding) heading, 113–115
reflex actions, control pressure, 7
Registration Certificate, 193
relative bearing, 118
remote communications outlets (RCOs), 254, **255**, 2474
reporting points, 262, 276, 279–280, 284
required basic equipment list, 184–18
required reports, 261–262, 402
respiration, 588–589, 603–604, **604, 605**
restricted airspace, 270–271, **271**, 274, 276
restrictions on aircraft performance, 66–68
rime ice, 150
roll, 3, 49
route planning, 159–162
runway alignment indicator lights (RAIL), 430
runway centerline lighting (RCLS), 428
runway centerline markings, 430
runway checks, 257
runway end identification lights (REIL), 430
runway length, 357–360, 357, 359
runway visibility value (RVV), 206, 435
runway visual range (RVR), 206, 428, 435–436, 458–459
runways
  on charts and maps, 352, **354**
  displaced thresholds, 359–360, 359
  jet runways, 360

saccadic movement of eyes, 583
safety belts/restraints, 185
satellites, GPS, 130
scale, on charts and maps, 271
scanning instruments, 24–25, 39, 43–50
  distractions, 44
  errors in instrument readouts, 47–48
  instrument fixation, 43, 85
  interpreting instrument readings, 44–45
  italicattitude, heading, altitudeitalic method, 43–44
  partial-panel (failure) procedures, 87
  proper technique, 45–47
  teaching how to scan, 43–44
sectional charts, 97–98, 147
segments of approach, 448
selective availability (SA), 139
sense indicator (*See* To/From indicator)
severe weather avoidance plan (SWAP), 200
SIGMETs, 198–200, 227
simplified directional facility (SDF), 272, 514, **515**

simplified short approach light system (SSALSR), 430
skid indicator, 148
skids, 35
sleep (*See also* fatigue), 594–596
slip indicator, 148
slip/skid indicator (*See also* turn and bank coordinator), 89, 148
slips, 35
snow, 218
spatial disorientation, 552–556, 562–568, 584–588, 598–600
squall lines, 218
stagnate hypoxia, 609
stalls, 80–81
  landing gear retraction, 81
  minimum controllable airspeed, 76
  recovery, 80–81
standard instrument departure (SID), 63, **64**, 65–68, 146, 159, 162, **166**, 267, 297–302, **298, 300, 301**
standard rate turns, 25–26, 34–36, 63
standard terminal arrival route (STAR), 63, 146, 163, 167, 267, 297–302, **298, 300, 301**, 410, **411**, 410
standard terminal procedures (TERPs), 312
static system, 17–18, **17**
station passage, VOR, 116
staying awake during night flight, 596–598
steep turns, 81–83
step climbs/descents, 28–30, **30**
straight-and-level flight, 6, 27–28, 36–38
  airspeed indicator, 53
  altimeters, 53
  artificial horizon, 53
  directional gyro, 53
  power setting, 53
  practice maneuver, 79
  practice pattern A, 73–74, **74**
  practice pattern B, 77–79, **78**
  standard-rate turns, 53
  trim, 53
  turn and bank indicator, 53
straight-in approach, 440–441, 453–454
stress, pilot, 623–631, 623
student/instructor relationship, 551–552
supplemental oxygen, 152, 190–191
supplemental type certificate (STC), 179
support instruments, 4–7, 5–6(*t*), 13(*t*)
symbols, on charts and maps, 268, 272, **273, 275, 277, 281**, 283–284, **285, 289, 290, 292**, 352–355, **353**
synopsis, in weather briefing, 221

TACAN stations, 272
tachometer, 76, 184, 185

tailwinds, 111
takeoff roll, 522
takeoff and departure, 51–52, 520, 522
  air route traffic control center (ARTCC),
    69
  air traffic control, 257–258
  airspeed indicator, 52
  attitude indicator setting, 51, 56
  brakes, 52
  clearance for IFR, 56, 69–71
  climb, 522
  climb attitude, 52
  climb gradients, 72
  climb profiles, 71–72
  climb speed, 52
  constant-rate climbs, 56–57
  control pressure, 522
  control towers, 69
  departure climbouts, 55–56
  Departure Control, 70
  departure procedures, 69–72
  directional gyro setting, 51, 52
  discrete clearance delivery, 69
  drift, 522
  elevation of airfield, 172–174, **173**
  Ground Control, 69, 259
  heading indicator, 522
  initial climb, 56
  instrument departure procedures, 72
  landing gear retraction, 81
  minimums, 360–362, **361**, 372, 375–383,
    376, 378–**382**, 520, 522
  obstacle clearance concerns, 62–65,
    299, 331, 475–**477**, 504
  power, 522
  practice, 522
  radio communications, 257–258
  rate of climb, 522
  restrictions on aircraft performance,
    66–68
  rolling engine run-up, 256–256
  runway checks, 257
  standard instrument departure (SID),
    63, **64**, 65–68, 146, 159, 162, **166**,
    267, 297–302, **298, 300, 301**
  takeoff roll, 522
  taxi checks, 256
  taxiing to takeoff, 51, 69
  throttle control, 52
  uncontrolled airports, 69
  vertical speed indicator, 52
  VFR conditions, 69
  visibility minimums, 204–205, 362
  weather minimums, 204–205
taxi checks, 256
taxiing, 51, 69
  rolling engine run-up, 256–256

taxiing (*Cont.*):
  runway checks, 257
  taxi checks, 256
taxiways, on charts and maps, 352, **354**
teardrop entry, holding pattern, 409,
  416–418, **417**
telephone communications facilities,
  274
Telephone Information Briefing Service
  (TIBS), 219–221
temperature, 197, 203–204
temperature gauge, 12, 185
terminal aerodrome forecasts (TAFs),
  214–217
terrain features, on charts and maps, 294,
  331, **332**
three-star fix, 136–137, **137, 138**
threshold crossing height (TCH), 340
throttle control, takeoffs, 52
thunderstorms, 149–151, 195, 196–200,
  211, 218, 219, 226, 538
time, 147
  estimated time enroute (ETE), 685
  estimated time of arrival (ETA), 685
  estimating time enroute, 236–237
  logging, 685
  required reports, 262, 402
  time/distance calculations, 90–93,
    108–109, **109**, 108
  time/distance checks, ADF, 119–120
  timing the approach, 469–470
  total time enroute, 237–238
time of useful consciousness (TUC),
  609–613
time zone boundaries, on charts and
  maps, 288
timing, in celestial navigation, 138–139
timing the approach, 469–470, 506
To/From indicator, VOR, 99, **100**
tornadoes, 218
touchdown zone elevation (TDZE),
  173–174, 335
touchdown zone lighting (TDZL), 428
Tower Control, 244
Tower Enroute Control (TEC), 162–163
tracking ADF, 123–126, **125**
tracking VOR, 100–102, 113
trailers, charts and maps, 370
training, student/instructor relationship,
  551–552
transition altitudes, 321
transitioning for approach, 327–329, **330**
transponders, 190–191, 247, 257
trim, 7–8, 33–34
  failure of instrument, 87
  steep turns, 82–83
  straight and level flight, 53

turbulence, 195, 199, 200–201, 211, 218,
　226–227, 538
turn-and-bank coordinator, 20–22, **21**,
　89, 148
　errors, 11
　failure of instrument, 87–88
　minimum controllable airspeed, 76
　standard rate turns with clock, 25–26,
　　34–36
　straight and level flight, 37–38, 53
　taxi checks, 256
turns, 6
turns and banks, 53
　airspeed indicator, 53
　altimeters, 53
　artificial horizon, 53
　bracketing, 115
　climbing tendency, 49
　control pressure, 35
　directional gyro, 53
　magnetic compass errors, 12–15
　practice maneuver, 77, 79
　skids, 35
　slips, 35
　standard rate turns, 25–26, 34–36, 53
　steep turns, 81–83
　trim, 33–34, 53
two-two-and-twenty rule, 40
type certificate data sheet (TCDS), 185

U.S. Standard for Terminal Instrument
　Procedures (TERPS), 449
unacceptable clearances, 253–254
uncontrolled airports, 69, 676–677
Universal Approach Chart Update
　notice, 313, 314–**317**
universal coordinated time (UTC/Zulu), 147
Universal Enroute Chart Update
　notice, 313, 314–**317**
unusual attitude, 83–85, 652–653
updrafts, 7, 25, 196
upslope fog, 203
UTC (*See* universal coordinated time)

vacuum system, pump driven, **19**, 19
variation in magnetic compass, 10
vectored approaches, 495
vectors to final approach, 526
vertical S maneuver, 77, **77**
vertical speed indicator (VSI), 4, 17, 19,
　20, 23–25, **24**, 37, 38
　"chasing the needle," 57, 115–116
　constant-rate climbs, 56–57
　descents, 54
　errors, 11
　minimum controllable airspeed, 76
　takeoffs, 52

vertigo, 552–556, 562–568, 584–588,
　598–600
VFR on top, required reports, 261, 402
Victor airways, 162
visibility, 195, 199, 204–206, 211, 217, 218,
　362, 429, 435–436, 539–540
　approach categories, 207–208, 429,
　　430–433, 458–459
vision, 616
　autokinesis illusion, 583
　dark adaptation of eyes, 580
　empty-field myopia, 577
　human eye, 578–584, **578**
　night blindness, 580–581
　night vision, 560–561, 576–584, **576**, 594
　saccadic movement of eyes, 583
visual and instrument correlation, 24–25,
　39, 43–50
visual approach, 436–438, 519–520, **521**
　charted visual approaches, 304–307,
　　305, 306
visual descent point (VDP), 336–337
visual meteorological conditions (VMC),
　537
void time clearance, 70, 254, 677
VOR, 58, 66, 90–93, 98, 158, 276, 429, 430
　16-point orientation test, 105–108, **106**
　ambiguity indicator (*See* To/From
　　indicator)
　approach, 481–484, **482, 485**, 659–661
　automatic direction finder (ADF)
　　orientation, 118–119
　bracketing, 650–651
　on charts and maps, 274, 284, 286, 287
　"chasing the needle," 115–116
　checks, 242–243, 246
　"cleared direct" on VOR, 113
　course deviation indicator (CDI), 92, 99,
　　99, 101–102
　course selector, 98
　discrete frequency, 636–639, **637, 638**
　distance measuring equipment (DME)
　　holds, 422–423, **423**
　equipment required for IFR, 148–149
　fixes, 278–279
　flags, 99
　frequency selector, 98
　global positioning system (GPS) vs., 102
　heading indicator errors, 102–104
　holding patterns, 651–652
　inbound turn example, 110–111, **110**
　interception mistakes, 112–113
　intersection holds, 421–422
　logbooks for VOR checks, 149
　omni bearing indicator (OBI), 92, 101
　omni bearing selector (OBS), 104, 245
　practice patterns, 116–117

VOR (*Cont.*):
   proficiency of technique, 104–105
   radial intercepts, 90–93, 109–111, **110**
   radials, 109–111, **110**
   reference (holding) heading, 113–115
   sense indicator (*See* To/From indicator)
   station passage, 116
   test facility, VOT, 148–149, 245–246
   time/distance calculations, 90–93,
      108–109, **109**
   To/From indicator, 99, **100**
   tracking and bracketing, 90–93,
      100–102, 113, 650–651
   VOR/DME approach, 488–490, **489**
   wind correction, 111–112, **114**
VOR test facility (VOT), 148–149, 245–246
VOR/DME approach, 488–490, **489**, 488
VORTAC, 383

warm fronts, 218
waypoints, 280, 290
weather, 157–158, 174–176, **175**, 195–196
   1-800-WX-BRIEF, 219–221
   abbreviated briefings, 234–236
   accidents, weather-related, 538–540,
      541, 542
   adverse conditions, in weather briefing,
      221, 224–225
   alternate airports, 208–209, **210**, 226,
      229–232, **232**
   approach categories, 207–208, 429,
      430–433
   ATC delays, 222
   briefings, 157–158, 174–176, **175**,
      198–200, 212, 221–226, 234–236,
      680–682
   computer services, 212–213
   current conditions, in weather briefing,
      221
   departure minimums, 204–205
   destination forecasts, in weather
      briefing, 221,225
   destination minimums, 205–208
   direct user access terminal service
      (DUATs), 213–214
   Enroute Flight Advisory Service (EFAS),
      639–642, **640**
   enroute forecasts, in weather briefing,
      221, 225
   equipment requirements vs., 190–191
   estimated climbout time, 228
   Flight Watch, 639–642, **640**
   forecast reliability, 217–219
   go/no-go decisions, 226–228
   instrument meteorological conditions
      (IMC), 436

weather (*Cont.*):
   METARs, 214–217
   minimums, 211, 228
   NOTAMs, 220, 222
   "one, two, three rule," 208
   one-call weather briefing/flight plan
      filing technique, 236
   personal minimums, 209–211
   Pilot Reports (PIREPs), 640
   Pilot's Automatic Telephone Weather
      Answering Service (PATWAS),
      219–221
   predicable vs. unpredictable changes,
      218–219
   preflight planning, 145
   required reports, 261, 402
   shorthand/abbreviations used, 222,
      223–**224**
   synopsis, in weather briefing, 221
   Telephone Information Briefing Service
      (TIBS), 219–221
   terminal aerodrome forecasts (TAFs),
      214–217
   transcribing, 221–222
   visibility minimums, 362
   visual meteorological conditions (VMC),
      537
   wind vs. ground speed, 228–229, **230,
      231**
   winds aloft, in weather briefing, 221, 225
weather briefings, 157–158, 174–176, **175**,
      198–200, 212, 221–226, 234–236,
      680–682
Weather Channel, 213
weather radar, 199
weather-related accidents, 538–540, 541,
      542
weight and balance data sheet, 193
who, who, where, what sequence,
      258–259
wind, 199–201
wind correction, 111–112, **114**, 121
   bracketing, 115, 123–126, **125**
   holding patterns, 414–415
wind drift, 121
wind shear, 25, 151, 197
wind vs. ground speed, 228–229, **230, 231**
winds aloft, in weather briefing, 221, 225
world aeronautical charts (WAC), 147

yaw, 3, 49

zenith, 135
zenith distance (ZD), 135
Zigdex, 268, **270**
Zulu time (*See* universal coordinated time)

About the Author

Thomas P. Turner is a commercial pilot who holds instructor ratings for instrument, single, and multiengine airplanes and a master's degree in aviation safety. The author of *Cockpit Resource Management*, Second Edition, and *Weather Patterns and Phenomena: A Pilot's Guide*, Second Edition, he speaks regularly to pilots, groups. His articles have appeared in 32 aviation publications.